CALCIUM AND
CELL FUNCTION

Volume I

CALMODULIN

MOLECULAR BIOLOGY

An International Series of Monographs and Textbooks

Editors: BERNARD HORECKER, NATHAN O. KAPLAN, JULIUS MARMUR, AND HAROLD A. SCHERAGA

A complete list of titles in this series appears at the end of this volume.

CALCIUM AND CELL FUNCTION

Volume I

CALMODULIN

Edited by

WAI YIU CHEUNG

Department of Biochemistry
St. Jude Children's Research Hospital
and University of Tennessee Center
for the Health Sciences
Memphis, Tennessee

ACADEMIC PRESS 1980
A Subsidiary of Harcourt Brace Jovanovich, Publishers
New York London Toronto Sydney San Francisco

ACADEMIC PRESS, INC.
111 Fifth Avenue, New York, New York 10003

United Kingdom Edition published by
ACADEMIC PRESS, INC. (LONDON) LTD.
24/28 Oval Road, London NW1 7DX

Library of Congress Cataloging in Publication Data
Main entry under title:

Calcium and cell function.

(Molecular biology series)
Vol. 1 edited by W. Y. Cheung.
Includes bibliographies and index.
1. Calcium–Physiological effect. 2. Calcium metabolism. 3. Cell physiology. I. Cheung, Wai Yiu. [DNLM: 1. Calcium. 2. Calcium–Binding proteins. QU55 C144]
QP535.C2C26 612'.3924 80–985
ISBN 0-12-171401-2 (v. 1)

PRINTED IN THE UNITED STATES OF AMERICA

80 81 82 83 9 8 7 6 5 4 3 2 1

Contents

List of Contributors

Numbers in parentheses indicate the pages on which the authors' contributions begin.

Robert S. Adelstein (167), Section on Molecular Cardiology, Cardiology Branch, National Heart, Lung, and Blood Institute, National Institutes of Health, Bethesda, Maryland 20205

James M. Anderson (201), Department of Biochemistry, University of Georgia, Athens, Georgia 30602

Lawrence S. Bradham (109), Department of Biochemistry, University of Tennessee Center for the Health Sciences, Memphis, Tennessee 38163

Harry Charbonneau (201), Department of Biochemistry, University of Georgia, Athens, Georgia 30602

Wai Yiu Cheung (1, 13, 79, 109, 273, 291), Department of Biochemistry, St. Jude Children's Research Hospital and University of Tennessee Center for the Health Sciences, Memphis, Tennessee 38101

Philip Cohen (183), Department of Biochemistry, Medical Sciences Institute, University of Dundee, Dundee DD1 4HN, Scotland

Milton J. Cormier (201), Department of Biochemistry, University of Georgia, Athens, Georgia 30602

E. Costa (253), Laboratory of Preclinical Pharmacology, National Institute of Mental Health, Saint Elizabeths Hospital, Washington, D.C. 20032

Paul Greengard (219), Department of Pharmacology, Yale University School of Medicine, New Haven, Connecticut 06510

I. Hanbauer (253), Section on Biochemical Pharmacology, Hypertension Endocrine Branch, National Heart, Lung, and Blood Institute, National Institutes of Health, Bethesda, Maryland 20205

Jeffrey F. Harper* (273), Department of Medicine, University of North Carolina School of Medicine, Chapel Hill, North Carolina 27514

* Present Address: Departments of Internal Medicine and Pharmacology, University of Texas Medical School at Houston, Houston, Texas 77025.

Thomas R. Hinds (127), Department of Pharmacology, University of Washington, Seattle, Washington 98195

Wieland B. Huttner (219), Department of Pharmacology, Yale University School of Medicine, New Haven, Connecticut 06510

Harold P. Jones (201), Department of Biochemistry, University of Georgia, Athens, Georgia 30602

Claude B. Klee (59, 167), Laboratory of Biochemistry, National Cancer Institute, National Institutes of Health, Bethesda, Maryland 20205

Steven N. Levine* (273), Department of Medicine, University of North Carolina, Chapel Hill, North Carolina 27514

Ying Ming Lin (79), Department of Chemistry, Tennessee State University, Nashville, Tennessee 37203

Richard O. McCann (201), Department of Biochemistry, University of Georgia, Athens, Georgia 30602

Howard Schulman (219), Department of Pharmacology, Stanford University School of Medicine, Stanford, California 94305

Rajendra K. Sharma (305), Department of Biochemistry, University of Manitoba, Winnipeg, Manitoba, Canada

Alton L. Steiner (273), Department of Medicine, University of North Carolina, Chapel Hill, North Carolina 27514

E. Ann Tallant (13), Department of Biochemistry, University of Tennessee Center for the Health Sciences, Memphis, Tennessee 38101

Stanley W. Tam (305), Department of Biochemistry, University of Manitoba, Winnipeg, Manitoba, Canada

Thomas C. Vanaman (41), Duke University Medical Center, Durham, North Carolina 27710

Frank F. Vincenzi (127), Department of Pharmacology, University of Washington, Seattle, Washington 98195

Robert W. Wallace (13, 273, 291), Department of Biochemistry, St. Jude Children's Research Hospital, Memphis, Tennessee 38101

Thomas L. Wallace (329), Department of Pharmacology, Medical College of Pennsylvania, Philadelphia, Pennsylvania 19129

Jerry H. Wang (305), Department of Biochemistry, University of Manitoba, Winnipeg, Manitoba, Canada

Benjamin Weiss (329), Department of Pharmacology, Medical College of Pennsylvania, Philadelphia, Pennsylvania 19129

John G. Wood (291), Department of Anatomy, University of Tennessee Center for the Health Sciences, Memphis, Tennessee 38163

* Present Address: Department of Medicine, Louisiana State University School of Medicine, Shreveport, Louisiana.

Preface

The role of the calcium ion in cell function has attracted the attention of investigators from almost every discipline of biological research for the last three decades. It is only within the past several years that its mode of action at the molecular level began to be unraveled. The calcium ion itself is inactive; its activity is mediated through a homologous class of calcium-binding proteins. The first protein to be identified as a mediator of calcium action was troponin C, whose function appears to be limited to the control of contraction in skeletal and cardiac muscles. In smooth muscle and nonmuscle cells, contraction is regulated by calmodulin. Calmodulin was discovered serendipitously as an activator of cyclic 3′,5′-nucleotide phosphodiesterase in the late 1960s in our laboratory. Ubiquitous in eukaryotes, this calcium-binding protein displays multifunctions controlling numerous key enzyme systems and cellular processes. Work from many laboratories subsequently demonstrated that calmodulin is a major mediator of Ca^{2+} functions in eukaryotes.

The chapters in this volume, written by leading investigators in the field, cover many calmodulin-regulated functions. As in any rapidly developing field, by the time the volume has been edited for production, new areas of investigation open up, and they deserve coverage. We hope that future volumes of this treatise will make up for the areas omitted from this volume, not by design but from the inability to keep up with the ever-increasing pace of development in the field.

The aim of this volume is to provide investigators, beginning as well as experienced, in all aspects of biological research easy access to the basic and critical information on the role of calmodulin in cell function.

I would like to thank all the contributors for their cooperative efforts which made the editing of this volume a gratifying experience. I am greatly indebted to Professor Edwin G. Krebs for his generous support and kind hospitality during my tenure as a Faculty Scholar, Josiah Macy, Jr., Foundation, in his laboratory of Molecular Pharmacology,

Howard Hughes Medical Institute, University of Washington. Much of the editing was completed there. The staff of Academic Press provided me with valuable assistance in all phases of the work. It has been a pleasure to work with them. Last, but by no means least, I am grateful to my wife and children for their love, devotion, and encouragement; to them I dedicate this volume.

Wai Yiu Cheung

Chapter 1

Calmodulin – An Introduction

WAI YIU CHEUNG

The discovery of calmodulin owes much to Professor Britton Chance, whose generous support and encouragement gave impetus to my early studies at the Johnson Research Foundation, University of Pennsylvania. In 1964, when I joined his department as a postdoctoral fellow, one of Dr. Chance's research interests was NADH oscillation in baker's yeast, a phenomenon strongly affected by cAMP (Chance *et al.*, 1965). At his suggestion, I began experiments to define the role of cAMP in the oscillation of this dinucleotide (Cheung, 1966a), but shortly after, my attention turned to regulatory mechanisms in the mammalian system. In a series of pioneering studies, Earl Sutherland and his co-workers had shown cAMP to be a mediator of several peptide hormones (for a review, see Robison *et al.*, 1971). I was particularly struck by the disparate activities of the two enzymes involved in the metabolism of cAMP. The activity of adenylate cyclase, which is responsible for the synthesis of cAMP, is generally one to two orders of magnitude lower than that of phosphodiesterase, an enzyme involved in the degradation of the cyclic nucleotide (Butcher and Sutherland, 1962; Sutherland *et al.*, 1962). Despite this discrepancy, tissue levels of cAMP increase sharply in response to hormone stimulation (Cheung and Williamson, 1965). The accumulation of cAMP in the face of relatively low synthetic activity suggested that the two enzymes are either

CALCIUM AND CELL FUNCTION, VOL. I

ISBN 0-12-171401-2

subject to different control mechanisms or are localized in different subcellular structures (Cheung, 1966b; Cheung and Salganicoff, 1967). Though oversimplified, this assumption proved a valid framework for my subsequent investigations.

I. DISCOVERY OF CALMODULIN

The regulating properties of phosphodiesterase were studied first in bovine brain, mainly because the tissue was readily available and contains a high level of the enzyme. My initial work indicated that phosphodiesterase is quite complex, displaying regulatory properties much like those of a metalloenzyme (Cheung, 1967a).

The method used to follow the activity of phosphodiesterase is outlined below.

$$\text{cAMP} \xrightarrow{\text{phosphodiesterase}} 5'\text{-AMP} \xrightarrow[\text{snake venom}]{5'\text{-nucleotidase}} \text{P}_i + \text{adenosine}$$

This assay (Butcher and Sutherland, 1962) can be completed with phosphodiesterase and 5′-nucleotidase incubated together (one stage), or it can be stopped at the 5′-AMP level and snake venom used for further incubation (two stages). Inorganic phosphate released by 5′-nucleotidase is measured colorimetrically. Generally, the amount of inorganic phosphate formed at the end of incubation does not depend on the type of assay procedure.

During attempts to purify phosphodiesterase for characterization studies, I noted that at certain stages of purification, the enzymatic activity varied greatly, depending on whether the one- or the two-stage procedure was used (Cheung, 1967b).

The effect of snake venom on phosphodiesterase activity is summarized in Table I. The results show that the crude enzyme had the same activity, whether preincubation was performed with or without snake venom. The purified enzyme, however, was several times more active with venom than without it. Thus, the venom, which is inactive toward cAMP, appeared to contain a factor that activates the purified enzyme. A systematic analysis disclosed that the effect of venom stems from an endogenous proteolytic activity. One explanation for the finding that the crude phosphodiesterase is fully active, whereas the purified enzyme depends on venom for optimum activity, is that a stimulatory factor originally present with phosphodiesterase dissociates from the enzyme during the course of purification (Cheung, 1967b). The results of a mixing experiment supported this impression.

Table II compares the activities of a crude and a purified phosphodies-

TABLE I

Effect of Snake Venom on the Activity of Phosphodiesterase[a,b]

Enzyme	Preincubation conditions activity (O.D./sample)	
	With venom	Without venom
Crude phosphodiesterase	0.625	0.621
"Purified" phosphodiesterase	0.476	0.061

[a] From Cheung (1967b).
[b] The two-stage procedure was used in the assay.

terase assayed separately with that of both enzymes assayed together. The activity of the mixture was more than twice the summed activities of the individual preparations. This experiment demonstrated the synergistic effect of mixing the enzyme preparations but did not reveal which of the two components contained the activating agent. The stimulation of purified phosphodiesterase by snake venom helped to clarify this question. Table III shows the effect of the venom on the activities of a crude and a purified phosphodiesterase. The venom activated the purified but not the crude enzyme. Again, the activity of the mixture was much greater than the combined activities of the individual preparations. Of particular interest was the finding that the sum of activities by the crude and the purified enzymes after their incubation with venom was comparable to the activity of the mixture. These observations prompted two conclusions. First, since the purified but not the crude enzyme was stimulated by venom, the

TABLE II

Comparison of Phosphodiesterase Activities in a Crude and a Partially Purified Preparation of Bovine Brain[a]

Fraction	Activity (nmole)
a. Crude phosphodiesterase	66
b. Purified phosphodiesterase	40
c. Crude + purified phosphodiesterase	260

[a] Phosphodiesterase was assayed by a two-stage procedure. A homogenate of bovine brain cerebra was dialyzed extensively against 20 m*M* Tris-Cl, pH 7.5, and was then centrifuged at 40,000 *g* for 20 min. An aliquot of the supernatant fluid was used as a source of the crude phosphodiesterase. The purified phosphodiesterase was prepared from a DEAE-cellulose column and was relatively inactive prior to activation. Activity is expressed as nanomoles of inorganic phosphate formed per 10 min. (From Cheung, 1970b.)

TABLE III

Effect of Snake Venom on the Activity of Crude and Purified Phosphodiesterase[a]

Fractions	Activity (nmole)
a. Crude phosphodiesterase	262
b. Crude phosphodiesterase + venom	279
c. Purified phosphodiesterase	80
d. Purified phosphodiesterase + venom	454
e. Crude + purified phosphodiesterase	658

[a] Phosphodiesterase was assayed by a two-stage procedure. When indicated, venom was present in the first stage of incubation. Activity is expressed as nanomoles of inorganic phosphate formed per 10 min. (From Cheung, 1970b.)

activating agent (or activator) was probably associated with the latter. Second, this activator was as effective as venom in stimulating the purifed enzyme (Cheung, 1970a,b).

I also considered the possibility that proteins in the venom stabilized some labile phosphodiesterase activity rather than stimulated the enzyme, but rejected this explanation because in my hands the enzyme proved to be quite stable; moreover, the presence of bovine serum albumin did not increase enzyme activity. I reasoned that the relative inactivity of the purified was caused by removal of an activator during the course of enzyme purification. Additional experiments demonstrated that the activator, now termed "calmodulin," was originally present in the crude homogenate and was removed from phosphodiesterase as purification proceeded. Fractions obtained early in the purification procedure were rich in calmodulin, whereas those obtained at a later stage were deficient. The purification of phosphodiesterase with a DEAE-cellulose column serves to illustrate this point. Before being placed on the column, the enzyme was fully active (Cheung, 1969, 1971a), but became relatively inactive after calmodulin was dissociated. Table IV shows the effect of calmodulin on the activity of purified phosphodiesterase. In this experiment, calmodulin—itself inactive—increased the activity of the purified enzyme more than fivefold. As before, the venom caused a pronounced stimulation. Figure 1 depicts the elution pattern of phosphodiesterase from a DEAE-cellulose column and shows the enzyme activity in its stimulated and nonstimulated states. These results substantiated the notion that purified phosphodiesterase is partially inactivated by the dissociation of calmodulin from the enzyme during purification. The crude enzyme remained fully active because it contained high concentrations of calmodu-

TABLE IV

Effect of Calmodulin on the Activity of Purified Phosphodiesterase[a]

Fractions	Activity (nmole)
a. Purified phosphodiesterase	56
b. Calmodulin	0
c. Purified phosphodiesterase + calmodulin	315
d. Purified phosphodiesterase + venom	306

[a] Phosphodiesterase was assayed by a two-stage procedure. When indicated, venom was present in the first stage of incubation. Calmodulin was a preparation from Fig. 1 that had been stored at −20°C for 6 months prior to use. Activity is expressed as nanomoles of inorganic phosphate formed per 10 min. (From Cheung, 1970b.)

lin; the purified enzyme, by contrast, was only partially active owing to a deficiency of the activator.

The nature of calmodulin was examined by use of trypsin, RNase, and DNase. Treatment with trypsin, but not with RNase or DNase, destroyed the activity of the activator, demonstrating that it is associated with a protein (Cheung, 1970a, 1971a).

The discovery of calmodulin as an activator of phosphodiesterase raised the question of its mode of action. Two experimental approaches were taken to resolve this issue. In one, we investigated the tissue, cellular, and subcellular localization of calmodulin as well as its developmental changes (Smoake *et al.*, 1974; Cheung *et al.*, 1975a). In the other, we purified the protein to apparent homogeneity, which permitted physical and chemical characterization. The concentration of calmodulin was much greater than that of phosphodiesterase and appeared in all tissues examined, even in those without Ca^{2+}-dependent phosphodiesterase activity. Moreover, the activity of calmodulin did not parallel that of phosphodiesterase during ontogentic development, indicating that the two proteins are under separate genetic regulation (Lynch *et al.*, 1975). Others extended this study and found calmodulin in all eukaryotes examined (for reviews, see Cheung, 1970a, 1980, and references therein). Collectively, these observations imply that besides the regulation of phosphodiesterase activity, calmodulin is involved in some basic cellular functions.

A prominent feature of calmodulin is its thermal stability and acidic nature, which we exploited to devise a simple procedure for purifying the protein from bovine brain to apparent homogeneity (Lin *et al.*, 1974). Another striking feature of calmodulin is its lack of tissue or species specificity.

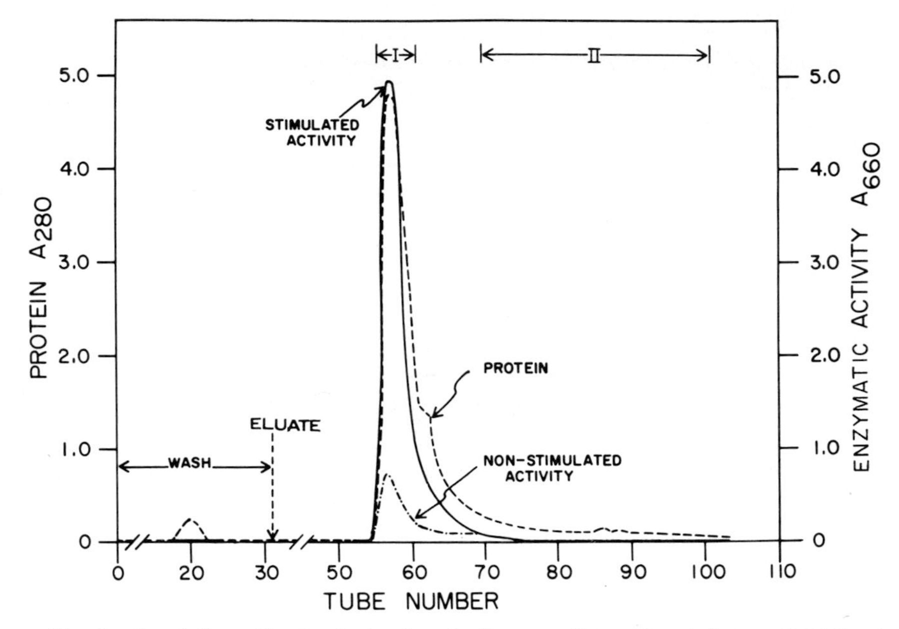

Fig. 1. Resolution of bovine brain phosphodiesterase from calmodulin on a DEAE-cellulose column. The column (2.5 × 42 cm) was equilibrated with 20 m*M* Tris-Cl (pH 7.5), and then charged with 240 mg of protein from a calcium phosphate eluate (Cheung, 1969). Each tube collected 10 ml. Approximately 300 ml of Tris-Cl was passed through the column as a wash, which contained no phosphodiesterase activity. The column was eluted with an exponential gradient generated from a lower reservoir containing 130 ml of 20m*M* Tris-Cl (pH 7.5) and an upper reservoir containing 1000 ml of 500 m*M* $(NH_4)2SO_4$ in 20 m*M* Tris-Cl (pH 7.5). Phosphodiesterase was assayed by the two-stage incubation, once with snake venom in the first stage of incubation to give the stimulated activity and again without venom to give the nonstimulated activity. Fraction II, which contained no phosphodiesterase activity, activated the enzyme and was the source of calmodulin used in Table IV. (From Cheung, 1971a.)

Although the regulation of phosphodiesterase by calmodulin appeared physiologically important, we were puzzled by the finding that calmodulin was always present in much higher concentration than the enzyme, especially in tissues without detectable calmodulin-dependent phosphodiesterase. This added to our lingering suspicion that calmodulin possessed other biological functions.

II. CALMODULIN AS A MULTIFUNCTIONAL REGULATORY PROTEIN

Teo and Wang (1973) first demonstrated that calmodulin is a Ca^{2+}-binding protein. This function was suggested indirectly in early studies on the metal requirement of phosphodiesterase. Mammalian phosphodiesterase was known to require divalent cations to express full activity. Kakiuchi and Yamazaki (1970) observed that calmodulin increased the sensitivity of phosphodiesterase to Ca^{2+}. Additionally, when purified to a stage at which sufficient calmodulin was retained for full activity, bovine brain phosphodiesterase was found to contain 1 mole of calcium per 250,000 g of enzyme (Cheung, 1971b). The question then was whether the divalent cation was associated with the enzyme or calmodulin, or both. The availability of pure calmodulin allowed experiments that resolved part of this question.

Measurements by atomic absorption spectrophotometry showed that calmodulin, as purified under our conditions, contained 1 mole of calcium (Lin *et al.*, 1974). This finding accounts for the calcium found in the partially purified phosphodiesterase. Using equilibrium dialysis, we found that calmodulin contains four Ca^{2+}-binding sites with dissociation constants ranging from 4 to 18 μM. Calmodulin from several different tissues was also found to have four Ca^{2+}-binding sites, although the dissociation constant showed considerable variation, probably because of different experimental conditions (Teo and Wang, 1973; Wolff *et al.*, 1977; Dedman *et al.*, 1977).

The subsequent finding that adenylate cyclase also requires calmodulin for maximal activity supported the notion that calmodulin has other functions (Brostrom *et al.*, 1975; Cheung *et al.*, 1975b). Even in brain, the level of calmodulin is higher than that of both adenylate cyclase and phosphodiesterase. In human erythrocytes, calmodulin is present in very high concentrations, yet there is neither detectable calmodulin-dependent phosphodiesterase nor adenylate cyclase (Smoake *et al.*, 1974). It was established in the early 1970s that the Ca^{2+}-ATPase of erythrocyte membrane requires a factor in the homolysate for activity (Bond and Clough, 1973; Luthra *et al.*, 1976). Recent studies have shown that the factor is identical to calmodulin (Gopinath and Vincenzi, 1977; Jarrett and Penniston, 1977). Within the past 2 years, numerous other calmodulin-regulated enzyme systems and cellular reactions have been reported (see following chapters). It now appears that calmodulin is the chief mediator of Ca^{2+} effects in eukaryotes. Several reviews on the central role of calmodulin in cellular regulation were recently published (Wolff and Brostrom, 1979; Cheung, 1980, Wang and Waisman, 1979; Klee *et al.*, 1980).

III. CALMODULIN AS A PROPER NAME

The fact that calmodulin regulates a wide variety of cellular functions and that research leading to the identification of these functions was done in many laboratories led to the use of different designations, for example, activator protein, modulator protein, calcium-dependent regulator (CDR), ATPase activator, phosphodiesterase-activating factor, and others. The use of different names to refer to the same protein caused considerable confusion in the literature. Many investigators felt that the adoption of a uniform name would be useful. We proposed "calmodulin" as the uniform name to denote that the protein is Ca^{2+}-modulated and that it also modulates the concentration of Ca^{2+} (Cheung *et al.*, 1978). An important consideration in this proposal was that calmodulin regulates the Ca^{2+}-ATPase in erythrocyte membranes, and since this enzyme is believed to be the calcium pump, calmodulin would regulate the concentration of Ca^{2+}. Subsequently, the protein was found to regulate the Ca^{2+}-ATPase associated with the synaptosome (Sobue *et al.*, 1979), the sarcoplasmic reticulum (Katz and Remtulla, 1978), and adipocyte (Pershadsingh *et al.*, 1980). Thus, the term "calmodulin" appears to reflect accurately the cellular functions of this versatile protein.

The response of the scientific community to the new designation was mixed. Many investigators expressed strong support, sharing our belief that the term appropriately reflects the functions of the protein. Others, who have used different terms, found it difficult to change. A few expressed reservations that, since all biological functions of calmodulin may not have been identified, a specific term could be premature. Despite some initial objections, the new term has gained wide acceptance in a relatively short time.

IV. EPILOGUE

Although calmodulin was discovered in the late 1960s, I did not publish a full account until later (Cheung, 1971a), as some aspects of the research were still incomplete. However, some of the basic features of calmodulin were known, and were reported in several meetings (Cheung, 1968, 1970a; Cheung and Jenkins, 1969; Cheung and Patrick, 1970). The following statements are taken from my early publications.

> "The fact that the crude phosphodiesterase is fully active while the 'purified' enzyme depends on venom for its optimal activity suggests that the stimulatory factor originally present with the enzyme must have dissociated during the course of its purification. Two lines of evidence support that such a dissociation has occurred. Firstly, a non-dialyzable substance has been obtained from the brain extract which shows no phosphodiesterase activity but is capable of activating the 'purified' enzyme. Secondly, the activity of a mixture of the crude and 'purified' enzyme is greater than the sum of the activities of the two enzymes assayed separately. Indeed, the activity of the mixture is comparable to the sum of the activities of the crude enzyme and the purified enzyme subsequent to its activation by the venom." (Cheung, 1967b)

"The factor was insensitive to deoxyribonuclease and ribonuclease, but was inactivated by tryptan. It was nondialyzable, heat stable, and appeared to be a protein." (Cheung, 1969)

Some investigators are not aware of these early findings, probably because the titles of these articles deal with the stimulation of phosphodiesterase by snake venom, and do not contain key words such as protein activator. These workers usually cite a subsequent note (Cheung, 1970b) as the first report on the discovery of calmodulin. The finding that phosphodiesterase requires calmodulin for maximum activity was soon confirmed in several laboratories (Kakiuchi and Yamazaki, 1970; Goren and Rosen, 1971; Uzunov and Weiss, 1972).

It might have been fortuitous that calmodulin was discovered during studies on cAMP metabolism, yet the function and metabolism of cAMP and calcium, two major regulators of cellular functions, have been closely linked since the emergence of the eukaryotes. Thus, it is perhaps not accidental that the abbreviation of the two are strikingly similar: cA and Ca. Indeed, in regulating the metabolism of both, calmodulin integrates the actions of both messengers (Cheung, 1980).

Calmodulin had a humble beginning. Discovered serendipitously as an activator of phosphodiesterase, it had to be studied intensively by many investigators before being established as an important mediator of Ca^{2+} action in eukaryotes. The following chapters not only describe aspects of calmodulin and its cellular functions but also provide numerous instances in which cA and Ca interact at the molecular and cellular levels.

This chapter merely serves as a personal account of how I remember the field of calmodulin has developed over the past decade or so, and, as such, probably depicts a biased view. Fortunately, many of the leading investigators describe their work in the following chapters. Others, who have also made important contributions, hopefully, will present their own accounts in future volumes of this treatise. Meanwhile, the serious reader is encouraged to consult the vast original literature.

ACKNOWLEDGMENTS

I am most grateful to Dr. Britton Chance, whose department gave birth to and nutured the early infancy of calmodulin. Without his generous support and unfailing encouragement, my work on calmodulin would not have been possible. I thank my colleagues, Y. M. Lin, Y. P. Liu, T. J. Lynch, J. A. Smoake, E. A. Tallant, and R. W. Wallace for their contributions to various phases of the work, and John Gilbert for editorial assistance. The research in my laboratory has been supported by grants from the USPHS and by ALSAC.

REFERENCES

Bond, G. H., and Clough, D. L. (1973). A soluble protein activator of (Mg^{2+} + Ca^{2+})-dependent ATPase in human red cell membranes. *Biochim. Biophys. Acta* **323,** 592–599.

Brostrom, C. O., Huang, Y. -C., Breckenridge, B. McL., and Wolff, D. J. (1975). Identification of a calcium binding protein as a calcium-dependent regulator of brain adenylate cyclase. *Proc. Natl. Acad. Sci. U.S.A.* **72,** 64–68.

Butcher, R. W., and Sutherland, E. W. (1962). Adenosine 3′,5′-phosphate in biological materials. 1. Purification and properties of cyclic 3′,5′-nucleotide phosphodiesterase and use of this enzyme to characterize adenosine 3′,5′-phosphate in human urine. *J. Biol. Chem.* **237,** 1244–1250.

Chance, B., Schoener, B., and Elsaesser, S. (1965). Metabolic control phenomena involved in damped sinusoidal oscillations of reduced diphosphoyridine nucleotide in a cell-free extract of *Saccharomyces carlsbergensis. J. Biol. Chem.* **240,** 3170–3181.

Cheung, W. Y. (1966a). Adenosine 3′,5′-phosphate and oscillations of DPNH in a cell-free extract of *Sccharomyces carlsbergensis. Biochim. Biophys. Acta* **115,** 235–239.

Cheung, W. Y. (1966b). Inhibition of cyclic nucleotide phosphodiesterase by adenosine 5′-triphosphate and inorganic pyrophosphate. *Biochem. Biophys. Res. Commun.* **23,** 214–219.

Chueng, W. Y. (1967a). Properties of cyclic 3′,5′-nucleotide phosphodiesterase from rat brain. *Biochemistry* **6,** 1079–1087.

Cheung, W. Y. (1967b). Cyclic 3′,5′-nucleotide phosphodiesterase: Pronounced stimulation by snake venom. *Biochem. Biophys. Res. Commun.* **29,** 478–482.

Cheung, W. Y. (1968). Activation of a partially inactive cyclic 3′,5′-nucleotide phosphodiesterase. *Fed. Proc., Fed. Am. Soc. Exp. Biol.* **27,** 783 (Abstr.).

Cheung, W. Y. (1969). Cyclic 3′,5′-nucleotide phosphodiesterase. Preparation of a partially inactive enzyme and its subsequent stimulation by snake venom. *Biochim. Biophys. Acta* **191,** 303–315.

Cheung, W. Y. (1970a). Cyclic nucleotide phosphodiesterase. *Adv. Biochem. Psychophormacol.* **3,** 51–65.

Cheung, W. Y. (1970b). Cyclic 3′,5′-nucleotide phosphodiesterase. Demonstration of an activator. *Biochem. Biophys. Res. Commun.* **38,** 533–538.

Cheung, W. Y. (1971a). Cyclic 3′,5′-nucleotide phosphodiesterase. Evidence for and properties of a protein activator. *J. Biol. Chem.* **246,** 2859–2869.

Cheung, W. Y. (1971b). Cyclic 3′,5′-nucleotide phosphodiesterase. Effect of divalent cations. *Biochim. Biophys. Acta* **242,** 395–409.

Cheung, W. Y. (1980). Calmodulin plays a pivotal role in cellular regulation. *Science* **207,** 19–27.

Cheung, W. Y., and Jenkins, A. (1969). Regulatory properties of cyclic 3′,5′-nucleotide phosphodiesterase. *Fed. Proc., Fed. Am. Soc. Exp. Biol.* **28,** 473 (Abstr.).

Cheung, W. Y., and Patrick, S. (1970). A protein activator of cyclic 3′,5′-nucleotide phosphodiesterase. *Fed. Proc., Fed. Am. Soc. Exp. Biol.* **29,** 602 (Abstr.).

Cheung, W. Y., and Salgonicoff, L. (1967). Cyclic 3′,5′-nucleotide phosphodiesterase. Localization and latent activity in rat brain. *Nature* (London) **214,** 90–91.

Cheung, W. Y., and Williamson, J. R. (1965). Kinetics of cyclic adenosine monophosphate changes in rat heart following epinephrine administration. *Nature* (*London*) **207,** 979–981.

Cheung, W. Y., Lin, Y. M., and Liu, Y. P. (1975a). Regulation of bovine brain cyclic 3′,5′-nucleotide phosphodiesterase by its protein activator. *In* "Cyclic Nucleotide in Disease" (B. Weiss, ed.), p. 321–350. University Press, Baltimore, Maryland.

Cheung, W. Y., Bradhom, L. S., Lunch, T. J., Lin, Y. M. W., and Tallant, E. A. (1975b). Protein activator of cyclic 3′,5′-nucleotide phosphodiesterase of bovine brain or rat brain also activates its adenylate cyclase. *Biochem. Biophys. Res. Commun.* **66,** 1055–1062.

Cheung, W. Y., Lynch T. J., and Wallace, R. W. (1978). An endogenous Ca^{2+}-dependent activator protein of brain adenylate cyclase and cyclic nucleotide phosphodiesterase. *Adv. Cyclic Nucleotide Res.* **9,** 233–251.

Dedman, J. R., Potter, J. D., Jackson, R. L., Johnson, D., and Means, A. R. (1977). Physiochemical properties of rat testis Ca^{2+}-dependent regulator protein of cyclic nucleotide phosphodiesterase. *J. Biol. Chem.* **252,** 8415–8422.

Gopinath, R. M., and Vincenzi, F. F. (1977). Phosphodiesterase protein activator mimics red blood cell cytoplasmic activator of (Ca^{2+} – Mg^{2+}) ATPase. *Biochem. Biophys. Res. Commun,* **77,** 1203–1209.

Goren, E. N., and Rosen, D. M. (1971). The effect of nucleotides and a nondialysable factor on the hydrolysis of cyclic AMP by a cyclic nucleotide phosphodiesterase from beef heart. *Arch. Biochem. Biophys.* **142,** 720–723.

Jarrett, H. W., and Penniston, J. T. (1977). Partial purification of the Ca^{2+} – Mg^{2+} ATPase activator from human erythrocytes: Its similarity to the activator of 3′:5′-cyclic nucleotide phosphodiesterase. *Biochem. Biophys. Res. Commun.* **77,** 1210–1216.

Kakiuchi, S., and Yamazaki, R. (1970). Calcium dependent phosphodiesterase activity and its activating factor (PAF) from brain. III. Studies on cyclic 3′,5′-nucleotide phosphodiesterase. *Biochem. Biophys. Res. Commun.* **41,** 1104–1110.

Katz, S., and Remtulla, M. A. (1978). Phosphodiesterase protein activator stimulates calcium transport in cardiac microsomal preparations enriched in sarcoplasmic reticulum. *Biochem. Biophys. Res. Commun.* **83,** 1373–1379.

Klee, C. B., Crouch, T. H. and Richman, P. (1980). Calmodulin. *Annu. Rev. Biochem.* **49,** 489–515.

Lin, Y. M., Liu, Y. P., and Cheung, W. Y. (1974). Cyclic 3′,5′-nucleotide phosphodiesterase. Purification, characterization, and active form of the protein activator from bovine brain. *J. Biol. Chem.* **249,** 4943–4954.

Luthra, M. G., Hildebrant, G. R., and Hanahan, D. J. (1976). Studies on an activator of the (Ca^{2+} + Mg^{2+})-ATPase of human erythrocyte membrane. *Biochim. Biophys. Acta* **419,** 164–179.

Lynch, T. J., Tallant, E. A., and Cheung, W. Y. (1975). Separate genetic regulation of cyclic nucleotide phosphodiesterase and its protein activator in cultured mouse fibroblasts. *Biochem. Biophys. Res. Commun.* **63,** 967–970.

Pershadsingh, H. A., Landt, M., and McDonald, J. M. (1980). High affinity ATP-dependent Ca^{2+} transport in adipocyte plasma membranes—stimulation by calmodulin. *Fed. Proc., Fed. Am. Soc. Exp. Biol.* **39,** 958.

Robison, G. A., Butcher, R. W., and Sutherland E. W. (1971). "Cyclic AMP," pp. 1–47. Academic Press, New York.

Smoake, J. A., Song, S. -Y., and Cheung, W. Y. (1974). Cyclic 3′,5′-nucleotide phosphodiesterase: Distribution and developmental changes of the enzyme and its protein activator in mammalian tissues and cells. *Biochim. Biophys. Acta* **341,** 402–411.

Sobue, K., Ichida, S., Yoshida, H., Yamazaki, R., and Kakiuchi, S. (1979). Occurrence of a Ca^{2+} and modulator protein. activatable ATPase in the synaptic plasma membranes of brain. *FEBS Lett.* **99,** 199–202.

Sutherland, E. W., Roll, T. W., and Menon, T. (1962). Adenyl cyclase. I. Distribution, preparation and properties. *J. Biol. Chem.* **237,** 1220–1227.

Teo, T. S., and Wang, J. H. (1973). Mechanism of activation of a cyclic adenosine 3′:5′-monophosphate phosphodiesterase from bovine heart by calcium ions. *J. Biol. Chem.* **248,** 5950–5955.

Uzunov, P., and Weiss, B. (1972). Separation of multiple molecular forms of cyclic adenosine-3′,5′-monophosphate phosphodiesterase in rat cerebellum by polyacrylamide gel electrophoresis. *Biochim. Biophys. Acta* **284,** 220–226.

Wang, J. H., and Waisman, D. M. (1979). Calmodulin and its role in the second messenger system. *Curr. Top. Cell. Regul.* **15,** 47–107.

Wolff, D. J., and Brostrom, C. O. (1979). Properties and functions of the Ca^{2+}-dependent regulator protein *Adv. Cyclic Neucleotide Res.* **11,** 28–88.

Wolff, D. J., Poirier, P. G., Brostrom, C. O., and Brostrom, M. A. (1977). Divalent cation binding properties of bovine brain Ca^{2+}-dependent regulator protein. *J. Biol. Chem.* **252,** 4108–4117.

Chapter 2

Assay, Preparation, and Properties of Calmodulin

ROBERT W. WALLACE
E. ANN TALLANT
WAI YIU CHEUNG

I. INTRODUCTION

Calmodulin, originally discovered as a calcium-dependent activator of cyclic nucleotide phosphodiesterase (Cheung, 1967, 1970; Kakiuchi and Yamazaki, 1970), is a multifunctional modulator protein, mediating the effects of Ca^{2+} in a variety of cellular reactions and processes (Cheung, 1980; Wolff and Brostrom, 1979). In addition to phosphodiesterase, calmodulin is known to regulate skeletal muscle phosphorylase kinase (Cohen *et al.*, 1978; Depaoli-Roach *et al.*, 1979), myosin light chain kinase

CALCIUM AND CELL FUNCTION, VOL. I

ISBN 0-12-171401-2

(Dabrowska and Hartshorne, 1978; Dabrowska *et al.*, 1978; Sherry *et al.*, 1978; Waisman *et al.*, 1978), NAD kinase in both plants (Anderson and Cormier, 1978) and animals (Epel *et al.*, 1980), phospholipase A_2 (Wong and Cheung, 1979), adenylate cyclase (C. O. Brostrom *et al.*, 1975, 1977; M. A. Brostrom *et al.*, 1976; Cheung *et al.*, 1975; Lynch *et al.*, 1977), guanylate cyclase (Nago *et al.*, 1979), Ca^{2+}-ATPase (Gopinath and Vincenzi, 1977; Jarrett and Penniston, 1977, 1978; Hanahan *et al.*, 1978; Lynch and Cheung, 1979; Niggli *et al.*, 1979; Sobue *et al.*, 1979), Ca^{2+} transport in erythrocytes (Hinds *et al.*, 1978; Larsen and Vincenzi, 1979) and sarcoplasmic reticulum (Katz and Remtulla, 1978), phosphorylation of membranes (Schulman and Greengard, 1978a,b), neurotransmitter release (DeLorenzo *et al.*, 1979), and the disassembly of microtubules (Marcum *et al.*, 1978; Nishida *et al.*, 1979). Moreover, there are numerous calmodulin-binding proteins whose functions have not been elucidated, suggesting that there may be additional roles of calmodulin (Wang and Desai, 1976, 1977; Klee and Krinks, 1978; Wallace *et al.*, 1978, 1979; LaPorte and Storm, 1978; Sharma *et al.*, 1978, 1979).

The mechanism by which calmodulin regulates enzymatic activity has been studied with phosphodiesterase (Teo and Wang, 1973; Kakiuchi *et al.*, 1973; Lin *et al.*, 1974), adenylate cyclase (Lynch *et al.*, 1977), and Ca^{2+}-ATPase (Lynch and Cheung, 1979). As depicted in the following scheme, Ca^{2+} is bound to calmodulin resulting in an active conformation. The calmodulin · Ca^{2+} complex (CaM* · Ca^{2+}) interacts with the apoenzyme (E) to give the active holoenzyme (CaM* · Ca^{2+} · E*).

$$\mathrm{CaM} + \mathrm{Ca}^{2+} \rightleftharpoons \mathrm{CaM}^* \cdot \mathrm{Ca}^{2+} \tag{1}$$

$$\underset{\text{(less active)}}{\mathrm{CaM}^* \cdot \mathrm{Ca}^{2} + E} \rightleftharpoons \underset{\text{(activated)}}{\mathrm{CaM}^* \cdot \mathrm{Ca}^{2+} \cdot \mathrm{E}^*} \tag{2}$$

Phosphorylase kinase contains 1 mole of calmodulin as an integral subunit; the enzyme is further stimulated by exogenous calmodulin (Cohen *et al.*, 1978; Depaoli-Roach *et al.*, 1979), presumably in a manner analogous to the scheme above. Myosin light chain kinase is of special interest; its activity is absolutely dependent on calmodulin (Dabrowska and Hartshorne, 1978).

In this chapter, we describe the general principles involved in the assay of calmodulin by the phosphodiesterase system and by a specific radioimmunoassay developed recently in our laboratory. The advantages of the two systems are compared. We next present the procedures currently available for the isolation of calmodulin from various tissues; the experimental details for preparing bovine brain calmodulin is given as an example. Finally, we summarize the salient properties of calmodulin.

II. MATERIALS

Freund's complete adjuvant was purchased from Miles Laboratories; sodium borohydride, 1,4-butanediol diglycidyl ether, PIPES buffer, Tris (hydroxymethyl) aminomethane (Tris), and dithiothreitol (DTT) were obtained from Sigma Chemical Co.; and 1-fluoro-2,4-dinitrobenzene (FDNB) was from Pierce Chemical Co. ^{3}H-cAMP (20 Ci/mmole) and ^{125}I (16 Ci/mmole) were obtained from Schwarz/Mann; ^{3}H-cAMP was purified by thin-layer chromatography on cellulose sheets (Eastman) with a solvent mixture of isopropanol, ammonium hydroxide, and water (7:1:2) (Lynch and Cheung, 1975). Fluphenazine · HC1 was a gift from E. R. Squibb and Sons, Inc. Acrylamide, *N,N'*-methylene-bisacrylamide (Bis), sodium dodecyl sulfate (SDS), and Coomassie brilliant blue were supplied by Bio-Rad Laboratories. All other reagents were of highest analytical grades.

Calmodulin was prepared either by a modification (Wallace and Cheung, 1979b) of a previous procedure (Lin *et al.*, 1974) or by the fluphenazine-Sepharose affinity technique based on the method of Charbonneau and Cormier (1979). Calcium-dependent phosphodiesterase from bovine brain was purified through the DEAE-cellulose stage (Cheung, 1971). The enzyme at this stage is calmodulin deficient and dependent upon exogenous calmodulin for full activity. The heat-labile calmodulin-binding protein ($CaM\text{-}BP_{80}$) was prepared from bovine brain (Wallace *et al.*, 1979). Goat anti-rabbit immunoglobulin F_c was a gift from Dr. William Walker, St. Jude Children's Research Hospital. Rabbit skeletal muscle troponin and its individual subunits were provided by Dr. Thomas Vanaman, Duke University.

III. ASSAY OF CALMODULIN

A. Preparation of Tissue Extract

Bovine brain cerebrum was obtained fresh from a local abattoir and cleaned of blood clots. The tissue, used fresh or after storage at −20°C, was homogenized in 2 volumes of buffer [50 m*M* Tris-HCl (pH 7.8), 3 m*M* $MgSO_4$, and 1 m*M* EGTA] in a Polytron (Brinkman) for three 1-min periods at a setting of 3. The homogenate was centrifuged at 100,000 *g* for 1 hr. The supernatant was heated rapidly to 95°C in a boiling water bath for 4.5 min; the denatured proteins were removed by centrifugation. The supernatant was dialyzed against 20 m*M* sodium phosphate buffer (pH 7.2) containing 0.15 *M* NaCl (NaCl/P_i). Appropriate fractions were measured for calmodulin either by the phosphodiesterase system or the radioimmunoassay.

Heat treatment of a tissue extract is a convenient means of removing extraneous proteins, and this treatment removes little or no [^{125}I] calmodulin that was added to monitor recovery (Wallace and Cheung, 1979b).

B. Assay of Calmodulin by Phosphodiesterase

The assay of calmodulin by phosphodiesterase is based on the stimulation of phosphodiesterase activity under specified conditions (Wallace *et al.*, 1978). One unit of calmodulin may be arbitrarily defined as the amount of protein giving half-maximum stimulation of phosphodiesterase activity. If a pure calmodulin is available, one can determine the absolute amount from a standard curve (Fig. 1).

Theoretically any of the calmodulin-dependent enzymes enumerated in Section I can be used to assay calmodulin. Phosphodiesterase has been widely used because the enzyme can be prepared easily in bulk (Cheung, 1971) and was the first responsive system to be studied (Cheung, 1967, 1970; Kakiuchi and Yamazaki, 1970).

Bovine brain contains several calmodulin-binding proteins which are capable of competing with phosphodiesterase for calmodulin (Wang and Desai, 1976, 1977; Klee and Krinks, 1978; Wallace *et al.*, 1978, 1979; La-Porte and Storm, 1978; Sharma *et al.*, 1978, 1979). In fact, all calmodulin-responsive enzymes are potential competitors. The presence of any of these proteins in a tissue extract will cause an underestimation of calmodulin. Figure 1 shows the effect of CaM-BP_{80} on the assay of calmodulin by the phosphodiesterase system. The presence of CaM-BP_{80} shifts the dose–response curve to the right, giving an apparently lower level of calmodulin in the extract.

C. Radioimmunoassay of Calmodulin

1. Radioiodination of Calmodulin

Calmodulin and DNB_3-calmodulin were iodinated according to a modification of the chloramine T procedure of Hunter and Greenwood (1962). The chloramine T iodination procedure proceeds in a strong oxidizing environment which would oxidize tryptophan or cysteine, but these residues are not present in calmodulin. The [^{125}I]calmodulin had a specific activity of 1560 Ci/mmole. DNB_3-calmodulin was labeled to approximately

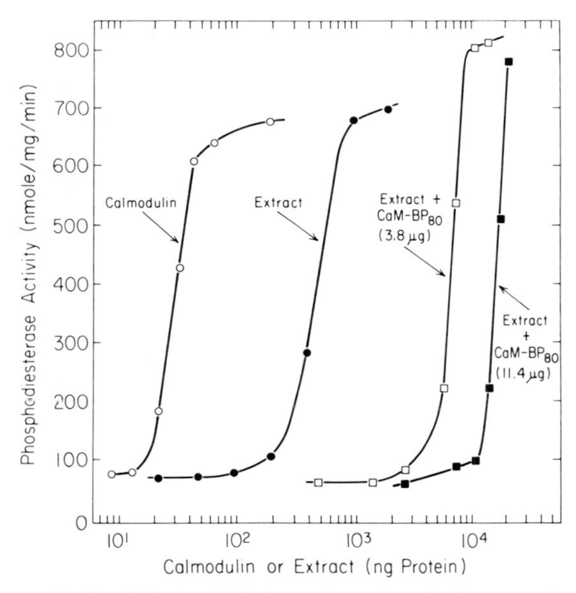

Fig. 1.. Assay of calmodulin by Ca^{2+}-dependent phosphodiesterase in the presence or absence of CaM-BP_{80}. The reaction mixture (0.1 ml) contained 40 m*M* Tris-HCl (pH 8.0), 5 m*M* $MgSO_4$, 50 μM $CaCl_2$, 10 μg calmodulin-deficient phosphodiesterase (Cheung, 1971), 2 m*M* ^{3}H-cAMP, an appropriate amount of calmodulin or heat-treated (95°C, 5 min) bovine brain extract, and the indicated amount of CaM-BP_{80}. The incubation was for 5 min at 30°C, and the assay was terminated by boiling for 1 min. 5′-AMP was converted to adenosine and inorganic phosphate by incubation with 20 μg snake venom (*Crotalus atrox*) for 10 min at 30°C. Adenosine was separated from unreacted cAMP by the addition of 1 ml of 33% solution of anion exchange resin AG1-X2 (Bio-Rad). The tubes were vortexed and centrifuged at low speed, and a 0.5 ml aliquot of the supernatant was removed for liquid scintillation counting. The anion exchange resin binds approximately 30% of the adenosine nonspecifically (Lynch and Cheung, 1975); the data have been corrected for this binding. Note that the presence of CaM-BP_{80} causes a shift in the response curve, causing an underestimation of tissue level of calmodulin. In the absence of CaM-BP_{80}, the brain extract contains 74 mg calmodulin/kg tissue. In the presence of 3.8 and 11.4 μg CaM-BP_{80} per reaction mixture, the level of calmodulin became 5.8 and 2.2 mg/kg, respectively. (From Wallace and Cheung, 1979b.)

the same extent. Calmodulin has two tyrosines, one of which is buried while the other is exposed to the solvent (Klee, 1977). The incorporation of 1 mole of ^{125}I per mole of calmodulin does not impair significantly its biological activity, as judged by its ability to stimulate phosphodiesterase. However, incorporation of 2 moles of ^{125}I per mole of calmodulin results in a marked loss of activity (Richman and Klee, 1978). In our experiments, an average of approximately 0.8 mole of ^{125}I was incorporated per mole of calmodulin.

2. *Preparation of Anti-Calmodulin Serum*

Calmodulin is a poor antigen, presumably because it is small and acidic and its primary structure is highly conserved. Several previous attempts to prepare an antibody against calmodulin in this laboratory have not been successful. The procedures included injecting rabbits with calmodulin adsorbed to poly (L-lysine) (Vunakis *et al.*, 1972), polymerized with ethyl chloroformate to form an insoluble complex (Avrameas and Ternynck, 1967), or covalently coupled to thyroglobulin (Skowsky and Fisher, 1972). In addition, the thyroglobulin–calmodulin complex was injected into chickens and frogs. In all cases, little or no antibody was produced which recognized the native calmodulin. However, incorporation of several dinitrophenyl groups into calmodulin render it highly antigenic, eliciting an antibody which recognizes the dinitrophenyl portion of the conjugate as well as the protein (Wallace and Cheung, 1979b).

As is indicated in Fig. 2, incorporation of an average of 3 moles of DNB per mole of calmodulin (DNB_3-calmodulin) has little effect on its ability to

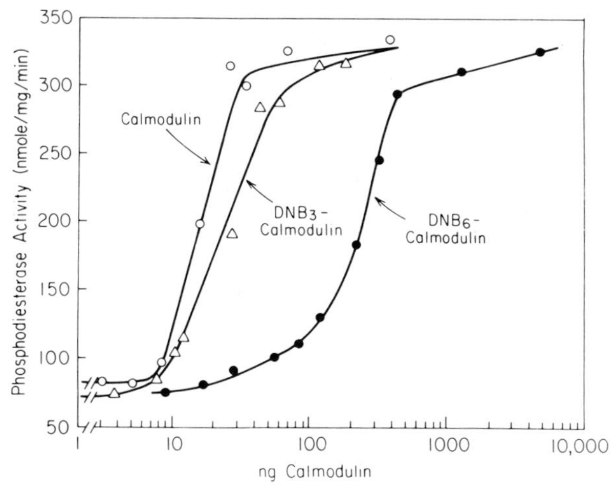

Fig. 2. Effect of dinitrophenylation on the biological activity of calmodulin. Dinitrophenylated calmodulin was prepared by combining 250 μl of calmodulin (4 mg/ml), 25 μl of 0.7 *M* $NaHCO_3$ and 50 μl 1-fluoro-2,4-dinitrobenzene (FDNB). The reaction mixture was vortexed for 15 or 60 min at room temperature and chromatographed on a Sephadex G-25 column (14 × 0.7 cm previously equilibrated with NaCl/Pi) to separate the DNB-calmodulin from the unreacted FDNB. The dinitrobenzene group (DNB) was determined by absorption at 365 nm using a molar extinction coefficient for DNB-L-lysine, $\epsilon = 1.74 \times 10^4$ (Little and Counts, 1969). Protein was determined by Lowry *et al.* (1951). Calmodulin was assayed by its ability to stimulate bovine brain phosphodiesterase as described in the legend to Fig. 1. The ratio of DNB/calmodulin is indicated by the subscripts. (From Wallace and Cheung, 1979b.)

stimulate phosphodiesterase. Increasing the hapten to 6 dinitrophenyl groups greatly reduces its ability to stimulate phosphodiesterase.

The DNB_3-calmodulin was used to immunize rabbits. Antibodies produced in the early stage of immunization recognize the derivatized and not the underivatized molecule, and appear specific for the dinitrophenyl moiety. With subsequent boosting, an antibody was produced which recognizes both the derivatized and the underivatized molecule. The antisera produced by day 72 was suitable for development of a radioimmunoassay for calmodulin (Wallace and Cheung, 1979b).

3. *Radioimmunoassay*

The radioimmunoassay for calmodulin utilizes the double antibody method in which bound [^{125}I]calmodulin is separated from the free by quantitative precipitation of the rabbit IgG with goat anti-rabbit immunoglobulin serum.

Figure 3A shows the binding of [^{125}I]calmodulin or ^{125}I-DNB_3-calmodulin to varying amounts of the antiserum. Consistent with the data in Fig. 2, the antibody had a much higher titer for ^{125}I-DNB_3-calmodulin than for [^{125}I]calmodulin. Binding of ^{125}I-DNB_3-calmodulin could be detected with 0.001 μl of antisera in the reaction mixture of 0.3 ml (300,000-fold dilution), whereas 0.10 μl antisera (3000-fold dilution) was necessary to detect the binding of [^{125}I]calmodulin, indicating the presence of high titers of antibody specific for the DNB moiety.

The binding of [^{125}I]calmodulin was blocked by preincubating the serum with unlabeled calmodulin or DNB_3-calmodulin (Fig. 3B). Fifty percent inhibition of binding of labeled antigen was obtained with 250 and 23 ng of calmodulin and DNB_3-calmodulin, respectively. The sensitivity of the assay was approximately 20 ng (1 pmole) for calmodulin or 2 ng (0.1 pmole) for DNB_3-calmodulin. The useful sample range lies between 20 and 2000 ng for calmodulin and 2 and 1000 ng for DNB_3-calmodulin. The sensitivity of the assay is comparable to the assay by phosphodiesterase, but the useful range is considerably greater. The sensitivity of both assays is sufficient for most tissue extracts, since they usually contain high levels of calmodulin.

Recently, Chafouleas *et al.* (1979) described a radioimmunoassay for calmodulin utilizing an antibody produced in sheep against the underivitized calmodulin. They claimed that their radioimmunoassay had a 300-fold greater sensitivity; however, direct comparison of inhibition curves for [^{125}I]calmodulin binding to the antisera indicates approximately a tenfold difference. The greater sensitivity of their assay probably results from a higher specific radioactivity of [^{125}I]calmodulin, the use of purified

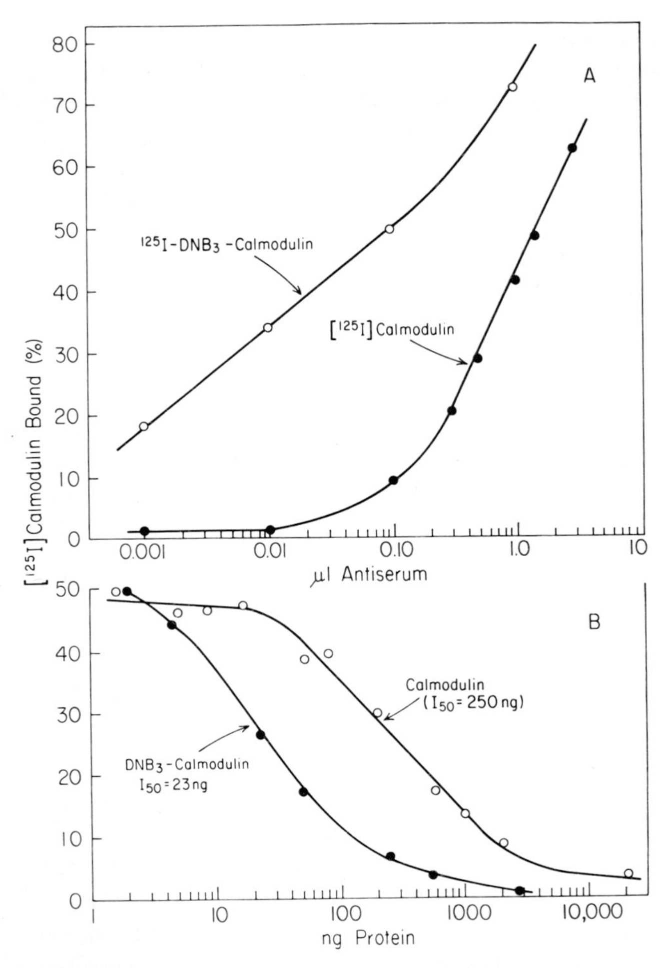

Fig. 3. Radioimmunoassay of calmodulin. (*A*) Binding of [^{125}I]calmodulin and ^{125}I-DNB$_3$-calmodulin to rabbit anti-calmodulin serum. The assay was conducted in BDS polymer tubes (12 × 75 mm) (Evergreen Scientific) and the reaction mixture (300 μl) contained NaCl/P$_i$, 0.02% sodium azide, 0.05% Triton X-100, and 3 m*M* EGTA. An appropriate amount of nonimmune rabbit serum was added so that each tube contained the equivalent of 3 μl of rabbit serum. [^{125}I]Calmodulin or ^{125}I-DNB$_3$-calmodulin (13,000 cpm, approximately 65 pg) was added, and the reaction mixture was incubated for 30 min at 30°C, and then 24 hr at 4°C, with constant shaking. Thirty microliters of undiluted goat anti-rabbit IgG serum was added and the reaction mixture was incubated overnight at 4°C. The precipitate was collected by centrifugation at 20,000 *g* for 20 min; the supernatant fluid was removed by aspiration, and the tube containing the IgG precipitate was counted in a Nuclear Chicago gamma spectrophotometer. The data, expressed as the mean of two determinations, were corrected for nonspecific binding to the tube and to IgG. The nonspecific binding was approximately 2% of the total counts. (*B*) Effect of various concentrations of calmodulin and DNB$_3$-calmodulin on the binding of [^{125}I]calmodulin to the antibody. The assay was conducted as de-

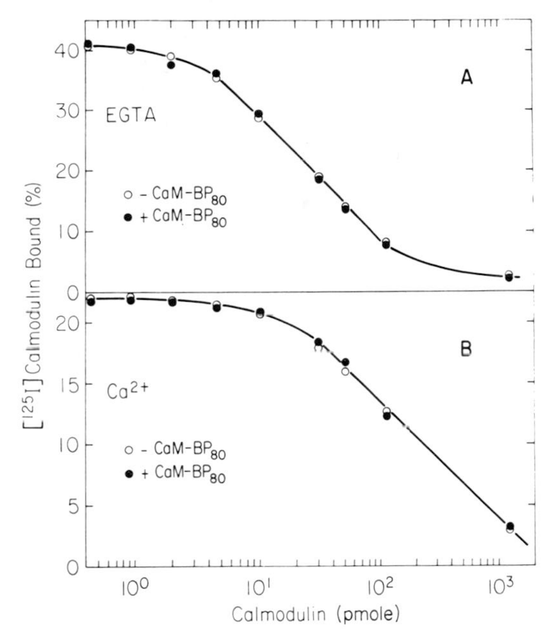

Fig. 4. Radioimmunoassay of calmodulin in the presence or absence of CaM-BP_{80}. The radioimmunoassay was conducted as described in the legend to Fig. 3B except the reaction mixture contained either 3 m*M* EGTA (A) or 0.3 m*M* Ca^{2+} (B). When present, the concentration of CaM-BP_{80} was 3.8 μg or 48 pmole/assay tube. (From Wallace and Cheung, 1979b.)

IgG in the assay system, or the use of native antigen in the development of antibodies.

Although CaM-BP_{80} interferes with the enzyme assay of calmodulin, resulting in its underestimation, it does not affect the radioimmunoassay, even in the presence of Ca^{2+} (Fig. 4). This suggests that the antigenic sites of calmodulin are accessible to the antibodies even in its ternary complex with CaM-BP_{80}. An alternative explanation is that the affinity of calmodulin for the antibody is greater than that for CaM-BP_{80}, and the long incubation period of the radioimmunoassay allows a new equilibrium in favor of the formation of the antigen-antibody complex. The radioimmunoassay is routinely conducted in the presence of EGTA. When EGTA is replaced

---

Fig. 3 (*continued*) scribed for (A) except for the following. Each tube contained 50 μl anti-calmodulin serum (diluted 50-fold with NaCl/P_i) to give the equivalent of 1 μl of undiluted serum in each tube. Different amounts of calmodulin or DNB_3-calmodulin (10^1-10^4 ng) were added, and the tubes were incubated for 30 min at 30°C before the addition of [^{125}I]calmodulin (13,000 cpm). Ten microliters of undiluted goat anti-rabbit IgG serum was used to precipitate the rabbit IgG. (From Wallace and Cheung, 1979b.)

with Ca^{2+}, the maximum binding of [^{125}I]calmodulin to the antisera is decreased and the binding curve shifted to the right. Calmodulin increases its helical structure upon binding Ca^{2+} (Liu and Cheung, 1976; Klee, 1977; Dedman *et al.*, 1977b; Wolff *et al.*, 1977), and the new conformation may decrease its avidity for the antibody, for reasons not yet apparent.

The antiserum exhibits a high degree of specificity toward calmodulin. Neither troponin nor its individual subunits—troponin I, troponin T, or troponin C—compete with [^{125}I]calmodulin for binding to the antibodies in the radioimmunoassay (Fig. 5). Troponin C is similar to calmodulin, having approximately 70% conservative and 50% direct sequence homology (Vanaman *et al.*, 1977). CaM-BP_{80} contains an 18,500-dalton subunit which binds Ca^{2+} (Wallace *et al.*, 1979; Klee *et al.*, 1979). None of these proteins compete with [^{125}I]calmodulin for binding to the antiserum, even at concentrations 1000-fold higher. These results demonstrate that the radioimmunoassay is highly specific for calmodulin.

D. Comparison of the Two Assays of Calmodulin

The assay of calmodulin by phosphodiesterase has been used widely. It is simple and easy to set up and the sensitivity is sufficient for measuring calmodulin in almost all tissues. The method appears accurate, provided that the tissue extract does not contain other calmodulin-binding proteins that interfere with the measurement. In this regard, the remarkable stability of calmodulin may be exploited to circumvent most, if not all, of the

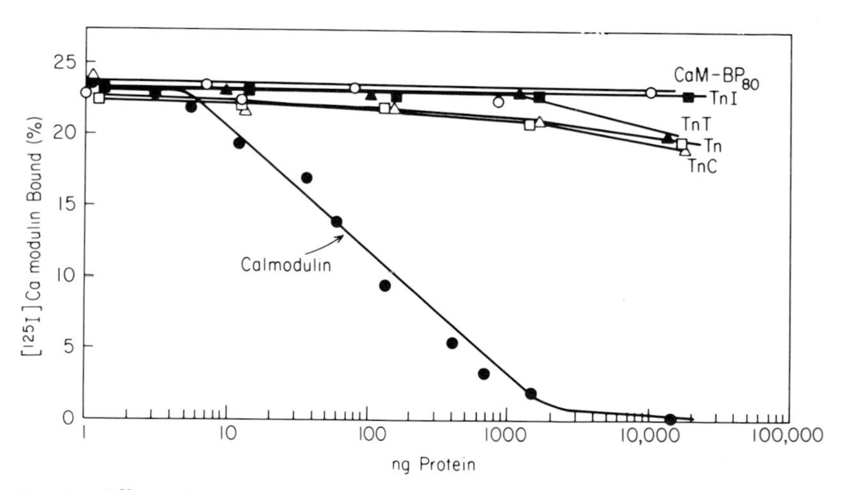

Fig. 5. Effect of various proteins on the binding of [^{125}I]calmodulin to anti-calmodulin serum. TnI, troponin I; TnT, troponin T; TnC, troponin C; Tn, troponin. (From Wallace and Cheung, 1979b.)

interfering proteins. Brief boiling of the tissue extract would denature the bulk of extraneous proteins which could be easily removed by centrifugation. Moreover, calmodulin withstands exposure to low pH and 8 *M* urea, conditions that denature many proteins. Thus, a judicious use of one or more of these harsh treatments on a tissue extract could allow an accurate quantitation of calmodulin in most tissues by the enzyme assay.

Compared to the assay of calmodulin by phosphodiesterase, the radioimmunoassay is time consuming and technically more complex. The main advantage is that it measures calmodulin on the basis of antigenic determinants rather than on bioligical activity, and would allow the study of protein turnover using immunoprecipitation techniques. In addition, the presence in tissue extracts of calmodulin-binding proteins that potentially interfere with the enzyme assay does not appear to affect the accuracy of the radioimmunoassay.

The assay of calmodulin by phosphodiesterase complements that by radioimmunoassay. With the increasing role of calmodulin in biological regulation, the availability of both biological and radioimmunoassay adds to the arsenal of tools at the disposal of the investigators.

IV. PREPARATION OF CALMODULIN

A variety of procedures have been described for preparing calmodulin from different tissues. In our laboratory, the protein was originally isolated from bovine brain by a technique which included heat denaturation of the bulk of extraneous protein in the tissue extract, followed by chromatography on diethylaminoethyl (DEAE)-cellulose and preparative polyacrylamide gel electrophoresis (Lin *et al.*, 1974). Other purification procedures avoid the heat denaturation step (Klee, 1977; Watterson *et al.*, 1976). Chromatography on a DEAE column is an effective step because calmodulin is much more acidic than most of the proteins in the tissue extract. Many procedures utilize this as a major step in the purification scheme, which affords milligram quantities of calmodulin 50–90% pure. As a final step in the protocol, other investigators used gel permeation (Teo *et al.*, 1973; Dedman *et al.*, 1977b), hydroxyapatite (Klee, 1977), and rechromatography on a DEAE column (Watterson *et al.*, 1976).

As properties of calmodulin are better understood, simpler and more efficient isolation procedures have been devised. Calmodulin binds troponin I in a calcium-dependent manner even in the presence of high concentrations of urea. Exploiting this unusual affinity, Grand *et al.* (1979) employed a troponin I-agarose affinity column as a major step in their purification of calmodulin from rabbit uterus.

Antipsychotic drugs of the phenothiazine type, such as trifluoperazine and chloropromazine, also bind calmodulin in the presence of Ca^{2+}, a property cleverly exploited to give a simple and efficient procedure to prepare calmodulin in high yields (Charbonneau and Cormier, 1979; Jamieson and Vanaman, 1979). The following section describes a protocol using the phenothiazine affinity column of Charbonneau *et al.* (1979; see also Cormier *et al,* this volume, Chapter 10).

Preparation of Calmodulin with a Fluphenazine-Sepharose Affinity Column

Fluphenazine is one of the psychoactive phenothiazine drugs having a hydroxyl group that can be coupled to an insoluble matrix. Sepharose-4B was activated by the bisoxrine procedure (Sunberg and Porath, 1974), and fluphenazine was coupled to the matrix essentially as described by Charbonneau and Cormier (1979). An experiment demonstrating the use of a fluphenazine affinity column to prepare calmodulin from bovine brain is presented i⁻ Fig. 6. In this experiment, an extract of bovine brain was loaded on the affinity column in the presence of Ca^{2+}, which allows calmodulin to bind to the immobilized fluphenazine. The bulk of extraneous proteins passed through the column. Washing the column with a buffer containing 0.5 *M* NaCl resulted in a small peak of protein that had no calmodulin activity. Elution with a buffer containing 10 m*M* EGTA released quantitatively the calmodulin bound to the column. The sample obtained from the column appeared homogeneous in SDS polyacrylamide gel electrophoresis (Fig. 7). Table I summarizes the quantitative data for the purification procedure. Calmodulin ws purified 250-fold in one step with a 72% yield. The specific activity of 3.7×10^4 units/mg protein is comparable to calmodulin isolated by conventional methods (Wallace *et al.,* 1979).

The use of a phenothiazine affinity column for the preparation of calmodulin appears superior to conventional methods; the procedure does not involve harsh conditions and is simple and efficient, giving a homogeneous product essentially by a single step. Moreover, the affinity column can be easily regenerated for repeated use.

V. PROPERTIES OF CALMODULIN

A. Physical and Chemical Properties

Calmodulin is a small, globular protein containing 148 amino acids with a molecular weight of 16,723 (Vanaman *et al.,* 1977). Some fifty of the

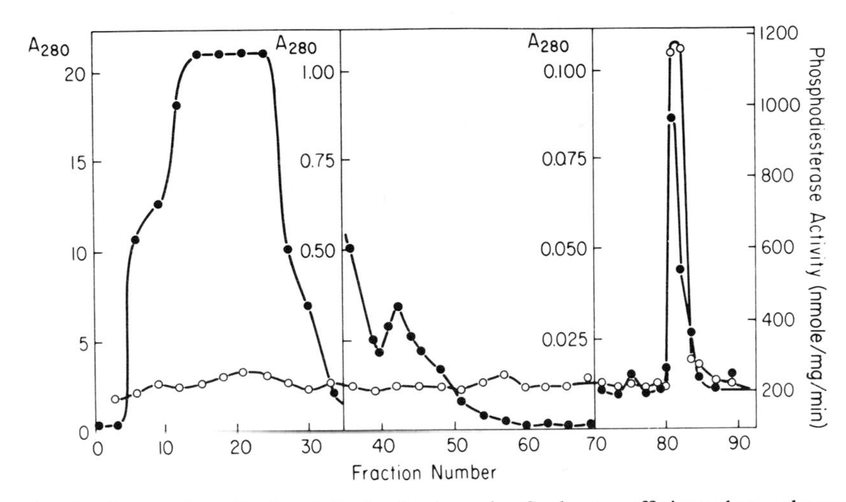

Fig. 6. Preparation of calmodulin by fluphenazine-Sepharose affinity column chromatography. One hundred grams of washed Sepharose were mixed with 100 ml of 0.6 *N* NaOH containing 3 mg/ml sodium borohydride and 100 ml of 1,4-butanediol diglycidyl ether. The suspension was incubated at 22°C for 8 hr with gentle stirring. The unreacted ether was removed by washing the activated Sepharose with 6 liters of water on a glass funnel. The activated Sepharose was combined with 200 ml of 0.1 *M* sodium carbonate (pH 11.0) containing 400 mg of fluphenazine. The suspension was incubated at 65°C for 48 hr with constant shaking. At the end of this period, the fluphenazine-Sepharose was washed in a glass funnel with 0.5 liter of 0.1 *M* sodium carbonate (pH 11.0), 1 liter of acetone, 0.5 liter of ethanol, and then 0.5 liter of 10 m*M* PIPES (pH 7.0). The Sepharose was poured into a column (2.6 × 29 cm) wrapped in aluminum foil and was equilibrated with 10 m*M* PIPES (pH 7.0). Fluphenazine is not soluble in the aqueous solution used for coupling; the coupling efficiency was not estimated. A tissue extract was prepared by homogenizing 100 g of bovine cerebra with 300 ml of 10 m*M* PIPES (pH 7.0) containing 1 m*M* EGTA and 0.02% NaN_3 for 2 min in a Waring blender. The homogenate was centrifuged at 10,000 g for 1 hr, and the supernatant fluid was made 2 m*M* with $CaCl_2$ and applied to the fluphenazine-Sepharose column at a flow rate of 1 ml/min. The column was washed with 100 ml of PIPES (pH 7.0) containing 1 m*M* $CaCl_2$ (fractions 1–35), and was eluted first with 350 ml of 10 m*M* PIPES (pH 7.0) containing 1 m*M* $CaCl_2$ and 0.5 *M* NaCl (fractions 36–70) and then with 250 ml of 50 m*M* Tris-HCl (pH 8.0) containing 0.5 *M* NaCl and 10 m*M* EGTA (fractions 71–95). Ten milliliter fractions were collected and aliquots (2 μl) were assayed for calmodulin with the phosphodiesterase system as described in the legend to Fig. 1. A_{280}, ●—●—●; phosphodiesterase activity, ○—○—○. The affinity column can be used repeatedly over a period of at least 6 months. The column was regenerated by washing with 1 liter 50 m*M* Tris (pH 8.0) containing 1.0 *M* NaCl and 10 m*M* EGTA followed by 1 liter of 10 m*M* PIPES (pH 7.0) containing 1 m*M* $CaCl_2$.

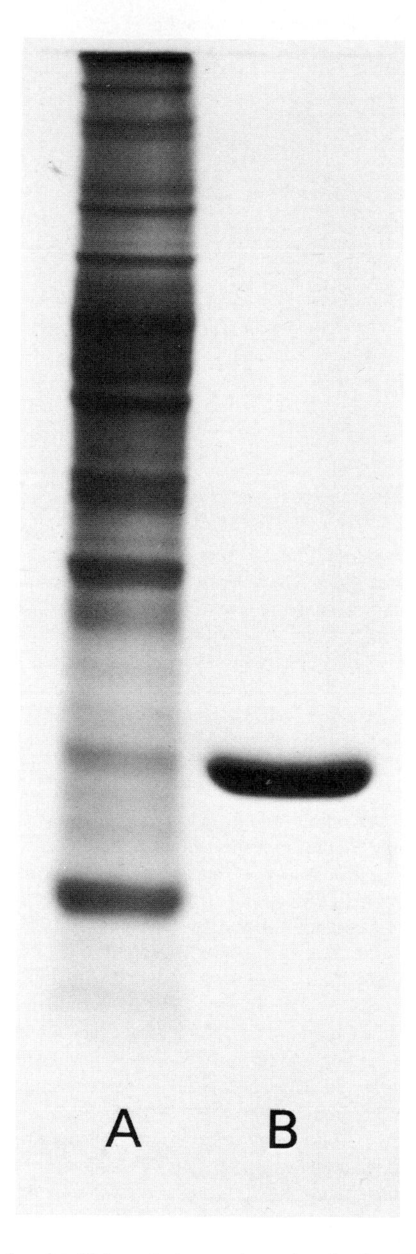

Fig. 7. Sodium dodecyl sulfate gel electrophoresis of calmodulin (A) 100 μg of bovine cerebral extract; (B) 15 μg of purified calmodulin from fractions 81–82 of Fig. 6. Electrophoresis was performed on a 10–20% linear acrylamide gradient gel using the buffer system described by Laemmli (1970). All buffers contained 1 m*M* EGTA; all samples contained 10 m*M* EGTA.

TABLE I

Purification of Calmodulin from Bovine Brain with a Fluphenazine-Sepharose Affinity Column[a]

Fraction	Total protein (mg)	Total activity (units)	Specific activity (units/mg protein)	Yield (%)	Purification (fold)
Supernatant	1736	2.59×10^5	1.49×10^2	100	1
Affinity column	5	1.86×10^5	3.72×10^4	72	260

[a] Calmodulin was prepared from 100 g of bovine cerebra. The overall yield of the procedure is approximately 50 mg of calmodulin per 1 kg of tissue.

amino acid residues are aspartate and glutamate accounting for an isoelectric point of about 4.3. (For a summary of physicochemical properties, see Klee, this volume, Chapter 4.) The molecule contains no cysteine or tryptophan and only 2 tyrosines and 8 phenylalanines (Lin *et al.*, 1974; Vanaman *et al.*, 1977). This composition of amino acids gives an unusual ultraviolet absorption spectra with peaks at 253, 259, 265, and 269 nm due to phenylalanine and at 277 nm due to tyrosine (Liu and Cheung, 1976; Watterson *et al.*, 1976; Klee, 1977). As expected, the extinction coefficient of calmodulin ($E^{1\%}_{1\,\mathrm{cm}} = 1.8$) is low compared to most other proteins (Watterson *et al.*, 1976). Calmodulin does not contain phosphate or carbohydrate, but does have a trimethylated lysine at position 115. Its amino terminus is acetylated (Vanaman *et al.*, 1977).

Calmodulin and troponin C exhibit approximately 70% conservative sequence homology and 50% direct sequence homology. Troponin C has a higher molecular weight than calmodulin (17,965 versus 16,723) due to an extra eight residues at its amino terminus (see Vanaman, this volume, Chapter 3).

Calmodulin contains four Ca^{2+}-binding sites with dissociation constants in the range of 4–18 μM (Teo and Wang, 1973; Lin *et al.*, 1974; Wolff *et al.*, 1977). In the presence of 1mM Mg^{2+}, one of the Ca^{2+} is displaced by Mg^{2+} giving a molecule with 3 Ca^{2+} and 1 Mg^{2+} (Wolff *et al.*, 1977). Binding of Ca^{2+} to calmodulin results in an increase in the α-helical content (Liu and Cheung, 1976; Wolff *et al.*, 1977; Klee, 1977). Most of the increase in helical content occurs upon binding of the first two molecules of Ca^{2+} (Klee, 1977). One of the two tyrosine residues, which is situated at the periphery, is further exposed to the solvent upon binding of Ca^{2+} (see Klee, this volume, Chapter 4). The Ca^{2+}-induced conformational changes in calmodulin have also been indicated by an increase in tyrosine fluorescence (Dedman *et al.*, 1977b), by an increase in refractoriness to cleavage by proteolytic enzymes (Ho *et al*, 1975; Liu and Cheung, 1976), and by nuclear magnetic resonance (Seamon, 1979).

One of the most striking characteristics of calmodulin is its stability. The protein may be exposed to 95°C or to a 9 M urea solution with retention of biological activity (Cheung, 1971). In addition, calmodulin retains its Ca^{2+}-dependent conformation and ability to bind troponin I in the presence of 8 *M* urea (Amphlett *et al.*, 1976).

B. Biological Properties

Calmodulin appears to be ubiquitous throughout the eukaryotes, lacking both tissue and species specificity. The protein has been isolated from a variety of sources including bovine brain (Lin *et al.*, 1974; Watterson *et al.*, 1976; Wolff *et al.*, 1977), bovine heart (Teo *et al.*, 1973), bovine uterus (Grand and Perry, 1978), bovine adrenal medulla (Brooks and Siegel, 1973; Kuo and Coffee, 1976), porcine brain (Wolff and Siegel, 1972; Klee, 1977), rat testis (Dedman *et al.*, 1977b), earthworm (Waisman *et al.*, 1975), sea urchin oocytes and sperm (Jones *et al.*, 1978), starfish oocytes (R. W. Wallace, L. Meijer, W. Y. Cheung, and D. Epel, unpublished data), pea seedlings (Anderson and Cormier, 1978), peanuts (Charbonneau and Cormier, 1979), and cottonseeds (Wallace and Cheung, 1979a). Calmodulin from all of these sources has similar properties and is capable of stimulating brain phosphodiesterase. Moreover, calmodulin from cottonseeds and starfish oocytes (R. W. Wallace, L. Meijer, W. Y. Cheung, and D. Epel, unpublished data) cross-reacts with the antibody directed against bovine brain calmodulin (Wallace and Cheung, 1979a); also calmodulin from slime mold (*Dictyostelium discoideum*), algae (*Chlamydomonas reinhardii*), rabbit polymorphonuclear leukocytes, and the electroplax of *Electrophorus electricus* cross-reacts with the antibody directed against rat testicular calmodulin (Chafouleas *et al.*, 1979). In addition, the amino acid sequences of calmodulin from bovine brain (Vanaman *et al.*, 1977), bovine uterus (Grand and Perry, 1978), and rat testis (Dedman *et al.*, 1978) are almost identical. The apparent similarity in the amino acid sequence of calmodulin from widely divergent sources suggests that the protein is ancient, with its primary structure highly conserved, an attribute not unexpected of a fundamental regulatory protein.

Comparison of the primary structure of calmodulin with that of other calcium-binding proteins indicates that calmodulin is a member of a homologous series of proteins which include cardiac and skeletal muscle troponin C, myosin light chains, the vitamin D-dependent calcium-binding protein, and parvalbumins (Barker *et al.*, 1977). All of these calcium-binding proteins may have evolved from a common ancestor through a process of gene duplication and point mutations. Calmodulin appears the most highly conserved member of this family of proteins. This may reflect functional constraints placed upon the protein's primary structure early in the evolutionary process.

The similarity of troponin C to calmodulin is most striking; the two proteins share some 50% direct and 70% conservative sequence homology (Vanaman *et al.*, 1977). Indeed, calmodulin substitutes effectively for troponin C in the activation of actomyosin ATPase (Amphlett *et al.*, 1976). Dedman *et al.* (1977a) reported that troponin C is one-six hundredth as active as calmodulin in stimulating brain phosphodiesterase. Wolff and Brostrom (1979), however, detected this trace activity only at concentrations of troponin C 2000-fold higher than calmodulin and attributed it to the contamination of troponin C with calmodulin.

One notable difference between calmodulin and troponin C is that the former is widely distributed and governs numerous divergent systems; troponin C appears limited to the skeletal and cardiac muscles. Another is that calmodulin contains a trimethylated lysine at position 115. It is not known whether the presence of this unusual amino acid is related to the versatility of calmodulin.

The enzymes which are regulated by calmodulin are fundamental for coordinating cellular activity. Calmodulin controls the degradation of cyclic nucleotides and other key enzymes and cellular processes as enumerated in Table II. The mechanism by which calmodulin regulates brain phosphodiesterase (Teo and Wang, 1973; Kakiuchi *et al.*, 1973; Lin *et al.*, 1974), adenylate cyclase (Lynch *et al.*, 1977), and Ca^{2+}-ATPase (Lynch and Cheung, 1979) has been studied in some detail. The binding of Ca^{2+} to calmodulin brings about a conformation change which allows the Ca^{2+}–calmodulin complex to interact with the apoenzyme to form a ternary complex, which is the active species [eq. (1) and (2) in Section I]. The mechanism depicted by Eq. (1) and (2) should not be regarded as the only mode by which calmodulin acts. Phosphorylase kinase (Cohen *et al.*, 1978; Depaoli-Roach *et al.*, 1979) contains calmodulin as an integral subunit which cannot be removed from the enzyme by EGTA. Lung phosphodiesterase may be another enzyme with a tightly bound calmodulin (Sharma and Wirch, 1979). In addition, calmodulin inhibits adenylate cyclase from Chinese hamster ovary cells (Evain *et al.*, 1979) and glioma cells (Brostrom *et al.*, 1976) under certain conditions.

Adenylate cyclase and phosphodiesterase are the only known enzymes involved in the metabolism of cAMP. In the brain (C. O. Brostrom *et al.*, 1975, 1977; M. A. Brostrom *et al.*, 1976; Cheung *et al.*, 1975; Lynch *et al.*, 1977) and adrenal gland (LeDonne and Coffee, 1979) calmodulin regulates the activities of both enzymes. In other tissues, including heart (Teo *et*

TABLE II

Calmodulin-Regulated Enzymes and Cellular Processes

Reaction or process	Reference
Cyclic nucleotide phosphodiesterase	Cheung, 1967, 1969, 1970a,b; Kakiuchi and Yamazaki, 1970
Adenylate cyclase	C. O. Brostrom *et al.*, 1975, 1977; M. A. Brostrom *et al.*, 1976; Cheung *et al.*, 1975; Lynch *et al.*, 1977
Guanylate cyclase	Nago *et al.*, 1979
Phosphorylase kinase	Cohen *et al.*, 1978; Depaoli-Roach *et al.*, 1979
Myosin light chain kinase	Waisman *et al.*, 1978; Dabrowska and Hartshorne, 1978; Dabrowska *et al.*, 1978; Sherry *et al.*, 1978; Yagi *et al.*, 1978
NAD kinase	Anderson and Cormier, 1978; Epel *et al.*, 1980.
Phospholipase A_2	Wong and Cheung, 1979
Ca^{2+}-ATPase	Gopinath and Vincenzi, 1977; Jarrett and Penniston, 1977, 1978; Hanahan *et al.*, 1978; Lynch and Cheung, 1979; Niggli *et al.*, 1979; Sobue *et al.*, 1979
Ca^{2+} transport in erythrocytes	Hinds *et al.*, 1978
Ca^{2+} transport in sarcoplasmic reticulum	Katz and Remtulla, 1978
Phosphorylation of membranes	Schulman and Greengard, 1978a,b
Neurotransmitter release	DeLorenzo *et al.*, 1979
Disassembly of microtubules	Marcum *et al.*, 1978; Nishida *et al.*, 1979

al., 1973; Kakiuchi *et al.*, 1975), liver, kidney (Kakiuchi *et al.*, 1975), and pituitary (Azhar and Menon, 1977), at least one form of phosphodiesterase is regulated by calmodulin. The activation of both the synthetic and degradative enzymes by calmodulin in neural tissues may allow a transient increase of cAMP by first activating adenylate cyclase (as Ca^{2+} passes through the plasma membrane) and then phosphodiesterase (as Ca^{2+} diffuses into the cytoplasm). Alternatively, since brain Ca^{2+}-dependent phosphodiesterase degrades cGMP faster than cAMP at micromolar concentrations, the sequential activation of adenylate cyclase and phosphodiesterase may result in an increase of cellular cAMP and a decrease of cellular cGMP. Another possibility is based on the observation that calmodulin is released from synaptosomal membranes upon phosphorylation of a membrane protein catalyzed by a cAMP-dependent protein kinase (Gnegy *et al.*, 1977). Activation of adenylate cyclase increases intracellular cAMP, which stimulates a cAMP-dependent phosphorylation of certain membrane components. This action releases calmodulin from the membrane compartment into the cytoplasm where it activates phosphodiesterase, returning cAMP to its prestimulated level. One uncertainty of this notion is that the concentration of calmodulin in the cytosol is invariably higher than that of phosphodiesterase, so that the enzyme activity appears more likely to be regulated by the cellular flux of Ca^{2+}.

Muscle contraction is governed by the force generated by the actin–myosin complex in response to the cellular level of Ca^{2+}. In skeletal and cardiac muscles the receptor of Ca^{2+} is troponin C, whereas in smooth muscle and nonmuscle cells the receptor is calmodulin. Calmodulin is absolutely required for the activity of myosin light chain kinase of smooth muscle or nonmuscle cells (Waisman *et al.*, 1978; Dabrowska and Hartshorne, 1978; Dabrowska *et al.*, 1978; Sherry *et al.*, 1978). The enzyme catalyzes the phosphorylation of a 20,000-dalton light chain of myosin. Phosphorylation of the light chain is obligatory for actomyosin ATPase activity.

Phosphorylase kinase controls the hydrolysis of glycogen. The enzyme is complex, consisting of four tetramers, $(\alpha\beta\gamma\delta)_4$; the δ subunit is calmodulin, tightly bound to the ternary structure. Phosphorylase kinase activity is further increased by an exogenous calmodulin (Cohen *et al.*, 1978; Depaoli-Roach *et al.*, 1979; Cohen, this volume, Chapter 9).

Calmodulin regulates a Ca^{2+}-dependent protein kinase found in a fraction of various tissues (Schulman and Greengard, 1978a,b). In synaptosomes the influx of Ca^{2+} following depolarization leads to the phosphorylation of two membrane proteins with molecular weights of 80,000 and 86,000 (Schulman and Greengard, 1978a). These polypeptides also serve as substrates for cAMP-dependent protein kinase (Sieghart *et al.*, 1979). DeLorenzo *et al.* (1979) noted that calmodulin-regulated protein phosphorylation may cause the release of neurotransmitters from synaptic vesicles. In addition, Yamauchi and Fujisawa (1979) showed that tryptophan 5-monooxygenase, a key enzyme in the biosynthesis of neurotransmitters, is activated by a calmodulin-induced phosphorylation.

Activity of NAD kinase in both pea seedling (Anderson and Cormier, 1978) and sea urchin oocyte (Epel, *et al.*, 1980) is governed by calmodulin. The kinase controls the ratio of NAD/NADP which determines the direction of certain metabolic pathways. One of the early events following fertilization of sea urchin oocytes is an increase of free Ca^{2+}, and a rapid conversion of NAD to NADP (Epel, 1964). Calmodulin is present at high levels in sea urchin oocyte (Head *et al.*, 1979) and is required for the NAD kinase for maximal activity (Epel *et al.*, 1980). The rapid increase of Ca^{2+} following fertilization of the egg may be related to the conversion of NAD to NADP.

Phospholipase A_2 is a key enzyme that governs the availability of arachidonic acid in many tissues. The level of free arachidonic acid is usually low, a rate-limiting step in the synthesis of endoperoxides, thromboxanes, prostaglandins, and prostacyclin. The enzyme depends on Ca^{2+} for activity; in human platelets, the effect of Ca^{2+} appears to be mediated through calmodulin (Wong and Cheung, 1979). It would be important to determine if the enzyme activity in other tissues is regulated through a similar mechanism.

Calmodulin regulates the activity of Ca^{2+}-ATPase of plasma membrane and endoplasmic reticulum (Gopinath and Vincenzi, 1977; Jarrett and Penniston, 1977, 1978; Hanahan *et al.*, 1978; Hinds *et al.*, 1978; Katz and Remtulla, 1978; Lynch and Cheung, 1979; Niggli *et al.*, 1979; Sobue *et al.*, 1979; Larsen and Vincenzi 1979). As intracellular Ca^{2+} increases, calmodulin activates the membrane Ca^{2+}-ATPase, which extrudes the Ca^{2+} into the extracellular space, lowering the Ca^{2+} concentrations to a steady state level, an excellent example of feedback control. Thus, calmodulin acts not only as a mediator of Ca^{2+} actions but also as a modulator of its intracellular level.

Using indirect immunofluorescence, Welsh *et al.* (1978) found that calmodulin-specific antibodies decorated the mitotic spindle during cell division of mouse fibroblast cells, human testis fibroblasts, rat sertoli cells, mouse mammary and mammary adenocarcinoma, mouse X human somatic cell hybrids, and chick liver fibroblasts. The fluorescence appeared most intense at the poles of the spindle. Agents such as Colcemid and N_2O which disrupt microtubule structure altered the distribution of fluorescence (Welsh *et al.*, 1979). In a follow-up experiment, the same group showed that calmodulin accelerated the disassembly of microtubules (Marcum *et al.*, 1978).

Wood *et al.* (1980) have localized calmodulin in mouse basal ganglia. They found that the protein is associated primarily with the postsynaptic densities (PSD) and the dendritic microtubules. This observation is consistent with the work of Grab *et al.* (1979) which shows that calmodulin is asociated with the PSD isolated by subcellular fractionation. The PSD also contains a Ca^{2+}-dependent protein kinase whose activity may be regulated by calmodulin (Grab and Siekvietz, 1979).

Harper *et al.* (1980) studied the distribution of calmodulin in rat liver, skeletal muscle, and adrenal gland. In liver calmodulin was associated with the nucleus, plasma membrane and glycogen granules. Skeletal muscle calmodulin was located at the I band and at the intermyofibrillar region probably in association with glycogen granules and the sarcoplasmic reticulum, respectively. In the adrenal gland, the endocrine state of the animal affects the distribution of calmodulin within the cell. Administration of ACTH to the animal greatly increased calmodulin in the nuclei, implying a role of calmodulin with certain nuclear functions.

In view of the extensive involvement of Ca^{2+} in cell functions, it would not be surprising that future studies will extend the role of calmodulin to other cellular processes.

VI. CONCLUDING REMARKS

Calmodulin has routinely been estimated by the phosphodiesterase system. The principle of the assay is based on the ability of calmodulin to stimulate the activity of the enzyme. In theory, all calmodulin-dependent proteins could interfere with the assay by competing with phosphodiesterase for calmodulin. The situation becomes particularly acute in determining calmodulin in a tissue extract. As is shown in Fig. 1, the presence of CaM-BP_{80} in the assay mixture causes an apparent underestimation of calmodulin.

Although the radioimmunoassay does not afford a greater sensitivity than the phosphodiesterase system, it has the advantage of detecting a wider span of sample concentrations. Moreover, the radioimmunossay, which measures calmodulin in the presence of EGTA, does not appear susceptible to interference by calmodulin-binding proteins, at least not by CaM-BP_{80}. In addition, the radioimmunoassay permits quantitation of calmodulin on the basis of antigenic determinants rather than biological activity, a feature potentially useful in determining calmodulin that may be biologically inactive.

Of the various procedures available for the purification of calmodulin, the protocol using phenothiazine affinity chromatography is obviously the method of choice. It is simple and efficient, giving a homogeneous calmodulin in essentially one step. Moreover, the affinity column can be used repeatedly after appropriate regeneration. The fact that calmodulin appears to be the only protein in the brain extract that binds to the fluphenazine affinity column with high affinity supports the contention by Weiss and Levin (1978) that the pharmacological action of the phenothiazine derivatives is through the inactivation of calmodulin.

Calcium ion has long been recognized as a key regulator of many physiological processes, such as muscle contraction, cell division and motility, neurotransmission, ion transport, and secretion (Kretsinger, 1979). It is increasingly clear that the actions of Ca^{2+} are mediated via a class of Ca^{2+}-binding proteins, of which calmodulin, the most widely distributed and versatile, appears to play a pivotal role in the coordination of basic cellular activities.

ACKNOWLEDGMENTS

This work was supported by an institutional Cancer Center Support Grant (CORE) CA 21765, by Project Grant NS 08059 and by ALSAC. R. W. Wallace is a recipient of a United States Public Health Service Fellowship AM 05689. We are grateful to Dr. William Walker and the Fineberg Packing Company for a generous supply of goat anti-rabbit immunoglobulin F_c sera and bovine brains, respectively. We thank Dr. M. Cormier and Dr. Thomas Vanaman for making manuscripts available to us prior to their publication.

REFERENCES

Amphlett, G. W., Vanaman, T. C., and Perry, S. V. (1976). Effect of the troponin C-like protein from bovine brain (brain modulator protein) on the Mg^{2+}-stimulated ATPase of skeletal muscle actomyosin. *FEBS Lett.* **72,** 163–168.

Anderson, J. M., and Cormier, M. J. (1978). Calcium-dependent regulator of NAD kinase in higher plants. *Biochem. Biophys. Res. Commun.* **84,** 595–602.

Avrameas, S., and Ternynck, T. (1967). Biologically active water-insoluble protein polymers: Their use for isolation of antigens and antibodies. *J. Biol. Chem.* **242,** 1651–1659.

Azhar, S., and Menon, K. M. J. (1977). Cyclic nucleotide phosphodiesterase from rat anterior pituitary: Characterization of multiple forms and regulation by protein activator and Ca^{2+}. *Eur. J. Biochem.* **731,** 73–82.

Barker, W. C., Ketcham, L. K., and Dayhoff, M. O. (1977). Evolutionary relationship among calcium-binding proteins. *In* "Calcium Binding Proteins and Calcium Function" (R. H. Wasserman, R. A. Corradino, E. Carafoli, R. S. Kretsinger, D. H. MacLennon, and F. L. Siegel, eds.), pp. 73–75, Elsevier/North-Holland Biomedical Press, Amsterdam.

Brooks, J. C., and Siegel, F.L. (1973). Purification of a calcium-binding phosphoprotein from beef adrenal medulla: Identity with one of two calcium-binding proteins of brain. *J. Biol. Chem.* **248,** 4189–4193.

Brostrom, C. O., Huang, Y. C., Breckenridge, B. McL., and Wolff, D. J. (1975). Identification of a calcium-binding protein as a calcium-dependent regulator of brain adenylate cyclase. *Proc. Natl. Acad. Sci. U.S.A.* **72,** 64–68.

Brostrom, C. O., Brostrom, M. A., and Wolff, D. J. (1977). Calcium-dependent adenylate cyclase from rat cerebral cortex: Reversible activation by sodium fluoride. *J. Biol. Chem.* **252,** 5677–5685.

Brostrom, M. A., Brostrom, C. O., Breckenridge, B. McL., and Wolff, D. J. (1976). Regulation of adenylate cyclase from glial tumor cells by calcium and a calcium-binding protein. *J. Biol. Chem.* **251,** 4744–4750.

Chafouleas, J. G., Dedman, J. R., Munjaal, R. P., and Means, A. R. (1979). Calmodulin: Development and application of a sensitive radioimmunoassay. *J. Biol. Chem.* **254,** 10262–10267.

Charbonneau, H., and Cormier, M. J. (1979). Purification of plant calmodulin by fluphenazine-sepharose affinity chromatography. *Biochem. Biophys. Res. Commun.* **90,** 1039–1047.

Cheung, W. Y. (1967). Cyclic 3′,5′-nucleotide phosphodiesterase: Pronounced stimulation by snake venom. *Biochem. Biophys. Res. Commun.* **29,** 478–482.

Cheung, W. Y. (1969). Cyclic 3′,5′-nucleotide phosphodiesterase: Preparation of a partially active enzyme and its subsequent stimulation by snake venom. *Biochem. Biophys. Acta* **191,** 303–315.

Cheung, W. Y. (1970a). Cyclic 3′,5′-nucleotide phosphodiesterase: Demonstration of an activator . *Biochem. Biophys. Res. Commun.* **38,** 533–538.

Cheung, W. Y. (1970b). Cyclic Nucleotide Phosphodiesterase *Adv. Biochem. Psychopharmacology* **3,** 51–65.

Cheung, W. Y. (1971). Cyclic 3′,5′-nucleotide phosphodiesterase: Evidence for and properties of a protein activator. *J. Biol. Chem.* **246,** 2859–2869.

Cheung, W. Y. (1980). Calmodulin plays a pivotal role in cellular regulation. *Science* **207,** 19–27.

Cheung W. Y., Bradham, L. S., Lynch, T. J., Lin, Y. M., and Tallant, E. A. (1975). Protein activator of cyclic 3′:5′-nucleotide phosphodiesterase of bovine or rat brain also activates its adenylate cyclase. *Biochem. Biophys. Res. Commun.* **66,** 1055–1062.

Cohen, P., Burchell, A., Foulkes, J. G., Cohen, P. T. W., Vanaman, T. C. and Nairn, A. L. (1978). Identification of the Ca^{2+}-dependent modulator protein as the fourth subunit of rabbit skeletal muscle phosphorylase kinase. *FEBS Lett.* **92,** 287–293.

Dabrowska, R., and Hartshorne, D. J. (1978). A Ca^{2+}-and modulator-dependent myosin light chain kinase from non-muscle cells. *Biochem. Biophys. Res. Commun.* **85,** 1352–1359.

Dabrowska, R., Sherry, J. M. F., Aromatorio, D. K., and Hartshorne, D. J. (1978). Modulator protein as a component of the myosin light chain kinase from chicken gizzard. *Biochemistry* **17,** 253–258.

Dedman, J. R., Potter, J. D., and Means, A. R. (1977a). Biological cross-reactivity of rat testis phosphodiesterase activator protein and rabbit skeletal muscle troponin-C. *J. Biol. Chem.* **252,** 2437–2440.

Dedman, J. R., Potter, J. D., Jackson, R. L., Johnson, J. D., and Means, A. R. (1977b). Physicochemical properties of rat testis Ca^{2+}-dependent regulator protein of cyclic nucleotide phosphodiesterase: Relationship of Ca^{2+}-binding, conformational changes, and phosphodiesterase activity. *J. Biol. Chem.* **252,** 8415–8422.

Dedman, J. R., Jackson, R. L., Schreiber, W. E., and Means, A. R. (1978). Sequence homology of the Ca^{2+}-dependent regulator of cyclic nucleotide phosphodiesterase from rat testis with other Ca^{2+}-binding proteins. *J. Biol. Chem.* **253,** 343–346.

DeLorenzo, R. J., Freedman, S. D., Yohe, W. B., and Maurer, S. C. (1979). Stimulation of Ca^{2+}-dependent neurotransmitter release and presynaptic nerve-terminal protein phosphorylation by calmodulin and a calmodulin-like protein isolated from synaptic vesicles. *Proc. Natl. Acad. Sci. U.S.A.* **76,** 1838–1842.

Depaoli-Roach, A. A., Gibbs, J. B., and Roach, P. J. (1979). Calcium and calmodulin activation of muscle phosphorylase kinase: Effect of tryptic proteolysis. *FEBS Lett.* **105,** 321–324.

Epel, D. (1964). A primary metabolic change of fertilization: Interconversion of pyridine nucleotides. *Biochem. Biophys. Res. Commun.* **17,** 62–68.

Epel, D., Wallace, R. W., and Cheung, W. Y. (1980). In preparation.

Evain, D., Klee, C., and Anderson, W. B. (1979). Chinese hamster ovary cell population density affects intracellular concentrations of calcium-dependent regulator and ability of regulator to inhibit adenylate cyclase activity. *Proc. Natl. Acad. Sci. U.S.A.* **76,** 3962–3966.

Gnegy, M. E., Nathanson, J. A., and Uzunov, P. (1977). Release of the phosphodiesterase activator by cyclic AMP-dependent ATP: Protein phosphotransferase from subcellular fractions of rat brain. *Biochim. Biophys. Acta* **497,** 75–85.

Gopinath, R. M., and Vincenzi, F. F. (1977). Phosphodiesterase protein activator mimics red blood cell cytoplasmic activator of (Ca^{2+} -Mg^{2+}) ATPase. *Biochem. Biophys. Res. Commun.* **77,** 1203–1209.

Grab, D. J., Berzins, K., Cohen, R. S., and Siekevitz, P. (1979). Presence of calmodulin in postsynaptic densities isolated from canine cerebral cortex. *J. Biol. Chem.* **254,** 8690–8696.

Grab, D. L., and Siekevitz, P. (1979). Calmodulin-dependent protein kinase activity in postsynaptic densities. *J. Cell Biol.* **83,** 603 (abstr.).

Grand, R. J. A., and Perry, S. V. (1978). The amino acid sequence of the troponin C-like protein (modulator protein) from bovine uterus. *FEBS Lett.* **92,** 137–142.

Grand, R. J. A., Perry, S. V., and Weeks, R. A. (1979). Troponin C-like proteins (calmodulins) from mammalian smooth muscle and other tissues. *Biochem. J.* **177,** 521–529.

Hanahan, D. J., Taverna, R. D., Glynn, D. D., and Ekholm, J. E. (1978). The interaction of Ca^{2+}/Mg^{2+} ATPase activator protein and Ca^{2+} with human erythrocyte membranes. *Biochem. Biophys. Res. Commun.* **84,** 1009–1015.

Harper, J. F., Cheung, W. Y., Wallace, R. W., Huang, H. L., Levine, S. N. and Steiner, A. L. (1980). Localization of calmodulin in rat tissues. *Proc. Natl. Acad. Sci. U.S.A.* **77,** 366–370.

Head, J. F., Mader, S., and Kaminer, B. (1979). Calcium-binding modulator protein from the unfertilized egg of the sea urchin *Arbacia punctulata*. *J. Cell. Biol.* **80,** 211–218.

Hinds, T. R., Larsen, F. L., and Vincenzi, F. F. (1978). Plasma membrane Ca^{2+} transport: Stimulation by soluble proteins. *Biochem. Biophys. Res. Commun.* **81,** 455–461.

Ho, H. C., Desai, R., and Wang, J. H. (1975). Effect of Ca^{2+} on the stability of the protein activator of cyclic nucleotide phosphodiesterase. *FEBS Lett.* **50,** 374–377.

Hunter, W. M., and Greenwood, F. C. (1962). Preparation of iodine-131 labelled human growth hormone of high specific activity. *Nature (Lond.)* **194,** 495–496.

Jamieson, G. A., and Vanaman, T. C. (1979). Calcium-dependent affinity chromatography of calmodulin on an immobilized phenothiazine. *Biochem. Biophys. Res. Commun.* **90,** 1048–1056.

Jarrett, H. W., and Penniston, J. T. (1977). Partial purification of the Ca^{2+}-Mg^{2+} ATPase activator from human erythrocytes: Its similarity to the activator of 3′:5′-cyclic nucleotide phosphodiesterase. *Biochem. Biophys. Res. Commun.* **77,** 1210–1216.

Jarrett, H. W., and Penniston, J. T. (1978). Purification of the Ca^{2+}-stimulated ATPase activator from human erythrocytes: Its membership in the class of Ca^{2+}-binding modulator proteins. *J. Biol. Chem.* **253,** 4676–4682.

Jones, H. P., Bradford, M. M., McRorie, R. A., and Cormier, M. J. (1978). High levels of a calcium-dependent modulator protein in spermatozoa and its similarity to brain modulator protein. *Biochem. Biophys. Res. Commun.* **82,** 1264–1272.

Kakiuchi, S., and Yamazaki, R. (1970). Calcium dependent phosphodiesterase activity and its activating factor (PAF) from brain. *Biochem. Biophys. Res. Commun.* **41,** 1104–1110.

Kakiuchi, S., Yamazaki, R., Teshima, Y., and Uenishi, K. (1973). Regulation of nucleoside cyclic 3′:5′-monophosphate phosphodiesterase activity from rat brain by a modulator and Ca^{2+}. *Proc. Natl. Acad. Sci. U.S.A.* **70,** 3526–3530.

Kakiuchi, S., Yamazaki, R., Teshima, Y., Uenishi, K., and Miyamoto, E. (1975). Multiple cyclic nucleotide phosphodiesterase activities from rat tissues and occurrence of a calcium-plus-magnesium-ion-dependent phosphodiesterase and its protein activator. *Biochem. J.* **146,** 109–120.

Katz, S., and Remtulla, M. A. (1978). Phosphodiesterase protein activator stimulates calcium transport in cardiac microsomal preparations enriched in sarcoplasmic reticulum. *Biochem. Biophys. Res. Commun.* **83,** 1373–1379.

Klee, C. B. (1977). Conformational transition accompanying the binding of Ca^{2+} to the protein activator of 3',5'-cyclic adenosine monophosphate phosphodiesterase. *Biochemistry* **16,** 1017–1024.

Klee, C. B., and Krinks, M. H. (1978). Purification of cyclic 3',5'-nucleotide phosphodiesterase inhibitory protein by affinity chromatography on activator protein coupled to sepharose. *Biochemistry* **17,** 120–126.

Klee, C. B., Crouch, T. H., and Krinks, M. H. (1979). Calcineurin: A calcium- and calmodulin-binding protein of the nervous system. *Proc. Natl. Acad. Sci. U.S.A.* **76,** 6270–6273.

Kretsinger, R. H. (1979). The informational role of calcium in the cytosol. *Adv. Cyclic Nucleotide Res.* **11,** 1–26.

Kuo, I. C. Y., and Coffee, C. J. (1976). Purification and characterization of a troponin-C-like protein from bovine adrenal medulla. *J. Biol. Chem.* **251,** 1603–1609.

Laemmli, U. K. (1970). Cleavage of structural proteins during the assembly of the head of bacteriophage T4. *Nature (London)* **227,** 680–684.

LaPorte, D. C., and Storm, D. R. (1978). Detection of calcium-dependent regulatory protein binding components using ^{125}I-labeled calcium-dependent regulatory protein. *J. Biol. Chem.* **253,** 3374–3377.

Larsen, F. L., and Vincenzi, F. F. (1979). Calcium transport across the plasma membrane: Stimulation by calmodulin. *Science* **204,** 306–309.

Le Donne, N. C., Jr., and Coffee, C. J. (1979). Properties of the bovine adrenal medulla adenyl cyclase. *Fed. Proc., Fed. Am. Soc. Exp. Biol.* **38,** 317 (abstr.).

Lin, Y. M., Liu, Y. P., and Cheung, W. Y. (1974). Cyclic 3':5'-nucleotide phosphodiesterase: Purification, characterization, and active form of an activator from bovine brain. *J. Biol. Chem.* **249,** 4943–4954.

Little, J. R., and Counts, R. B. (1969). Affinity and heterogeneity of antibodies induced by ϵ-2,4-dinitrophenylinsulin. *Biochemistry* **8,** 2729–2736.

Liu, Y. P., and Cheung, W. Y. (1976). Cyclic 3':5'-nucleotide phosphodiesterase: Ca^{2+} confers more helical conformation to the protein activator. *J. Biol. Chem.* **251,** 4193–4198.

Lowry, O. H., Rosebrough, N. J., Farr, A. L., and Randall, R. J. (1951). Protein measurement with the Folin phenol reagent. *J. Biol. Chem.* **193,** 265–275.

Lynch, T. J., and Cheung, W. Y. (1975). Underestimation of cyclic 3',5'-nucleotide phosphodiesterase activity by a radioisotopic assay using an anionic-exchange resin. *Anal. Biochem.* **67,** 130–138.

Lynch, T. J., and Cheung, W. Y. (1979). Human erythrocyte Ca^{2+} -Mg^{2+}-ATPase: Mechanism of stimulation by Ca^{2+}. *Arch. Biochem. Biophys.* **194,** 165–170.

Lynch, T. J., Tallant, E. A., and Cheung, W. Y. (1977). Rat brain adenylate cyclase: Further studies on its stimulation by a Ca^{2+}-binding protein. *Arch. Biochem. Biophys.* **182,** 124–133.

Marcum, J. M., Dedman, J. R., Brinkley, B. R., and Means, A. R. (1978). Control of microtubule assembly-disassembly by calcium-dependent regulator protein. *Proc. Natl. Acad. Sci. U.S.A.* **75,** 3771–3775.

Nago, S., Suzuki, Y., Watanabe, Y. and Nozawa, Y. (1979). Activation by a calcium-binding protein of guanylate cyclase in *Tetrahymena pyriformis*. *Biochem. Biophys. Res. Commun.* **90,** 261–268.

Niggli, V., Penniston, J. T., and Carafoli, E. (1979). Purification of the (Ca^{2+}-Mg^{2+})-ATPase from human erythrocyte membranes using a calmodulin affinity column. *J. Biol. Chem.* **254,** 9955–9958.

Nishida, E., Kumagai, H., Ohtsuki, I., and Sakai, H. (1979). The interaction between calcium-dependent regulator protein of cyclic nucleotide phosphodiesterase and microtubule proteins. I. Effect of calcium-dependent regulator protein on the calcium sensitivity of microtubule assembly. *J. Biochem.* **85,** 1257–1266.

Richman, P. G., and Klee, C. B. (1978). Interaction of ^{125}I-labeled Ca^{2+}-dependent regulator protein with cyclic nucleotide phosphodiesterase and its inhibitory protein. *J. Biol. Chem.* **253,** 6323–6326.

Schulman, H., and Greengard, P. (1978a). Stimulation of brain protein phosphorylation by calcium and an endogenous heat-stable protein. *Nature (London)* **271,** 478–479.

Schulman, H., and Greengard, P. (1978b). Ca^{2+}-dependent protein phosphorylation system in membranes from various tissues, and its activation by "calcium-dependent regulator." *Proc. Natl. Acad. Sci. U.S.A.* **75,** 5432–5436.

Seamon, K. (1979). Cation dependent conformations of brain Ca^{2+}-dependent regulator protein detected by nuclear magnetic resonance. *Biochem. Biophys. Res. Commun.* **86,** 1256–1265.

Sharma, R. K. and Wirch, E. (1979). Ca^{2+}-dependent cyclic nucleotide phosphodiesterase from rabbit lung. *Biochem. Biophys. Res. Commun.* **91,** 338–344.

Sharma, R. K., Wirch, E., and Wang, J. H. (1978). Inhibition of Ca^{2+}-activated cyclic nucleotide phosphodiesterase reaction by a heat-stable inhibitor protein from bovine brain. *J. Biol. Chem.* **253,** 3575–3580.

Sharma, R. K., Desai, R., Waisman, D. M., and Wang, J. H. (1979). Purification and subunit structure of bovine brain modulator binding protein. *J. Biol. Chem.* **254,** 4276–4282.

Sherry, J. M. F., Gorecka, A., Aksoy, M. O., Dabrowska, R., and Hartshorne, D. J. (1978). Role of calcium and phosphorylation in the regulation of the activity of gizzard myosin. *Biochemistry* **17,** 4411–4418.

Seighart, W., Forn, J., and Greengard, P. (1979). Ca^{2+} and cyclic AMP regulate phosphorylation of same two membrane-associated proteins specific to nerve tissue. *Proc. Natl. Acad. Sci. U.S.A.* **76,** 2475–2479.

Skowsky, W. R., and Fisher, D. A. (1972). The use of thyroglobulin to induce antigenicity to small molecules. *J. Lab. Clin. Med.* **80,** 134–144.

Sobue, K., Ichida, S., Yoshida, H. Yamazaki, R., and Kakiuchi, S. (1979). Occurrence of a Ca^{2+}- and modulator protein-activatable ATPase in the synaptic plasma membranes of brain. *FEBS Lett.* **99,** 199–202.

Sundberg, L., and Porath, J. (1974). Preparation of adsorbents for biospecific affinity chromatography. I. Attachment of group-containing ligands to insoluble polymers by means of bifunctional oxiranes. *J. Chromatogr.* **90,** 87–98.

Teo, T. S., and Wang, J. H., (1973). Mechanism of activation of a cyclic adenosine 3′:5′-monophosphate phosphodiesterase from bovine heart by calcium ions: Identification of the protein activator as a Ca^{2+} binding protein. *J. Biol. Chem.* **248,** 5950–5955.

Teo, T. S., Wang, T. H., and Wang, J. H. (1973). Purification and properties of the protein activator of bovine heart cyclic adenosine 3′,5′-monophosphate phosphodiesterase. *J. Biol. Chem.* **248,** 588–595.

Vanaman, T. C., Sharief, F., and Watterson, D. M. (1977). Structural homology between brain modulator protein and muscle TnCs. *In* "Calcium Binding Proteins and Calcium Function" (R. H. Wasserman, R. A. Corradino, E. Carafoli, R. H. Kretsinger, D. H. MacLennan, and F. L. Siegel, eds.), pp. 107–116, Elsevier/North-Holland Biomedical Pres, Amsterdam.

Vunakis, H. V., Wasserman, E., and Levine, L. (1972). Specificities of antibodies to morphine. *J. Pharmacol. Exp. Ther.* **180,** 514–522.

Waisman, D. M., Stevens, F. C., and Wang, J. H. (1975). The distribution of the Ca^{2+}-dependent protein activator of cyclic nucleotide phosphodiesterase in invertebrates. *Biochem.Biophys. Res. Commun.* **65,** 975–982.

Waisman, D. M., Singh, T. J. and Wang, J. H. (1978). The modulator-dependent protein kinase: A multifunctional protein kinase activatable by the Ca^{2+}-dependent modulator protein of the cyclic nucleotide system. *J. Biol. Chem.* **253,** 3387–3390.

Wallace, R. W., and Cheung, W. Y. (1979a). Antibody against bovine brain calmodulin cross-reacts with cottonseed calmodulin. *Fed. Proc., Fed. Am. Soc. Exp. Biol.* **38,** 478 (abstr.).

Wallace, R. W., and Cheung, W. Y. (1979b). Calmodulin: Production of an antibody in rabbit and development of a radioimmunoassay. *J. Biol. Chem.* **254,** 6564–6571.

Wallace, R. W., Lynch, T. J., Tallant, E. A., and Cheung, W. Y. (1978). An endogenous inhibitor protein of brain adenylate cyclase and cyclic nucleotide phosphodiesterase. *Arch. Biochem. Biophys.* **187,** 328–334.

Wallace, R. W., Lynch, T. J., Tallant, E. A., and Cheung, W. Y. (1979). Purification and characterization of an inhibitor protein of brain adenylate cyclase and cyclic nucleotide phosphodiesterase. *J. Biol. Chem.* **254,** 377–382.

Wang, J. H., and Desai, R. (1976). A brain protein and its effect on the Ca^{2+}- and protein modulator-activated cyclic nucleotide phosphodiesterase. *Biochem. Biophys. Res. Commun.* **72,** 926–932.

Wang, J. H., and Desai, R. (1977). Modulator binding protein: Bovine brain protein exhibiting the Ca^{2+}-dependent association with the protein modulator of cyclic nucleotide phosphodiesterase. *J. Biol. Chem.* **252,** 4175–4184.

Watterson, D. M., Harrelson, W. G., Keller, P. M., Jr., Sharief, F., and Vanaman, T. C. (1976). Structural similarities between the Ca^{2+}-dependent regulatory proteins of 3′:5′-cyclic nucleotide phosphodiesterase and actomyosin ATPase. *J. Biol. Chem.* **251,** 4501–4513.

Weiss, B., and Levin, R. M. (1978). Mechanism for selectively inhibiting the activation of cyclic nucleotide phosphodiesterase and adenylate cyclase by antipsychotic agents. *Adv. Cyclic Nucleotide Res.* **9,** 285–303.

Welsh, M. J., Dedman, J. R., Brinkley, B. R., and Means, A. R. (1978). Calcium-dependent regulator protein: Localization in mitotic apparatus of eukaryotic cells. *Proc. Natl. Acad. Sci. U.S.A.* **15,** 1867–1871.

Welsh, M. J., Dedman, J. R., Brinkley, B. R., and Means, A. R. (1979). Tubulin and calmodulin: Effects of microtubules and microfilament inhibitors on localization in the mitotic apparatus. *J. Cell. Biol.* **81,** 624–634.

Wolff, D. J., and Brostrom, C. O. (1979). Properties and functions of the calcium-dependent regulator protein. *Adv. Cyclic Nucleotide Res.* **11,** 27–88.

Wolff, D. J., and Siegel, F. L. (1972). Purification of a calcium-binding phosphoprotein from pig brain. *J. Biol. Chem.* **247,** 4180–4185.

Wolff, D. J., Poirier, P. G., Brostrom, C. O., and Brostrom, M. A. (1977). Divalent cation binding properties of bovine brain Ca^{2+}-dependent regulator protein. *J. Biol. Chem.* **252,** 4108–4117.

Wong, P. Y. K., and Cheung, W. Y. (1979). Calmodulin stimulates human platelet phospholipase A_2. *Biochem. Biophys. Res. Commun.* **90,** 473–480.

Wood, J. G., Wallace, R. W., Whitaker, J. N., and Cheung, W. Y. (1980). Immunocytochemial localization of calmodulin and a heat-labile calmodulin binding protein in basal ganglia of mouse brain. *J. Cell Biol.* **84,** 66–76.

Yagi, K., Yazawa, M., Kakiuchi, S., Ohshima, M., and Uenishi, K. (1978). Identification of an activator protein for myosin light chain kinase as the Ca^{2+}-dependent modulator protein. *J. Biol. Chem.* **253,** 1338–1340.

Yamauchi, T., and Fujisawa, H. (1979). Activation of tryptophan 5-monooxygenase by calcium-dependent regulator protein. *Biochem. Biophys. Res. Commun.* **90,** 28–35.

Chapter 3

Structure, Function, and Evolution of Calmodulin

THOMAS C. VANAMAN

I. INTRODUCTION

The divalent cation, Ca^{2+}, has been shown in recent years to be a major regulatory signal for such diverse cellular processes as endo- and exocytosis, intermediary metabolism, motility, and proliferation (for review, see Berridge, 1975; Rasmussen and Goodman, 1977; Greengard, 1978). These processes are regulated by Ca^{2+} through a limited set of structurally related calcium-dependent regulatory proteins including, among others, calmodulins, muscle troponin Cs, and myosin light chains. Of these, only calmodulin has a broad distribution within the cell and throughout different tissues and species. In addition, calmodulin alone is a multifunctional regulatory protein which activates, in a calcium-dependent manner, a number of the enzymes involved in these physiological processes as noted elsewhere in this volume.

The mechanism through which calmodulin acts is understood in some

CALCIUM AND CELL FUNCTION, VOL. I

ISBN 0-12-171401-2

detail. It possesses four equivalent sites which specifically bind calcium with a dissociation constant of between 1 and 10 μM (Potter *et al.*, 1977). Calcium binding induces significant conformational changes in calmodulin leading to greater helical content and a more stable structure (see Klee, this volume, Chapter 4). These calcium-induced changes in structure produce sites on the surface of calmodulin through which it can interact with and activate the enzymes which it regulates. With the exception of phosphorylase kinase (see Cohen, this volume, Chapter 9) the association between calmodulin and its regulated enzymes is readily reversible, occurring only in the presence of saturating calcium concentrations (Watterson and Vanaman, 1976).

Although first studied in higher animals, it is now clear that calmodulin is a constituent of most, if not all, eukaryotic cells. Its broad phylogenetic distribution and regulatory activities have led us and other investigators to propose that calmodulin may be the primary intracellular receptor for Ca^{2+} signals. This chapter will discuss our current knowledge concerning the structure of calmodulin, its conservation during eukaryotic evolution, the structural features of calmodulin which endow it with its special functional capabilities, and how these structural and functional properties are related to other calcium binding proteins, particularly the muscle troponin Cs.

II. GENERAL PHYSICOCHEMICAL PROPERTIES OF CALMODULINS

While calmodulin was first detected as an activator of cyclic nucleotide phosphodiesterase (PDE) (Cheung, 1970; Kakiuchi *et al.*, 1970), the protein was purified and characterized originally because of its unique physicochemical properties and abundance in neurosecretory tissues (Schneider, 1973). Our initial studies concerned the troponin C (TnC)-like properties of this acidic brain protein (Vanaman *et al.*, 1975) while other laboratories purified and characterized it as the bovine brain (Lin *et al.*, 1974) and heart (Teo and Wang, 1973) PDE activators. The identity of these two independent isolates of calmodulin was subsequently demonstrated for both the bovine brain (Watterson *et al.*, 1976) and heart (Stevens *et al.*, 1976) proteins.

As shown in Tables I and II, the physicochemical properites, calcium-binding abilities, and amino acid compositions of calmodulins are quite similar to the muscle TnCs as we (Vanaman *et al.*, 1975; Watterson *et al.*, 1976), Stevens and co-workers (1976) and others (Kuo and Coffee, 1976; Dedman *et al.*, 1977) have noted. Detailed sequence analyses, discussed

TABLE I
Comparison of Physicochemical Properties of CaM and TnCs

Physical properties	Bovine brain CaM	Skeletal TnC	Cardiac TnC
MW (sequence)	16,680[a]	17,923[b]	18,459[c]
p*I*	4.2	4.2	~4
uv spectrum	Phenylalanine fine spectrum	Phenylalanine fine spectrum	Phenylalanine fine spectrum
Ca^{2+} binding	4 Sites	4 Sites	3 Sites

[a] Taken from Watterson *et al.* (1980a).
[b] Taken from Collins *et al.* (1977).
[c] Taken from Van Eerd and Takahashi (1976).

later in this chapter, proved unequivocally that these similarities result from substantial sequence homology (Vanaman *et al.*, 1977).

The physicochemical properties listed in Table I for calmodulins and TnCs are distinctive for this family of calcium-binding proteins which also includes parvalbumins, myosin light chains, and other proteins (Barker *et al.*, 1978). Their relatively small size and high net negative charge (p*I* ~ 4) make them easily distinguishable from most other proteins by gel electrophoretic analyses (Watterson *et al.*, 1976). Another distinctive characteristic shared by calmodulins, TnCs, and parvalbumins is their uv absorption spectrum showing a number of absorption peaks between 276 and 250 nm. This spectrum results from high phenylalanine and low tyrosine contents and the lack of tryptophan. The low molar extinction coefficients of tyrosine and phenylalanine give calmodulin solutions a very low absorption at wavelengths normally used for protein estimation (Watterson *et al.*, 1976) [bovine brain calmodulin has an $E_{1 mg/ml, 276\ nm} = 0.18$ (Watterson *et al.*, 1976).

Although calmodulins and muscle TnCs are quite similar in composition for those residues dictating gross physicochemical properties, a number of important and distinctive differences do exist in contents of specific amino acids. For example, Table II show that both calmodulins and TnCs have high contents of acidic amino acids, methionine and phenylalanine; low contents of proline and histidine; and no tryptophan. However, calmodulin alone is devoid of cysteine and more importantly possesses a single fully trimethylated lysyl residue, not found in TnCs (Watterson *et al.*, 1976; Vanaman *et al.*, 1976; Jackson *et al.*, 1977). The 3:1 ratio of threonine to serine in calmodulin is also quite different from that of TnCs where these residues are present in equal amounts. As discussed more fully in

TABLE II

Comparison of Amino Acid Composition of Calmodulins and TnC from Various Sources

	Calmodulins			
Amino acid	Bovine brain[a] (residues/molecule)	*Renilla reniformis*[b] (residues/molecule)	*Tetrahymena pyriformis*[c] (moles/16,700 g)	Muscle TnC Rabbit skeletal[d] (residues/molecule)
Lysine	7	8	7.72	9
Trimethyllysine[e]	1	1	(e)	0
Histidine	1	1	2.3	1
Arginine	6	6	5.4	7
Aspartic Acid	23	23	23.5	22
Threonine	12	12	10.6	6
Serine	4	5	4.7	7
Glutamic acid	27	25	26.5	31
Proline	2	2	2.3	1
Glycine	11	11	12.1	13
Alanine	11	10	10.7	13
Valine	7	7	6.2	7
Methionine	9	9	8.1	10
Isoleucine	8	8	9.2	10
Leucine	9	9	13.6	9
Tyrosine	2	1	1.0	2
Phenylalanine	8	9	8.5	10
½ Cystine	0	0	0	1
Tryptophan	0	0	0	0

[a] From Vanaman *et al.* (1977), and Watterson *et al.* (1980a).

[b] From F. Sharief, H. P. Jones, M. J. Cormier, and T. C. Vanaman, unpublished observations.

[c] From Jamieson *et al.* (1979).

[d] From Collins *et al.* (1977).

[e] Trimethyllysine detected as a shoulder on the lysine peak by double column methodology is reported together with lysine.

the following sections, these unique compositional characteristics have been demonstrated for all homogeneous preparations of calmodulin studied to date including those from animals, plants, and protozoa. They thus can be used to distinguish calmodulins from other calcium-binding proteins of this family, even the closely related TnCs. Together with its unique calcium-dependent regulatory activities, they form a set of criteria for unequivocally identifying a protein as a calmodulin.

III. STRUCTURAL STUDIES OF CALMODULIN

A. Vertebrate Calmodulins

The results of detailed sequence analyses of four vertebrate calmodulins and one invertebrate calmodulin coupled with detailed comparative studies of calmodulins from a number of other sources have demonstrated that its structure is very highly conserved in metozoans. Comparison of calmodulin's primary structure to those known for other calcium-binding proteins has also allowed the tentative identification of several important features of the linear sequence (see Fig. 1). The sequence shown in Fig. 1 is that of bovine brain calmodulin (Vanaman *et al.*, 1977; Watterson *et al.*, 1980a) arranged so as to depict the regions and residues predicted, based on the criteria of Kretsinger (1976), to be involved in forming its four helix-loop-helix calcium-binding site structures. The loops consist of 12 amino acids residues, six of which have oxygen-containing side chains assumed to contribute oxygen ligands (solid lines) to calcium at the vertices of an octahedron (Kretsinger, 1976). These six bonding residues alternate in the sequence as shown, most often being interspersed with glycine or other small side chain amino acids. The glycyl residue found at the apex (between bonding positions 3 and 4) of each loop has been proposed to be especially important, allowing the loop to make a sharp bend at this position (Kretsinger, 1976). Eight residues on either side of each loop are shown in a staggered configuration to indicate that they could form two helices as found for the calcium binding sites in parvalbumin (Kretsinger, 1976). Little of the linear sequence of calmodulin is found outside these predicted helix-loop-helix domains-only the 11 amino terminal residues and three 8–9 residue interdomain sequences assumed to be hinge regions.

Sequence studies have also been reported for vertebrate calmodulins from two other sources, bovine uterine smooth muscle (Grand and Perry, 1978) and rat testis (Dedman *et al.*, 1978). The sequence shown in Fig. 1 is identical to that reported by Grand and Perry (1978) for the bovine uterus protein except for the amidation states of the aspartyl residues at two positions, residues 24 and 97. Together, these bovine calmodulin sequences differ at a number of positions from the sequence reported for the calmodulin from rat testis (Dedman *et al.*, 1978), largely in amide assignments. However, there is one placement difference, the inversion in the rat testis protein of alanine-57 and asparagine-60 in the bovine protein. Such an inversion of these two residues between the rat and bovine proteins seems unlikely for two important reasons. First, the amino acid sequence of this

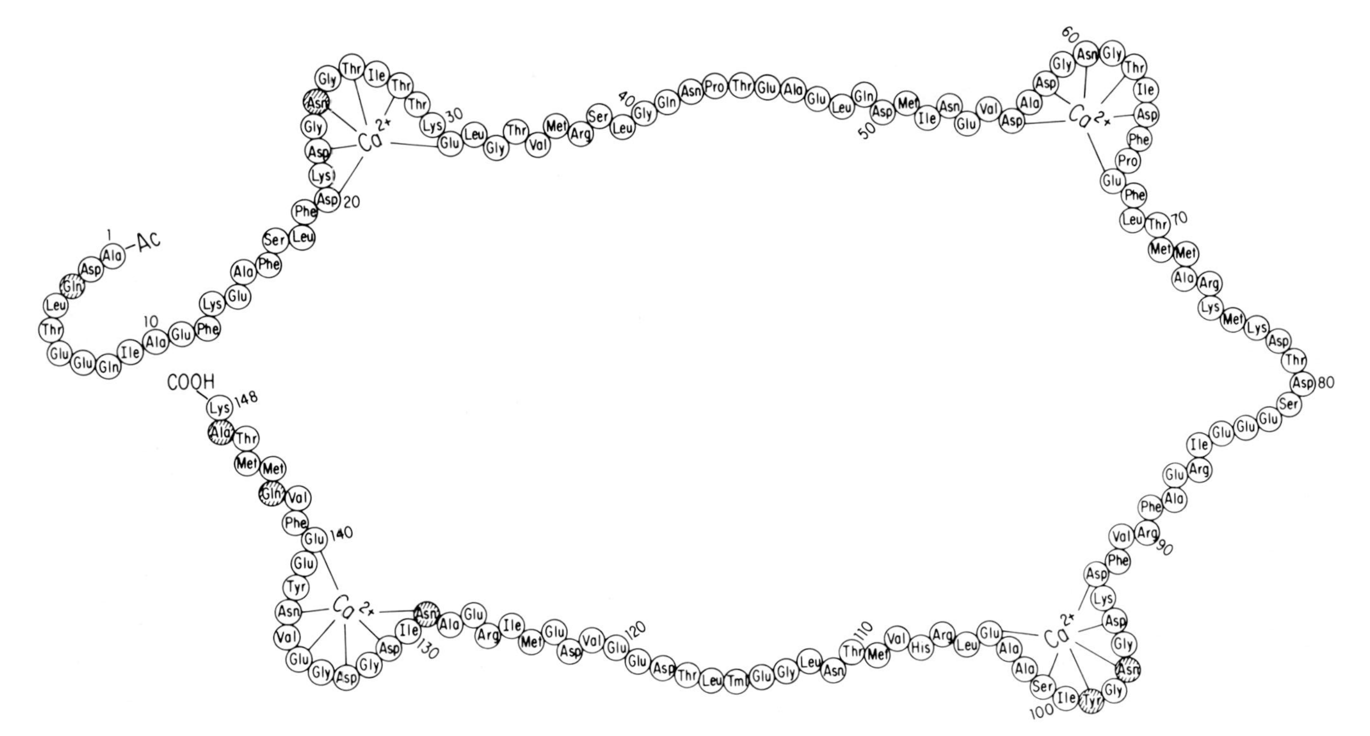

Fig. 1. Sequence of bovine brain and *Renilla reniformis* calmodulins. The sequence shown is that of the bovine brain protein determined by Watterson *et al.* (1980). *Renilla reniformis* protein sequence (Sharief *et al.*, 1980) is identical except for those residues hatched in the figure as discussed in the text. The sequence is arranged to depict the putative helix–loop–helix regions forming each of the four calcium binding sites based on the E-F hand model of Kretsinger (1976). Predicted calcium liganding residues are indicated by the solid lines in each loop; predicted helix forming residues adjacent to each loop are staggered.

region of calmodulin from the sea invertebrate, *Renilla reniformis,* discussed below is identical to that of the bovine proteins. Second, position 60 (bovine, Asn; rat, Ala) is a predicted liganding residue in the second calcium binding domain (see Fig. 1).

Detailed comparative studies other than sequence analysis have been performed on a number of vertebrate calmodulins. As noted in Section II, Stevens and co-workers (1976) found no differences between bovine heart and brain calmodulins based on studies including tryptic peptide mapping. Similarly, the amino acid compositions, physicochemical properties, functional activities, and tryptic peptide maps of bovine, porcine, rabbit, rat, and chicken brain calmodulins have been shown to be indistinguishable (Vanaman *et al.,* 1976; Watterson *et al.,* 1980b). The calmodulins from various other tissues (Kuo and Coffee, 1976; Jarrett and Penniston, 1978; Grand *et al.,* 1979) as well as transformed chick embryo fibroblasts (Van Eldik and Watterson,1979) and the murine cell line $P388D_1$ (Jamieson and Vanaman, 1979) also appear to be undistinguishable from the brain proteins by similar criteria.

B. Invertebrate Calmodulins

In view of the very highly conserved structures of vertebrate calmodulins, we have recently (Vanaman and Sharief, 1979) begun investigating the structure of calmodulins from more divergent species. Figure 1 *also* presents the sequence of calmodulin from the coelenerate *Renilla reniformis* isolated as described by Cormier and co-workers (Jones *et al.,* 1979). A set of tryptic peptides was prepared from performic acid-oxidized *Renilla* calmodulin, and their complete sequences established by a combination of direct sequence analysis and analysis of thermolysin-produced subfragments. These tryptic peptide sequences were identical to their counterparts in the bovine brain protein except for 7 residues (shaded in Fig. 1) as discussed briefly below.

The amino terminal tryptic peptide of *Renilla* calmodulin was identified because of its blocked amino terminus, the only such tryptic peptide isolated. The amino acid composition of this blocked peptide had one less glutamic acid residue than the corresponding peptide from the bovine protein. Thermolysin digestion products of this peptide had sequences identical to those previously found in similar studies of the bovine protein except for the blocked amino terminal subfragment isolated as a dipeptide containing Ala and Asx. The tripeptide Ac-Ala-Asp-Gln was isolated from this sequence in the bovine brain protein (Watterson *et al.,* 1980a). It was therefore concluded that the glutaminyl residue at position 3 in the bovine protein was missing in *Renilla reniformis* calmodulin.

The other differences in sequence between the *Renilla* and bovine brain calmodulins were established by direct residue identification. These included the following differences in the *Renilla* protein: Phe replacing Tyr-99, Asp replacing Asn-129, Lys replacing Gln-142, and Ser replacing Ala-147 in the sequence shown in Fig. 1. In addition, the *Renilla* protein has aspartic acid residues not asparagines, at positions 60 and 97 just as assigned in the bovine uterine smooth muscle protein (Grand and Perry, 1978).

The very limited number of apparent differences between the vertebrate and invertebrate calmodulins is striking particularly in view of their position and character: (1) Three of the five major differences are clustered in the amino terminal 3 and carboxyl terminal 6 residues; (2) the lack of glutamine at position 3 in the *Renilla* protein is unlikely to have any significant effect on function as it occurs in a region of the sequence coming before the start of the helix-loop-helix forming residues of the first calcium-binding domain; (3) the interchange of Ala-147 in bovine calmodulin with a seryl residue represents a conservative replacement. In addition, the two other differences which occur between bovine and *Renilla* calmodulins, in the fourth calcium-binding domain (Asn-129 and Gln-143 in the bovine protein being Asp and Lys, respectively, in the *Renilla* protein), are also conservative since they are paired replacements which maintain the exact charge character of this region of the sequence.

The presence of a phenylalanyl residue in place of Tyr-99 in the bovine protein might appear to be a quite significant change as this position is predicted to be involved in oxygen coordinated calcium liganding in the third domain loop structure. However Phe occurs in the equivalent position in chicken skeletal muscle TnC and in parvalbumins (Barker *et al.*, 1978). In addition, TnCs and parvalbumins contain arginyl and lysyl residues in the equivalent position in other calcium-binding loops (Barker *et al.*, 1978).

In the case of parvalbumins, the main chain carbonyl oxygen has been assigned as the liganding species at this position by X-ray crystallographic studies (Moews and Kretsinger, 1975). By homology, this is also assumed to be the case in muscle TnCs (Kretsinger and Barry 1975). Therefore, the presence of a phenylalanyl residue at this position would not be expected to alter calcium binding substantially.

The results presented here concerning the conserved structures of vertebrate and invertebrate calmodulins is in general agreement with recent studies of other invertebrate calmodulins, including those from the earthworm, *Lumbricus terrestris* (Waisman *et al.*, 1978), and the nematode, *Caenorhabditis elegans* (Schachat *et al.*, 1980). It is extremely important to note, however, that *C. elegans* has a troponin C-like protein (TnCLP) in addition to a calmodulin (Schachat *et al.,* 1980). The physicochemical properties of these two *C. elegans* proteins are so similar that their resolution by gel electrophoresis or ion exchange chromatography is difficult. Both proteins bind to troponin I-Sepharose affinity resins in a calcium-dependent manner. However, they are clearly distinct proteins as judged by amino acid compostion, CNBr peptide map, and functional activity as the TnCLP does not serve as a calcium-dependent activator of PDE. Studies are currently in progress to examine the potential role of these two proteins in providing both thick and thin filament calcium-dependent regulation of motility in *C. elegans* (F. Schachat, personal communication).

C. Protozoan and Plant Calmodulins

A number of reports have appeared showing the presence of proteins with calmodulin-like properties in plants (Anderson and Cormier, 1979; Wallace and Cheung, 1979; Van Eldik *et al.,* 1980) and protozoa (Kuznicki *et al.,* 1979; Nagao *et al.,* 1979; Maihle and Satir, 1979; Jamieson *et al.,* 1979). However, only limited information is available concerning their detailed structural and functional properties. Recently, we have purified calmodulin from the protozoan ciliate *Tetrahymena pyriformis* and have examined its properties in detail (Jamieson *et al.,* 1979). The amino acid composition of *Tetrahymena* calmodulin (Table II) has many of the general features previously determined for animal calmodulins, including the absence of tryptophan and cysteine and the presence of trimethyllysine. However, this composition is clearly less rigidly conserved than those of the animal proteins. In addition, *Tetrahymena* calmodulin is smaller (MW$\simeq$15,000). and differs from inverterate and vertebrate calmodulins in at least 2 out of 7 CNBr cleavage products as judged by gel electrophoretic analysis. Despite these detectable differences in structure, *Tetrahymena* calmodulin retains all the calcium-dependent activities of the metozoan proteins thus far examined (Jamieson *et al.,* 1979), including activation of bovine brain PDE, formation of complexes with rabbit skeletal muscle troponin I, and binding of phenothiazines. More definitive information concerning structural differences between *Tetrahymena,* animal, and plant calmodulins will be obtained from detailed comparison of their amino acid sequences.

IV. SIMILARITIES BETWEEN CALMODULIN AND TROPONIN C

As noted in Section I, calmodulin provides myosin-linked calcium-dependent regulation of the actomyosin contractile apparatus of most animal tissues through activation of myosin light chain kinase (Dabrowska *et al.*, 1978; Yazawa and Yagi, 1977; Nairn and Perry, 1979; Hathaway and Adelstein, 1979; Scordilis and Adelstein 1978). However, striated muscles have an additional, specialized regulatory system, the troponin complex, which provides calcium-dependent regulation of the actomyosin system through interaction with actin thin filaments as described by Ebashi and co-workers (1967).* Troponin (Tn) is a tightly associated, equimolar complex of three polypeptides: troponin I (TnI) and troponin T (TnT) which interact with actin and tropomyosin, respectively, and the calcium-binding subunit, troponin C (TnC).

Conformational changes induced in TnC upon calcium binding increase the strength of its interaction with TnI and TnT, resulting in a structural rearrangement of the troponin—tropomyosin—actin thin filament complex, removing the steric interference to actin—myosin interaction (Hitchcock *et al.*, 1973). TnC therefore acts as an essentially nondissociating calcium-dependent activator of striated muscle actomyosin contractile activity.

As noted earlier, numerous reports have appeared showing the TnC-like physicochemical and functional properties of calmodulins. Comparisons as in Fig. 2 of the amino acid sequences of animal calmodulins to those determined for skeletal (Collins *et al.*, 1977) and cardiac (Van Erd and Takahashi, 1976) muscle TnCs clearly demonstrates their close relationship. When the sequence of bovine brain calmodulin (Fig. 2, line B) is aligned with residue 8 in rabbit skeletal muscle TnC (Fig. 2, line C) and residue 9 in the bovine cardiac muscle protein (Fig. 2, line A), maximum homology is maintained throughout the three protein sequences by introducing only two small gaps in the calmodulin sequence. The first of these is a single residue gap, placed here at a position corresponding to Cys-35 in bovine cardiac muscle TnC, which is also required in this region of the sequences of all skeletal muscle TnCs (Barker *et al.*, 1978) for proper alignment to the bovine cardiac muscle protein. The second is a gap of three residues required to align the calmodulin sequence to both TnCs. This gap is placed between residues 80 and 81 in the calmodulin sequence (position 90 in the Fig. 2 alignments), in the center of the putative hinge sequence between domains 2 and 3 (see Fig. 1).

The number of positions at which all three proteins have either identical (Fig. 2, solid boxes) or functionally conserved (Fig. 2, dashed boxes) residues is quite striking. As shown in Table III, the sum of these homologous

* Recent studies (Mikawa *et al.*, 1978) suggest that a similar but distinct regulatory system, termed leiotonin, may provide actin-linked regulation in other tissues such as smooth muscle. However, further evidence is required in order to prove this fact unequivocally.

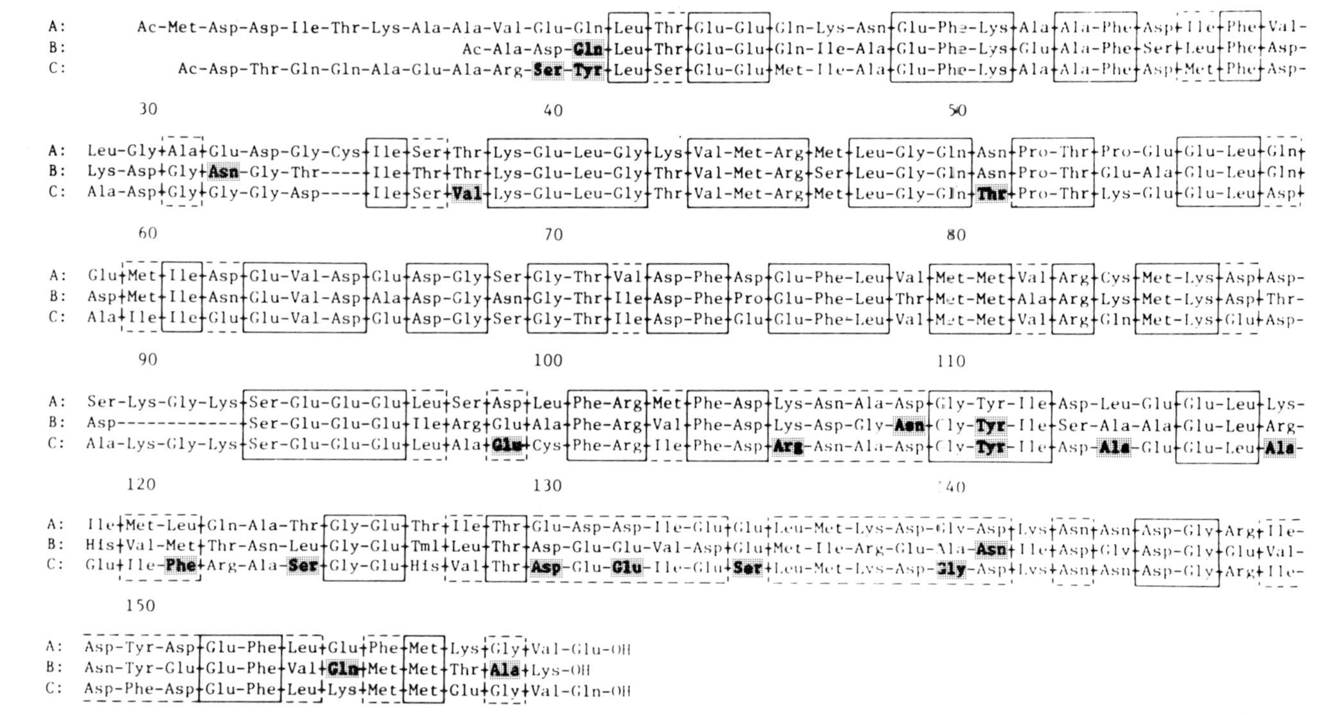

Fig. 2. Homology between calmodulin and muscle TnCs. The amino acid sequences of bovine cardiac [sequence A (Van Eerd and Takahashi, 1976)] and rabbit skeletal [sequence C (Collins *et al.*, 1977)] muscle TnCs are aligned with that of bovine brain calmodulin [sequence B (Watterson *et al.*, 1980a)] as discussed in the text. Solid boxes indicate positions at which all three proteins have identical residues, dashed boxes conserved residues. Gaps are indicated by a dashed line. The shaded residues in the rabbit skeletal muscle TnC sequence (C) indicate positions at which it differs from that of the chicken muscle protein (Wilkinson, 1976). Shaded residues in the calmodulin sequence (B) differ in the *Renilla reniformis* protein (Sharief *et al.*, 1980), and are shown as hatched residues in Fig. 1. (Adapted from Watterson *et al.*, 1980a.)

TABLE III

Comparison of Calmodulin and Muscle TnCs[a]

Proteins compared	Identities	Conserved	Sum	Percent
CaM versus cardiac TnC	74	40	114	76
CaM versus skeletal TnC	77	33	110	74
Skeletal TnC versus cardiac TnC	103	23	126	85
All three	63	38	101	68

[a] Taken from the sequences shown in Fig. 2; 148 positions compared.

positions in all three proteins is 68%, only slightly less than that obtained when calmodulin is compared to either TnC alone. It should be noted that TnCs, unlike calmodulin, occur in at least two distinctly different tissue-specific types represented by the sequences of the cardiac and skeletal muscle proteins shown here. Only a limited number of differences exist between the sequences of rabbit and chicken fast skeletal muscle TnCs (Fig. 2, line C, shaded residues). However, even these differences are greater than those found for vertebrate and invertebrate calmodulins (Fig. 2, line B, shaded residues).

Additional evidence strongly supports the view that calmodulins and TnCs share a common evolutionary ancestry. Collins (1976) first noted that the amino acid sequence of rabbit skeletal muscle TnC was composed of four homologous domains. The amino acid sequence of bovine brain calmodulin also possesses internal homology (Vanaman *et al.*, 1977) as shown in Fig. 3. Although all four domains are related in sequence, the level of homology is greatest when the first domain is aligned with the third domain and the second domain is aligned with the fourth. Of the 33 residues compared in each pair of domains, 18 residues are identical and 6 are conservative replacements between domains 1 and 3. There are 13 identical residues and 13 conservative replacements between domains 2 and 4. This level of internal homology is greater even than that observed within the muscle TnCs (Barker *et al.*, 1978).

From the considerations presented above, it is now clear that the entire super family of calcium-binding proteins have arisen from a common genetic ancestry. Two successive tandem duplications of a gene encoding a single domain containing protein of 30–40 residues gave a two-domain molecule. This, in turn, was duplicated to give a four-domain precursor from which this family was then derived. The sequence of calmodulin most closely resembles that predicted for the original four-domained protein (Barker *et al.*, 1978; Goodman *et al.*, 1980).

The structural similarities between calmodulin and muscle TnCs noted

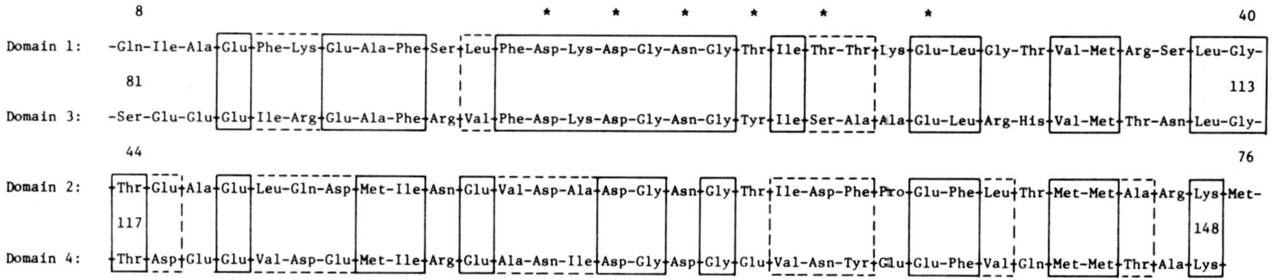

```
8                                            *    *    *    *    *       *                                   40
Domain 1:  -Gln-Ile-Ala-Glu-Phe-Lys-Glu-Ala-Phe-Ser-Leu-Phe-Asp-Lys-Asp-Gly-Asn-Gly-Thr-Ile-Thr-Thr-Lys-Glu-Leu-Gly-Thr-Val-Met-Arg-Ser-Leu-Gly-
           81                                                                                                           113
Domain 3:  -Ser-Glu-Glu-Glu-Ile-Arg-Glu-Ala-Phe-Arg-Val-Phe-Asp-Lys-Asp-Gly-Asn-Gly-Tyr-Ile-Ser-Ala-Ala-Glu-Leu-Arg-His-Val-Met-Thr-Asn-Leu-Gly-

           44                                                                                                            76
Domain 2:  -Thr-Glu-Ala-Glu-Leu-Gln-Asp-Met-Ile-Asn-Glu-Val-Asp-Ala-Asp-Gly-Asn-Gly-Thr-Ile-Asp-Phe-Pro-Glu-Phe-Leu-Thr-Met-Met-Ala-Arg-Lys-Met-
           117                                                                                                      148
Domain 4:  -Thr-Asp-Glu-Glu-Val-Asp-Glu-Met-Ile-Arg-Glu-Ala-Asn-Ile-Asp-Gly-Asp-Gly-Glu-Val-Asn-Tyr-Glu-Glu-Phe-Val-Gln-Met-Met-Thr-Ala-Lys-
```

Fig. 3. Internal homology in calmodulin. The four domains of bovine brain calmodulin are aligned to give maximum homology as discussed in the text. Starred residues are those predicted to provide oxygen ligands for calcium coordination. Numerals refer to the residue position in the linear sequence shown in Fig. 1. Solid boxes denote identities, dashed boxes conserved residues. (Taken from Watterson *et al.*, 1980a, by permission.)

here account for the fact that calmodulin possesses limited TnC like functional properties (Dedman *et al.*, 1977; Amphlett *et al.*, 1976). However, it is clear that calmodulin possesses numerous unique regulatory activities not shared by muscle TnCs to any significant extent. The most obvious structural differences between calmodulins and TnCs which might correlate with calmodulin-specific functions are (1) the region containing the single trimethyllysyl residue found uniquely in calmodulins and (2) the greater variability in the sequences of the C-terminal region of these proteins encompassing their entire fourth domains (see Fig. 2). To date, no exact role has been established for trimethyllysine in calmodulin. However, detailed studies (Perry *et al.*, 1979) of bovine brain calmodulin and rabbit skeletal muscle TnC interaction with TnI suggest that a region of amino acid sequence encompassing at least domain 3 in both proteins represents a site for specific TnI interaction. The sequence -Ser-Glu-Glu-Glu- (positions 93–96 in the cardiac protein sequence shown in Fig. 2) is particularly noteworthy as this sequence is conserved in all TnCs and calmodulins thus far studied (Barker *et al.*, 1978; Goodman *et al.*, 1980). It seems likely that this sequence may form a portion of the TnI recognition site in this domain, and may represent a portion of the site for calcium-dependent interaction between calmodulin and other proteins.

V. CONCLUDING REMARKS

From the studies presented here, it is clear that calmodulin is a protein of highly conserved structure which is present in all eukaryotes thus far examined. Although calmodulin has not been found in prokaryotes, structurally related proteins will almost certainly be found in these organisms.

The rigidly conserved structure of calmodulin no doubt is due to the multiple specific interactions which it must make with both small molecules and proteins in order to properly regulate cellular activity. The enzyme systems which calmodulin regulates are so essential to cellular function that any significant alteration in calmodulin structure would be lethal to the organism. In fact, it would appear that tissue-specific forms of calmodulin cannot be tolerated, since even the closely related muscle TnCs, (which might be considered more highly evolved, specialized forms of calcium-dependent regulator) possess no significant calmodulin-like activity.

The second messenger function of calcium is now well established as being at least as important as that of the cyclic nucleotides. The calcium–calmodulin complex regulates a set of specific enzymes, including protein kinases which allow for specific biochemical responses to external stimuli. In contrast, cyclic nucleotide-dependent kinases have very broad specificity more commensurate with a pleiotropic cellular response. Sets of calmodulin-regulated enzymes, whose presence in a given tissue could be genetically determined, would appear to be best suited for providing tissue and function-specific stimulus–response systems in eukaryotic cells.

ACKNOWLEDGMENTS

This work was supported by NIH Grant NS10123. The author wishes to thank Mr. Gordon Jamieson and Dr. Fred Schachat for critical reading of this manuscript and Ms. Ann Allen for its preparation.

REFERENCES

Amphlett, G. W., Vanaman, T. C., and Perry, S. V. (1976) Effect of the troponin C-like protein from bovine brain (brain modulator protein) on the Mg^{2+}-stimulated ATPase of skeletal muscle actomyosin. *FEBS Lett.* **72,** 163–168.

Anderson, J. M., and Cormier, M. J. (1979). Isolation of calcium-dependent modulator protein from higher plants and fungi. *Fed. Proc., Fed. Am. Soc. Exp. Biol.* **38,** 478.

Barker, W. C., Ketcham, L. K., and Dayhoff, M. O. (1978). Contractile system proteins. *In* "Atlas of Protein Sequence and Structure" (M. O. Dayhoff, ed.) pp. 273–283. Nat. Biomed. Res. Found., Washington, D.C.

Berridge, M. J. (1975). The interaction of cyclic nucleotides and calcium in the control of cellular activity. *Adv. Cyclic Nucleotide Res.* **6,** 1–98.

Cheung, W.-Y. (1970). Cyclic 3′,5′-nucleotide phosphodiesterase - Demonstration of an activator. *Biochem. Biophys. Res. Commun.* **38,** 533–538.

Collins, J. H. (1976). Structure and evolution of troponin C and related proteins. *Symp. Soc. Expl. Biol.* **30,** 303–334.

Collins, J. H., Greaser, M. L., Potter, J. D., and Horn, M. J. (1977). Determination of the amino acid sequence of troponin C from rabbit skeletal muscle. *J. Biol. Chem.* **252,** 6356–6362.

Dabrowska, R., Sherry, J. M. F., Aramatoria, D. K., and Hartshorne, D. J. (1978). Modulator protein as a component of the myosin light chain kinase from chicken gizzard. *Biochemistry* **17,** 253–258.

Dedman, J. R., Potter, J. D., Jackson, R. L., Johnson, J. D., and Means, A. R. (1977). Physicochemical properties of rat testis Ca^{2+}-dependent regulator protein of cyclic nucleotide phosphodiesterase. Relationship of Ca^{2+}-binding, conformational changes, and phosphodiesterase activity. *J. Biol. Chem.* **252,** 8415–8422.

Dedman, J. R., Jackson, R. L., Schreiber, W. E., and Means, A. R. (1978). Sequence homology of the Ca^{2+}-dependent regulator of cyclic nucleotide phosphosiesterase from rat testis with other Ca^{2+}-binding proteins. *J. Biol. Chem.* **253,** 343–346.

Ebashi, S., Ebashi, F., and Kodama, A. (1967). Troponin as the Ca^{++}-receptive protein in the contractile system. *J. Biochem.* (*Tokyo*) **62,** 137–138.

Goodman, M., Pechère, J. -F., Haiech, J., and Demaille, J. G. (1979). Evolutionary diversification of structure and function in the family of intracellular calcium-binding proteins. *J. Mol. Evol.* **13,** 331–352.

Grand, R. J. A., and Perry, S. V. (1978). The amino acid sequence of the troponin C-like protein (modulator protein) from bovine uterus. *FEBS Lett.* **92,** 137–142.

Grand, R. J. A., Perry, S. V. and Weeks, R. A. (1979). Troponin C-like proteins (calmodulins) from mammalian smooth muscle and other tissues. *Biochem. J.* **177,** 521–529.

Greengard, P. (1978). Phosphorylated proteins as physiological effectors. *Science* **199,** 146–152.

Hathaway, D. R., and Adelstein, R. S. (1979). Human platelet myosin light-chain kinase requires the calcium-binding protein calmodulin for activity. *Proc. Natl. Acad. Sci. U.S.A.* **76,** 1653–1657.

Hitchcock, S. E., Huxley, H. E., and Szent-Gÿorgi, A. G. (1973). Calcium sensitive binding of troponin to actin-tropomyosin: A two-site model for troponin action. *J. Mol. Biol.* **80,** 825–836.

Jackson, R. L., Dedman, J. R., Schreiber, W. E., Knapp, R., Bhatnagar, P. K., and Means, A. R. (1977). Identification of ε-N-Trimethyllysine in a rat testis calcium-dependent regulatory protein of cyclic nucleotide phosphodiesterase. *Biochem. Biophys. Res. Commun.* **77,** 723–729.

Jamieson, G. A., and Vanaman, T. C. (1979). Isolation and characterization of calmodulin from a murine macrophage-like cell line. *Fed. Proc., Fed. Am. Soc. Exp. Biol.* **38,** 478.

Jamieson, G. A., Jr., Vanaman, T. C., and Blum, J. J. (1979). Presence of calmodulin in *Tetrahymena*. *Proc. Natl. Acad. Sci. U.S.A.* **76,** 6471–6475.

Jarrett, H. W., and Penniston, J. T. (1978). Purification of the Ca^{2+}-stimulated ATPase activator from human erythrocytes. Its membership in the class of Ca^{2+}-binding modulator proteins. *J. Biol. Chem.* **253,** 4676–4682.

Jones, H. P., Matthews, J. C., and Cormier, M. J. (1979). Isolation and characterization of Ca^{2+}-dependent modulator protein from the marine invertebrate *Renilla reniformis*. *Biochemistry* **18,** 55–60.

Kakiuchi, S., Yamazaki, R., and Nakajimi, H. (1970). Properties of a heat-stable phosphodiesterase activating factor isolated from brain extracts-Studies on cyclic 3′,5′-nucleotide phosphodiesterase II. *Proc. Jpn. Acad.* **46,** 589–592.

Kretsinger, R. H. (1976). Calcium binding proteins. *Annu. Rev. Biochem.* **45,** 239–266.

Kretsinger, R. H., and Barry, C. D. (1975). The predicted structure of the calcium binding component of troponin. *Biochim. Biophys. Acta* **405,** 40–52.

Kuo, I. C. Y., and Coffee, C. J. (1976). Purification and characterization of a troponin-C like protein from adrenal medulla. *J. Biol. Chem.* **251,** 1603–1609.

Kuznicki, J., Kuznicki, L., and Drabikowski, W. (1979). Ca^{2+}-binding modulator protein in protozoa and myxomycete. *Cell Biol. In. Rep.* **3,** 17–23.

Lin, Y. M., Liu, Y. P., and Cheung, W. Y. (1974). Cyclic 3′:5′-Nucleotide phosphodiesterase. Purification, characterization and active form of the protein activator from bovine brain. *J. Biol. Chem.* **249,** 4943–4954.

Maihle, N. J., and Satir, B. H. (1979). Indirect immunoflourescent localization of calmodulin in *Paramecium tetraurelia*. *J. Protozool.* **26,** 10a.

Mikawa, T., Nonomura, Y., Hirata, M., Ebashi, S., and Kakiuchi, S. (1978). Involvement of an acidic protein in regulation of smooth muscle contraction by the tropomyosin-leiotonin system. *J. Biochem* (*Tokyo*) **84,** 1633–1636.

Moews, P. C., and Kretsinger, R. H. (1975). Refinement of the structure of carp muscle calcium-binding parvalbumin by model building and difference fourier analysis. *J. Mol. Biol.* **91,** 201–227.

Nagao, S., Sukuki, Y., Watanabe, Y., and Nozawa, Y. (1979). Activation by a calcium-binding protein in guanylate cyclase in *Tetrahymena pyriformis*. *Biochem. Biophys. Res. Commun.* **90,** 261–268.

Nairn, A. C., and Perry, S. V. (1979). Calmodulin and myosin-light-chain kinase of rabbit fast skeletal muscle. *Biochem. J.* **179,** 89–97.

Perry, S. V., Grand, R. J. A., Nairn, A. C., Vanaman, T. C., and Wall, C. A. (1979). Calcium-binding proteins and the regulation of contractile activity. *Biochem. Soc. Trans.* **7,** 619–622.

Potter J. D., Johnson, J. D., Dedman, J. R., Schreiber, W. E., Mandel, F., Jackson, R. L., and Means, A. R. (1977). Calcium-binding proteins: Relationship of binding structure, conformation and biological function. *In* "Calcium Binding Proteins and Calcium Function" (R. H. Wasserman, R. A. Corradino, E. Carafoli, R. H. Kretsinger, D. A. MacClennan, and F. L. Siegel, eds.) pp. 239–250. Am. Elsevier, New York.

Rasmussen, H., and Goodman, D. B. P. (1977). Relationships between calcium and cyclic nucleotides in cell activation. *Physiol. Rev.* **57,** 421–509.

Schachat, F. H., Bronson, D. D., Jamieson, G. A., Jr., and Vanaman, T. C. (1980). Characterization and roles of calmodulin and a second calcium regulatory protein from *Caenorabditis elegans*. *In* "Calcium Binding Proteins and Calcium Function (F. L. Seigel, E. Carafoli, R. H. Kretsinger, D. H. MacLennan, and R. H. Wasserman, eds.). Elsevier/North Holland, New York (in press).

Sharief, F. S., Jones, H. P., Cormier, M. J., and Vanaman, T. C. (1980). In preparation.

Schneider, D. J. (1973). Studies of Nervous System Proteins. In "Proteins of the Nervous System" D. J. Schneider, R. H. Angelletti, R. A. Bradshaw, A. Grasso, and B. W. Moore, eds.), pp.81–83. Raven, New York.

Scordilis, S. P., and Adelstein, R. S. (1978). A comparative study of the myosin light chain kinases from myoblast and muscle sources - Studies on the kinases from proliferative rat myoblasts in culture, rat thigh muscle, and rabbit skeletal muscle. *J. Biol. Chem.* **253,** 9041–9048.

Stevens, F. C., Walsh, M., Ho, H. C., Teo, T. S., and Wang, J. H. (1976). Comparison of calcium-binding proteins. Bovine heart and brain protein activators of cyclic nucleotide phosphodiesterase and skeletal muscle troponic C*J. Biol. Chem.* **251,** 4495–4500.

Teo, T. S., and Wang, J. H. (1973). Mechanism of activation of a cyclic adenosine 3′:5′-monophosphate phosphodiesterase from bovine heart by calcium ions - Identification of the protein activator as a Ca^{2+} binding protein. *J. Biol. Chem.* **248,** 588–595.

Vanaman, T. C., and Sharief, F. (1979). Structural properties of calmodulins from divergent eucaryotic organisms. *Fed. Proc., Fed. Am. Soc. Exp. Biol.* **38,** 788.

Vanaman, T. C. Harrelson, W. G., and Watterson, D. M. (1975). Studies on a troponin C like Ca^{2+} binding protein from brain. *Fed. Proc., Fed. Am. Soc. Exp. Biol.* **34,** 307.

Vanaman, T. C., Sharief, F., Awramik, J. L., Mendel, P. A., and Watterson, D. M. (1976). Chemical and biological properties of the ubiquitous troponin-C like protein from non-muscle tissues, a multifunctional Ca^{2+}-dependent regulatory protein. *In* "Contractile Systems in Non-Muscle Tissues" (S. V. Perry, A. Margreth, and R. S. Adelstein, eds.), pp. 165–176. Elsevier/North-Holland Biomedical Press, New York.

Vanaman, T. C., Sharief, F., and Watterson, D. M. (1977). Structural homology between brain modulator protein and muscle TnCs. In "Calcium Binding Proteins and Calcium Function" (R. H. Wasserman, R. A. Corradino, E. Carafoli, R. H. Kretsinger, D. A. MacClennan, and F. L. Siegel, eds.). pp. 107–116. Am. Elsevier, New York.

Van Eerd, J. -P., and Takahashi, K. (1976). Determinations of the complete amino acid sequence of bovine cardiac troponin C. *Biochemistry* **15,** 1171–1180.

Van Eldik, L. J., Grossman, A. R., Iverson, D. B., and Watterson, D. M. (1980). Isolation and characterization of calmodulin from spinach leaves and *in vitro* translation mixtures. *Proc. Natl. Acad. Sci. U.S.A.* **77,** 1912–1916.

Van Eldik, L. J., and Watterson, D. M. (1979). Characterization of a calcium-modulated protein from transformed chicken fibroblasts. *J. Biol. Chem.* **254,** 10250–10255.

Waisman, D. M., Stevens, F. C., and Wang, J. H. (1978). Purification and characterization of a Ca^{2+}-binding protein in *Lumbricus terrestris*. *J. Biol. Chem.* **253,** 1106–1113.

Wallace, R. M., and Cheung, W. Y. (1979). Antibody against bovine brain calmodulin cross-reacts with cotton seed calmodulin. *Fed. Proc., Fed. Am. Soc. Exp. Biol.* **38,** 478.

Watterson, D. M., and Vanaman, T. C. (1976). Affinity chromatography purification of a cyclic nucleotide phosphodiesterase using immobilized modulator protein, a troponin C-like protein from brain. *Biochem. Biophys. Res. Commun.* **73,** 40–46.

Watterson, D. M., Harrelson, W. G., Jr., Keller, P. M., Sharief, F., and Vanaman, T. C. (1976) Structural similarities between the Ca^{2+}-dependent regulatory proteins of 3′,5′-cyclic nucleotide phosphodiesterase and actomyosin ATPase. *J. Biol. Chem.* **251,** 4501–4513.

Watterson, D. M., Sharief, F. S., and Vanaman, T. C. (1980a). The complete amino acid sequence of the Ca^{2+}-dependent modulator protein (Calmodulin) of bovine brain. *J. Biol. Chem.* **255,** 962–975.

Watterson, D. M., Mendel, P. A., and Vanaman, T. C. (1980b). Comparison of calcium-modulated proteins from vertebrate brains. *Biochemistry* **19,** 2672–2676.

Wilkinson, J. M. (1976). The amino acid sequence of troponin C from chicken skeletal muscle. *FEBS Lett.* **70,** 254–256.

Yazawa, M., and Yagi, K. (1977). A calcium-binding subunit of myosin light chain kinase. *J. Biochem (Tokyo)* **82,** 287–289.

Chapter 4

Calmodulin: Structure–Function Relationships

Claude B. Klee

I. INTRODUCTION

Calmodulin (CaM), a heat- and acid-stable protein, was discovered as an activator of cyclic nucleotide phosphodiesterase (Cheung, this volume, Chapter 1) and was identified as a Ca^{2+}-binding protein by Teo and Wang (1973). Its similarity to troponin C (TnC) another Ca^{2+}-binding protein, was first recognized by Wang *et al.*, (1975) and has been confirmed by analysis of the sequence of the two proteins (Watterson *et al.*, 1976). Calmodulin is a member of a specific group of Ca^{2+}-binding proteins defined by Kretsinger (1977) as "EF hand"- containing proteins which are believed to be derived from a common ancestor (Kretsinger, 1980; Goodman *et al.*, 1979). In contrast to others of these Ca^{2+}-binding proteins, such as troponin C, parvalbumins, S-100 protein, and intestinal Ca^{2+}-binding protein, which are both tissue- and species-specific in structure

CALCIUM AND CELL FUNCTION, VOL. I

ISBN 0-12-171401-2

and perform unique functions, calmodulin is a ubiquitous protein whose almost perfect structural conservation throughout evolution (Watterson *et al.*, 1980) must reflect its multifunctional roles (for reviews, see Wolff and Brostrom, 1979, Wang and Waisman, 1979; Cheung, 1980; Means and Dedman, 1980; Klee *et al.*, 1980).

II. CALMODULIN STRUCTURE

The primary structure of calmodulins isolated from several vertebrate and invertebrate organisms as well as plants is the subject of another chapter (Vanaman, this volume, Chapter 3). The remarkable similarity between the various calmodulins and the known ubiquity of the protein reflects its ability to mediate the control of the activity of a large number of cellular processes by Ca^{2+} acting as a second messenger (Rasmussen, 1970). It also accounts for the fact that the physical and chemical properties of all calmodulins described currently are nearly identical. As shown in Fig. 1, calmodulin is believed to consist of four Ca^{2+}-binding domains (or "EF hands") (Vanaman *et al.*, 1977) each of which is composed of two α-helical regions separated by a Ca^{2+}-binding loop formed by appropriately spaced amino acid residues which serve as the Ca^{2+} ligands (Kretsinger, 1980). The four Ca^{2+}-binding domains are very similar, particularly domains I and III and domains II and IV. The relative positions of hydrophobic residues, particularly methionyl and phenylalanyl residues, are identical in each set of two domains. No two domains are totally identical, however, as evidenced by the relative position of the polar and neutral amino acids. The Ca^{2+}-binding loops themselves are different, and the particular characteristics of the fourth loop have been emphasized by Vogt *et al.* (1979). These different domains could fulfill different functions and are probably responsible for the unique Ca^{2+}-binding properties of calmodulin.

The blocked amino-terminal peptide and the three peptides connecting the Ca^{2+} binding domains are also different from each other and highly preserved throughout evolution suggesting that they are functionally important. The third connecting peptide (residues 113–120) contains the single trimethyllysyl residue (at position 115) (Watterson *et al.*, 1976, 1980; Jackson *et al.*, 1977; Miyake and Kakiuchi, 1978) which is characteristic of all calmodulins. Its role in the biological activity of the protein has not yet been elucidated. In view of the extreme conservation of calmodulin, as opposed to other Ca^{2+}-binding proteins of similar structure which perform specific functions, one can assume that the whole calmodulin molecule plays a role in one or the other of its many functions. One of the major

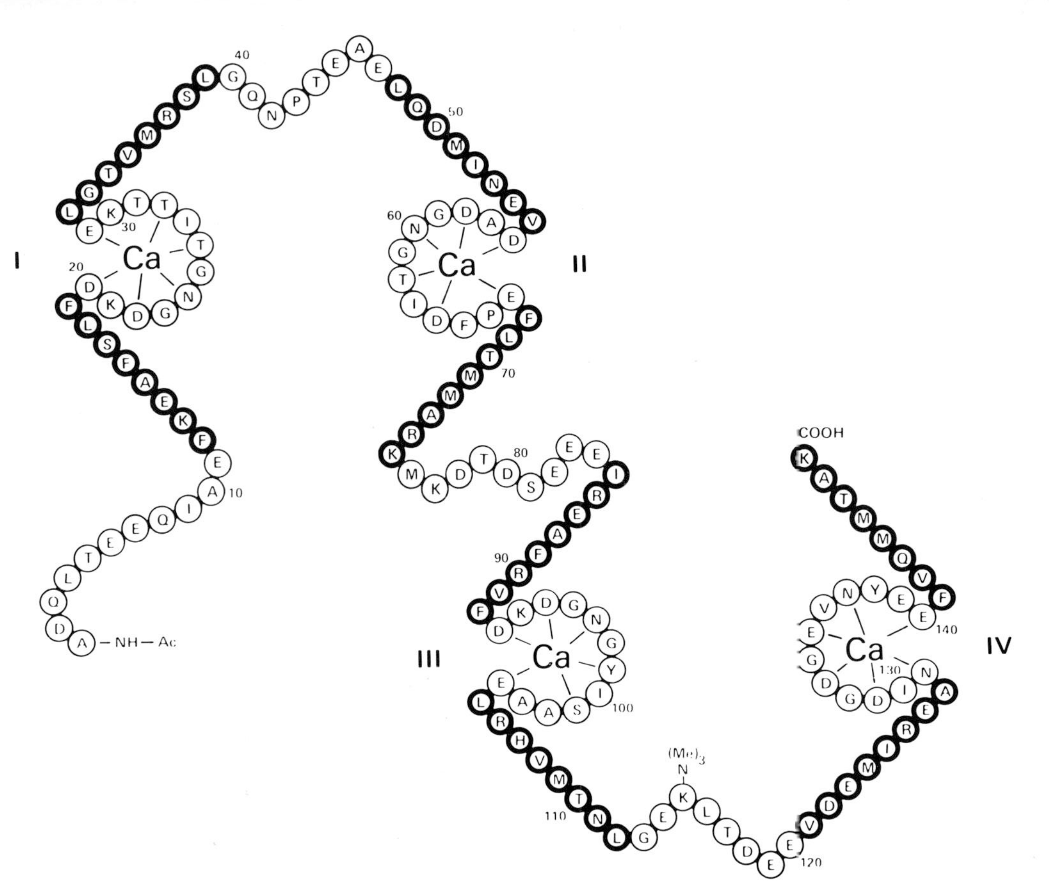

Fig. 1. Sequence of calmodulin. The sequence of bovine brain calmodulin (Watterson *et al.*, 1980) is shown utilizing the one-letter code for amino acid residues. A, Ala; D, Asp; E, Glu; F, Phe; G, Gly; H, His; I, Ile; K, Lys; L, Leu; M, Met; N, Asn; P, Pro; Q, Gln; R, Arg; S, Ser; T, Thr; V, Val; Y, Tyr. The four proposed Ca^{2+}-binding domains (see text) with the stretches of α-helix (darker circles) are indicated (from Klee *et al.*, 1980).

questions to be answered will therefore be: are different regions of calmodulin recognized by different enzymes or proteins or is the same calmodulin interacting site recognized by a specific calmodulin-binding region shared by several proteins?

III. PHYSICOCHEMICAL PROPERTIES

A. Hydrodynamic Properties

The physicochemical properties of calmodulins isolated from mammalian tissues in several laboratories are summarized in Table I. The molecular weight based on the amino acid sequence (16,723) of bovine brain calmodulin (Watterson *et al.*, 1980) is very similar to that determined by electrophoresis in the presence of sodium dodecyl sulfate (SDS), although in the latter case abnormally high values are obtained when chelating agents are present during electrophoresis (Grab *et al.*, 1979; Klee *et al.*, 1979b; Burgess *et al.*, 1980). This behavior is characteristic of several Ca^{2+}-binding proteins and is not understood, but presumably reflects either Ca^{2+} binding (Burgess *et al.*, 1980) or chelator binding (Haiech *et al.*, 1979). Calmodulin exists as a monomer under native conditions in the presence or absence of Ca^{2+}. Because it is a highly acidic protein (isoelectric point of 3.9-4.3), calmodulin migrates on gel filtration, at low ionic strength, with an anomalously large apparent Stokes radius (Dedman *et al.*, 1977). The frictional coefficient indicates that calmodulin is a compact, globular protein which contains a large degree of secondary structure even in the absence of Ca^{2+}. Indeed optical rotatory dispersion and circular dichroic (CD) studies have shown that the Ca^{2+}-free protein exhibits a fairly high content of α-helix (28–45%) and β-pleated sheet (15–20%) (Liu and Cheung, 1976; Klee, 1977a; Wolff *et al.*, 1977; Dedman *et al.*, 1977).

B. Spectroscopic Properties

The amino acid composition of calmodulins is characterized by the absence of tryptophan and a low tyrosine : phenylalanine ratio. Thus, the protein displays a characteristic and unusual uv-absorption spectrum (Watterson *et al.*, 1976; Klee, 1977a). A maximum at 276 nm and a shoulder at 282 nm correspond to the tyrosine spectrum. The multiple peaks at 253, 258.5, 265 and 268.5 nm are characteristic of the fine structure of the absorption band of phenylalanine. Calmodulins isolated from invertebrate species, such as *Renilla reniformis* (Jones *et al.*, 1979; T. Vanaman, per-

TABLE I

Physicochemical Properties of Calmodulin

Property	+ Ca^{2+}	+ EGTA/EDTA
Molecular weight		
Amino acid sequence	16,723[a]	
Sedimentation equilibrium	18,700[b]	18,700[b]
	15,000[c]	17,800[d]
SDS-gel electrophoresis	16,500[e]	19,000[e]
$s_{20,w}$ (S)	1.87[b]	1.83[b]
	1.85[c]	
	2.00[f]	
D (cm/s)	1.09×10^{-6}[c]	
$\bar{v}$ (ml/g)	0.707[b]	0.712[b]
Stokes radius (Å)		
Calculated from $\bar{v}$, $s_{20,w}$, M_r	20.9	21.4
Frictional coefficient	1.20[c,b]	1.23[b]
Isoelectric point	3.9–4.3[c,g,h]	
$\varepsilon_{276nm}^{1\%}$	1.8[d,i]	2.0[i]
θ_{222} (deg cm²/dmole)	−15,000[i]	−12,000[i]
	−18,200[g]	−14,700[g]
	−16,000[j]	−11,000[j]

[a] Watterson *et al.*, 1980.
[b] Crouch and Klee, 1980.
[c] Lin *et al.*, 1974.
[d] Watterson *et al.*, 1976.
[e] Klee *et al.*, 1979b.
[f] Teo *et al.*, 1973.
[g] Dedman *et al.*, 1977.
[h] Stevens *et al.*, 1976.
[i] Klee, 1977a.
[j] Wolff *et al.*, 1977.

sonal communication), scallop and sea anemone (Yasawa *et al.*, 1980), *Octopus vulgaris* (K. Seamon, personal communication), and *Tetrahymena pyziformis* (Jamieson *et al.*, 1979), contain only a single tyrosyl residue at position 138 and have accordingly an even lower absorption coefficient at 276 nm than do most calmodulins. The near-uv circular dichroic (CD) spectrum of calmodulin reflects the asymmetric environment of its aromatic residues (Wolff *et al.*, 1977; Walsh *et al.*, 1979; Crouch and Klee, 1980). In the presence of 0.1 *M* salt the CD spectrum is characterized by the presence of three negative peaks at 279, 261, and 268 nm which indicates that both tyrosyl and phenylalanyl residues are in an asymmetrical environment. The relative ellipticities of these bands are strongly dependent on salt concentration, and at low ionic strength (in the absence of divalent cations) the tyrosyl residues show a small positive ellipticity. More recent nuclear magnetic resonance (NMR) studies have allowed a more precise analysis of the environment of the aromatic residues and also that of the trimethyllysyl and histidinyl residues (Seamon, 1980).

C. Ca^{2+} Binding to Calmodulin

Studies of the interaction of calmodulin with Ca^{2+} generally demonstrate the existence of four high-affinity Ca^{2+}-binding sites ($K_d = 10^{-5}$ to 10^{-6} *M*) per mole of calmodulin (Table II) in agreement with the four Ca^{2+}-binding domains postulated on the basis of the amino acid sequence (Van-

TABLE II

Ca^{2+}-binding Properties of Calmodulin

Experimental conditions	Number of Ca^{2+}-binding sites	Dissociation constants
25 m*M* Tris-HCl, 25 m*M* imidazole, 3 m*M* Mg^{2+}; 25°C[a]	3	10^{-6} *M* (1)[g], 1.2 × 10^{-5} *M* (2)
25 m*M* Tris-HCl, pH 8.0; 4°C[b]	4	3.5 × 10^{-6} *M* (3), 1.8 × 10^{-5} *M* (1)
100 m*M* Imidazole, pH 7.0; 24°C[c]	4	1.1 × 10^{-6} *M* (2), 8.6 × 10^{-4} *M* (2)
10 m*M* Tris, pH 7.4; 25°C[d]	4	2 × 10^{-7} *M* (3), 10^{-6} *M* (1)
10 m*M* Tris, pH 7.4, 1 m*M* Mg^{2+}; 25°C[d]	3	3 × 10^{-6} *M* (3)
10 m*M* HEPES, pH 6.5, 100 m*M* KCl 0.1 m*M* EGTA; 4°C[e]	4	2.4 × 10^{-6} *M* (4)
10 m*M* HEPES, pH 7.5, 100 m*M* KCl, 25°C[f]	4	3.3 × 10^{-6} *M* (1), 10^{-6} *M* (1), 8.6 × 10^{-6} *M* (1), 2 × 10^{-5} *M* (1)
10 m*M* HEPES, pH 7.5, 100 m*M* KCl, 3 m*M* Mg^{2+}; 25°C[f]	4	5 × 10^{-6} *M* (1), 4 × 10^{-6} *M* (1), 2.5 × 10^{-5} *M* (1), 4 × 10^{-5} *M* (1)

[a] Teo and Wang, 1973.
[b] Lin *et al.*, 1974.
[c] Watterson *et al.*, 1976.
[d] Wolff *et al.*, 1977.
[e] Dedman *et al.*, 1977.
[f] Crouch and Klee, 1980.
[g] The numbers in parenthesis indicate the number of sites with the corresponding dissociation constant.

aman *et al.*, 1977). With one exception (Dedman *et al.*, 1977), all studies report two classes of sites which have different affinity for Ca^{2+} or a negative cooperativity between the sites. A more recent study indicates a positive cooperative interaction between the two first sites (Crouch and Klee, 1980). The number of sites in each class is still the subject of some uncertainty as are the values of the binding constants. These discrepancies in the values of the binding constants are probably the result of the different experimental conditions, such as pH, temperature, and ionic strength, which have been used in different laboratories. For example, the significantly higher values of the binding constants reported by Wolff *et al.*, (1977) may be due to the low ionic strength used by these authors. The affinity of calmodulin for Ca^{2+} is greatly diminished by increased salt concentration (J. Haiech, personal communication). Monovalent cations affect the structure of calmodulin (Section III, B) and may thereby affect its Ca^{2+}-binding properties. Alternatively monovalent cations may compete with Ca^{2+} at one or more of the Ca^{2+} sites. Divalent cations other than Ca^{2+} have been reported to interact with calmodulin (Teo and Wang, 1973; Lin *et al.*, 1974). One of the most efficient competitors Mn^{2+} (with an affinity which is about one-tenth that of Ca^{2+}), is believed to interact with the Ca^{2+} sites and to have very similar effects on the structure and biological activity of calmodulin as does Ca^{2+} (Wolff *et al.*, 1977). The role of Mg^{2+} is less well understood. At millimolar concentrations, it decreases the affinity of the Ca^{2+} sites for Ca^{2+} competitively and even completely suppresses one of the sites (Wolff *et al.*, 1977), but it fails to activate calmodulin or to induce the characteristic Ca^{2+}-dependent structural changes (Wolff *et al.*, 1977; Klee, 1977a; Richman and Klee, 1979; Seamon, 1980).

D. Ca^{2+}-Induced Conformational Changes

Upon binding of Ca^{2+}, calmodulin undergoes a specific conformational transition which affects its α-helix content (Liu and Cheung, 1976; Klee, 1977a; Dedman *et al.*, 1977; Wolff *et al.*, 1977) and the microenvironment of several amino acid residues without significantly affecting its hydrodynamic properties (and therefore its gross size and shape) (Crouch and Klee, 1980). Monovalent cations and Mg^{2+} have a less marked effect on the α-helix content (Wolff *et al.*, 1977; Richman and Klee, 1979). The 10–20% increase in α-helix content which accompanies Ca^{2+} binding to calmodulin reflects a large conformational change which has been studied by a number of different techniques.

The presence of Ca^{2+} greatly decreases the susceptibility of calmodulin to trypsin digestion and changes the initial sites of bond cleavage as well (Drabikowski *et al.*, 1977). In the presence of Ca^{2+}, trypsin cleaves first at

Lys-77 (located between the second and the third Ca^{2+}-binding domain) and generates two polypeptides. The amino-terminal half of the molecule (domains I and II) is apparently rapidly degraded further, its electrophoretic mobility is not greatly affected by Ca^{2+}, and is probably not highly structured; the carboxy-terminal half of the molecule (domains III and IV) accumulates under these conditions, interacts with Ca^{2+}, and undergoes a typical Ca^{2+}-dependent conformational transition as analyzed by tyrosine fluorescence (Drabikowski *et al.*, 1977; Walsh *et al.*, 1977). On the other hand, in the presence of EGTA, different cleavage sites by trypsin have been identified and proteolysis now occurs at much lower trypsin concentrations. The most susceptible bond (at Arg-106) is cleaved to yield calmodulin (1–106) which contains the two first domains and a fraction of the third. This polypeptide is subsequently cleaved at Arg-90. Both peptides 1–106 and 1–90 appear to interact with Ca^{2+} (Walsh *et al.*, 1977). Since the peptide 1–77 which does not contain the connecting peptide (residues 76–84) fails to interact with Ca^{2+} it is possible that the connecting peptide (residues between domains II and III) plays an important role in maintaining the conformational stability of the binding domains I and II. In accord with this hypothesis a pronounced decrease in affinity for Ca^{2+} is observed after modification of methionyl residues 71, 72, and 76 by *N*-chlorosuccinimide (Walsh and Stevens, 1978). Note too that, in the absence of Ca^{2+}, the third binding domain of calmodulin is apparently unfolded since residues 106 and 90 both of which are located in α-helical regions of this postulated structure are susceptible to proteolysis. Thus, limited proteolysis studies suggest that upon binding of Ca^{2+} large structural changes occur in the third Ca^{2+}-binding domain of calmodulin.

Spectroscopic methods have provided a large body of evidence that the fourth Ca^{2+}-binding domain also undergoes a major transition upon binding of Ca^{2+}. Among the observed effects are the appearance of a characteristic and Ca^{2+}-specific uv-difference spectrum and increased ellipticity at 279, 268, and 261 nm (Wolff *et al.*, 1977; Klee, 1977a; Yasawa *et al.*, 1978) an increased tyrosine fluorescence (Wang *et al.*, 1975; Dedman *et al.*, 1977) and perturbation of proton nuclear magnetic resonances (Seamon, 1980). The two tyrosyl residues of calmodulin are located in the third and fourth Ca^{2+}-binding loops. Tyr-99 (in the third loop) has a normal pK_a and reacts with tetranitromethane and *N*-acetylimidazole both in the presence and absence of Ca^{2+} (Klee, 1977a; Richman and Klee, 1978; Richman, 1978). Tyr-99 is therefore believed to be exposed to solvent and does not appear to be modified by Ca^{2+} binding. On the other hand, Tyr-138 (in the fourth Ca^{2+}-binding loop) has an abnormal pK_a, and in the absence of Ca^{2+} reacts with *N*-acetylimidazole but not with tetranitromethane. It is therefore believed to be exposed to solvent but that the proximity of negative charges may increase the pK_a and prevent reaction with tetranitromethane. Upon addition of Ca^{2+}, this residue loses its ability to react with *N*-acetylimizadole, and the pK_a of its nitrated derivative is increased. Thus Tyr-138 becomes more buried as a result of the Ca^{2+}-induced conformational change (Richman, 1978; Richman and Klee, 1979). Seamon (1980) has reached similar conclusions from nuclear magnetic resonance (NMR) studies. This increase in hydrophobicity in the environment of Tyr-138 is somewhat surprising in view of the fact that this residue is responsible for the Ca^{2+}-induced negative uv-difference spectrum characteristic of calmodulin (Klee, 1977a; Yasawa *et al.*, 1978). This conclusion has been confirmed by the recent observation that calmodulin from lower invertebrates which lacks Tyr-99 but contains Tyr-138 exhibits a similar uv difference spectrum (Yasawa *et al.*, 1980; K. Seamon, personal communication). It was also shown by NMR (Seamon, 1980) that a slow conformational transition between two stable states of the protein is responsible for the perturbation of the environments of Tyr-138, Tyr-99 a specific phenylalanyl residue, and the single trimethyllysyl residue.

Spectroscopic studies have revealed that the Ca^{2+}-dependent conformational transition of calmodulin occurs in discrete, distinguishable steps. The increase in α-helix content as determined by ellipticity measurements at 220 nm is almost complete upon binding of two Ca^{2+} per mole (Klee, 1977a). Similarly the uv absorption changes accompanying the modification of the environment of Tyr-138 and the near-uv CD changes are almost complete in the calmodulin $\cdot Ca_2{}^{2+}$ complex, and exhibit a sigmoidicity which indicates a positive cooperativity between the two first Ca^{2+}-binding sites (Klee, 1977a; Crouch and Klee, 1980). Using nitrated calmodulin McCubbin *et al.* (1979) have recently shown two distinct transitions in the CD changes both in the near- and far-uv regions. Similarly, flurescence changes upon Ca^{2+} binding to calmodulin (Dedman *et al.*, 1977) are essentially complete upon binding of 1–2 Ca^{2+} per mole. Nuclear magnetic resonance studies have confirmed and further defined these multiple transitions (Seamon, 1980). The first transition upon binding of Ca^{2+} affects both tyrosyl residues, a specific phenylalanine, and the trimethyllysyl residue via a slow conformational change which affects a large portion of the calmodulin molecule. Within this transition, the ortho protons of Tyr-138 also exhibit a fast exchange which could reflect a more localized modification due to Ca^{2+} binding at the fourth domain. Finally, binding of two additional Ca^{2+} induces other specific changes in the environment of Tyr-138 and the phenylalanyl residues in a faster structural change. The data could be explained by specific structural changes occurring upon random binding of Ca^{2+} at any of the four domains (Dedman *et al.*, 1977). However, all four sites are structurally different, and exhibit different affinities for Ca^{2+} (Crouch and Klee, 1980), and the NMR data (Seamon, 1980) are more consistent with an ordered sequential binding of Ca^{2+} to calmodulin. Thus it seems that at least four different conformations of Ca^{2+}·calmodulin complexes exist which could be recognized by different proteins. Calmodulin would thereby translate a quantitative Ca^{2+} signal into qualitatively different cellular responses (Klee *et al.*, 1980).

The mechanism of activation of several enzymes by calmodulin occurs by a two-step mechanism (for a review, see Wolff and Brostrom, 1979). At least two of these calmodulin-dependent enzymes, cyclic nucleotide phosphodiesterase and myosin light chain kinase, have been shown to require the calmodulin·Ca_3^{2+} or calmodulin·Ca_4^{2+} complex for activation (Wolff *et al.*, 1977; Crouch and Klee, 1980; Stull *et al.*, 1980). Furthermore, interaction with these enzymes induces a strong cooperativity in Ca^{2+} dependence which produces a very sharp Ca^{2+} threshold for activation (Crouch and Klee, 1980). On the other hand, phosphorylase kinase has been shown to interact with calmodulin free of Ca^{2+} (Cohen *et al.*, 1978), although Ca^{2+} is still required for increased activity. It will be interesting to learn if this Ca^{2+} dependence of activation differs from that of enzymes which only interact with Ca^{2+}·calmodulin complexes.

IV. INTERACTION OF CALMODULIN WITH ITS TARGET PROTEINS

Calmodulin interacts with its target proteins in at least three different ways as shown in Table III. The majority of the calmodulin-regulated proteins, such as cyclic nucleotide phosphodiesterase, calcineurin (CaM-BP_{80}), myosin kinases, and (Ca^{2+} − Mg^{2+}) ATPase of the erythrocyte plasma membrane form tight and specific Ca^{2+}-dependent complexes with calmodulin ($K_d \leqq 10^{-8}$ *M*). Some other proteins (troponin I, tubulin, and phosphorylase kinase) interact not only with calmodulin but also with troponin C. Troponin I and phosphorylase kinase can form both Ca^{2+}-dependent and -independent complexes with calmodulin. The interesting dual interaction of calmodulin and phosphorylase kinase is discussed by Cohen in Chapter 9 of this volume. In an analogous fashion troponin I, which under physiological conditions interacts with troponin C, can also form Ca^{2+}-dependent (C. B. Klee and P. G. Richman, unpublished results) and -dependent stoichiometric complexes with calmodulin (Amphlett *et al.*, 1976; Dedman *et al.*, 1977; Drabikowski et al., 1977). Calmodulin has also been reported to be associated with membrane fractions from which it can only be dissociated by treatments known to alter drastically the membrane structure (Kakiuchi *et al.*, 1978).

TABLE III

Interaction of Calmodulin with Its Target Proteins

	Subunits (s) M_r	Structure	Ca^{2+} requirement[a]	$K_d(M)$ TNC	$K_d(M)$ CaM[b]
Cyclic nucleotide phosphodiesterase[c]	59,000	Dimer $(\alpha)_2$	+		10^{-9}
Heat-labile calmodulin-binding protein (CaM-BP_{80} or calcineurin)[d]	61,000 (A); 15,000 (B)	Heterodimer (A · B)	+		6×10^{-9}
Myosin light chain kinases[e]	80,000–130,000	Monomer	+		10^{-9}
$(Ca^{2+} - Mg^{2+})$ erythrocyte ATPase[f]	125,000		+		
Troponin I[g]	24,000	Monomer	+/−	ND[c]	2×10^{-6}
Tubulin[h]	55,000		+	10^{-4}	10^{-4}
Phosphorylase kinase[i]	145,000 (α); 128,000 (β)	Heterohexadecamer $(\alpha,\beta,\gamma,\delta)_4$	+	10^{-6}	10^{-8}
	40,000 (γ); 16,500 (δ)		−		ND
Membrane protein kinase[j]			+		10^{-9}

[a] The Ca^{2+} requirement applies only to calmodulin enzyme interaction and not to enzyme activation.

[b] The K_d corresponds to the concentration of calmodulin needed to obtain 50% of the maximal extent of stimulation of the enzyme. In the case of calcineurin or troponin I, K_d was calculated from the concentration-dependent inhibition of the stimulation of phosphodiesterase by calmodulin assuming a one-to-one complex of calmodulin and the appropriate protein. The K_d for tubulin is estimated on the basis of the concentration of calmodulin or TnC needed to obtain 50% inhibition of tubulin polymerization.

[c] ND, not determined (very tight complex).

[c] Teshima and Kakiuchi, 1974; Lin *et al.*, 1975; LaPorte *et al.*, 1979; Klee *et al.*, 1979a; Morrill *et al.*, 1979; Klee, 1977b; Sharma *et al.*, 1980.

[d] Wang and Desai, 1977; Klee and Krinks, 1978; Wallace *et al.*, 1979; see also Chapter 17.

[e] Dabrowska *et al.*, 1978; Yasawa and Yagi, 1978; Nairn and Perry, 1979; Conti and Adelstein, 1980; Adelstein and Klee, 1980; Walsh *et al.*, 1980; see also Chapter 11.

[f] Niggli *et al.*, 1979; Gopinath and Vincenzi, 1977.

[g] Amphlett *et al.*, 1976; Drabikowski *et al.*, 1977.

[h] Marcum *et al.*, 1978; Kumagai and Nishida, 1979.

[i] Cohen *et al.*, 1978; Cohen *et al.*, 1979.

[j] LePeuch *et al.*, 1979.

The mechanism by which this single protein can take part in so many interactions is still unknown. It has become clear during the last 2 years that enzymes which form tight Ca^{2+}-dependent complexes with calmodulin [such as cyclic nucleotide phosphodiesterase, myosin light chain kinases, and erythrocyte (Ca^{2+} – Mg^{2+}) ATPase] do not share a common calmodulin binding subunit. These three enzymes have been purified to homogeneity (see Table III) and have been shown to be composed of only a single type of subunit which carries both the catalytic activity and the ability to bind calmodulin. Furthermore, the three types of enzyme have molecular weights which are significantly different from one another. They could, however, share a calmodulin-binding domain, a region of the polypeptide sequence which exerts an inhibitory effect on the enzymatically active site which is relieved upon binding of calmodulin. Cheung (1971) first reported that limited proteolysis causes an irreversible activation of cyclic nucleotide phosphodiesterase which renders the enzyme independent of Ca^{2+} and calmodulin. We have now shown that this activation is accompanied by the cleavage of a 22,000 dalton peptide originating from the NH_2—or the COOH—terminal portion of the catalytic subunit of the enzyme. The native enzyme, a dimer composed of two identical 59,000 dalton polypeptides, is converted in the presence of Ca^{2+} and calmodulin into a tetrameric protein composed of two catalytic subunits and two regulatory subunits, calmodulins, as schematized in Fig. 2 (LaPorte *et al.*, 1979; Krinks and Klee, 1980) The proteolyzed enzyme loses its

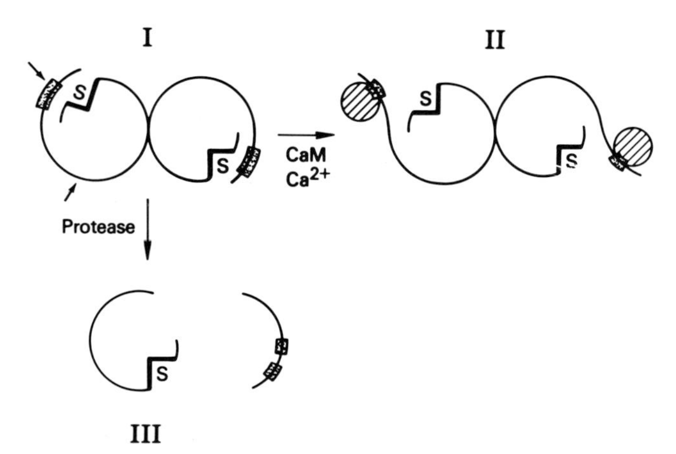

Fig. 2. Schematic representation of the activation of cyclic nucleotide phosphodiesterase by calmodulin and proteases. S indicates the substrate, cAMP or cGMP; the stippled area corresponds to the calmodulin-binding site; the hatched circle represents calmodulin; the small arrows indicates the sites of cleavage by trypsin (from Klee, 1980).

ability to bind calmodulin and to dimerize as well. Activation by proteolysis and calmodulin have very similar effects on the enzyme. Both forms of the enzyme have identical kinetic properties, a high specific activity of 200–350 units/mg and a K_m value of $100 \mu M$ for cAMP, and 40–60 units/mg and a K_m value of $10 \mu M$ for cGMP. The only difference is in the Ca^{2+} dependence, which is observed only for the native holoenzyme (Krinks and Klee, 1980). The low level of basal activity exhibited by the native enzyme which does not obey Michaelis–Menten kinetics may be the result of a small contamination by degraded enzyme. These results can be taken as evidence that binding of calmodulin, like limited proteolysis, results in the displacement of an inhibitory domain. A similar mechanism could explain the activation of phosphodiesterase by phospholipids (Wolff and Brostrom, 1976; Pichard and Cheung, 1977) The interaction of calmodulin with adenylate cyclase (Brostrom *et al.*, 1975; Cheung *et al.*, 1975) is more complex and may require a regulatory factor (Toscano *et al.*, 1979).

Interaction of calmodulin with similar calmodulin binding domains on different proteins could explain the inhibition of any specific calmodulin-dependent enzyme by another calmodulin-binding protein (Wang, Sharma, and Tam, Chapter 15, this volume). Alternatively, the inhibitory effect could indicate that the formation of calmodulin complexes with a particular group of calmodulin-binding proteins prevents the interaction of calmodulin with another calmodulin-binding protein interacting at a different site.

Attempts to isolate ternary complexes of calmodulin with two calmodulin-binding proteins, [the heat-labile calmodulin-binding protein (calcineurin) and troponin I] by cross-linking experiments have so far been unsuccessful (C. B. Klee, unpublished results). Troponin I, which inhibits the activation of cyclic nucleotide phosphodiesterase by calmodulin with a low affinity ($K_d = 2 \times 10^{-6}\ M$), also forms a stoichiometric, one-to-one, complex with calmodulin. The formation of this complex prevents the interaction with calcineurin. The existence of ternary complexes of calmodulin with its target proteins has been postulated in the case of phosphorylase kinase (Cohen, *et al.*, 1978) and tubulin (Kumagai and Nishida, 1979), but these remain to be demonstrated unequivocally. These postulated complexes constitute the only presumptive evidence for the alternative idea that calmodulin does indeed contain different sites that interact with different proteins.

Although the site(s) of interaction of calmodulin with its target enzymes has not been identified preliminary evidence indicates that the site that binds troponin I is located in the second connecting peptide between residues 78 and 106 since both tryptic peptides between residues 78–148, and

1–106 can interact with troponin I (Drabikowski *et al.*, 1977; Walsh *et al.*, 1977). A more detailed study of this interaction site is presented in Chapter 3, this volume, by Vanaman. The interacting site with cyclic nucleotide phosphodiesterase could also be located in the same part of the molecule. Peptide 1–106 but not peptide 1–90 preserves some ability to activate the enzyme although with a large decrease in affinity (200-fold). Peptide 78–148, on the other hand, is devoid of significant activity (Walsh *et al.*, 1977). It appears that if the same site can indeed interact with different proteins, the requirements for the interaction with some target enzymes are more stringent than that of others (Walsh and Stevens, 1979). Calmodulin isolated from vertebrate sources is apparently less efficient in the activation of myosin kinase than in the activation of cyclic nucleotide phosphodiesterase (Yasawa *et al.*, 1980). These differences may be due to the fact that different proteins recognize slightly different conformations. As mentioned in Section III, the conformation of the portion of calmodulin containing the two last Ca^{2+}-binding domains is greatly affected by the degree of Ca^{2+} occupancy. We also know that troponin I and phosphorylase kinase can interact with calmodulin in the absence of Ca^{2+}, whereas phosphodiesterase (Wolff *et al.*, 1977; Crouch and Klee, 1980) and myosin light chain kinase (Stull *et al.*, 1980) require almost full occupancy of the calmodulin Ca^{2+}-binding sites.

V. CONCLUSIONS

Despite its structural similarity to many other Ca^{2+}-binding proteins, calmodulin is functionally unique in that it interacts with a great many target proteins. Thus, calmodulin has been shown to interact specifically with such proteins as cyclic nucleotide phosphodiesterase, myosin light chain kinases, the (Ca^{2+} – Mg^{2+}) ATPase of the erythrocyte plasma membrane, and the heat-labile calmodulin-binding protein (calcineurin). The structural basis for the versatility of calmodulin has not yet been fully elucidated and could be the result of specific and localized features of calmodulin such as the trimethyllysyl residue. It seems more likely, however, that different parts of the calmodulin molecule may be important for different functions, which would explain the extreme conservation of the protein and at the same time the lack of sensitivity of a given function to some chemical modifications (Walsh and Stevens, 1977, 1978; Richman and Klee, 1978). Although there is no direct evidence at present linking different domains of the calmodulin molecule with different target proteins, it is attractive to speculate that the many conformations induced by binding of Ca^{2+} may generate different interaction sites for recognition by different proteins. Indeed the calmodulin · Ca^{2+} complexes are reognized by different enzymes: phosphorylase kinase under physiological conditions can interact with calmodulin in the absence of Ca^{2+} and a large number of other calmodulin-dependent enzymes interact only in the presence of Ca^{2+}. This Ca^{2+}-induced generation of recognition sites would allow a concerted modulation of the Ca^{2+} signal. The role of calmodulin as a mediator of the second messenger is analogous to that of cAMP-dependent protein kinase as a mediator of cAMP signal (Wang and Waisman, 1979) and in a analogous fashion allows a coordinated regulation of various cellular processes. Furthermore, since the enzymes controlling cAMP metabolism and synthesis are under calmodulin control in many tissues, the two second messenger systems are directly coupled by the action of this extraordinarily versatile protein.

ACKNOWLEDGMENTS

I thank many of my co-workers in this field who gracefully communicated manuscripts to me prior to publication and Ms. May Liu and Anahid Ayrandjian for their excellent editorial assistance.

REFERENCES

Adelstein, R. S., and Klee, C. B. (1980). In preparation.

Amphlett, G. W., Vanaman, T. C., and Perry, S. V. (1976). Effect of troponin C-like protein from bovine brain (brain modulator protein) on the Mg^{2+} stimulated ATPase of skeletal muscle actomyosin. *FEBS Lett.* **72,** 163–168.

Brostrom, C. O., Huang, Y. C., Breckenridge, B., McL., and Wolff, D. J. (1975). Identification of calcium binding protein as a calcium-dependent regulator of brain adenylate cyclase. *Proc. Natl. Acad. Sci. U.S.A.* **72,** 64–68.

Burgess, W. H., Jemiolo, D. K., and Kretsinger, R. H. (1980). Interaction of calcium and calmodulin in the presence of sodium dodecyl sulfate. *Biochim. Biophys. Acta.* (in press) (Submitted for publication.)

Cheung, W. Y. (1971). Cyclic 3′,5′-nucleotide phosphodiesterase: Evidence for and properties of a protein activator. *J. Biol. Chem.* **246,** 2859–2869.

Cheung, W. Y. (1980). Calmodulin plays a pivotal role in cellular regulation. *Science* **207,** 19–27.

Cheung, W. Y., Bradham, L. S., Lynch, T. J., Lin, Y. M., and Tallant, E. A. (1975). Protein activator of cyclic 3′,5′ nucleotide phosphodiesterase of bovine rat brain activates its adenylate cyclase, *Biochem. Biophys. Res. Commun.* **66,** 1055–1062.

Cohen, P., Burchell, A., Foulkes, J. G., Cohen, P. T. W., Vanaman, T. C., and Nairn, A. (1978). Identification of the Ca^{2+}-dependent modulator protein as the fourth subunit of rabbit skeletal muscle phosphorylase kinase. *FEBS Lett.* **92,** 287–293.

Cohen, P., Picton, C., and Klee, C. B. (1979). Activation of phosphorylase kinase from rabbit skeletal muscle by calmodulin and troponin. *FEBS Lett.* **104,** 25–30.

Conti, M. A., and Adelstein, R. S. (1980). Mechanism of regulation of smooth muscle myosin kinase by phosphorylation *Fed. Proc., Fed. Am. Soc. Exp. Biol.* **39,** 1569–1573.

Crouch, T. H., and Klee, C. B. (1980). Positive cooperative binding of calcium to bovine brain calmodulin. *Biochemistry* **19,** 3692–3697.

Dabrowska, R., Sherry, J. M. F., Aromatorio, D. K., and Hartshorne, D. J. (1978). Modulator protein as a component of the myosin light-chain kinase from chicken gizzard. *Biochemistry* **17,** 253–258.

Dedman, J. R., Potter, J. D., Jackson, R. L., Johnson, J. D., and Means, A. R. (1977). Physicochemical properties of rat testis Ca^{2+}-dependent regulator protein of cyclic nucleotide phosphodiesterase. *J. Biol. Chem.* **252,** 8415–8422.

Drabikowski, W., Kuznicki, J., and Grabarek, Z. (1977). Similarity in Ca^{2+}-induced changes between troponin C and protein activator of 3′:5′-cyclic nucleotide phosphodiesterase and their tryptic fragments. *Biochim. Biophys. Acta* **485,** 124–133.

Goodman, M., Pechère, J. F., Haiech, J., and Demaille, J. G. (1979). Evidence for the Darwinian evolution in the diversification of structure and function in the family of intracellular calcium-binding proteins. *J. Mol. Evol.* **13,** 331–352.

Gopinath, R. M., and Vincenzi, F. F. (1977). Phosphodiesterase protein activator mimics red blood cell cytoplasmic activator of ($Ca^{2+} - Mg^{2+}$) ATPase. *Biochem. Biophys. Res. Commun.* **77,** 1203–1209.

Grab, D. J., Berzins, K., Cohen, R. S., and Siekevitz, P. (1979). Presence of calmodulin in postsynaptic densities isolated from canine cerebral cortex. *J. Biol. Chem.* **254,** 8690–8696.

Haiech, J., Derancourt, J., Pechère, J. F., and Demaille, J. (1979). Magnesium and calcium binding to parvalbumins: Evidence for differences between parvalbumins and an explanation of their relaxing function. *Biochemistry* **18,** 2752–2758.

Jackson, R. L., Dedman, J. R., Schreiber, W. E., Bhatnagar, P. K., Knapp, R. D., and Means, A. R. (1977). Identification of ε-*N*-trimethyllysine in a rat testis calcium-dependent regulatory protein of cyclic nucleotide phosphodiesterase. *Biochem. Biophys. Res. Commun.* **77,** 723–729.

Jamieson, G. A., Jr., Vanaman, T. C., and Blum, J. J. (1979). Presence of calmodulin in tetrahymena. *Proc. Natl. Acad. Sci. U.S.A.* **76,** 6471–6475.

Jones, P. H., Matthews, J. C., and Cormier, M. J. (1979). Isolation and characterization of Ca^{2+}-dependent modulator protein from the marine invertebrate *Renilla reniformis. Biochemistry* **18,** 55–60.

Kakiuchi, S., Yamasaki, R., Teshima, Y., Uenishi, K., Yasuda, S., Kashiba, A., Sobue, K., Ohshima, M., and Kakajima, T. (1978). Membrane-bound protein modulator and phosphodiesterase. *Adv. Cyclic Nucleotide Res.* **9,** 253–263.

Klee, C. B. (1977a). Conformational transition accompanying the binding of Ca^{2+} to the protein activator of 3′,5′-cyclic adenosine monophosphate phosphodiesterase. *Biochemistry* **16,** 1017–1024.

Klee, C. B. (1977b). Interactions of Ca^{2+}-dependent activator protein and troponin C with cyclic AMP phosphodiesterase and its inhibitor. *In* "Proceedings of the Third US-USSR Joint Symposium on Myocardial Metabolism," DHEW Publication no. (NIH) 78–1457, pp. 83–91.

Klee, C. B. (1980). Calmodulin: the coupling factor of the two second messengers Ca^{2+} and cAMP. FMI-EMBO workshop on Protein Phosphorylation and Bioregulation, Karger, Basel (in press).

Klee, C. B., and Krinks, M. H. (1978). Purification of cyclic 3′,5′-nucleotide phosphodiesterase inhibitory protein by affinity chromatography on activator protein coupled to Sepharose. *Biochemistry* **17,** 120–126.

Klee, C. B., Crouch, T. H., and Krinks, M. H. (1979a). Subunit structure and catalytic properties of bovine brain Ca^{2+}-dependent cyclic nucleotide phosphodiesterase. *Biochemistry* **18,** 722–729.

Klee, C. B., Crouch, T. H., and Krinks, M. H. (1979b). Calcineurin: A calcium- and calmodulin-binding protein of the nervous system. *Proc. Natl. Acad. Sci. U.S.A.* **76,** 270–273.

Klee, C. B., Crouch, T. H., and Richman, P. G. (1980). Calmodulin. *Annu. Rev. Biochem.* **49,** 489–515.

Kretsinger, R. H. (1977). Evolution of the informational role of calcium in eukaryotes. *In* "Calcium Binding Proteins and Calcium Function" (R. Wasserman, ed.), pp. 63–72. Am. Elsevier, New York.

Kretsinger, R. H. (1980). Structure and evolution of calcium modulated proteins. *Crit. Rev. Biochem.,* (in press).

Krinks, M. H., and Klee, C. B. (1980) Proteopylic Activation of Ca^{2+}-dependent cyclic nucleotide phosphodiesterase *Fed. Proc.* **39,** (6) 1624.

Kumagai, H., and Nishida, E. (1979). The interactions between calcium-dependent regulator protein of cyclic nucleotide phosphodiesterase and microtubule proteins. *J. Biochem.* (*Tokyo*) **85,** 1267–1274.

LaPorte, D. C., Toscano, W. A., Jr., and Storm, D. R. (1979). Cross-linking of ^{125}I-labeled calcium dependent regulatory protein to the Ca^{2+}-sensitive phosphodiesterase. *Biochemistry* **18,** 2820–2825.

LePeuch, C. J., Haiech, J., and Demaille, J. G. (1979). The concerted regulation of cardiac sarcoplasmic reticulum calcium transport by cAMP-dependent and calcium calmodulin-dependent phosphorylations. *Biochemistry* **18,** 5150–5157.

Lin, Y. M., Liu, Y. P., and Cheung, W. Y. (1974). Cyclic nucleotide phosphodiesterase purification, characterization and active form of the protein activator from bovine brain. *J. Biol. Chem.* **249,** 4943–4954.

Lin, Y. M., Liu, Y. P., and Cheung, W. Y. (1975). Cyclic 3′,5′-nucleotide phosphodiesterase: Ca^{2+}-dependent formation of bovine brain enzyme-activator complex. *FEBS Lett.* **49,** 356–360.

Liu, Y. P., and Cheung, W. Y. (1976). Cyclic 3′:5′-nucleotide phosphodiesterase Ca^{2+} confers more helical conformation to the protein activator. *J. Biol. Chem.* **251,** 4193–4198.

Means, A. R., and Dedman, J. R. (1980). Calmodulin-an intracellular calcium receptor. *Nature* (*London*) **285,** 73–77.

McCubbin, W. D., Hincke, M. T., and Kay, C. M. (1979). The utility of the nitrotyrosine chromophore as a spectroscopic probe in troponin C and modulator protein. *Can. J. Biochem.* **57,** 15–20.

Marcum, J. M., Dedman, J. R., Brinkley, B. R., and Means, A. R. (1978). Control of microtubule assembly-disassembly by calcium-dependent regulator protein. *Proc. Natl. Acad. Sci. U.S.A.* **75,** 3771–3775.

Miyake, M., and Kakiuchi, S. (1978). Calcium-dependent activator protein of cyclic nucleotide phosphodiesterase from rat and bovine brain; presence of *N*-trimethyllysine residue. *Brain Res.* **139,** 378–380.

Morrill, M. E., Thompson, S. T., and Stellwagen, E. (1979). Purification of a cyclic nucleotide phosphodiesterase from bovine brain using blue dextran-Sepharose chromatography. *J. Biol. Chem.* **254,** 4371–4374.

Nairn, A. C., and Perry, S. V. (1979). Calmodulin and myosin light-chain kinase of rabbit fast skeletal muscle. *Biochem. J.* **179,** 89–97.

Niggli, V., Penniston, J. T., and Carafoli, E. (1979). Purification of the ($Ca^{2+} - Mg^{2+}$)-AT-

Pase from human erythrocyte membranes using a calmodulin affinity column. *J. Biol. Chem.* **254,** 9955–9958.

Pichard, A. L., and Cheung, W. Y. (1977). Cyclic 3′,5′-nucleotide phosphodiesterase stimulation of bovine brain cytoplasmic enzyme by lysophosphatidyl choline. T. *Biol. Chem.* **252,** 4872–4875.

Rasmussen, H. (1970). Cell communication, calcium ion and cyclic adenosine monophosphate. *Science* **170,** 404–412.

Richman, P. G. (1978). Conformation-dependent acetylation and nitration of the protein activator of cyclic adenosine 3′,5′-monophosphate phosphodiesterase. Selective nitration of tyrosine residue 138. *Biochemistry* **17,** 3001–3005.

Richman, P. G., and Klee, C. B. (1978). Conformation-dependent nitration of the protein activator of cyclic adenosine 3′,5′-monophosphate phosphodiesterase. *Biochemistry* **17,** 928–935.

Richman, P. G., and Klee, C. B. (1979). Specific perturbation by Ca^{2+} of tyrosyl residue 138 of calmodulin. *J. Biol. Chem.* **254,** 5372–5376.

Seamon, K. (1980). Ca^{2+} – Mg^{2+}-dependent conformational states of calmodulin as determined by nuclear magnetic resonance. *Biochemistry* **19,** 207–215.

Sharma, R. K., Wang T. H., Wirch, E., and Wang, J. H. (1980). Purification and properties of bovine brain calmodulin-dependent cyclic nucleotide phosphodiesterase. *Fed. Proc.* **39,** 1624.

Stevens, F. C., Walsh, M., Ho, H. C., Teo, T. S., and Wang, J. H. (1976). Comparison of calcium binding proteins: Bovine heart and brain protein activators of cyclic nucleotide phosphodiesterase and rabbit skeletal muscle troponin C. *J. Biol. Chem.* **251,** 4495–4500.

Stull, J. T., Manning, D. R., High, C. W., and Blumenthal, D. K. (1980). Phosphorylation of contractile proteins in heart and skeletal muscle. *Fed. Proc., Fed. Am. Soc. Exp. Biol.* **39,** 1552–1557.

Teo, T. S., and Wang, J. H. (1973). Mechanism of activation of a cyclic adenosine 3′:5′-monophosphate phosphodiesterase from bovine heart by calcium ions. *J. Biol. Chem.* **248,** 5950–5955.

Teo, T. S., Wang, J. H., and Wang, J. H. (1973). Purification and properties of the protein activator of bovine heart cyclic adenosine 3′,5′-monophosphate phosphodiesterase. *J. Biol. Chem.* **248,** 588–595.

Teshima, Y., and Kakiuchi, S. (1974). Mechanism of stimulation of Ca^{2+} plus Mg^{2+}-dependent phosphodiesterase from rat cerebral cortex by the modulator protein and Ca^{2+}. *Biochem. Biophys. Res. Commun.* **56,** 489–495.

Toscano, W. A., Westcott, K. E., LaPorte, D. C., and Storm, D. R. (1979). Evidence for a dissociable protein subunit required for calmodulin stimulation of brain adenylate cyclase. *Proc. Natl. Acad. Sci. U.S.A.* **76,** 5582–5586.

Vanaman, T. C., Sharief, F., and Watterson, D. M. (1977). Structural homology between brain modulator protein and muscle TnCs. *In* "Calcium Binding Proteins and Calcium Function" (R. Wasserman, ed.), pp. 63–72. Am. Elsevier, New York.

Vogt, H. P., Strassburger, W., Wollmer, A., Fleischhauer, J. Bullard, B., and Mercola, D. (1979). Calcium binding by troponin C and homologs is correlated with the position and linear density of "β-turns forming" residues. *J. Theor. Biol.* **76,** 297–310.

Wallace, R. W., Lynch, T. J., Tallant, E. A., and Cheung, W. Y. (1979). Purification and characterization of an inhibitor protein of brain adenylate cyclase and cyclic nucleotide phosphodiesterase. *J. Biol. Chem.* **254,** 377–382.

Walsh, M., and Stevens, F. C. (1977). Chemical modification studies on the Ca^{2+}-dependent protein modulator of cyclic nucleotide phosphodiesterase. *Biochemistry* **16,** 2742–2749.

Walsh, M., and Stevens, F. C. (1978). Chemical modification studies on the Ca^{2+}-dependent protein modulator: The role of methionine residues in activation of cyclic nucleotide phosphodiesterase. *Biochemistry* **17,** 3924–3928.

Walsh, M., and Stevens, F. C. (1979). Preparation, characterization and properties of a novel triple-modified derivative of the Ca^{2+}-dependent protein modulator. *Can. J. Biochem.* **56,** 420–429.

Walsh, M., Stevens, F. C., Kuznicki, J., and Drabikowski, W. (1977). Characterization of tryptic fragments obtained from bovine brain protein modulator of cyclic nucleotide phosphodiesterase. *J. Biol. Chem.* **252,** 7440–7443.

Walsh, M., Stevens, F. C., and Oikawa, K., and Kay, C. M. (1979). Circular dichroism studies of native and chemically modified Ca^{2+}-dependent protein modulator. *Can. J. Biochem.* **57,** 267–277.

Walsh, M. P., Vallet, B., Cavadore, J. C., and Demaille, J. G. (1980). Homologous calcium binding proteins in the activation of skeletal, cardiac and smooth muscle myosin light chain kinases. *J. Biol. Chem.* **255,** 335–337.

Wang, J. H., and Desai, R. (1977). Modulator-binding protein: Bovine brain protein exhibiting the Ca^{2+}-dependent association with the protein modulator of cyclic nucleotide phosphodiesterase. *J. Biol. Chem.* **252,** 4175–4184.

Wang, J. H., and Waisman, D. M., (1979). Calmodulin and its role in the second messenger system. *Curr. Top. Cell. Regul.* **15,** 47–107.

Wang, J. H., Teo, T. S., Ho, H. C., and Stevens, F. C. (1975). Bovine heart protein activator of cyclic nucleotide phosphodiesterase. *Adv. Cyclic Nucleotide Res.* **5,** 179–194.

Watterson, D. M., Harrelson, W. G., Jr., Keller. P. M., Sharief, F., and Vanaman, T. C. (1976). Structural similarities between the Ca^{2+}-dependent regulatory proteins of 3′:5′-cyclic nucleotide phosphodiesterase and actomyosin ATPase. *J. Biol. Chem.* **251,** 4501–4513.

Watterson, D. M., Sharief, F., and Vanaman, T. C. (1980). The complete amino acid sequence of the Ca^{2+}-dependent modulator protein (calmodulin) of bovine brain. *J. Biol. Chem.* **255,** 462–475.

Wolff, D. J., and Brostrom, C. O. (1976). Ca^{2+}-dependent cyclic nucleotide phosphodiesterase from brain: Identification of phospholipids as calcium-independent activators. *Arch. Biochem. Biophys.* **173,** 720–731.

Wolff, D. J., and Brostrom, C. O. (1979). Properties and functions of the Ca^{2+}-dependent regulator protein. *Adv. Cyclic Nucleotide Res.* **11,** 28–88.

Wolff, D. J., Poirier, P. G., Brostrom, C. O., and Brostrom, M. A. (1977). Divalent cation binding properties of bovine brain Ca^{2+}-dependent regulator protein. *J. Biol. Chem.* **252,** 4108–4117.

Yasawa, M., and Yagi, K. (1978). Purification of modulator-deficient myosin light chain kinase by modulator protein-Sepharose affinity chromatography. *J. Biochem. (Tokyo)* **84,** 1259–1265.

Yasawa, M., Kuwayama, H., and Yagi, K. (1978). Modulator protein as a Ca^{2+}-dependent activator of rabbit skeletal myosin light-chain kinase. *J. Biochem. (Tokyo)* **84,** 1253–1258.

Yasawa, M., Sakuma, M., and Yagi, K. (1980). Calmodulins from muscles of marine invertebrates, scallop and sea anemone. Comparison with calmodulins from rabbit skeletal muscle and pig brain. *J. Biochem. (Tokyo)* **87,** 1313–1320.

Chapter 5

Ca^{2+}-Dependent Cyclic Nucleotide Phosphodiesterase

YING MING LIN
WAI YIU CHEUNG

I. INTRODUCTION

The recognition of Ca^{2+} as a mediator, in addition to cyclic AMP and cyclic GMP, of many physiological processes (Berridge, 1975; Rasmussen *et al.*, 1975) reflects the complex nature of biological regulation. Perhaps because of the multitude and diversity of cellular processes, more than a single regulator is necessary. The coordination and interaction of the reg-

CALCIUM AND CELL FUNCTION, VOL. I

ISBN 0-12-171401-2

ulators, Ca^{2+}, cyclic AMP, and cyclic GMP, maintain a subtle balance of the physiological states in response to external stimuli.

Calmodulin was originally discovered by Cheung (1967b, 1969b, 1970a) as an activator of cyclic 3′,5′-nucleotide phosphodiesterase during purification of the enzyme from bovine brain. Kakiuchi and co-workers (Kakiuchi and Yamazaki, 1970; Kakiuchi *et al.*, 1971) resolved rat brain phosphodiesterase by Sepharose 6B column chromatography into a Mg^{2+}-dependent and a Ca^{2+} plus Mg^{2+}-dependent activities. They subsequently found that only the Ca^{2+} plus Mg^{2+}-dependent activity was activated by a protein similar to calmodulin. The nature of the activation of the Ca^{2+}-dependent phosphodiesterase by Ca^{2+} and calmodulin was studied by Kakiuchi *et al.* (1973), Teo and Wang (1973), Lin *et al.* (1974b), and Brostrom and Wolff (1976). After its successful purification (Teo *et al.*, 1973; Teo and Wang, 1973; Lin *et al.*, 1974b), calmodulin was identified to be a Ca^{2+}-binding protein. A Ca^{2+}-binding phosphoprotein purified from pig brain by Wolff and Siegel (1972) was later found to be identical to calmodulin (Wolff and Brostrom, 1974). The further demonstration of Ca^{2+}-dependent association of phosphodiesterase and calmodulin (Teshima and Kakiuchi, 1974; Lin *et al.*, 1975; Wang *et al.*, 1975) led to the elucidation of the mechanism of phosphodiesterase activation by calmodulin.

Calmodulin has now been purified and studied in many other laboratories (Watterson *et al.*, 1976; Wolff *et al.*, 1977; Klee, 1977; Dedman *et al.*, 1977). A number of other Ca^{2+}-dependent enzymes or processes have also been found to be activated by calmodulin (see other chapters of this volume). In addition, a heat-labile (Wang and Desai, 1977; Klee and Krinks, 1978; Richman and Klee, 1978b; Wallace *et al.*, 1979) and a heat-stable (Sharma *et al.*, 1978a,b) protein have recently been found to inhibit specifically Ca^{2+}-dependent phosphodiesterase activity and to undergo Ca^{2+}-dependent association with calmodulin (Wang *et al.*, Chapter 15, this volume).

Calcium-dependent phosphodiesterase is one of the multiple forms of cyclic 3′,5′-nucleotide phosphodiesterase, an enzyme discovered by Sutherland and Rall (1958). The enzyme is widely distributed. Both cytosol and membrane particles contain phosphodiesterase, and their distribution varies with the types of cells and species of animals (Cheung, 1970b). Phosphodiesterase is the only enzyme that catabolizes cAMP and cGMP, thereby regulating the tissue levels of these cyclic nucleotides (Robison *et al.*, 1971). In certain cases, a change of tissue cyclic nucleotide levels involves a direct interaction of the enzyme with a hormone (Wells and Hardman, 1977; Strada and Thompson, 1978).

The introduction of a simple and sensitive method for assaying phosphodiesterase with a radioactive substrate (Brooker *et al.*, 1968; Thompson and Appleman, 1971a) enabled the detection of abnormal kinetics of phosphodiesterase and facilitated the identification of multiple forms of the enzyme. The different forms of phosphodiesterase possess distinct physical as well as kinetic properties. Furthermore, these forms may be selectively inhibited by certain types of pharmacological agents (Weiss *et al.*, 1974; Weiss, 1975; Amer and Kreighbaum, 1975; Weiss and Levin, 1978). A selective inhibition of phosphodiesterase may be a useful basis in developing new drugs (Chasin and Harris, 1976).

This chapter deals primarily with the Ca^{2+}-dependent phosphodiesterase, particularly its regulation involving Ca^{2+} and calmodulin. A section devoted to the discussion of the multiple forms of phosphodiesterase is included to provide a broader overview. General aspects of phosphodiesterase (Cheung, 1970b; Appleman *et al.*, 1973; Wells and Hardman, 1977) have been reviewed.

II. ASSAY

The presence of Ca^{2+}-dependent phosphodiesterase is indicated by an increase in enzyme activity when assayed in the presence of Ca^{2+} and calmodulin, as compared with the basal activity obtained in the presence of EGTA. The Ca^{2+}-dependent phosphodiesterase activity is the difference between the total and the basal activities. All tissue extracts examined contain high levels of calmodulin (Smoake *et al.*, 1974; Waisman *et al.*, 1975), and exogenous calmodulin is usually not necessary for maximum phosphodiesterase activity. However, exogenous calmodulin is always required for purified preparations from which calmodulin has been partially or totally removed.

Brain is an excellent source for calmodulin. A crude calmodulin may simply be prepared by boiling a high-speed supernatant fluid of a tissue extract (Cheung, 1970a). A highly purified calmodulin may be obtained by passing a boiled extract through a DEAE-cellulose column (Lin *et al.*, 1974a). Calmodulin, being acidic, is adsorbed to the column, while the bulk of the contaminating proteins passes through the column. Homogeneous calmodulin preparations have been obtained from different laboratories by different methods (Teo and Wang, 1973; Lin *et al.*, 1974a,b; Watterson *et al.*, 1976; Klee, 1977; Wolff *et al.*, 1977; Dedman *et al.*, 1977). A recent procedure exploiting the high affinity of calmodulin for trifluoroperazine linked to an immobile matrix offers a simple and highly efficient method for isolating calmodulin (Cormier *et al.*, Chapter 10, and Wallace *et al.*, Chapter 2, this volume).

Several methods have been developed for assaying phosphodiesterase: (1) isolation and determination of radioactive 5′-nucleotides from a reaction mixture by paper chromatography (Cheung, 1967a); (2) monitoring the decrease in optical density at 265 nm as 5′-AMP is converted to 5′-IMP with 5′-adenylic acid deaminase as an auxillary enzyme (Drummond and Perrott-Yee, 1961); (3) spectrophotometric determination of NADH by coupling to myokinase, pyruvate kinase, and pyruvate dehydrogenase (Cheung, 1966); (4) titration of protons released as a result of conversion from cAMP (one negative charge) to 5′-AMP (two negative charges) (Cheung, 1969a); (5) measurement by the luciferin–luciferase technique of ATP quantitatively converted from 5′-AMP with myokinase (Weiss *et al.*, 1972); (6) colorimetric measurement of inorganic phosphate generated from hydrolysis of the 5′-nucleotide with 5′-nucleotidase (Butcher and Sutherland, 1962); and (7) separation of nucleoside quantitatively converted from 5′-nucleotide with 5′-nucleotidase by an anion exchange resin (Thompson and Appleman, 1971a).

Colorimetric measurement had been widely used because of its simplicity. However, the method lacks the sensitivity for measuring enzymatic activity at micromolar substrate levels and has largely been replaced by the more sensitive resin method of Thompson and Appleman. A slight drawback of the resin method is the considerable high background arising from an incomplete adsorption of cyclic nucleotides and the nonspecific binding of purine bases (Lynch and Cheung, 1975; Boudreau and Drummond, 1975). Thompson *et al.* (1979) described a modified procedure with reduction of nonspecific binding.

A wide range of substrate concentrations have been employed for the determination of phosphodiesterase activity. Since the rate of an enzyme reaction varies with the substrate concentrations unless a saturated substrate level is maintained, the activities determined at different substrate concentrations should not be compared directly. In addition, one form of phosphodiesterase may hydrolyze cAMP more than cGMP at one substrate concentration, and the reverse may be true at another substrate concentration (Beavo *et al.*, 1970; Kakiuchi *et al.*, 1973; Lin *et al.*, 1974b; Ho *et al.*, 1976). Thus, substrate concentration is an important parameter in reporting phosphodiesterase activity.

In general, assay of phosphodiesterase at a lower substrate concentration has an inherently higher sensitivity, but the initial velocity should be used to obtain accurate results. A tissue extract usually contains more than one form of phosphodiesterase, and the high K_m form of the enzyme is more easily detected with high substrate concentrations.

III. PREPARATION AND PURIFICATION

Calmodulin is an acidic protein (p*I* = 4.3) with a molecular weight of 16,700 (Teo *et al.*, 1973; Lin *et al.*, 1974b; Watterson *et al.*, 1980). The association of Ca^{2+}-dependent phosphodiesterase with calmodulin requires the presence of Ca^{2+} (see Section IX, B). These properties have been exploited for preparing the enzyme essentially free from calmodulin.

Using gel filtration column chromatography, Kakiuchi and his co-workers (1971, 1972, 1975) resolved phosphodiesterase from several rat tissues into Mg^{2+}-dependent and Ca^{2+} plus Mg^{2+}-dependent activities. A single peak of Ca^{2+}-dependent phosphodiesterase, free from calmodulin, was obtained. A similar technique was used for analysis of the interaction between the enzyme and calmodulin (Teshima and Kakiuchi, 1974; Lin *et al.*, 1974b). An effective separation of the two proteins occurred only when EGTA was included in the buffer to chelate free Ca^{2+}.

Uzunov and Weiss (1972) employed preparative polyacrylamide gel electrophoresis to separate multiple forms of phosphodiesterase from rat brain. They found that two out of six distinct activity peaks depended on calmodulin for full activity. However, a similar technique used by Cheung and Lin (1974) for purification of phosphodiesterase from bovine brain resulted in loss of Ca^{2+}-dependent activity. The loss of Ca^{2+}-dependent activity might be due to the low stability of phosphodiesterase in a highly purified state. Others also experienced loss of Ca^{2+}-dependent phosphodiesterase activity in a highly purified state (Ho *et al.*, 1977) or during separation of the enzyme with isoelectric focusing (Pledger *et al.*, 1974).

A widely used method for preparing Ca^{2+}-dependent phosphodiesterase devoid of calmodulin is that employing a DEAE-cellulose column. Two peaks of Ca^{2+}-dependent activity are usually obtained (see Table I). Since elution of proteins from a DEAE-cellulose column depends on the ionic strength of the elution medium, gradients of various salts have been used.

Ho *et al.* (1977) purified some 5000-fold a Ca^{2+}-dependent phosphodiesterase from bovine heart with a yield of 3.5%. The purified phosphodiesterase was activated sixfold by calmodulin; a less purified preparation was activated ninefold. A key step in the purification scheme utilized the higher affinity of the enzyme–calmodulin complex toward the DEAE-cellulose. This caused a shift of elution position of the enzyme and removal of a bulk of the contaminants. Though highly purified, the preparation was not yet homogeneous, and attempts for further purification has resulted in great loss (90%) of the activity.

A recent report (Morrill *et al.*, 1979) described a successful three-step purification of Ca^{2+}-dependent phosphodiesterase from bovine brain, with a yield of 11%. The enzyme preparation, after ammonium sulfate fractionation, was twice subjected to Blue dextran-Sepharose column chromatography. The two steps of Blue dextran-Sepharose column chromatography achieved a 1000-fold purification of the enzyme. The purified enzyme was apparently homogeneous, and was stimulated twofold by calmodulin. Gel filtration chromatography and SDS polyacrylamide gel electrophoresis show that the enzyme has a molecular weight of 126,000, consisting of two identical subunits with a molecular weight of 59,000.

Watterson and Vanaman (1976) applied a calmodulin affinity column for purifying Ca^{2+}-dependent phosphodiesterase. The enzyme purified from bovine brain exhibited three major bands with molecular weights of 60,000, 40,000, and 18,000 and a number of minor bands, as analyzed with SDS polyacrylamide gel electrophoresis.

Using DEAE-cellulose, Sephadex G-200, and hydroxyapatite chromatography in addition to affinity chromatography, Klee *et al.* (1979) obtained an apparently homogeneous Ca^{2+}-dependent phosphodiesterase from bovine brain. The enzyme, purified 5000-fold with a 1.4% yield, contained subunits of 61,000, 59,000 and 15,000 daltons. Only the 59,000 subunit displayed enzyme activity. The 61,000 and 15,000 subunits appeared identical to those of a heat-labile calmodulin-binding protein (CaM-BP_{80}) (Klee and Krinks, 1978; Wallace *et al.*, 1979; Wang and Desai, 1977; see also Wang *et al.* Chapter 15, this volume) on the basis of molecular weight.

LaPorte *et al.* (1979) achieved an apparently homogeneous Ca^{2+}-dependent phosphodiesterase from bovine heart by adopting the method of Ho *et al.* (1977) followed by a calmodulin affinity chromatography. The enzyme was purified 13,750-fold with a 10% yield and was activated tenfold by calmodulin. The heart enzyme was a monomeric protein with a molecular weight of 57,000.

IV. MULTIPLE MOLECULAR FORMS OF PHOSPHODIESTERASE

Earlier studies of phosphodiesterase from bovine brain (Cheung, 1969a) and frog erythrocyte (Rosen, 1970) demonstrated the presence of two forms of phosphodiesterase. Using a sensitive assay method, Thompson and Appleman (1971a,b) detected three forms of phosphodiesterase in rat brain as well as several other rat tissues. The existence of multiple forms of phosphodiesterase has since been confirmed in numerous laboratories.

Thompson and Appleman showed that with the exception of liver, several rat tissues contained three forms of phosphodiesterase: an exclusion fraction, a high molecular weight (400,000) fraction, and a low molecular weight (200,000) fraction. The exclusion fraction, probably a particulate phosphodiesterase, and the low molecular weight fraction exhibited high affinity for cyclic AMP. The high molecular weight fraction possessed a higher affinity for cyclic GMP than cyclic AMP. Rat liver contained only the high molecular weight fraction. Further study with DEAE-cellulose chromatography (Russell *et al.*, 1973), however, shows that rat liver also contained three forms of phosphodiesterase.

Calcium-dependent phosphodiesterase partially purified from bovine brain (Lin *et al.*, 1975) has been resolved on a Sephadex G-200 column into two Ca^{2+}-dependent activity fractions. One fraction excluded from the column had a molecular weight larger than 400,000. The molecular weight of the other fraction was estimated to be 170,000 and 230,000 for the free enzyme and the calmodulin-complexed enzyme, respectively. A corresponding molecular weight of 150,000 and 200,000 for rat brain (Teshima and Kakiuchi, 1974) and of 155,000 and 230,000 for bovine heart (Wang *et al.*, 1975) have also been obtained with similar techniques.

In contrast to the findings of Thompson and Appleman (1971a,b), Kakiuchi *et al.* (1975) found varying patterns of phosphodiesterase from several rat tissues by Sepharose 6B chromatography. A Ca^{2+}-dependent phosphodiesterase with a molecular weight of 150,000 was found to be a predominant form in many tissues examined.

Several studies on multiple phosphodiesterase forms employed electrophoresis. Monn and co-workers (Monn and Christiansen, 1971; Monn *et al.*, 1972) identified as many as seven different forms of phosphodiesterase in rat and rabbit tissues using starch gel electrophoresis and activity staining. Brain contained four forms, whereas the other tissues contained from one to three forms of phosphodiesterase. All seven forms of phosphodiesterase hydrolyzed both cAMP and cGMP. Campbell and Oliver (1972) obtained essentially similar results by assaying sliced gels after analytical polyacrylamide gel electrophoresis of phosphodiesterase.

Many studies on the multiple forms of phosphodiesterase used DEAE-cellulose column chromatography, which resolved the enzyme into three forms (Hidaka *et al.*, 1977; Donnelly, 1977; Taniguchi *et al.*, 1978; Sakai *et al.*, 1977), designated as forms I, II, and III according to the order of emergence from the column. At micromolar substrate concentration forms I and II hydrolyze cGMP faster than cAMP (Ho *et al.*, 1976). In contrast form III hydrolyzes cAMP at a faster rate at all substrate concentrations. Forms I and II generally have a lower K_m value for cGMP than for cAMP, and the reverse is true for form III. Form II may be further distinguished from form I in that it has a K_m value for cAMP comparable to cGMP and that cGMP stimulates the hydrolysis of cAMP (Hidaka *et al.*,

1977). Forms I and II in most cases are Ca^{2+} dependent, and the extent of activation varies with the tissue.

Brain generally has a more complex phosphodiesterase pattern. Hidaka *et al.* (1977) resolved human brain phosphodiesterase into four activity peaks by DEAE-cellulose column. Peaks II, III, and IV appear comparable to forms I, II, and III of other mammalian phosphodiesterases, respectively.

It is difficult to correlate with certainty the individual forms of phosphodiesterase obtained by two different separation methods. Separation of multiple forms by DEAE-cellulose chromatography and by electrophoresis are based on differences in charge, and forms I, II, and III from DEAE-cellulose chromatography may correspond to the three common forms obtained by starch gel electrophoresis.

The designation of different forms of phosphodiesterase is primarily based on the physical behavior by a particular separation method. Similarity in physical and kinetic properties may be found for the same form of phosphodiesterase from different species, but significant differences may exist in different tissues obtained from the same species. For instance, the extent of Ca^{2+}-dependent activation of phosphodiesterase is different, and the same form of phosphodiesterase from different tissues may have one or two K_m values. One possible explanation for the apparent discrepancies is that the enzyme is contaminated by other forms of phosphodiesterase; another is that tissue protease (see Section VII) decreases the sensitivity of phosphodiesterase to calmodulin (Cheung, 1970a). Sakai *et al.* (1978) showed that the molecular forms of phosphodiesterase may be altered by lysosome protease when phosphodiesterase is prepared from a lysosome-rich tissue, especially after freezing and thawing of the tissue. Calcium-dependent phosphodiesterase invariously was found less responsive to Ca^{2+} after purification (Cheung and Lin, 1974; Ho *et al.*, 1977; Pledger *et al.*, 1974). Moreover, conditions of enzyme purification may affect the properties of the different enzyme forms (Van Inwegen *et al.*, 1976), giving rise to differences between the same form of phosphodiesterase from different tissues.

The conversion of one form of phosphodiesterase into another has been noted. Using Sepharose 4B chromatography, Cheung and Lin (1974) resolved phosphodiesterase partially purified from bovine brain into two activity peaks. When an aged rather than a fresh sample was used, the activity corresponding to the higher molecular weight increased at the expense of the activity corresponding to the lower molecular weight. The conversion was prevented by prior incubation of the aged sample with β-mercaptoethanol. Aggregation and deaggregation caused by a change of the ionic strength of the medium (Tisdal, 1975; Schröder and Rickenberg, 1973) or of the enzyme concentration (Pichard and Cheung, 1976) have also been reported.

V. DISTRIBUTION

Some of the tissues known to contain Ca^{2+}-dependent phosphodiesterase are listed in Table I. These phosphodiesterases are cytosolic in origin. Two forms of Ca^{2+}-dependent phosphodiesterase are found in many tissues, classified as forms I and II according to the order of emergence in a DEAE-cellulose column. In cases in which a separation technique other than DEAE-cellulose chromatography was used or the existence of only a single form of Ca^{2+}-dependent phosphodiesterase was reported, the entry as form I or II is arbitrary. The extent of Ca^{2+}-dependent activation and the separation technique used are also included in the table, which is intended to give qualitative information on the tissue distribution. A quantitative estimation of the tissue level of Ca^{2+}-dependent phosphodiesterase is difficult because of the presence of multiple forms and the varying extent of Ca^{2+} activation of each form. Based on a single form of Ca^{2+}-dependent phosphodiesterase isolated by Sepharose 6B chromatography, Kakiuchi *et al.* (1975) estimated the level of Ca^{2+}-dependent phosphodiesterase in several rat tissues in the following order: cerebral cortex, kidney, cerebellum, heart, and liver. Several tissues, including sea urchin sperm (Wells and Garbers, 1976), bovine pineal gland (Sankaran *et al.*, 1978), and rat superior cervical ganglion (Boudrea and Drummond, 1975), appear to lack Ca^{2+}-dependent phosphodiesterase. Lower forms of animals contain little or no Ca^{2+}-dependent phosphodiesterase (Waisman *et al.*, 1975).

Kakiuchi *et al.* (1978) reported Ca^{2+}-dependent phosphodiesterase in a particulate fraction of rat liver, kidney, and heart. These experiments did not exclude the possibility that the particulate fraction may be contaminated by an enzyme from the cytosol.

A number of investigators studied phosphodiesterase during ontogenetic development; only a few examined specifically the Ca^{2+}-dependent activity. Tanigawa and Shimoyama (1976) found that in the early embryonic state of the chick embryo nearly all phosphodiesterases were Ca^{2+} dependent, and the activity progressively decreased until little or no Ca^{2+}-dependent activity in the adult. In contrast, Singer *et al.* (1978) showed that early in development rabbit fetal brain, heart, kidney, liver, and lung had the mature array of phosphodiesterase, except that in liver and lung one form became Ca^{2+}-dependent late in gestation.

TABLE I

K_m Values of Ca^{2+}-Dependent Phosphodiesterase

Tissues	K_m (μM)				References
	Form I		Form II		
	cAMP	cGMP	cAMP	cGMP	
Human					
Heart	ND[a]	2.0, 10	ND	15	Hidaka *et al.* (1977)
Lung	ND	0.9, 3.1	15	29	Hidaka *et al.* (1977)
Kidney	ND	3.0	90	27	Hidaka *et al.* (1977)
Liver	ND	0.65	44	10	Hidaka *et al.* (1977)
Skeletal muscle	ND	1.5, 20	50	24	Hidaka *et al.* (1977)
Brain	0.27, 0.55	0.11, 1.25	ND	ND	Hidaka *et al.* (1977)
Rabbit					
Sinoatrial node	8.3	2.0	1.2, 50	20, 20	Taniguchi *et al.* (1978)
Atria	1.3, 25	5.0, 50	1.6, 20	25, 33	Taniguchi *et al.* (1978)
Ventricle	0.1, 20	1.3, 10	3.0, 50	22, 25	Taniguchi *et al.* (1978)
Porcine Aorta	40–100	1–4			Wells *et al.* (1975)
Brain	4000/180[b]	200/8			Brostrom and Wolff (1976)
Rat					
Kidney	ND	1.3			Kakiuchi *et al.* (1975)
Heart	ND	1.7			Kakiuchi *et al.* (1975)
Liver	ND	5			Kakiuchi *et al.* (1975)
Brain	ND	2			Kakiuchi *et al.* (1975)
Brain	ND	ND	3.1/1.7	1.3/0.63	Sakai *et al.* (1977)
Brain	4.9/4.2	4.5/2			Kakiuchi *et al.* (1973)
Brain	350/80	5–9/5–9			Uzunov *et al.* (1976)
Bovine					
Heart	220–520	1, 15/3	ND	ND	Donnelly (1977)
Heart	1200–1700 / 160–270	210–330 / 7–11			Ho *et al.* (1976)
Brain	150/150	9/9			Morrill *et al.* (1979)
Brain	310/100	ND			Klee *et al.* (1979)

[a] ND, not determined.

[b] Value for nonactivated state/value for activated state.

VI. GENERAL PROPERTIES

Calcium-dependent phosphodiesterase in crude preparations is stable and may be stored at −20°C or below for months without a significant loss of activity (Cheung, 1970b). Upon storage, a purified brain enzyme increased its molecular weight (Cheung, 1970b; Cheung and Lin, 1974). The change was reversed by incubation with β-mercaptoethanol. Upon storage or during purification, a decrease in Ca^{2+}-dependent activation also occurred (Cheung and Lin, 1974; Ho *et al.*, 1977; Kakiuchi *et al.*, 1975), probably due to the action of proteolytic enzymes (Cheung, 1971b; Cheung and Lin, 1974). For tissues rich in lysosomes (Sakai *et al.*, 1978), care should be taken during enzyme preparation to minimize the effect of lysosomal protease by avoiding freezing and thawing and by the addition of suitable protease inhibitors.

In the purified state, Ca^{2+}-dependent phosphodiesterase appears more labile. Severe loss of Ca^{2+}-dependent activity has been reported for several purified preparations (Goren and Rosen, 1972; Cheung and Lin, 1974; Ho *et al.*, 1977). Ho *et al.* (1977) compared the stabilities of a highly purified and a less purified bovine heart preparation. They found that at 55°C phosphodiesterase in the highly purified state was inactivated much faster than that in the less purified one. Calcium and calmodulin stabilized the enzyme in the highly purified preparation, but destabilized it in the less purifed one. At 0°C, inclusion of EGTA to remove Ca^{2+} caused a rapid loss of phosphodiesterase activity in the highly purified preparation. The inactivation was retarded by the addition of bovine serum albumin. In contrast, Morrill *et al.* (1979) noted that no apparent loss of the enzyme activity was observed when a homogeneous enzyme was kept at 4°C for 4 weeks or at 30°C for 40 min in the absence of calmodulin and the presence of 2 m*M* EGTA.

Calcium-dependent phosphodiesterase requires Mn^{2+} or Mg^{2+} for maximum activity (Cheung, 1969b; Kakiuchi and Yamazaki, 1970; Teo and Wang, 1973). The enzyme is activated by calmodulin, and its activation in the presence of Mg^{2+} requires Ca^{2+} (Kakiuchi *et al.*, 1971; Teo and Wang, 1973). Calcium is not required for the activation when Mg^{2+} is replaced by Mn^{2+} (Lin *et al.*, 1974b). Tissue concentration of Mn^{2+} is one to two orders of magnitude lower than that of Ca^{2+}, and Ca^{2+} appears likely to be the physiological cation responsible for regulating the activity of phosphodiesterase. Besides Mn^{2+}, Co^{2+} may replace Mg^{2+} (Cheung, 1967a, 1971a) and Sr^{2+}, Ba^{2+}, and Co^{2+} may replace Ca^{2+} (Kakiuchi *et al.*, 1972; Teo and Wang, 1973). However, the dissociation constant of the enzyme for these ions are at least one order of magnitude higher than that for Ca^{2+} (Teo and Wang, 1973).

Phosphodiesterase exhibits specificity for cyclic 3′,5′-nucleotides with a purine base (Cheung, 1970b); cyclic 3′,5′-UMP is the only pyrimidine nucleotide known to be hydrolyzed to a significant degree. Hardman and Sutherland (1965) reported that a phosphodiesterase from beef and dog hearts preferred cUMP as a substrate. It is possible that a separate enzyme is responsible for the hydrolysis of cUMP. Calcium-dependent phosphodiesterase catalyzes the hydrolysis of cAMP faster than cGMP at millimolar concentration but slower at micromolar concentrations (Kakiuchi *et al.*, 1973; Lin *et al.*, 1974b; Ho *et al.*, 1976; Brostrom *et al.*, 1976). The extent of Ca^{2+}-dependent activation is invariably higher at micromolar than at millimolar substrate concentrations; the extent of stimulation apparently varies with different tissues (see Table I).

VII. KINETIC PARAMETERS

Brooker *et al.* (1968) demonstrated anomalous kinetics of phosphodiesterase in a crude rat brain preparation. Based on the observation that a Lineweaver–Burk plot gave a downward curvature, they suggested the presence of either more than one form of the enzyme or an enzyme with negative cooperativity. The anomalous kinetics has since been shown with most, if not all, crude phosphodiesterases. The failure to detect similar kinetic anomaly in earlier studies may be attributed to the use of less sensitive assay methods. Although different types of kinetic plots have been employed, including Lineweaver–Burk, Hofstee, and Hanes, a similar, nonlinear curve profile has always been obtained with the same type of plot used (Lin and Cheung, 1975). On the basis of theoretical considerations, Lin and Cheung (1975) suggested that the nonlinear curves indicate the presence of either multiple forms of an enzyme, each form exhibiting typical Michaelis–Menten kinetics, or one or more forms of an enzyme with multiple nonidentical and noninteracting sites. The theoretical analysis by Russell *et al.* (1972), on the other hand, suggests an enzyme with negative cooperativity. The existence of multiple forms of phosphodiesterase has been noted in most tissues (see Section IV).

The apparent K_m values for phosphodiesterase in a crude preparation that exhibits a nonlinear curve have often been obtained by extrapolating the two apparently linear regions of the kinetic plots. Depending on the activity contribution of each form of phosphodiesterase, the K_m values so obtained may have considerable error.

In classic Micahelis–Menten kinetics, the K_m value, a constant characteristic of an enzyme, represents the substrate concentration at which the initial velocity of the enzyme is half-maximal. The K_m value also provides information as to what concentration range within which the substrate may regulate the enzyme activity. When a single enzyme possesses two apparent K_m values, they no longer convey the conventional meaning, and perhaps represent an operational term whose significance remains unclear. Thus, in handling the unconventional kinetics of phosphodiesterase, one needs to consider the various kinetic parameters arising from variations in assay conditions and the extent of contamination by multiple enzyme forms.

The K_m values for Ca^{2+}-dependent phosphodiesterase are compiled in Table II; presumably they represent a single form of a partially purified Ca^{2+}-dependent phosphodiesterase. When available, the values for the nonactivated and activated states are included.

Calcium-dependent phosphodiesterase generally has a lower K_m value for cGMP than for cAMP, and it has a higher V_{max} for cAMP (Kakiuchi *et al.*, 1973; Brostrom *et al.*, 1976; Ho *et al.*, 1976). In the activated state, the K_m value ranges from 0.1 to 300 μM for cAMP and from 0.1 to 70 μM for cGMP. The K_m values for phosphodiesterase partially purified from bovine heart and from brain agree fairly well. Two forms of Ca^{2+}-dependent phosphodiesterase in most mammalian tissues have been discussed earlier. Form I and II are distinguished mainly on the basis of their elution pattern on a DEAE-cellulose column. Form II has a comparable K_m for both cAMP and cGMP. Form I has a higher K_m for cAMP than for cGMP. Further, cGMP at low concentrations (micromolar) stimulated cAMP hydrolysis by form II (Hidaka *et al.*, 1977; Taniguchi *et al.*, 1978). In contrast, at a higher substrate concentration, cAMP inhibits the hydrolysis of cGMP and vice versa, in a manner that the K_i of one nucleotide as an inhibitor of the hydrolysis of the other is comparable to the K_m of each nucleotide as a substrate, suggesting that the two nucleotides compete for the same substrate site.

The effect of calmodulin on kinetic parameters of phosphodiesterase has also been studied; the results are mixed. An increased V_{max} without a change in the K_m has been found for the enzyme from rat brain with cAMP as the substrate (Cheung, 1971b; Weiss *et al.*, 1974; Donnelly, 1977; Pichard and Cheung, 1976) and for that from bovine brain with cAMP or cGMP as the substrate (Morrill *et al.*, 1979). In contrast, a decreased K_m, without a change in the V_{max}, has been found for a porcine brain enzyme with either cAMP or cGMP as the substrate (Brostrom and Wolff, 1976). In addition, a rat brain enzyme exhibits a decreased K_m with no change in the V_{max} for cAMP, and an increased V_{max} with no change in the K_m (Uzunov *et al.*, 1976).

TABLE II

Extent of Activation of Ca^{2+}-Dependent Phosphodiesterase by Calmodulin

Tissues	Activation (fold)				Separation methods	References
	Form I		Form II			
	Cyclic AMP	Cyclic GMP	Cyclic AMP	Cyclic GMP		
Human						
Heart	2.3 (0.4)[a]	3.5 (0.4)	1.2 (0.4)	2.0 (0.4)	DEAE-cellulose	Hidaka *et al.* (1977)
Lung	1.8 (0.4)	1.3 (0.4)	1.5 (0.4)	2.5 (0.4)	DEAE-cellulose	Hidaka *et al.* (1977)
Kidney	1.0 (0.4)	1.0 (0.4)	1.9 (0.4)	2.7 (0.4)	DEAE-cellulose	Hidaka *et al.* (1977)
Liver	1.0 (0.4)	2.3 (0.4)	1.3 (0.4)	1.8 (0.4)	DEAE-cellulose	Hidaka *et al.* (1977)
Skeletal muscle	2.3 (0.4)	3.8 (0.4)	1.0 (0.4)	2.6 (0.4)	DEAE-cellulose	Hidaka *et al.* (1977)
Brain	5.6 (0.4)	10 (0.4)	2.9 (0.4)	5.5 (0.4)	DEAE-cellulose	Hidaka *et al.* (1977)
Rabbit sinoatrial	2.0 (0.4)	1.4 (0.4)	1.9 (0.4)	3.7 (0.4)	DEAE-cellulose	Taniguchi *et al.* (1978)
Porcine brain	1 (V_{max})	1 (V_{max})			Partial purification	Brostrom and Wolff (1976)
Rat						
Aorta	1.4–2.15 (1)	1.5–2.2 (1)	1.4–2.2 (1)	1.4–2.2 (1)	DEAE-cellulose	Donnelly (1978)
Heart	1.7–2.4 (1)	1.4–2.6 (1)	1.5–2.4 (1)	1.4–2.6 (1)	DEAE-cellulose	Donnelly (1978)
Brain	1.4 (1)	1.0 (1)	8.4 (1)	4.2 (1)	DEAE-cellulose	Sakai *et al.* (1977)
Brain	2 (200)	ND	6 (200)	ND	Polyacrylamide gel electrophoresis	Uzunov and Weiss (1972)
Brain	4.5 (0.4)	9.6 (0.4)			Sepharose 6B chromatography	Kakiuchi *et al.* (1973)
	7.1 (V_{max})	5.3(V_{max})			Sepharose 6B chromatography	Kakiuchi *et al.* (1973)
Bovine						
Heart	1.7 (V_{max})	2.0 (V_{max})	ND	ND	DEAE-cellulose	Donnelly (1977)
Heart	13.4 (20)	5.8 (20)			Partial purification	Ho *et al.* (1976)
	5 (V_{max})	1 (V_{max})			Partial purification	Ho *et al.* (1976)
Brain	2 (V_{max})	2 (V_{max})			Purification	Morrill *et al.* (1979)
Brain	3.6 (200)	ND			Purification	Klee *et al.* (1979)
	6 (V_{max})	ND				
Brain	10 (500)	ND			Purification	LaPorte et al. (1979)

[a] Values in parentheses indicate the substrate concentration (micromolar).

VIII. EFFECTORS

Compounds with diverse structure and pharmacological activities affect phosphodiesterase activity, a subject reviewed elsewhere (Amer and Kreighbaum, 1975; Chasin and Harris, 1976). Whether the pharmacological effects of these compounds are related to their action on phosphodiesterase remains an open question. Weiss and Levin (1978) showed that trifluoroperazine, an antidepressant, inhibits the calmodulin-supported phosphodiesterase activity; they postulated that the effect may be related to the neuroleptic activity of the drug (see also Weiss and Wallace, this volume, Chapter 16).

EGTA, a Ca^{2+} chelator, is another compound that specifically inhibits the calmodulin-dependent enzyme activity (Kakiuchi and Yamazaki, 1970). It exerts its effect by removing Ca^{2+} that is essential for the activity of calmodulin. Recently, Davis and Daly (1978) showed that K^+ inhibits selectively the basal activity of Ca^{2+}-dependent phosphodiesterase from rat cerebrum. At 100 μM, the cation inhibits essentially all the basal, but not the calmodulin-stimulated, activity. These authors argued that since the intracellular K^+ concentration is in the range of 90 to 120 μM, the enzyme is essentially inactive unless activated by calmodulin.

A heat-labile calmodulin-binding protein with a molecular weight of 80,000 (CaM-BP_{80}) (Wang and Desai, 1976, 1977; Klee and Krinks, 1978; Richman and Klee, 1978b; Wallace *et al.*, 1978, 1979) and a heat-stable (Sharma *et al.*, 1978a,b) calmodulin-binding protein with a molecular weight of 70,000 (CaM-BP_{70}) from brain was found to specifically inhibit the Ca^{2+}-dependent activity, apparently by competing with the enzyme for calmodulin. The biological activity of the two proteins has not been identified; they may be other calmodulin-regulated enzymes (Wang *et al.*, this volume, Chapter 15).

Certain compounds are known to activate phosphodiesterase and are usually specific for the Ca^{2+}-dependent activity, for example, ammonium ions and imidazole (Butcher and Sutherland, 1962; Nair, 1966; Ho *et al.*, 1976; Cheung, 1967b, 1971b; Cheung and Lin, 1974; Sakai *et al.*, 1977). Proteolytic enzymes, such as trypsin, chymotrypsin, pronase, and *B. subtilis* alkaline protease, activate Ca^{2+}-dependent phosphodiesterase by partial proteolysis. The activated phosphodiesterase decreased in its molecular size and lost its response to calmodulin (Cheung and Lin, 1974; Sakai *et al.*, 1977). In contrast to these endopeptidases, exopeptidases such as carboxypeptidase and leucine amino peptidase, did not show significant effect on phosphodiesterase activity (Sakai *et al.*, 1977).

Certain phospholipids (Bublitz, 1973; Wolff and Brostrom, 1976; Pichard and Cheung, 1977) and a vitamin E derivative, α-tocopherylphosphate (Sakai, *et al.*, 1977), also activate Ca^{2+}-dependent phosphodiesterase. Among several phospholipids tested, lysolecithin appears most potent (Wolff and Brostrom, 1976; Pichard and Cheung, 1977). Similar to the effect of calmodulin, that of lysolecithin was immediate and reversible. Both agents rendered the enzyme more thermal labile, decreased the energy of activation, and increased the V_{max} of phosphodiesterase without affecting the K_m for cAMP. However, the activation by lysolecithin was reversed by Lubrol-P, a nonionic detergent, but not by EGTA, suggesting that the mode of activation is different from that by calmodulin.

IX. Ca^{2+}-DEPENDENT ACTIVATION BY CALMODULIN

Calcium-dependent phosphodiesterase requires Mg^{2+} for the catalytic activity and Ca^{2+} for its activation by calmodulin (Kakiuchi *et al.*, 1971; Teo and Wang, 1973; Lin *et al.*, 1974b). The simultaneous presence of Ca^{2+} and calmodulin is necessary for the activation of the enzyme, and one agent increases the sensitivity of the enzyme to the other. For example, at 0.4 μg of calmodulin, 100 μM Ca^{2+} is required to reach half-maximal stimulation of phosphodiesterase, whereas at 2 μg of calmodulin, only 3 μM Ca^{2+} is necessary. On the other hand, at 100 μM Ca^{2+}, 2 μg of calmodulin achieves full activation, whereas at 20 μM Ca^{2+}, 40 μg of calmodulin is required (Lin *et al.*, 1974b).

The Ca^{2+}-dependent activation of phosphodiesterase by calmodulin is immediate and reversible (Cheung, 1971b; Wolff and Brostrom, 1974; Lin *et al.*, 1975). The cellular level of calmodulin is generally much higher than phosphodiesterase, and the cellular flux of Ca^{2+} in response to stimuli appears important in regulating the activity of the enzyme.

A. Ca^{2+} and Calmodulin Interaction

To account for the requirement of both Ca^{2+} and calmodulin for the activation of Ca^{2+}-dependent phosphodiesterase, a number of mechanisms may be considered. The initial interaction may be the binding of any two of Ca^{2+}, calmodulin, and phosphodiesterase, or the simultaneous binding of all three components. One of these possibilities was amenable to experimental verification when homogeneous calmodulin became available. Using calmodulin from bovine heart, Teo and Wang (1973) demonstrated that strong binding occurs between Ca^{2+} and calmodulin. A Scatchard analysis of the equilibrium binding of Ca^{2+} and calmodulin indicated one binding site with a K_d (dissociation constant) of 2.9 μM and two to three binding sites with a K_d of 12 μM. Using bovine brain camodulin, Lin *et al.* (1974b) found a total of four Ca^{2+}-binding sites with K_ds ranging from 3.5 to 18 μM. The binding of Ca^{2+} was inhibited by Mn^{2+}, Co^{2+}, Zn^{2+}, and Sr^{2+}. These ions could partially replace Ca^{2+} for the activation of phosphodiesterase (Kakiuchi *et al.*, 1972; Teo and Wang, 1973). At 500 μM Mg^{2+} did not significantly affect the binding of Ca^{2+} to calmodulin; the K_d for Mg^{2+} was at least ten times higher than that for Ca^{2+}. Other investigators (Brostrom and Wolff, 1974; Dedman *et al.*, 1977; Klee, 1977; Wolff *et al.*, 1977; Watterson *et al.*, 1976) showed that calmodulin possessed four Ca^{2+} binding sites, the K_ds ranging from 1 to 12 μM. In addition to Ca^{2+}, Wolff *et al.* (1977) examined the binding of Mg^{2+} and Mn^{2+} to bovine brain calmodulin. They found one set of three binding sites with K_ds of 0.2, 1.2, and 140 μM, and another set of one binding site with a K_d of 1, 4, and 20 μM for Ca^{2+}, Mn^{2+}, and Mg^{2+}, respectively. Considering that the free concentration of Mg^{2+} in brain is about 1 mM and that the free intracellular concentrations of Ca^{2+} of the unstimulated and stimulated cells are approximately 0.1 and 1 μM, respectively, they suggested that calmodulin exists as a tetramagnesium complex under normal conditions. Upon stimulation of the cells, increase in free Ca^{2+} causes the replacement of Mg^{2+} by Ca^{2+} at the three sites that have a higher affinity for Ca^{2+} and results in conversion of a tetramagnesium complex into tricalcium–monomagnesium complex, which is thought to be the active species of calmodulin for the activation of phosphodiesterase.

Although apparent discrepancies for the K_ds from different laboratories exist, the different values are within the range of the K_a (half-maximal stimulation) of phosphodiesterase. The closeness of the two constants supports the notion that the binding of Ca^{2+} to calmodulin is closely related to the activation of phosphodiesterase.

The observation that calmodulin is a Ca^{2+}-binding protein led to the suggestion that the active calmodulin is a Ca^{2+}–calmodulin complex. To support this hypothesis, many investigators have looked into conformational change of calmodulin upon binding Ca^{2+}. That a conformational change of calmodulin does occur in the presence of Ca^{2+} is indicated by enhancement of fluorescence emission intensity (Wang *et al.*, 1975), increased resistance to proteolytic inactivation (Ho *et al.*, 1976; Liu and Cheung, 1976), increased helical structure estimated by optical rotatory dispersion or circular dicroism (Liu and Cheung, 1976; Wolff *et al.*, 1977; Klee, 1977; Dedman *et al.*, 1977), and the alteration of the environment of one tyrosine residue as reflected from a shift of pK value (Klee, 1977), decreased intensity of tyrosine absorption band (Klee, 1977), and a change of reactivity toward chemical modifications (Richman and Klee,

1978a; Richman, 1978). Both tyrosine residues of calmodulin have been modified without affecting the activity of calmodulin (Walsh and Stevens, 1978; Richman and Klee, 1978a; Richman, 1978). Although calmodulin modified with *N*-chlorosuccinimide did not affect the extent of Ca^{2+}-induced increase of helical content, the protein lost its high affinity Ca^{2+}-binding sites and its ability to activate phosphodiesterase (Walsh and Stevens, 1978). Contents of α-helix may represent only gross conformation, while numerous minor variations with the same gross conformation represented by them may exist. One such variation may be optimal for the activation of phosphodiesterase.

B. Ca^{2+}-Dependent Association of Calmodulin and Phosphodiesterase

A criterion that the Ca^{2+}–calmodulin complex is the active species is that an interaction of Ca^{2+}–calmodulin with phosphodiesterase occurs and results in the activation of phosphodiesterase. Though direct proof is lacking, the qualitative similarity in the dependence on Ca^{2+} and calmodulin for the formation of the Ca^{2+}–calmodulin complex and the activation of phosphodiesterase strongly suggests that Ca^{2+}–calmodulin complex is the species that interacts with phosphodiesterase.

The interaction of calmodulin and phosphodiesterase has been investigated with gel filtration column chromatography (Teshima and Kakiuchi, 1974; Lin *et al.*, 1975). In the presence of Ca^{2+}, phosphodiesterase forms a complex with calmodulin as indicated by the following: (1) the enzyme eluted from the column does not require exogenous calmodulin for full activity; (2) calmodulin is associated with the enzyme; and (3) the elution profile of the enzyme is shifted to one that corresponds to the calmodulin–enzyme complex. The formation of the calmodulin–enzyme complex requires Ca^{2+}, since the complex did not occur in the presence of EGTA, which chelates Ca^{2+}. The elution profile of the calmodulin–enzyme complex suggests that the complex consisted of 1 mole of phosphodiesterase and 2 moles of calmodulin. In contrast, Richman and Klee (1978b) and LaPorte *et al.* (1979) reported a molar ratio of 1 to 1, based on analysis of the complex by cross-linking agents.

The Ca^{2+}-dependent association of calmodulin and the enzyme indicates that either the calmodulin–phosphodiesterase complex or the Ca^{2+}–calmodulin–phosphodiesterase complex is the active enzyme. If Ca^{2+} merely serves as a catalyst for complex formation, the activation of phosphodiesterase would not depend on Ca^{2+} concentration. Since the activation of phosphodiesterase does depend on Ca^{2+} concentration, the Ca^{2+}–calmodulin–phosphodiesterase complex is most likely the active form.

C. Mechanism of Activation

The simplest model for the activation of phosphodiesterase by calmodulin (Lin *et al.*, 1974b) consistent with experimental observations may be described in the following scheme:

Ca^{2+} + calmodulin ⇌ Ca^{2+}-calmodulin* (active)
Ca^{2+}–calmodulin* (active) + phosphodiesterase (less active) ⇌
Ca^{2+}–calmodulin*–phosphodiesterase* (active)

According to this scheme, hormonal or electrical stimulation of the cell increases the cellular concentration of Ca^{2+} from 10^{-7} to 10^{-6} or 10^{-5} *M*. The increase of cellular Ca^{2+} enables calmodulin to assume the active Ca^{2+}–calmodulin species which then interacts with the apoenzyme of phosphodiesterase to produce the activated holoenzyme. Lowering of cellular Ca^{2+} (uptake by sarcoplasmic reticulum and mitochondria, or extrusion through the plasma membrane by Ca^{2+}-ATPase) dissociates the ternary complex, returning the enzyme activity to its steady state level. Thus, phosphodiesterase activity is controlled by the cellular flux of Ca^{2+}. Similar models have been proposed by other workers (Teshima and Kakiuchi, 1974; Wolff and Brostrom, 1974; Wang *et al.*, 1975).

The stoichiometry of the individual components in the complexes has not been defined. Calmodulin possesses four Ca^{2+}-binding sites; based on a K_a value of 2 to 8 μM for the Ca^{2+} activation of phosphodiesterase, it is likely that only the high-affinity sites are essential for activity. Alternatively, Wolff *et al.* (1977) suggested that the tricalcium–monomagnesium–calmodulin complex is the active calmodulin.

The molar ratio of calmodulin and phosphodiesterase in the active Ca^{2+}–calmodulin–phosphodiesterase complex has been estimated to be either 2 to 1 (Teshima and Kakiuchi, 1974; Lin *et al.*, 1975) or 1 to 1 (Richman and Klee, 1978; LaPorte *et al.*, 1979). The number of Ca^{2+} present in the active enzyme complex is not known.

A comparison of the stimulation of phosphodiesterase by calmodulin with that by trypsin may shed light on the molecular structure of phosphodiesterase. Activation of phosphodiesterase by calmodulin proceeds stoichiometrically and reversibly, whereas that by trypsin proceeds catalytically and irreversibly. Activation by trypsin reduces the molecular weight of phosphodiesterase and disables it to interact with and respond to calmodulin. Stimulation by either agent produces an activated enzyme with comparable activities. Figure 1 depicts a speculative scheme on the molecular events associated with the stimulation of phosphodiesterase by both agents. Binding of calmodulin to phosphodiesterase is believed to cause the enzyme to assume a new conformation, with the catalytic site more accessible to the substrate. On the other hand, limited trypsinization

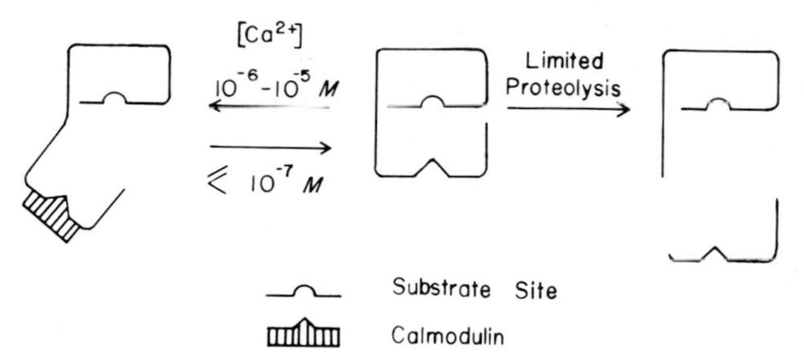

Fig. 1. A schematic illustration of the activation of phosphodiesterase by calmodulin and by trypsin. Binding of calmodulin renders phosphodiesterase to assume a new conformation with the catalytic site more accessible to the substrate. Dissociation of the ternary complex returns the enzyme to its prestimulated conformation. The process is reversible, and the equilibrium is controlled by the cellular flux of Ca^{2+}. Limited trypsinization releases a polypeptide from phosphodiesterase. Removal of the polypeptide exposed the substrate site, and the result is phenomenally comparable to the effect of calmodulin—and increase in V_{max} with little or no change in the K_m for cAMP. The cleavage of phosphodiesterase by trypsin is not reversible and is probably not physiological. The botched rectangle represents calmodulin; the semi circle represents the substrate site.

of phosphodiesterase releases a polypeptide from the latter, with a similar consequence—an enzyme with a more exposed substrate site. Although the scheme in Fig. 1 assumes that the polypeptide released from phosphodiesterase contains the calmodulin-binding domain, this is not a necessary condition; other alternatives are equally possible.

D. The Role of Ca^{2+} in Phosphodiesterase Regulation

The notion that the activity of Ca^{2+}-dependent phosphodiesterase is regulated by Ca^{2+} flux, first suggested by Kakiuchi *et al.* (1972), is supported by several lines of evidence: (1) the stimulation of phosphodiesterase requires micromolar Ca^{2+}, which is close to its physiological level (Hodgkin and Keynes, 1957); (2) the intracellular Ca^{2+} concentration fluctuates in response to stimuli (Rasmussen *et al.*, 1975); and (3) the response of phosphodiesterase to Ca^{2+} is swift (Wolff and Brostrom, 1974; Lin *et al.*, 1975) and is compatible with the rapid catabolism of cAMP (Kakiuchi and Rall, 1968a,b). Tissue levels of calmodulin are generally higher than those of phosphodiesterase; the enzyme activity is likely to be limited by the availability of Ca^{2+}. A possibility may exist that the enzyme activity is regulated by a change in the concentration of calmodulin available to phosphodiesterase. Gnegy *et al.* (1976) observed that a purified cAMP-dependent protein kinase, in the presence of cAMP and ATP, stimulated from a particulate fraction of rat brain and adrenal medulla the release of calmodulin. The released calmodulin may regulate the activity of the cytosolic phosphodiesterase (see also Hanbauer and Costa, this volume, Chapter 12). This notion, however, fails to take into consideration the general finding that, in the cytosolic compartment, the concentration of calmodulin is considerably higher than that of phosphodiesterase.

The degradation of intracellular cAMP and cGMP may involve other forms of phosphodiesterase. The contribution of individual forms of phosphodiesterase to the hydrolysis of either cAMP or cGMP depends not only on the relative concentrations of these forms and their affinities for the substrates but also the concentrations of the substrates and the various effectors. Although some tissues contain predominantly the Ca^{2+}-dependent form, exceptions have been noted (see Section V). In lower forms of animals, Ca^{2+}-dependent phosphodiesterase has not been detected (Waisman *et al.*, 1975). Calcium-dependent phosphodiesterase is considered by some workers to be a cGMP phosphodiesterase because of its higher affinity for cyclic GMP (Kakiuchi *et al.*, 1973). However, the Ca^{2+}-dependent form may become dominant in hydrolysing AMP when the concentration of the latter is markedly elevated as a result of stimulation of the tissue.

Besides Ca^{2+}-dependent phosphodiesterase, Ca^{2+}-dependent adenylate cyclase also exists in the brain (Brostrom *et al.*, 1975; Cheung *et al.*, 1975) and glial tumor cells (Brostrom *et al.*, 1976). The stimulation of adenylate cyclase by calmodulin appears to follow a mechanism similar to that of phosphodiesterase (C. O. Brostrom *et al.*, 1975; Cheung *et al.*, 1975; Lynch *et al.*, 1976; M. A. Brostrom *et al.*, 1976; Bradham and Cheung, this volume, Chapter 6). A number of other enzymes have now been found to be activated by calmodulin (see other chapters) and may well be regulated in a similar manner.

The fact that both adenylate cyclase and phosphodiesterase are activated by calmodulin appears puzzling. Two hypotheses have been put forth. Cheung *et al.* (1975) proposed that influx of Ca^{2+} across the plasma membrane into the cells causes a sequential activation of first adenylate cyclase in the membrane and then phosphodiesterase in the cytosol, resulting in a transient accumulation of cAMP. Brostrom *et al.* (1975) postulated that calmodulin stimulates the synthesis of cAMP and a concurrent degradation of cGMP.

Some of the experiments on the effects of Ca^{2+} on cyclic nucleotide levels in intact tissues are relevant to our discussion of both hypotheses. For instance, under different conditions, both Ca^{2+}-stimulated accumulation (Shimizu *et al.*, 1970) and Ca^{2+}-stimulated degradation (Schultz, 1975) of cyclic AMP in brain have been observed. In contrast, only Ca^{2+}-

activated cyclic GMP synthesis in smooth muscle has been reported (Schultz and Hardman, 1975). These observations, however, do not provide conclusive evidence to support either of the hypotheses. The tissue levels of cAMP are probably the result of complex interactions involving multiple forms of phosphodiesterase and the various modulators of the enzymes.

X. CONCLUDING REMARKS

Substantial progress has been made in recent years in phosphodiesterase research. Following successful purification of calmodulin and elucidation of the mechanism of its activation of phosphodiesterase, a complete purification of a Ca^{2+}-dependent phosphodiesterase, which had long resisted many attempts, has been achieved recently. The availability of a homogeneous enzyme will facilitate further exploration of the properties and molecular structure of the enzyme.

Apparent discrepancies on certain aspects of phosphodiesterase from different laboratories have appeared. Most of the discrepancies are probably due to different experimental conditions, heterogeneity of the enzyme preparation, or instability of the enzyme. Proteolysis alters the enzyme and may be related to the instability and generation of multiple forms of the enzyme (Cheung, 1970b; Cheung and Lin, 1974; Sakai *et al.,* 1978). Other factors cause interconversion of the different enzyme forms (Tisdale, 1975; Pichard and Cheung, 1976). Future efforts should be directed to sort out the physiologically relevant from laboratory artifacts.

ACKNOWLEDGMENTS

The work from our laboratories were supported by USPHS Grant NS 08059 and RR 08092, and by ALSAC.

REFERENCES

Amer, M. S., and Kreighbaum, W. E. (1975). Cyclic nucleotide phosphodiesterase: Properties, activators, inhibitors, structure-activity relationship and possible role in drug development. *J. Pharm. Sci.* **64,** 1–37.

Appleman, M. M., Thompson, W. J., and Russell, T. R. (1973). Cyclic nucleotide phosphodiesterase. *Adv. Cyclic Nucleotide Res.* **3,** 65–98.

Beavo, J. A., Hardman, J. G., and Sutherland, E. W. (1970). Hydrolysis of cyclic guanosine and adenosine 3′,5′-monophosphates by rat and bovine tissues. *J. Biol. Chem.* **245,** 5649–5655.

Berridge, M. J. (1975). Interaction of cyclic nucleotide and calcium in the control of cellular activity. *Adv. Cyclic Nucleotide Res.* **6,** 1–98.

Boudreau, R. J., and Drummond, G. I. (1975). The effect of Ca^{2+} on cyclic nucleotide phosphodiesterases of superior cervical ganglion. *J. Cyclic Nucleotide Res.* 219–228.

Brooker, G., Thomas, L. J., Jr., and Appleman, M. M. (1968). The assay of adenosine 3′,5′-monophosphate and guanosine 3″,5′-monophosphate in biological materials by enzymatic radioisotopic displacement. *Biochemistry* **2,** 4177–4181.

Brostrom, C. O., and Wolff, D. J. (1976). Calcium-dependent cyclic nucleotide phosphodiesterase from brain: Comparison of adenosine 3′,5′-monophosphate and guanosin 3′,5′-monophosphate as substrates. *Arch. Biochem. Biophys.* **172,** 301–311.

Brostrom, C. O., Huang, Y. -C., Breckenridge, B. M., and Wolff, D. J. (1975). Identification of calcium binding protein as a calcium dependent regulator of brain adenylate cyclase. *Proc. Natl. Acad. Sci. U.S.A.* **72,** 64–68.

Brostrom, M. A., Brostrom, C. O., Breckenridge, B. M., and Wolff, D. J. (1976). Regulation of adenylate cyclase from glial tumor cells by calcium and a calcium-binding protein. *J. Biol. Chem.* **251,** 4744–4750.

Bublitz, C. (1973). Effects of lipids on cyclic-nucleotide phosphodiesterase. *Biochem. Biophys. Res. Commun.* **52,** 173–179.

Butcher, R. W., and Sutherland, E. W. (1962). Adenosine 3′,5′-phosphate in biological materials. I. Purification and properites of cyclic 3′,5′-nucleotide phosphodiesterase and use of this enzyme to characterize adenosine 3′,5′-phosphate in human urine. *J. Biol. Chem.* **237,** 1244–1250.

Campbell, M. T., and Oliver, I. T. (1972). 3′,5′-Cyclic-nucleotide phosphodiesterase in rat tissues. *Eur. J. Biochem.* **28,** 30–37.

Chasin, M., and Harris, D. N. (1976). Inhibitors and activators of cyclic nucleotide phosphodiesterase. *Adv. Cyclic Nucleotide Res.* **7,** 225–264.

Cheung, W. Y. (1966). Inhibition of cyclic nucleotide phosphodiesterase by adenosine 5′-phosphate and inorganic phosphate. *Biochem. Biophys. Res. Commun.* **23,** 214–219.

Cheung, W. Y. (1967a). Properties of cyclic 3′,5′-nucleotide phosphodiesterase from rat brain. *Biochemistry* **6,** 1079–1087.

Cheung, W. Y. (1967b). Cyclic 3′,5′-nucleotide phosphodiesterase: Pronounced stimulation by snake venom. *Biochem. Biophys. Res. Commun.* **29,** 478–482.

Cheung, W. Y. (1969a). Cyclic 3′,5′nucleotide phosphodiesterase: A continous titrimetric assay. *Anal. Biochem.* **28,** 182–191.

Cheung, W. Y. (1969b). Cyclic 3′,5′-nucleotide phosphodiesterase: Preparation of a partially inactive enzyme and its subsequent stimulation by snake venom. *Biochim, Biophys. Acta* **191,** 303–315.

Cheung, W. Y. (1970a). Cyclic 3′,5′-nucleotide phosphodiesterase: Demonstration of an activator. *Biochem. Biophys. Res. Commun.* **38,** 533–537.

Cheung, W. Y. (1970b). Cyclic nucleotide phosphodiesterase. *Adv. Biochem. Pharmacol.* **3,** 51–66.

Cheung, W. Y. (1971a). Cyclic 3′,5′-nucleotide phosphodiesterase. Effect of divalent cations. *Biochim. Biophys. Acta* **242,** 395–409.

Cheung, W. Y. (1971b). Cyclic nucleotide phosphodiesterase: Evidence for and properties of a protein activator. *J. Biol. Chem.* **246,** 2859–2869.

Cheung, W. Y., and Lin, Y. M. (1974). Purification and characterization of cyclic 3′,5′-nucleotide phosphodiasterase from bovine brain. *In* "Methods in Enzymology" (J. G. Hardman, and B. W. O'Malley, eds.), Vol. 38, Part C, pp. 223–239. Academic Press, New York.

Cheung, W. Y., Bradham, L. S., Lynch, T. J., Lin, Y. M., and Tallant, E. A. (1975). Protein activator of cyclic 3′,5′-nucleotide phosphodiesterase of bovine or rat brain also activates its adenylate cyclase. *Biochem. Biophys. Res. Commun.* **66,** 1055–1062.

Davis, C. W., and Daly, J. W. (1978). Calcium-dependent 3′,5′-cyclic nucleotide phosphodiesterase. Inhibition of basal activity at physiological levels of potassium ions. *J. Biol. Chem.* **253,** 8683–8686.

Dedman, J. R., Potter, J. D., Jackson, R. L., Johnson, J. D., and Means, A. R. (1977). Physicochemical properties of rat testis Ca^{2+}-dependent regulator protein of cyclic nucleotide phosphodiesterase. *J. Biol. Chem.* **252,** 8415–8422.

Donnelly, T. E., Jr. (1977). Properties of the activator-dependent cyclic nucleotide phosphodiesterase from bovine heart. *Biochim. Biophys. Acta* **480,** 193–203.

Donnelly, T. E., Jr. (1978). Lack of altered cyclic nucleotide phosphodiesterase activity in aorta and heart at the spontaneously hypertensive rat. *Biochim. Biophys. Acta* **542,** 245–252.

Drummond, G. I., and Perrott-Yee, S. (1961). Enzymatic hydrolysis of adenosine 3′,5′-phosphoric acid. *J. Biol. Chem.* **236,** 1126–1129.

Gnegy, M. E., Costa, E., and Uzunor, P. (1976). Regulation of transynaptically elicited increase of 3′,5′-cyclic AMP by endogenous phosphodiesterase activator. *Proc. Natl. Acad. Sci. U.S.A.* **73,** 352–355.

Goren, E. N., and Rosen, O. M. (1972). Purification and properties of a cyclic nucleotide phosphodiesterase from bovine heart. *Arch. Biochem. Biophys.* **153,** 384–397.

Hardman, J. G., and Sutherland, E. W. (1965). A cyclic 3′,5′-nucleotide phosphodiesterase from heart with specificity for uridine 3′,5′-phosphate. *J. Biol. Chem.* **240,** 3704–3705.

Hidaka, H., Yamaki, T., Ochiai, Y., Asano, T., and Yamabe, H. (1977). Cyclic 3′,5′-nucleotide phosphodiesterase determined in various human tissues by DEAE-cellulose chromatography. *Biochim. Biophys. Acta* **484,** 398–407.

Ho, H. C., Teo, T. S., Desai, R., and Wang, J. H. (1976). Catalytic and regulatory properties of two forms of bovine heart cyclic nucleotide phosphodiesterase. *Biochim. Biophys. Acta* **429,** 461–473.

Ho, H. C., Wirch, E., Stevens, F. C., and Wang, J. H. (1977). Purification of a Ca^{2+}-activatable cyclic nucleotide phosphodiesterase from bovine heart by specific interaction with its Ca^{2+} dependent modulator protein. *J. Biol. Chem.* **252,** 43–50.

Hodgkin, A. L., and Keynes, R. D. (1957). Movement of labeled calcium in squid giant axons. *J. Physiol. (London)* **138,** 253–281.

Kakiuchi, S., and Rall, T. W. (1968a). The influence of chemical agents on the accumulation of adenosine 3′,5′-phosphate in slices of rabbit cerebellum. *Mol. Pharmacol.* **4,** 367–378.

Kakiuchi, S., and Rall, T. W. (1968b). Studies on adenosine 3′,5′-phosphate in rabbit cerebral cortex. *Mol. Pharmacol.* **4,** 379–388.

Kakiuchi, S., and Yamazaki, R. (1970). Stimulation of activity of cyclic 3′,5′-nucleotide phosphodiesterase by calcium ion. *Proc. Jpn. Acad.* **46,** 387–392.

Kakiuchi, S., Yamazaki, R., and Teshima, Y. (1971). Cyclic 3′,5′-nucleotide phosphodiesterase. IV. Two enzymes with different properties from brain. *Biochem. Biophys. Res. Commun.* **42,** 968–974.

Kakiuchi, S., Yamazaki, R., and Teshima, Y. (1972). Regulation of brain phosphodiesterase activity: Ca^{2+} plus Mg^{2+}-dependent phosphodiesterase and its activating factor from rat brain. *Adv. Cyclic Nucleotide Res.* **1,** 455–478.

Kakiuchi, S., Yamazaki, R., and Uenishi, K. (1973). Regulation of nucleotide cyclic 3′,5′-monophosphate phosphodiesterase activity from rat brain by a modulator and Ca^{2+}. *Proc. Natl. Acad. Sci. U.S.A.* **70,** 3520–3530.

Kakiuchi, S., Yamazaki, R., Teshima, Y., Uenish, K., and Miyamoto, E. (1975). Multiple cyclic nucleotide phosphodiesterase activities from rat tissues and occurrence of a calcium-plus-magnesium-ion-dependent phosphodiesterase and its protein activator. *Biochem. J.* **146,** 109–120.

Kakiuchi, S., Yamazaki, R., Teshima, Y., Uenishi, K., Yasuda, S., Kashiba, A., Sobue, K., Ohshima, M., and Nakajima, T. (1978) Membrane-bound protein modulator and phosphodiesterase. *Adv. Cyclic Nucleotide Res.* **9,** 253–264.

Klee, C. B. (1977). Conformational transition accompanying the binding of Ca^{2+} to the protein activator of 3′,5′-cyclic adenosine monophosphate phosphodiesterase. *Biochemistry* **16,** 1017–1024.

Klee, C. B., and Krinks, M. H. (1978). Purification of cyclic 3′,5′-nucleotide phosphodiesterase inhibitory protein by affinity chromatography on activator protein coupled to Sepharose. *Biochemistry* **17,** 120–126.

Klee, C. B., Crouch, T. H., and Krinks, M. H. (1979). Subunit structure and catalytic properties of bovine brain Ca^{2+}-dependent cyclic nucleotide phosphodiesterase. *Biochemistry* **18,** 722–729.

LaPorte, D. C., Toscano, W. A., Jr., and Storm, D. R. (1979). Cross-linking of iodine-125 labelled calcium-dependent regulatory protein to the Ca^{2+}-sensitive phosphodiesterase purified from bovine heart. *Biochemistry* **18,** 2821–2825.

Lin, Y. M., and Cheung, W. Y. (1975). Cyclic 3′,5′-nucleotide phosphodiesterase. A theoretical account of its anomalous kinetics in terms of multiple isozymes. *Int. J. Biochem.* **6,** 271–280.

Lin, Y. M., Liu, Y. P., and Cheung, W. Y. (1974a). Purification and characterization of protein activator of cyclic nucleotide phosphodiesterase from bovine brain. *In* "Methods in Enzymology" (J. G. Hardman, and B. W. O'Malley, eds.), Vol. 38, Part C, pp. 223–239. Academic Press, New York.

Lin, Y. M., Liu, Y. P., and Cheung, W. Y. (1974b). Cyclic 3′,5′-nucleotide phosphodiesterase. Purification, characterization and active form of the protein activator from bovine brain. *J. Biol. Chem.* **249,** 4943–4594.

Lin, Y. M., Liu, Y. P., and Cheung, W. Y. (1975). Cyclic 3′,5′-nucleotide phosphodiesterase. Ca^{2+}-dependent formation of bovine brain enzyme–activator complex. *FEBS Lett.* **49,** 356–360.

Liu, Y. P. and Cheung, W. Y. (1976). Cyclic 3′,5′-nucleotide phosphodiesterase. Ca^{2+} confers more helical conformation to the protein activator. *J. Biol. Chem.* **251,** 4193–4198.

Lynch, T. J., and Cheung, W. Y. (1975). Underestimation of cyclic 3′,5′-nucleotide phosphodiesterase activity by a radioisotopic assay using an anion exchange resin. *Anal. Chem.* **67,** 130–138.

Lynch, T. J., Tallant, E. A., and Cheung, W. Y. (1976). Ca^{2+}-dependent formation of brain adenylate cylcase-protein activator complex. *Biochem. Biophys. Res. Commun.* **68,** 616–625.

Monn, E., and Christiansen, R. O. (1971). Adenosine 3′,5′-monophosphate phosphodiesterase: Multiple molecular forms. *Science* **173,** 540–541.

Monn, E., Desautel, M., and Christiansen, R. O. (1972). Highly specific testicular adenosine 3′,5′-monophosphate phosphodiesterase associated with sexual maturation. *Endocrinology* **91,** 716–720.

Morrill, M. E., Thompson, S. T., and Stellwagen, E. (1979). Purification of a cyclic nucleotide phosphodiesterase from bovine brain using blue dextran-Sepharose chromatography. *J. Biol. Chem.* **254,** 4371–4374.

Nair, K. G. (1966). Purification and properties of 3′,5′-cyclic nucleotide phosphodiesterase from dog heart. *Biochemistry* **5,** 150–157.

Pichard, A. -L., and Cheung, W. Y. (1976). Cyclic nucleotide phosphodiesterase. Interconvertible multiple forms and their effects on enzyme activity and kinetics. *J. Biol. Chem.* **251,** 5726–5737.

Pichard, A. L., and Cheung, W. Y. (1977). Cyclic 3′,5′-nucleotide phosphodiesterase. Stimulation of bovine brain cytoplasmic enzyme by lysophosphatidylcholine *J. Biol. Chem.* **252,** 4872–4875.

Pledger, W. J., Stancel, G. M., Thompson, W. J., and Strada, S. J. (1974). Separation of multiple forms of cyclic nucleotide phosphodiesterase from rat brain by isoelectrofocusing. *Biochim. Biophys. Acta* **370,** 242–248.

Rasmussen, H., Jensen, P., Lake, W., Friedman, N., and Goodman, D. B. P. (1975). Cyclic nucleotides and cellular acalcium metabolism. *Adv. Cyclic Nucleotide Res.* **5,** 375–414.

Richman, P. G. (1978). Conformation-dependent nitration and acetylation of the protein activator of cyclic 3′,5′-monophosphate phosphodiesterase. Selective nitration of tyrosine residue 138. *Biochemistry* **17,** 3001–3005.

Richman, P.G., and Klee, C. B. (1978a). Conformational-dependent nitration of the protein activator of cyclic 3′,5′-nucleotide phosphodiesterase. *Biochemistry* **17,** 928–934.

Richman, P. G., and Klee, C. B. (1978b). Interaction of ^{125}I labelled Ca^{2+}-dependent regulator protein with cyclic nucleotide phosphodiesterase and its inhibitory protein. *J. Biol. Chem.* **253,** 6323–6326.

Robison, G. A., Butcher, R. W., and Sutherland, E. W. (1971). "Cyclic AMP." Academic Press, New York.

Rosen, O. M. (1970). Preparation and properties of a cyclic 3′,5′-nucleotide phosphodiesterase isolated from frog erythrocytes. *Arch. Biochem. Biophys.* **137,** 435–441.

Russell, T. R., Thompson, W. J., Schneider, F. W., and Appleman, M. M. (1972). 3′,5′-cyclic adenosine monophosphate phosphodiesterase: Negative cooperativity. *Proc. Natl. Acad. Sci. U.S.A.* **69,** 1791–1795.

Russell, T. R., Terasaki, W. L., and Appleman, M. M. (1973). Separate phosphodiesterase for the hydrolysis of cyclic adenosine 3′,5′-monophosphate and cyclic guanosine-monophosphate in rat liver. *J. Biol. Chem.* **248,** 1334–1340.

Sakai, T., Yamanaka, H., Tanaka, R., Makine, H., and Kasi, H. (1977). Stimulation of cyclic nucleotide phosphodiesterase from rat brain by activator protein, proteolytic enzymes and a vitamin E derivative. *Biochim. Biophys. Acta* **483,** 121–134.

Sakai, T., Makino, H., and Tanaka, R. (1978). Increased activity of cyclic AMP phosphodiesterase from frozen-thawed rat liver. *Biochim. Biophys. Acta* **522,** 477–490.

Sankaran, K., Hanbauer, I., and Lovenberg, W. (1978). Heat-stable low molecular weight form of phosphodiesterase from bovine pineal gland. *Proc. Natl. Acad. Sci. U.S.A.* **75,** 3188–3191.

Schröder, J., and Rickenberg, H. V. (1973). Partial purification and properties of the cyclic AMP and the cyclic GMP phosphodiesterase of bovine liver. *Biochim. Biophys. Acta* **302,** 50–63.

Schultz, G., and Hardman, J. G. (1975). Regulation of cyclic GMP levels in durtus deferens of the rat. *Adv. Cyclic Nucleotide Res.* **5,** 339–351.

Schultz, J. (1975). Cyclic adenosine 3′,5′-monophosphate in guinea pig cerebral cortical slices: Possible regulation of phosphodiesterase activity by cyclic 3′,5′-monophosphate and calcium. *J. Neurochem.* **24,** 496–501.

Sharma, R. K., Wirch, E., and Wang, J. H. (1978a). Inhibition of Ca^{2+}-activated cyclic nucleotide phosphodiesterase reaction by a heat-stable inhibitor protein from bovine brain. *J. Biol. Chem.* **253,** 3575–3580.

Sharma, R. K., Desai, R., Thompson, T. R., and Wang, J. H. (1978b). Purification of the heat stable inhibitor protein of the Ca^{2+}-activated cyclic nucleotide phosphodiesterase by affinity chromatography. *Can. J. Biochem.* **56,** 598–604.

Shimizu, H., Creveling, C. R., and Daly, J. (1970). Cyclic adenosine 3′,5′-monophosphate formation in brain slices: Stimulation by batrachotoxin, ouabain, veratridine, and potassium ions. *Mol. Pharmacol.* **6,** 184–188.

Singer, A. L., Dunn, A., and Appleman, M. M. (1978). Cyclic nucleotide phosphodiesterase and protein activator in fetal rabbit tissues. *Arch. Biochem. Biophys.* **187,** 406–413.

Smoake, J. A., Song, S. -Y., and Cheung, W. Y. (1974). Cyclic 3′,5′-nucleotide phosphodiesterase. Distribution and developmental changes of the enzyme and its protein activator in mammalian tissues and cells. *Biochim. Biophys. Acta* **341,** 402–411.

Strada, S. J., and Thompson, W. J. (1978). Multiple forms of cyclic nucleotide phosphodiesterases: Anomalies or biological regulators? *Adv. Cyclic Nucleotide Res.* **9,** 265–283.

Sutherland, E. W., and Rall, T. W. (1958). Fractionation and characterization of a cyclic adenine ribonucleotide and characterization of a cyclic adenine ribonucleotide formed by tissue particles. *J. Biol. Chem.* **232,** 1077–1091.

Tanigawa, Y., and Simoyama, M. (1976). Reciprocal changes in Ca^{2+}/protein activator-dependent cyclic AMP phosphodiesterase. *Biochem. Biophys. Res. Commun.* **73,** 19–24.

Taniguchi, T., Fujisawa, M., Lee, J. J., and Hidaka, H. (1978). Cyclic 3′,5′-nucleotide phosphodiesterase of rabbit sinoatrial node. *Biochim. Biophys. Acta* **522,** 465–476.

Teo, T. W., and Wang, J. H. (1973). Mechanism of activation of a cyclic adenosine 3′,5′-monophosphate phosphodiesterase from bovine heart by calcium ions. *J. Biol. Chem.* **247,** 5950–5955.

Teo, T. W., Wang, T. H., and Wang, J. H. (1973). Purification and the properties of the protein activator of bovine heart cyclic adenosine 3′,5′-monophosphate phosphodiesterase. *J. Biol. Chem.* **248,** 588–595.

Teshima, Y., and Kakiuchi, S. (1974). Mechanism of stimulation of Ca^{2+} plus Mg^{2+}-dependent phosphodiesterase from rat cerebral cortex by the modulator protein and Ca^{2+}. *Biochem. Biophys. Res. Commun.* **56,** 489–495.

Thompson, W. J., and Appleman, M. M. (1971a). Multiple cyclic nucleotide phosphodiesterase activities from rat brain. *Biochemistry* **10,** 311–316.

Thompson, W. J., and Appleman, M. M. (1971b). Characterization of cyclic nucleotide phosphodiesterases of rat tissues. *J. Biol. Chem.* **246,** 3145–3150.

Thompson, W. J., Epstein, P. M., and Strada, S. J. (1979). Purification and characterization of high-affinity cyclic adenosine monophosphate phosphodiesterase from dog kidney. *Biochemistry* **18,** 5228–5237.

Tisdale, M. J. (1975). Characterization of cyclic 3′,5′-monophosphate phosphodiesterase from Walker Carcinoma sensitive and resistant to bifunctional alkylating agent. *Biochim. Biophys. Acta* **397,** 134–143.

Uzunov, P., and Weiss, B. (1972). Separation of multiple forms of cyclic adenosine 3′,5′-monophosphate phosphodiesterase in rat cerebellum by polyacrylamide gel electrophoresis. *Biochim. Biophys. Acta* **284,** 220–226.

Uzunov, P., Gnegy, M. E., Revuelta, A., and Costa, E. (1976). Regulation of the high K_m cyclic nucleotide phosphodiesterase of adrenal medulla by the exogenous calcium-dependent-protein activator. *Biochem. Biophys. Res. Commun.* **70,** 132–138.

Van Inwegen, R. G., Pledger, W. J., Strada, S. J., and Thompson, W. J. (1976). Characterization of cyclic nucleotide phosphodiesterase with multiple separation techniques. *Arch. Biochem. Biophys.* **175,** 700–709.

Waisman, D., Stevens, F. C., and Wang, J. H. (1975). The distribution of the Ca^{2+}-dependent protein activator of cyclic nucleotide phosphodiesterase in invertebrates. *Biochem. Biophys. Res. Commun.* **65,** 975–982.

Wallace, R. W., Lynch, T. J., Tallant, E. A., and Cheung, W. Y. (1978). An endogenous inhibitor protein of brain adenylate cyclase and cyclic nucleotide phosphodiesterase. *Arch. Biochem. Biophys.* **187,** 328–334.

Wallace, R. W., Lynch, T. J., Tallant, E. A., and Cheung, W. Y. (1979). Purification and characterization of inhibitor protein of brain adenylate cyclase and cyclic nucleotide phosphodiesterase. *J. Biol. Chem.* **254,** 377–382.

Walsh, M., and Stevens, F. C. (1978). Ca^{2+}-dependent protein modulator. The role of methionine residues in the activation of cyclic nucleotide phosphodiesterase.*Biochemistry* **17,** 3924–3928.

Wang, J. H., and Desai, R. (1976). A brain protein and its effect on the Ca^{2+}- and protein modulator-activated cyclic nucleotide phosphodiesterase. *Biochem. Biophys. Res. Commun.* **72,** 926–932.

Wang, J. H., and Desai, R. (1977). Modulator binding protein. Bovine brain protein exhibiting the Ca^{2+}-dependent association with the protein modulator of cyclic nucleotide phosphodiesterase. *J. Biol. Chem.* **252,** 4175–4184.

Wang, J. H., Teo, T. S., Ho, H. C., and Stevens, F. C. (1975). Bovine heart protein activator of cyclic nucleotide phosphodiesterase. *Adv. Cyclic Nucleotide Res.* **5,** 179–194.

Watterson, D. M., Harrelson, W. G., Jr., Keller, P. M., Sharief, F., and Vanaman, T. C. (1976). Structural similarities between the Ca^{2+}-dependent regulatory proteins of 3′,5′-cyclic nucleotide phosphodiesterase and actomyosin ATPase. *J. Biol. Chem.* **251,** 4501–4513.

Watterson, D. M., and Vanaman, T. C. (1976). Affinity chromatography purification of a cyclic nucleotide phosphodiesterase using immobilized modulator protein, a troponin c-like protein from brain. *Biochem. Biophys. Res. Commun.* **73,** 40–46.

Watterson, D. M., Sharief, F., and Vanaman, T. C. (1980). The complete amino acid sequence of calmodulin of bovine brain. *J. Biol. Chem.* **255,** 962–975.

Weiss, B. (1975). Differential activation and inhibition of the multiple forms of cyclic nucleotide phosphodiesterase. *Adv. Cyclic Nucleotide Res.* **5,** 195–211.

Weiss, B., and Levin, R. M. (1978). Mechanism for selectively inhibiting the activation of cyclic nucleotide phosphodiestease and adenylate cyclase by antipsychotic agents. *Adv. Cyclic Nucleotide Res.* **8,** 285–304.

Weiss, B., Lehne, R., and Strada, S. (1972). Rapid micro-assay of adenosine 3′,5′-monophosphate phosphodiesterase activity. *Anal. Biochem.* **45,** 222–235.

Weiss, B., Fertel, R., Figlin, R., and Uzunov, P. (1974). Selective alteration of the activity of the multiple forms of cyclic 3′,5′-AMP phosphodiesterase of rat cerebrum. *Mol. Pharmacol.* **10,** 615–626.

Wells, J. N., and Garbers, D. L. (1976). Nucleoside 3′,5′-monophosphate phosphodiesterase in sea urchin sperm. *Biol. Reprod.* **15,** 46–53.

Wells, J. N., and Hardman, J. G. (1977). Cyclic nucleotide phosphodiesterase. *Adv. Cyclic Nucleotide Res.* **8,** 119–143.

Wells, J. N., Wu, Y. J., Baird, C. E., and Hardman, J. G. (1975). Phsophodiesterase from porcine coronary arteries. Inhibition of separated forms by xanthines, papaverine, and cyclic nucleotides. *Mol. Pharmacol.* **11,** 775–783.

Wolff, D. J., and Brostrom, C. O. (1974). Calcium-binding phosphorprotein from pig brain: Identification as a calcium-dependent regulator of brain cyclic nucleotide phosphodiesterase. *Arch. Biochem. Biophys.* **163,** 349–358.

Wolff, D. J., and Brostrom, C. O. (1976). Calcium-dependent cyclic nucleotide phosphodiesterase from brain: Identification of phospholipids as calcium-independent activators. *Arch. Biochem. Biophys.* **173,** 720–731.

Wolff, D. J., and Siegel, F. L. (1972). Purification of a calcium-binding phosphorprotein from pig brain. *J. Biol. Chem.* **247,** 4180–4185.

Wolff, D.J., Porrier, P. G., Brostrom, C. O., and Brostrom, M. A. (1977). Divalent cation binding properties of bovine brain Ca^{2+}-dependent regulator protein. *J. Biol. Chem.* **252,** 4108–4117.

Chapter 6

Calmodulin-Dependent Adenylate Cyclase

LAURENCE S. BRADHAM
WAI YIU CHEUNG

I. ADENYLATE CYCLASE AND DIVALENT CATIONS

A. Introduction

Since its initial isolation and characterization (Sutherland *et al.,* 1962), adenylate cyclase [ATP pyrophosphate-lyase (cyclizing), EC 4.6.1.1] has been found to be subject to regulation by a wide variety of agents. These include hormones of diverse structure, fluoride ions, nucleotides such as GTP and its synthetic structural analogs, cholera toxin, and a variety of pharmacological agents. A complete review of all of these aspects of adenylate cyclase regulation is beyond the scope of this chapter. In keeping with the theme of this book, we have limited ourselves to the role of diva-

CALCIUM AND CELL FUNCTION, Vol. I

ISBN 0-12-171401-2

lent cations, and more specifically, to the role of Ca^{2+} and calmodulin in the regulation of adenylate cyclase. In the course of the discussion which follows, references will be made to the effects on the enzyme of some of the substances listed above without citation of all of the original literature. For more detailed discussion of these aspects of adenylate cyclase, the reader is referred to various other review articles (Birnbaumer *et al.*, 1970; Perkins, 1973; Birnbaumer, 1973; Rodbell *et al.*, 1975; Cuatrecasas *et al.*, 1975; Gill, 1978; Kebabian, 1976; Lefkowitz and Williams, 1978; Levitzki, 1978; Maguire *et al.*, 1977; Ross *et al.*, 1978a; Moss and Vaughn, 1979).

B. Requirement for Mg^{2+} or Mn^{2+}

The activity of adenylate cyclase is dependent upon the presence of divalent cations in the assay medium. It is generally believed that the cation serving this function in the intact cell is Mg^{2+}. The concentration of Mg^{2+} required for maximal activity of isolated adenylate cyclase is variable, depending upon the source of the enzyme and the assay conditions. In general, maximal activity is observed when the concentration of Mg^{2+} exceeds that of ATP (Birnbaumer *et al.*, 1969; Drummond and Duncan, 1970; Drummond *et al.*, 1971; Severson *et al.*, 1972; Pohl *et al.*, 1971; Bradham, 1972; Perkins, 1973; Finn *et al.*, 1975; Johnson and Sutherland, 1973). As an example, maximal activity of brain adenylate cyclase in the presence of an ATP-regenerating system is obtained at 5 m*M* Mg^{2+} and 1 m*M* ATP (Perkins, 1973). The effect of the excess Mg^{2+} is an increase in the V_{max} of the enzyme, a finding that has led to the postulation of an allosteric binding site for Mg^{2+} and a role for the cation as a regulator of adenylate cyclase (Birnbaumer *et al.*, 1970; de Haen, 1974; Garbers and Johnson, 1975; Londos and Preston, 1977).

The divalent cation requirement of adenylate cyclase can be fulfilled by Mn^{2+} which is usually more effective than Mg^{2+}, and a lower concentration of Mn^{2+} produces a higher V_{max} (Drummond *et al.*, 1971; Perkins, 1973; Johnson and Sutherland, 1973; Pilkis and Johnson, 1974; Burke, 1970; Herman *et al.*, 1976; Londos and Preston, 1977; Neer, 1979). Adenylate cyclase from some sources appears to be active only in the presence of Mn^{2+} (Flawia and Torres, 1972; Ross *et al.*, 1978b; Neer, 1978b). Maximal adenylate cyclase activity is observed at a concentration of Mn^{2+} exceeding that of ATP (Drummond *et al.*, 1971; Flawia and Torres, 1972; Johnson and Sutherland, 1973; Neer, 1979). The concentration of Mn^{2+} required for maximal activity of brain adenylate cyclase is in the range of 1 to 3 m*M* (Perkins, 1973; Neer, 1979), and the enzyme activity is two- to threefold greater than that in the presence of 5 mM Mg^{2+}. At higher concentrations of Mn^{2+} there is a slight inhibition of the activity of the enzyme, but it is still higher with Mn^{2+} than with Mg^{2+}. The effects of the two cations are not additive (Neer, 1979). Although MnATP is a better substrate for adenylate cyclase *in vitro,* the physiological substrate is presumably MgATP. In addition, Mg^{2+} may also have a regulatory role.

C. Inhibition by Ca^{2+}

With adenylate cyclase from many tissues, Ca^{2+} cannot replace Mg^{2+} or Mn^{2+}; in fact, Ca^{2+} inhibits the enzyme assayed in the presence of Mg^{2+} (Taunton *et al.,* 1969; Birnbaumer *et al.,* 1969; Drummond and Duncan, 1970; Bradham *et al.,* 1970; Perkins and Moore, 1971; Burke, 1970; Hepp *et al.,* 1970; Severson *et al.,* 1972; Johnson and Sutherland, 1973; Steer and Levitzki, 1975; Steer *et al.,* 1976; Finn *et al.,* 1975; Stolc, 1977). In general, the concentration of Ca^{2+} that produces maximal inhibition is on the order of 10^{-3} M in the presence of millimolar Mg^{2+}. The brain enzyme, for example, is maximally inhibited by 1 mM Ca^{2+} at 5 mM Mg^{2+} (see below). The inhibition results in a decrease in V_{max} with little or no effect on the apparent K_m for ATP. There is evidence that Ca^{2+} competes with Mg^{2+} (Drummond and Duncan, 1970; Steer and Levitzki, 1975). It has been proposed that Ca^{2+} inhibition of adenylate cyclase may be important in the intracellular regulation of the enzyme (Steer and Levitzki, 1975; Steer *et al.,* 1976; Stolc, 1977).

When the brain enzyme is assayed with Mn^{2+} rather than Mg^{2+}, the inhibition by Ca^{2+} is greatly reduced. Particulate brain adenylate cyclase, for example, is inhibited by 35 and 80% when assayed with 2 mM Mn^{2+} and 5 mM Mg^{2+}, respectively (L. S. Bradham, unpublished data). Johnson and Sutherland (1973) obtained similar results with detergent-dispersed cyclase from rat brain. They found that the enzyme was almost completely inhibited by 1 mM Ca^{2+} in the presence of 8 mM Mg^{2+}, but there was no effect with 8 mM Mn^{2+}. At a higher concentration of Ca^{2+} (10 mM), however, some inhibition (40%) was observed. Perhaps, as suggested by Steer and Levitzki (1975), Mn^{2+} and Ca^{2+} may compete for the same allosteric binding site on the enzyme.

II. CALMODULIN-DEPENDENT ADENYLATE CYCLASE

A. Introduction

It is clear, from the preceeding discussion, that the activity of adenylate cyclase is affected by a number of divalent cations. Although inhibited by millimolar Ca^{2+}, there is considerable evidence that the enzyme isolated from brain and from cultured cells derived from this organ requires micromolar Ca^{2+} for full activity. As described in the following sections, the effect of Ca^{2+} on brain adenylate cyclase is mediated through calmodulin.

B. Requirement of Brain Adenylate Cyclase for Ca^{2+}

1. Reversible Inhibition of Brain Adenylate Cyclase by EGTA

The initial observation that brain adenylate cyclase requires divalent cations other than Mg^{2+} was made by Bradham *et al.* (1970) using a membrane fraction prepared from beef cerebral cortex. Addition of the chelating agent ethylene glycol-bis(β-aminoethyl ether) N,N'-tetraacetic acid (EGTA) to an assay mixture inhibited the enzyme. In the presence of 5 m*M* Mg^{2+}, the enzyme activity was decreased 60% by 25 μM EGTA. The high concentration of Mg^{2+} precluded the possibility that its chelation by EGTA accounted for the enzyme inhibition. Calcium ion, equimolar to EGTA, completely reversed the inhibition. Strontium ion was almost as effective as Ca^{2+}; Ba^{2+} was ineffective. A typical assay mixture contained approximately 30 μM Ca^{2+} as a contaminant. This value correlated well with the concentration of EGTA (25 to 50 μM) required for maximal inhibition of the enzyme.

The inhibition by EGTA was noncompetitive with respect to the substrate MgATP (Bradham, 1972). The chelator decreased the V_{max} but did not significantly affect the apparent K_m. Calcium ion, at a concentration well below that of EGTA, effectively reversed the inhibition. EGTA had no effect on either the recovery of added cyclic AMP or on the formation of inorganic phosphate from ATP, suggesting that the inhibition did not result from an effect of EGTA on cyclic 3′,5′-nucleotide phosphodiesterase or on ATPase contaminating the enzyme preparation.

Similar results were subsequently obtained by Johnson and Sutherland (1973), Perkins (1973), MacDonald (1975), and von Hungen and Roberts (1973). The last two groups reported virtually identical results using particulate preparations from beef and rat brain, respectively. Adenylate cyclase was partially inhibited by 25 to 50 μM EGTA, and the inhibition was reversed by equimolar Ca^{2+}. At higher concentrations of Ca^{2+}, the enzyme was inhibited.

Johnson and Sutherland (1973) investigated the effect of chelating agents on the activity of detergent-dispersed adenylate cyclase prepared from rat cerebellar membranes. Maximal inhibition of basal activity (more than 90%) was obtained in the presence of 100 μM EGTA and 8 m*M* Mg^{2+}. Little inhibition was seen when Mg^{2+} was replaced by Mn^{2+}. The effect of several cations on reversing the inhibition by EGTA was examined; Ca^{2+}, Mn^{2+}, Ni^{2+}, Sr^{2+}, Zn^{2+}, Fe^{2+}, and Cu^{2+} were all effective to various extents.

The degree of inhibition of adenylate cyclase by EGTA was dependent upon the concentration of enzyme protein (Johnson and Sutherland, 1973). At a low concentration of the chelator, inhibition was reversed by increasing the concentration of enzyme protein, suggesting that EGTA was chelating some enzyme-bound metal ions. An attempt was made to remove the ions by preincubation of the enzyme with EGTA and passage of the mixture through a column of Sephadex G-25. The eluted enzyme retained its original specific activity and was still inhibited 50% by 30 μM EGTA, indicating that the metal ions required for activity were not removed by this treatment.

2. *Identity of Metal Ions Required for Brain Adenylate Cyclase Activity*

The results discussed above suggest that a cation other than Mg^{2+} is required for maximal activity of brain adenylate cyclase. Presumably the ion is Ca^{2+}, since it effectively reverses the inhibition by EGTA. EGTA forms stable complexes with a number of divalent cations. The stability constant of Mg-EGTA is $10^{5.4}$, whereas that of Ca-EGTA is $10^{10.9}$ (Holloway and Reilley, 1960), a difference of five orders of magnitude. Assuming that this difference is maintained under the conditions of adenylate cyclase assay, the chelation of Ca^{2+} by EGTA would account for the inhibition of the enzyme.

Other cations form even more stable complexes with EGTA (Holloway and Reilley, 1960). The stability constant of Mn-EGTA is $10^{12.3}$, an affinity some 25 times greater than that of Ca^{2+} for EGTA. The inhibition of brain adenylate cyclase by EGTA is completely reversed by equimolar Mn^{2+} (Johnson and Sutherland, 1973; MacDonald, 1975; Neer, 1979). Neer (1979) has suggested that the physiological cation is Mn^{2+} rather than Ca^{2+}. This conclusion is based upon an observation that Mn^{2+} is more effective than Ca^{2+} in reversing the inhibition by EGTA.

We have performed similar experiments and have found no difference in the effectiveness of Ca^{2+} and Mn^{2+} in reversing EGTA inhibition, provided that the concentration of divalent cation added does not exceed that of the chelating agent (L. S. Bradham, unpublished data). Neer used metal ions in excess of EGTA to reverse inhibition. An excess of Ca^{2+} inhibits adenylate cyclase, whereas Mn^{2+} stimulates the enzyme; this could account for the apparent discrepancy. Since EGTA lacks absolute specificity as a chelating agent, results of this type cannot be used as the sole criterion for establishing the identity of the metal ion required by adenylate cyclase. The fact that its effect is mediated through calmodulin, however, strongly suggests that the metal ion is Ca^{2+}.

C. Requirement of Brain Adenylate Cyclase for Calmodulin

1. Demonstration of the Requirement for Calmodulin

The inhibitory effect of EGTA on adenylate cyclase may be explained by the chelation of Ca^{2+} and the dissociation of calmodulin from the enzyme. This was demonstrated by Brostrom *et al.* (1975) and Cheung *et al.* (1975) using essentially identical experimental procedures. Detergent-solubilized adenylate cyclase prepared from porcine brain (Brostrom *et al.*, 1975) or beef and rat brain (Cheung *et al.*, 1975) was chromatographed on an anionic exchange column to resolve the enzyme from calmodulin; the latter, being acidic, was retained by the column whereas adenylate cyclase emerged unretarded. The eluted enzyme had little activity unless assayed in the presence of Ca^{2+} and calmodulin.

2. Reversible Binding of Calmodulin to Adenylate Cyclase

The reversible binding of calmodulin to the apoenzyme of adenylate cyclase in response to Ca^{2+} was demonstrated by Lynch *et al.* (1976) by gel filtration (Sephadex G-200) using a detergent-solubilzed adenylate cyclase. The enzyme, with a molecular weight of 300,000 (Neer, 1978a), emerged from the column in the void volume and was clearly separated from calmodulin, which has a molecular weight of 16,700. When the experiment was performed in the presence of EGTA, no calmodulin was associated with the enzyme, and maximum adenylate cyclase activity was dependent upon exogenous calmodulin. In contrast, when the experiment was repeated in the presence of Ca^{2+}, some of the calmodulin was associated with the enzyme which had comparable activity in the presence or absence of added calmodulin. These experiments indicated that the apoenzyme of adenylate cyclase formed a complex with calmodulin in the presence of Ca^{2+} to produce an active holoenzyme.

Evidence showing that formation of the holoenzyme was rapid and reversible was provided by kinetic experiments using adenylate cyclase deprived of calmodulin (Cheung *et al.*, 1975) or Ca^{2+} (Lynch *et al.*, 1976). Addition of calmodulin to the apoenzyme assayed in the presence of Ca^{2+} produced an immediate increase in the rate of cyclic AMP formation. A similar result was obtained when Ca^{2+} was added to Ca^{2+}-deficient adenylate cyclase supplemented with calmodulin. On the other hand, the addition of EGTA to the reaction mixtures containing the holoenezyme resulted in an immediate return of the rate of cyclic AMP formation to the basal level. Since the active species of the enzyme is the calmodulin–adenylate cyclase complex, the change in the rate of cyclic AMP synthesis reflects the rapid and reversible formation of the enzyme–calmodulin complex.

Although experiments performed with detergent-solubilized adenylate cyclase are informative, it is possible that treatment with detergent modifies the properties of the enzyme. Lynch *et al.* (1977) showed, however, that membrane-bound adenylate cyclase also requires calmodulin for activity. This was demonstrated by repeated washing of a particulate rat brain adenylate cyclase with buffers containing EGTA. The washing removed some of the endogenous calmodulin, rendering the enzyme responsive to exogenous calmodulin. A complete removal of endogenous calmodulin was accomplished by including a low concentration of detergent in the washing medium.

The removal of endogenous calmodulin produced a decrease in the V_{max} with no appreciable change in the apparent K_m for ATP. Reconstitution of the holoenzyme by exogenous calmodulin and Ca^{2+} resulted in an increase in V_{max}. The reconstituted enzyme was stimulated by NaF and by 5′-guanylyl imidodiphosphate [Gpp(NH)p], and the effect of these modulators was additive to that of calmodulin.

3. *Evidence for the Existence of Two Forms of Adenylate Cyclase*

MacDonald (1975) studied the effect of F^- on brain adenylate cyclase and noted that the unstimulated enzyme was markedly inhibited (84%) by EGTA, whereas the F^--stimulated enzyme was only partially inhibited (60%). The F^--stimulated activity, moreover, appeared to be comparable regardless of the presence or absence of EGTA. Similar results were obtained by Lynch *et al.* (1977) using calmodulin-depleted particulate adenylate cyclase from brain. They found that the increased activity produced by F^- or by Gpp(NH)p was approximately the same in the presence or absence of exogenous calmodulin. These data suggest that stimulation of adenylate cyclase by calmodulin is independent of that by F^- and Gpp(NH)p.

Using particulate adenylate cyclase from rat brain, Brostrom *et al.* (1977, 1978) found that 80% of the basal adenylate cyclase was calmodulin dependent. The dependent and independent forms displayed differential thermal stability, the independent form could be inactivated by preincubation of the enzyme at 37°C. Following heat inactivation and removal of endogenous calmodulin, the enzyme activity was almost totally dependent upon exogenous calmodulin which stimulated the enzyme as much as 30-fold.

The dependent and independent forms of adenylate cyclase were both activated by F^- and guanine nucleotides. Activation of the independent form was greater (four- to sixfold) than that of the dependent form (50 to 100%) so that, in the activated state, each form constituted approximately half of the total enzyme activity. Activation of the independent form by Gpp(NH)p was greater than that by GTP, whereas the two nucleotides had essentially the same potency on the dependent form (Brostrom *et al.*, 1978). Pretreatment of the enzyme with F^- in the absence of calmodulin and Ca^{2+} resulted in the activation of only the independent form. The dependent form could be activated by pretreatment with F^-, calmodulin, and Ca^{2+}, and the increased activity was totally dependent upon calmodulin (Brostrom *et al.*, 1977).

The dependent form of the enzyme was inhibited by chlorpromazine, but the independent form was not affected. Chlorpromazine and other phenothiazine derivatives also inhibit calmodulin-dependent cyclic nucleotide phosphodiesterase (Levin and Weiss, 1976; Wolff and Brostrom, 1976). Inhibition is competitive with respect to calmodulin and has been attributed to binding of the drug to calmodulin (Levin and Weiss, 1977; Weiss and Wallace, this volume, Chapter 16).

Westcott *et al.* (1979) resolved brain adenylate cyclase into calmodulin-dependent and -independent forms using a calmodulin affinity column. After preliminary removal of endogenous calmodulin, approximately 20% of detergent-solubilized brain adenylate cyclase was adsorbed to the affinity column. The remaining activity emerged in the void volume of the column and did not require exogenous calmodulin for activity. The dependent form was eluted from the column by EGTA, and its activity was stimulated fourfold by exogenous calmodulin and Ca^{2+}. In agreement with the results of Wallace *et al.* (1978), this form of the enzyme was inhibited by a heat-labile calmodulin-binding protein (CaM-BP_{80}).

4. Demonstration of Calmodulin-Dependent Adenylate Cyclase in Intact Cells

The experiments cited above demonstrate the physical reality of two forms of brain adenylate cyclase. Brain tissue, however, is heterogeneous, and the two forms of the enzyme may originate from different cell types. Brostrom *et al.* (1976, 1979) studied the properties of adenylate cyclase in C-6 glioma cells, which are known to have many of the characteristics of normal glial cells. The glioma enzyme contained both calmodulin-dependent and independent forms (Brostrom *et al.*, 1976). Norepinephrine stimulated the enzyme, and calmodulin did not affect the response or the sensitivity of the enzyme to the catecholamine.

For demonstration of the dependency of glioma adenylate cyclase on

Ca^{2+}, cells were washed with buffered saline containing 1 m*M* EGTA. This treatment reduced the Ca^{2+} content by approximately 80%. Norepinephrine caused a sixfold increase in the rate of cyclic AMP formation in these cells. This rate, however, was more than doubled following restoration of intracellular Ca^{2+}. Conversely, the rate of cyclic AMP formation diminished sharply when Ca^{2+}-restored cells were transferred to a medium containing EGTA. These results implied that at least part of the adenylate cyclase activity in the intact cells was Ca^{2+} dependent.

A number of β-adrenergic agonists stimulated adenylate cyclase of glioma cells, and the activity of the Ca^{2+}-dependent enzyme was inhibited by β-adrenergic antagonists, such as propanolol, alprenolol, and sotalol. Calcium ion depletion had no effect on either the number of β-adrenergic receptors or on the binding of agonists to the receptor (Brostrom *et al.*, 1979).

Verapamil, which inhibits the cellular influx of Ca^{2+} (Singh *et al.*, 1978), blocked the stimulation of Ca^{2+}-dependent adenylate cyclase activity when added to the cells prior to, but not after, restoration of intracellular Ca^{2+}. This suggested that the cyclase activity was regulated by Ca^{2+} inside the cell.

Although direct confirmation could not be obtained, various experiments suggested that the effect of Ca^{2+} on the accumulation of cyclic AMP in C-6 glioma cells was mediated by calmodulin. For example, trifluoperazine inhibited the Ca^{2+}-dependent adenylate cyclase in Ca^{2+}-restored cells. This phenothiazine, like chlorpromazine, binds to calmodulin and renders it biologically inactive (Levin and Weiss, 1977).

D. Mechanism of Calmodulin Stimulation

Based on a wide range of experimental evidence, the sequence of events associated with the stimulation of adenylate cyclase by calmodulin may be depicted by the following equations:

$$Ca^{2+} + \text{CaM (inactive)} \rightleftharpoons Ca^{2+} - \text{CaM* (active)} \quad (1)$$

$$Ca^{2+} - \text{CaM*} + \text{E (inactive)} \rightleftharpoons Ca^{2+} - \text{CaM*} - \text{E* (active)} \quad (2)$$

The binding of Ca^{2+} to calmodulin (CaM) induces a change in the conformation of the protein facilitating its binding to the apoenzyme (E) of adenylate cyclase. Although there is no direct evidence of a conformational change in the enzyme as a result of binding calmodulin, indirect evidence is provided by the finding that the holoenzyme is thermally more stable. Calmodulin also complexes with Mn^{2+}, but the affinity of calmodulin for Mn^{2+} is tenfold less (Lin *et al.*, 1974; Wolff *et al.*, 1977). The cellular content of Mn^{2+}, moreover, is lower than Ca^{2+}. It is likely, therefore, that Ca^{2+} is the physiological cation whose cellular flux, in response to stimuli, regulates the activity of adenylate cyclase. A similar mechanism is believed to regulate the activity of Ca^{2+}-dependent phosphodiesterase (Lin and Cheung, this volume, Chapter 5).

E. Functional Considerations

The functional relationship between calmodulin-dependent adenylate cyclase and calmodulin-dependent phosphodiesterase is unclear. The influx of Ca^{2+} through the plasma membrane or the release of membrane-bound Ca^{2+} in response to a stimulus may activate adenylate cyclase, leading to an increase in the intracellular level of cyclic AMP. The increased concentration of Ca^{2+} in the cytosol may then activate the calmodulin-dependent phosphodiesterase, restoring the concentration of cyclic AMP to its steady state level. The sequential activation of the two enzymes would produce a transient increase in cyclic AMP.

Alternatively, sequential stimulation of adenylate cyclase and phosphodiesterase may occur by translocation of calmodulin from the membrane to the cytosol. Calmodulin is associated with synaptic membranes isolated from brain (Gnegy *et al.*, 1977a) and is released under certain conditions by some neurohormones (Revuelta *et al.*, 1976). In membrane fractions isolated from striatum or adrenal medulla, calmodulin was released upon phosphorylation of certain membrane proteins by a cyclic AMP-dependent protein kinase (Gnegy *et al.*, 1976a,b). Neither calmodulin nor adenylate cyclase was phosphorylated under these conditions. The release of calmodulin from the membrane preparation was accompanied by a decrease in the sensitivity of adenylate cyclase to dopamine (Costa *et al.*, 1977; Gnegy *et al.*, 1976b). Conversely, *in vivo* experiments showed that an increase in the membrane content of calmodulin resulted in enhanced sensitivity of the enzyme to the neurohormone (Gnegy *et al.*, 1977b,c). These workers further showed that following chronic treatment of rats with haloperidol the calmodulin content in striatal membranes increased as did the sensitivity of the striatal cyclase to dopamine. The biochemical changes were accompanied by behavioral symptoms typical of withdrawal from the drug.

On the basis of these experiments, Costa *et al.* (1977) proposed that the intracellular concentration of cyclic AMP may be self-regulated. They visualize that an increase in cyclic AMP following hormonal stimulation causes the phosphorylation of certain membrane proteins leading to the release of calmodulin. Calmodulin released into the cytosol then activates Ca^{2+}-dependent phosphodiesterase returning the concentration of cyclic AMP to the basal level. The decrease of calmodulin in the membrane diminishes the sensitivity of adenylate cyclase to further stimulation by the hormone (Hanbauer and Costa, this volume, Chapter 12). One difficulty with this hypothesis is that the concentration of calmodulin in the cytosol is usually more than sufficient to fully activate phosphodiesterase (Cheung, 1971; Smoake *et al.*, 1974).

F. Ca^{2+}-Dependent Adenylate Cyclase in Other Tissues

In addition to brain, adenylate cyclase in other tissues may also be Ca^{2+} dependent. An example is the Ca^{2+} requirement for the stimulation by ACTH of adenylate cyclase from adipocytes (Birnbaumer and Rodbell, 1969; Bar and Hechter, 1969a) or adrenal cortex (Bar and Hechter, 1969b; Lefkowtiz *et al.*, 1970; Kelly and Koritz, 1971; Finn *et al.*, 1975). Adenylate cyclase of both tissues was inhibited by EGTA when measured in the presence of ACTH. Inhibition of the adrenal enzyme was reversed by the inclusion of Ca^{2+} equimolar to EGTA (Bar and Hechter, 1969a; Lefkowitz *et al.*, 1970). There was no inhibition of the adipocyte enzyme by EGTA when assayed in the presence of epinephrine and glucagon (Birnbaumer and Rodbell, 1969). Indeed, the stimulatory effect of these hormones was significantly increased by the presence of EGTA, which had no effect on the basal or F^--stimulated activity.

A biphasic response to Ca^{2+} was observed with adenylate cyclase prepared from isolated bladder epithelial cells (Bockaert *et al.*, 1972). After treatment with EGTA and extensive washing of the particles to remove the chelator, Ca^{2+} was required for the stimulation of the enzyme by oxytocin. A maximal response was observed in the range of 10^{-6} to 10^{-5} M Ca^{2+}; at higher concentrations, the cation was inhibitory.

The activity of adenylate cyclase of renal medulla is also dependent upon Ca^{2+} (Campbell *et al.*, 1972). Both the basal- and vasopressin-stimulated activities were inhibited by EGTA, and the inhibition was reversed by equimolar Ca^{2+}. Higher Ca^{2+} concentrations inhibited the enzyme.

Finally the adenylate cyclases of parotid gland (Franks *et al.*, 1974) and of liver (Hepp *et al.*, 1970) have been found to be Ca^{2+} dependent. The parotid enzyme was inhibited by EGTA when it was assayed alone or in the presence of NaF or isoproterenol. Inhibition by the chelating agent was reversed by Ca^{2+}. The response of the liver enzyme to glucagon was not affected by EGTA, but the stimulation by NaF was partially inhibited.

One possible explanation of these results is that Ca^{2+} is required for the binding of hormones to their respective receptors. The adipocyte membrane has receptors for a number of hormones, but only the stimulatory effect of ACTH is dependent on Ca^{2+}. Calcium ion may be required for the binding of ACTH to its receptor but not for the binding of epinephrine and glucagon. The data of Lefkowitz *et al.* (1970) and Campbell *et al.* (1972) demonstrated, however, that Ca^{2+} is not required for the binding of ACTH and vasopressin to adrenal or renal membranes.

Gnegy *et al.* (1976a) have shown that calmodulin is bound to membranes of adrenal medulla and that transsynaptic stimulation of the organ results in a release of calmodulin into the cytosol. By analogy with the system in brain, the membrane-bound calmodulin may be required for the activity of the adrenal cyclase. LeDonne and Coffee found that the adrenal enzyme is inhibited by trifluoperazine, suggesting that it might be calmodulin dependent (C. J. Coffee, personal communication).

It is also conceivable that adenylate cyclase in certain tissues contains an integral calmodulin that is not dissociated by EGTA, as in the case of phosphorylase kinase (Cohen, this volume, Chapter 9). Such an enzyme would respond to Ca^{2+} but not necessarily to exogenous calmodulin. This possibility deserves exploration in the future.

III. CONCLUDING REMARKS

The regulation of adenylate cyclase is complex, involving a number of membrane components. In addition to hormone receptors, there are other proteins that are important. One is calmodulin which confers Ca^{2+} sensitivity upon the brain enzyme; other proteins may be involved in the activation by F^- and by guanine nucleotides (Bradham, 1977; Ross and Gilman, 1977; Sahyoun *et al.*, 1978; Hebdon *et al.*, 1978; Ross *et al.*, 1978a, 1978b; Howlett *et al.*, 1979; Pfeuffer and Helmreich, 1975; Pfeuffer, 1977; Cassel and Pfeuffer, 1978; Johnson *et al.*, 1978; Renart *et al.*, 1979). The size of the entire multimeric complex required for full adenylate cyclase activity in hepatic membranes appears to be large (Schlegel *et al.*, 1979), and the mechanism by which these components interact to express adenylate cyclase is, at present, far from being fully understood.

Even with brain adenylate cyclase, the mechanism of stimulation by calmodulin may be more complex than that described here. Toscano *et al.* (1979) found that a calmodulin-independent form of brain adenylate cyclase was converted to a dependent form when the enzyme was reconstituted with another protein factor. Moss and Vaughn (1977) reported that the activation of brain adenylate cyclase by cholera toxin required both calmodulin and GTP. It is too early to assess the significance of these findings, but they do emphasize the complexity of the enzyme.

In conclusion, it appears likely that calmodulin and Ca^{2+} play a central role in the regulation of brain adenylate cyclase. Whether they are of equal importance in controlling the activity of adenylate cyclase in other tissues remains to be determined.

ACKNOWLEDGMENTS

The authors are indebted to Drs. M. A. Brostrom and D. R. Storm for providing data prior to their publication. The work from our laboratories were supported by USPHS Grants NS-12122 and NS-08509 and by ALSAC.

REFERENCES

Bar, H. P., and Hechter, O. (1969a). Adenyl cyclase and hormone action. I. Effects of adrenocorticotropic hormone, glucagon, and epinephrine on the plasma membrane of rat fat cells. *Proc. Natl. Acad. Sci. U.S.A.* **66,** 350–356.

Bar, H. P., and Hechter, O. (1969b). Adenyl cyclase and hormone action. III. Calcium requirement for ACTH stimulation of adenyl cyclase. *Biochem. Biophys. Res. Commun.* **35,** 681–686.

Birnbaumer, L. (1973). Hormone-sensitive adenylyl cyclases. Useful models for studying hormone receptor functions in cell-free systems. *Biochim. Biophys. Acta* **300,** 129–138.

Birnbaumer, L., and Rodbell, M. (1969). Adenyl cyclase in fat cells. II. Hormone receptors. *J. Biol. Chem.* **244,** 3477–3482.

Birnbaumer, L., Pohl, S. L., and Rodbell, M. (1969). Adenyl cyclase in fat cells. I. Properties and the effects of adrenocorticotropin and fluoride. *J. Biol. Chem.* **244,** 3468–3476.

Birnbaumer, L., Pohl, S. L., Krans, H. M. J., and Rodbell, M. (1970). The actions of hormones on the adenyl cyclase system. *In* "Role of Cyclic AMP in Cell Function" (P. Greengard and E. Costa, eds.), pp. 185–208. Raven, New York.

Bockaert, J., Roy, C., and Jard, S. (1972). Oxytocin-sensitive adenylate cyclase in frog bladder epithelial cells. Role of calcium, nucleotides, and other factors in hormone stimulation. *J. Biol. Chem.* **247,** 7073–7081.

Bradham, L. S. (1972). Comparison of the effects of Ca^{2+} and Mg^{2+} on the adenyl cyclase of beef brain. *Biochim. Biophys. Acta* **276,** 434–443.

Bradham, L. S. (1977). Fluoride activation of rat brain adenylate cyclase: The requirement for a protein co-factor. *J. Cyclic Nucleotide Res.* **3,** 119–128.

Bradham, L. S., Holt, D. A., and Sims, M. (1970). The effect of Ca^{2+} on the adenyl cyclase of calf brain. *Biochim. Biophys. Acta* **201,** 250–260.

Brostrom, C. O., Huang, Y., Breckenridge, B. M., and Wolff, D. J. (1975). Identification of a calcium-binding protein as a calcium-dependent regulator of brain adenylate cyclase. *Proc. Natl. Acad. Sci. U.S.A.* **72,** 64–68.

Brostrom, C. O., Brostrom, M. A., and Wolff, D. J. (1977). Calcium-dependent adenylate cyclase from rat cerebral cortex. Reversible activation by sodium fluoride. *J. Biol. Chem.* **252,** 5677–5685.

Brostrom, M. A., Brostrom, C. O., Breckenridge, B. M., and Wolff, D. J. (1976). Regulation of adenylate cyclase from glial tumor cells by calcium and calcium-binding protein. *J. Biol. Chem.* **251,** 4744–4750.

Brostrom, M. A., Brostrom, C. O., and Wolff, D. J. (1978). Calcium-dependent adenylate cyclase from rat cerebral cortex: Activation by guanine nucleotides. *Arch. Biochem. Biophys.* **191,** 341–350.

Brostrom, M. A., Brostrom, C. O., and Wolff, D. J. (1979). Calcium dependence of hormone-stimulated cAMP accumulation in intact glial tumor cells. *J. Biol. Chem.* **254,** 7548–7557.

Burke, G. (1970). Effects of cations and oubain on thyroid adenyl cyclase. *Biochim. Biophys. Acta* **220,** 30–41.

Campbell, B. J., Woodward, G., and Borberg, V. (1972). Calcium-mediated interactions between the antidiuretic hormone and renal plasma membranes. *J. Biol. Chem.* **247,** 6167–6173.

Cassel, D., and Pfeuffer, T. (1978). Mechanism of cholera toxin action: Covalent modification of the guanyl nucleotide-binding protein of the adenylate cyclase system. *Proc. Natl. Acad. Sci. U.S.A.* **75,** 2669–2673.

Cheung, W. Y. (1971). Cyclic 3′,5′-nucleotide phosphodiesterase. Evidence for and properties of a protein activator. *J. Biol. Chem.* **246,** 2859–2869.

Cheung, W. Y., Bradham, L. S., Lynch, T. J., Lin, Y. M., and Tallant, E. A. (1975). Protein activator of cyclic 3′,5′-nucleotide phosphodiesterase of bovine or rat brain also activates its adenylate cyclase. *Biochem. Biophys. Res. Commun.* **66,** 1055–1062.

Costa, E., Gnegy, M., Revuelta, A., and Uzunov, P. (1977). Regulation of dopamine-dependent adenylate cyclase by a Ca^{2+}-binding protein stored in synaptic membranes. *Biochem. Physchopharmacol.* **16,** 403–408.

Cuatrecasas, P., Hollenberg, M. D., Chang, K., and Bennett, V. (1975). Hormone receptor complexes and their modulation of membrane function. *Recent Prog. Horm. Res.* **31,** 37–94.

de Haen, C. (1974). Adenylate cyclase. A new kinetic analysis of the effects of hormones and fluoride ion. *J. Biol. Chem.* **249,** 2756–2762.

Drummond, G. I., and Duncan, L. (1970). Adenyl cyclase in cardiac tissue. *J. Biol. Chem.* **245,** 976–983.

Drummond, G. I., Severson, D. L., and Duncan, L. (1971). Adenyl cyclase. Kinetic properties and nature of fluoride and hormone stimulation. *J. Biol. Chem.* **246,** 4166–4173.

Finn, F. M., Montibeller, J. A., Ushijima, Y., and Hofmann, K. (1975). Adenylate cyclase system of bovine adrenal plasma membranes. *J. Biol. Chem.* **250,** 1186–1192.

Flawia, M. M., and Torres, H. N. (1972). Adenylate cyclase activity in *Neurospora crassa.* II. Kinetics. *J. Biol. Chem.* **247,** 6880–6883.

Franks, D. J., Perrin, L. S., and Malamud, D. (1974). Calcium ion: A modulator of parotid adenylate cyclase activity. *FEBS Lett.* **42,** 267–270.

Garbers, D. L., and Johnson, R. A. (1975). Metal and metal-ATP interactions with brain and cardiac adenylate cyclases. *J. Biol. Chem.* **250,** 8449–8456.

Gill, D. M. (1978). Mechanism of action of cholera toxin. *Adv. Cyclic Nucleotide Res.* **8,** 85–118.

Gnegy, M. E., Costa, E., and Uzunov, P. (1976a). Regulation of transsynaptically elicted increases of 3′,5′-cyclic AMP by endogenous phosphodiesterase activator. *Proc. Natl. Acad. Sci. U.S.A.* **73,** 352–355.

Gnegy, M. E., Uzunov, P., and Costa, E. (1976b). Regulation of dopamine stimulation of striatal adenylate cyclase by an endogenous Ca^{++}-binding protein. *Proc. Natl. Acad. Sci., U.S.A.* **73,** 3887–3890.

Gnegy, M. E., Nathanson, J. A., and Uzunov, P. (1977a). Release of the phosphodiesterase activator by cyclic AMP-dependent ATP-protein phosphotransferase from subcellular fractions of rat brain. *Biochim. Biophys. Acta* **497,** 75–85.

Gnegy, M. E., Uzunov, P., and Costa, E. (1977b). Participation of an endogenous Ca^{++}-binding protein activator in the development of drug-induced supersensitivity of striatal dopamine receptors. *J. Pharmacol. Exp. Ther.* **202,** 558–564.

Gnegy, M. E., Lucchelli, A., and Costa, E. (1977c). Correlation between drug induced supersensitivity of dopamine dependent striatal mechanisms and the increase in striatal content of Ca^{2+} regulated protein activator of cAMP phosphodiesterase. *Naunyn-Schmiedeberg's Arch. Pharmacol.* **301,** 121–127.

Hebdon, M., Le Vine, H., III, Sahyoun, N., Schmitges, C. J., and Cuatrecasas, P. (1978). Properties of the interaction of fluoride- and guanylyl-5′-imidodiphosphate-regulatory proteins with adenylate cyclase. *Proc. Natl. Acad. Sci. U.S.A.* **75,** 3693–3697.

Hepp, E. D., Edel, R., and Wieland, O. (1970). Hormone action on liver adenyl cyclase activity. *Eur. J. Biochem.* **17,** 171–177.

Herman, C. A., Zahlen, W. L., Doak, G. A., and Campbell, B. J. (1976). Bull sperm adenylate cyclase: Localization and partial characterization. *Arch. Biochem. Biophys.* **177,** 622–629.

Holloway, J. H., and Reilley, C. N. (1960). Metal chelate stability constants of aminopolycarboxylate ligands. *Anal. Chem.* **32,** 249–256.

Howlett, A. C., Sternweiss, P. C., Macik, B. A., van Arsdale, P. M., and Gilman, A. G. (1979). Reconstitution of catecholamine-sensitive adenylate cyclase. Association of a regulatory component of the enzyme with membranes containing the catalytic protein and β-adrenergic receptors. *J. Biol. Chem.* **254,** 2287–2295.

Johnson, G. L., Kaslow, H. R., and Bourne, H. R. (1978). Genetic evidence that cholera toxin substrates are regulatory components of adenylate cyclase. *J. Biol. Chem.* **253,** 7120–7123.

Johnson, R. A., and Sutherland, E. W. (1973). Detergent-dispersed adenylate cyclase from rat brain. Effects of fluoride, cations, and chelators. *J. Biol. Chem.* **248,** 5114–5121.

Kebabian, J. W. (1978). Dopamine-sensitive adenylyl cyclase: A receptor mechanism for dopamine. *Adv. Biochem. Psychopharmaco.* **19,** 131–154.

Kelly, L. A., and Koritz, S. B. (1971). Bovine adrenal cortex adenyl cyclase and its stimulation by adrenocorticotropic hormone and NaF. *Biochim. Biophys. Acta* **237,** 141–155.

Lefkowitz, R. J., and Williams, L. T. (1978). Molecular mechanisms of activation and desentitization of adenylate cyclase coupled β-adrenergic receptors. *Adv. Cyclic Nucleotide Res.* **9,** 1–17.

Lefkowitz, R. J., Roth, J., and Pastan, I. (1970). Effects of calcium on ACTH stimulation of the adrenal: Separation of hormone binding from adenyl cyclase activation. *Nature (London)* **228,** 864–866.

Levin, R. M., and Weiss, B. (1976). Mechanism by which psychotropic drugs inhibit adenosine cyclic 3′,5′-monophosphate phosphodiesterase of brain. *Mol. Pharmacol.* **12,** 581–589.

Levin, R. M., and Weiss, B. (1977). Binding of trifluoperazine to the calcium-dependent activator of cyclic nucleotide phosphodiesterase. *Mol. Pharmacol.* **13,** 690–697.

Levitzki, A. (1978). The mode of coupling of adenylate cyclase to hormone receptors and its modulation of GTP. *Biochem. Pharmacol.* **27,** 2081–2088.

Lin, Y. M., Liu, Y. P., and Cheung, W. Y. (1974). Cyclic 3′,5′-nucleotide phosphodiesterase. Purification, characterization, and active form of the protein activator from bovine brain. *J. Biol. Chem.* **249,** 4943–4954.

Londos, C., and Preston, M. S. (1977). Activation of the hepatic adenylate cyclase system by divalent cations. A reassessment. *J. Biol. Chem.* **252,** 5957–5961.

Lynch, T. J., Tallant, E. A., and Cheung, W. Y. (1976). Ca^{++}-dependent formation of brain adenylate cyclase-protein activator complex. *Biochem. Biophys. Res. Commun.* **68,** 616–625.

Lynch, T. J., Tallant, E. A., and Cheung, W. Y. (1977). Rat brain adenylate cyclase. Further studies on its stimulation by Ca^{++}-binding protein. *Arch. Biochem. Biophys.* **182,** 124–133.

MacDonald, I. A. (1975). Differentiation of fluoride stimulated and non-fluoride stimulated components of beef brain adenylate cyclase by calcium ions, ethyleneglycol-bis(β-aminoethyl ether) *N,N'*-tetraacetic acid and Triton X-100. *Biochim. Biophys. Acta* **397,** 244–253.

Maquire, M. E., Ross, E. M., and Gilman, A. G. (1977). β-Adrenergic receptor. Ligand binding properties and the interaction with adenylyl cyclase. *Adv. Cyclic Nucleotide Res.* **8,** 1–83.

Moss, J., and Vaughan, M. (1977). Choleragen activation of solubilized adenylate cyclase: Requirement for GTP and protein activator for demonstration of enzymatic activity. *Proc. Natl. Acad. Sci. U.S.A.* **74,** 4396–4400.

Moss, J. and Vaughan, M. (1979). Activation of adenylate cyclase by choleragen. *Ann. Rev. Biochem.* **48,** 581–600.

Neer, E. J. (1978a). Size and detergent binding of adenylate cyclase from bovine cerebral cortex. *J. Biol. Chem.* **253,** 1498–1502.

Neer, E. J. (1978b). Physical and functional properties of adenylate cyclase from mature rat testis. *J. Biol. Chem.* **253,** 5808–5812.

Neer, E. J. (1979). Interaction of soluble brain adenylate cyclase with manganese. *J. Biol. Chem.* **254,** 2089–2096.

Perkins, J. P. (1973). Adenyl cyclase. *Adv. Cyclic Nucleotide Res.* **3,** 1–64.

Perkins, J. P., and Moore, M. (1971). Adenyl cyclase of rat cerebral cortex. *J. Biol. Chem.* **246,** 62–68.

Pfeuffer, T. (1977). GTP-binding proteins in membranes and the control of adenylate cyclase activity. *J. Biol. Chem.* **252,** 7224–7234.

Pfeuffer, T., and Helmreich, E. J. M. (1975). Activation of pigeon erythrocyte membrane adenylate cyclase by guanyl nucleotide analogues and separation of a nucleotide binding protein. *J. Biol. Chem.* **250,** 867–876.

Pilkis, S. J., and Johnson, R. A. (1974). Detergent dispersion of adenylate cyclase from partially purified rat liver plasma membranes. *Biochim. Biophys. Acta* **341,** 388–395.

Pohl, S. L., Birnbaumer, L., and Rodbell, M. (1971). The glucagon-sensitive adenyl cyclase system in plasma membranes of rat liver. Properties. *J. Biol. Chem.* **246,** 1849–1856.

Renart, M. F., Ayanoglu, G., Mansour, J. M., and Mansour, T. E. (1979). Fluoride and guanosine nucleotide activated adenylate cyclase from *Fasciola hepatica:* Reconstitution after inactivation. *Biochem. Biophys. Res. Commun.* **89,** 1146–1153.

Revuelta, A., Uzunov, P., and Costa, E. (1976). Release of phosphodiesterase activator from particulate fractions of cerebellum and striatum by putative neurotransmitters. *Neurochem. Res.* **1,** 217–227.

Rodbell, M., Lin, M. C., Salomon, Y., Londos, C., Harwood, J. P., Martin, B. R., Rendell, M., and Berman, M. (1975). Role of adenine and guanine nucleotides in the activity and response of adenylate cyclase systems to hormone: Evidence for multiple transition states. *Adv. Cyclic Nucleotide Res.* **5,** 3–29.

Ross, E. M., and Gilman, A. G. (1977). Reconstitution of catecholamine-sensitive adenylate cyclase activity: Interaction of solubilized components with receptor-replete membranes. *Proc. Natl. Acad. Sci. U.S.A.* **74,** 3715–3719.

Ross, E. M., Haga, T., Howlett, A. C., Schwang, J., Schleifer, L. S., and Gilman, A. G. (1978a). Hormone-sensitive adenylate cyclase: Resolution and reconstitution of some components necessary for regulation of the enzyme. *Adv. Cyclic Nucleotide Res.* **9,** 1–17.

Ross, E. M., Howlett, A. C., Ferguson, K. M., and Gilman, A. G. (1978b). Reconstitution of hormone-sensitive adenylate cyclase activity with resolved components of the enzyme. *J. Biol. Chem.* **253,** 6401–6412.

Sahyoun, N., Schmitges, C. J., Le Vine, H., III, and Cuatrecasas, P. (1978). Molecular resolution and reconstitution of the Gpp(NH)p and NaF sensitive adenylate cyclase system. *Life Sci.* **21,** 1857–1864.

Schlegel, W., Kempner, E. S., and Rodbell, M. (1979). Activation of adenylate cyclase in hepatic membranes involves interactions of the catalytic unit with multimeric complexes of regulatory proteins. *J. Biol. Chem.* **254,** 5168–5176.

Severson, D. L., Drummond, G. I., and Sulakhe, P. V. (1972). Adenylate cyclase in skeletal muscle. Kinetic properties and hormonal stimulation. *J. Biol. Chem.* **247,** 2949–2958.

Singh, B. N., Ellrodt, G., and Peter, C. T. (1978). Verapamil: A review of its pharmacological properties and therapeutic use. *Drugs* **15,** 169–197.

Smoake, J. A., Song, S.-Y., and Cheung, W. Y. (1974). Cyclic 3′,5′-nucleotide phosphodiesterase. Distribution and developmental changes of the enzyme and its protein activator in mammalian tissues and cells. *Biochim. Biophys. Acta* **341,** 402–411.

Steer, M. L., and Levitzki, A. (1975). The control of adenylate cyclase by calcium in turkey erythrocyte ghosts. *J. Biol. Chem.* **250,** 2080–2084.

Steer, M. L., Baldwin, C., and Levitzki, A. (1976). Preparation and characterization of hormone-sensitive, resealed erythrocyte ghosts. *J. Biol. Chem.* **251,** 4930–4935.

Stolc, V. (1977). Mechanism of regulation of adenylate cyclase activity in human polymorphonuclear leukocytes by calcium, guanosyl nucleotides, and positive effectors. *J. Biol. Chem.* **252,** 1901–1907.

Sutherland, E. W., Rall, T. W., and Maxwell, M. (1962). Adenyl cyclase. I. Distribution, preparation, and properties. *J. Biol. Chem.* **237,** 1220–1227.

Taunton, C. D., Roth, J., and Pastan, I. (1969). Studies on the adrenocorticotropic hormone-activated adenyl cyclase of a functional adrenal tumor. *J. Biol. Chem.* **244,** 247–253.

Toscano, W. A., Jr., Westcott, K. R., LaPorte, D. C., and Storm D. R. (1979). Evidence for a dissociable protein subunit required for calmodulin stimulation of brain adenylate cyclase. *Proc. Natl. Acad. Sci., U.S.A.* **76,** 5582–5586.

von Hungen, K., and Roberts, S. (1973). Catecholamine and Ca^{2+} activation of adenylate cyclase systems in synaptosomal fractions from rat cerebral cortex. *Nature (London) New Biol.* **242,** 58–60.

Wallace, R. W., Lynch, T. J., Tallant, E. A., and Cheung, W. Y. (1978). An endogenous inhibitor protein of brain adenylate cyclase and cyclic nucleotide phosphodiesterase. *Arch. Biochem. Biophys.* **187,** 328–334.

Westcott, K. R., LaPorte, D. C., and Storm, D. R. (1979). Resolution of adenylate cyclase sensitive and insensitive to Ca^{2+} and calcium-dependent regulatory protein (CDR) by CDR-Sepharose affinity chromatography. *Proc. Natl. Acad. Sci. U.S.A.* **76,** 204–208.

Wolff, D. J., and Brostrom, C. O. (1976). Calcium-dependent cyclic nucleotide phosphodiesterase from brain: Identification of phospholipids as calcium-independent activators. *Arch. Biochem. Biophys.* **173,** 720–731.

Wolff, D. J., Poirier, P. G., Brostrom, C. O., and Brostrom, M. A. (1977). Divalent cation binding properties of bovine brain Ca^{2+}-dependent regulator protein. *J. Biol. Chem.* **252,** 4108–4117.

Chapter 7

Calmodulin and Plasma Membrane Calcium Transport

FRANK F. VINCENZI
THOMAS R. HINDS

CALCIUM AND CELL FUNCTION, VOL. I

ISBN 0-12-171401-2

I. MEMBRANE Ca^{2+} TRANSPORT

A. Introduction

A substantial body of evidence demonstrates that membranes of living cells actively transport Ca^{2+}. We intend to review some of this evidence as well as recent evidence which implicates calmodulin (CaM) in the regulation of the plasma membrane Ca^{2+} pump. If, as suggested below, CaM participates in the regulation of the plasma membrane Ca^{2+} pump, then CaM participates in the regulation of how much free Ca^{2+} is available for intracellular Ca^{2+}-binding proteins, including CaM itself; an idea implicit in the name coined by Cheung *et al.* (1978). Calmodulin will be used to denote the Ca^{2+}-binding protein previously known by various names, including phosphodiesterase activator (Cheung, 1971), calcium-dependent regulator (CDR; Brostrom *et al.,* 1975), modulator protein (Wang and Desai, 1977), and ATPase activator (Gopinath and Vincenzi, 1977; Luthra *et al.,* 1976; Jarrett and Penniston, 1978).

At the outset we want to establish some terminology and general frame of reference. We will use the symbol "Ca^{2+}" to denote ionic calcium and "calcium" to denote total calcium or calcium for which the bonding state is indeterminate. We will use the commonly accepted term "pump" to mean the process by which membranes utilize energy and which results in the movement of certain species against their electrochemical gradients. We consider processes by which the energy available for the movement of one species down its electrochemical gradient is coupled to the movement of another species up its electrochemical gradient to be an exchange rather than a pump (e.g., $Na^{+}-Ca^{2+}$ exchange across the plasma membrane) (Baker, 1972).

The extracellular fluid of most organisms contains about 1 mM Ca^{2+}, yet there is evidence that erythrocyte (red blood cell, RBC) cytoplasmic $[Ca^{2+}]$ is maintained in the range of 10^{-7} (Simons, 1976b) or less. Thus, a concentration gradient for Ca^{2+} of approximately 10,000-fold exists across the plasma membrane. Considering that cells maintain a trans(plasma)-membrane potential negative on the inside (Jay and Burton, 1969), it is clear that the electrochemical gradient is even greater than the concentration gradient. This electrochemical gradient is maintained at the cost of energy liberated by enzymatic hydrolysis of ATP.

Kretsinger (1977) advanced the notion that the sole function of Ca^{2+} in the cytosol is to transmit information. Thus, Ca^{2+} can be viewed as an intracellular second messenger. We would like to think of Ca^{2+} as the "first second messenger," in contrast to cyclic AMP, which obviously evolved later than Ca^{2+} (Penzias, 1979). The basic notion of this treatise, i.e., calcium and cell function, is that the binding of calcium to certain proteins (e.g., troponin C or CaM) can regulate specific cellular functions. A major role of the Ca^{2+} pump is to regulate how much calcium is available to these Ca^{2+}-binding proteins. In turn, this indirectly regulates cellular functions. Other chapters in this volume will emphasize the roles of CaM in the mediation of the effects of intracellular Ca^{2+}. This chapter will emphasize that CaM (with the Ca^{2+} pump) also participates in the termination of the effects of intracellular Ca^{2+}.

Evidence for plasma membrane Ca^{2+} transport will be reviewed with emphasis on the human erythrocyte. Other plasma membranes will be considered only briefly and mainly by inference. The role(s) of CaM in the regulation of the Ca^{2+} pump in health and disease will also be considered. Several closely related, excellent reviews recently appeared. Kretsinger (1976) presented an excellent review on Ca^{2+}-binding proteins. Carafoli and Crompton (1978) reviewed the general question of the regulation of intracellular Ca^{2+}, and Roufogalis (1980) reviewed the regulation of Ca^{2+} translocation across the RBC membrane. The review by Roufogalis is an excellent up-to-date statement on the regulation of the RBC Ca^{2+} pump. Lew and Ferriera (1978) considered passive Ca^{2+} transport and Ca^{2+} buffering in the RBC. We will try to emphasize the possible relationships between calmodulin and the plasma membrane Ca^{2+} pumps in general, but recognize that the RBC is the only unambiguous model available at this time.

B. Ca^{2+} Transport by the Human Erythrocyte

The human RBC has been used as a model to study the regulation of intracellular Ca^{2+}. RBCs have unique advantages which make them useful. They may be obtained in large quantities, and they are available from patients with interesting inherited and/or acquired diseases. Furthermore, RBCs lack organelles, and the major cytoplasmic protein (hemoglobin) is easily removed from the plasma membrane. RBCs are easily separated from other blood cells and can be fractionated into preparations which are enriched in "young" or "old" RBCs (Danon and Marikovsky, 1964).

For laboratories interested in membranes and transport, the RBC has been exceedingly useful. Pure plasma membranes can be readily prepared in large quantities from RBCs (Dodge *et al.*, 1963), and we use such membranes as a bioassay system for CaM. Because RBCs lack organelles there is no question of contamination with other membranes. A distinct advantage is, therefore, that one can study plasma membrane processes in relatively "clean" preparations.

Carafoli and Crompton (1978) considered the total surface area of membranes capable of transporting Ca^{2+} in various cells. In liver and heart cells, the plasma membrane constitutes only 11.4 and 0.8% of the total surface area, respectively. The relative physiological role of mitochondrial, sarcoplasmic reticular, and plasma membranes in rapid and/or long-term regulation of intracellular Ca^{2+} in non-erythrocytes is an important, if yet unresolved, question. At least in the RBC, it is clear that the plasma membrane bears the full responsibility for eventualy removing Ca^{2+} from the cytosol. It should be noted that RBCs have a finite, if relatively low, permeability to Ca^{2+} (Schatzmann and Vincenzi, 1969; Lew and Ferreira, 1978), but they do not continue to accumulate Ca^{2+}, in spite of a massive inward electrochemical gradient. The Ca^{2+} which enters the cell must eventually be extruded through the plasma membrane. An obvious limitation in the study of these processes is that "membranes" prepared from liver and heart cells, etc., are usually heterogeneous. In short, if one wishes unambiguously to study plasma membrane transport in the regulation of intracellular Ca^{2+}, then the RBC is the cell of choice at this time. We have, therefore, chosen to concentrate our studies on human RBCs. It is intended that such studies will yield information on plasma membrane Ca^{2+} transport as a mechanism of control of Ca^{2+} available to regulatory proteins and, thus, the regulation of Ca^{2+}-dependent cellular functions.

A disadvantage of limiting the study of Ca^{2+} transport to the plasma membrane is that rapid Ca^{2+} transients, such as may be achieved and/or modified by organelles (Carafoli and Crompton, 1978), are not present in RBCs. These transients are probably involved mainly in specialized functions of excitable cells.

The lack of organelles in RBCs is an advantage from the point of view of being able experimentally to manipulate the cells for transport purposes. For example, when RBCs are exposed to a short period of hypotonic shock, they become "leaky" to molecules which are normally impermeant, including hemoglobin. However, unlike what might be anticipated based on a simple notion of osmotic rupture, it is possible to promote "resealing" of the RBCs (now called resealed RBC ghosts) by restoring isotonicity. The technique is called reversible hemolysis. Factors involved in the opening and resealing of resealed RBC ghosts have been studied by Bodemann and Passow (1972), among others. The cells lose a good deal of hemoglobin during the osmotic shock and are pale, hence the term ghosts. The main point about resealed RBC ghosts is that during the osmotic stress the extracellular and intracellular compartments approach equilibrium. Thus, one may rapidly load RBCs with various substances that might never be present or that would enter RBCs only very slowly. Loading is achieved by placing the substances to be loaded in the hypotonic medium. RBCs have been loaded by this method with various substances including Ca^{2+} buffers (Simons, 1976a) and selected proteins (DeLoach and Ihler, 1977).

As noted above, the ability to load RBCs makes them useful for transport work. In 1966, Schatzmann, using the reversible hemolysis technique, provided the first evidence for active transport of Ca^{2+} across the RBC membrane. Subsequently, Schatzmann and co-workers (Schatzmann and Vincenzi, 1969; Schatzmann, 1973, 1975) as well as other laboratories (Lee and Shin, 1969; Olson and Cazort, 1969) confirmed the existence of a Ca^{2+} pump in the RBC membrane. Transport occurs against an electrochemical gradient and is dependent on ATP (Schatzmann and Vincenzi, 1969; Vincenzi, 1971). Other nucleoside triphosphates which give rise to ATP also support Ca^{2+} transport (Olson and Cazort, 1969). Other means of loading RBCs with Ca^{2+} have included metabolic depletion (Dunn, 1974) and the use of the Ca^{2+} ionophore A23187 (Sarkadi *et al.*, 1976). The latter method is especially attractive because of its rapid and relatively specific effect.

Another interesting and extremely useful feature of RBCs in the study of membrane transport is that they can be inverted to form inside-out vesicles (Kant and Steck, 1972; Steck, 1974a). This has allowed various workers to demonstrate the assymetry of the Ca^{2+} pump. For the Ca^{2+} pump to operate, ATP and Mg^{2+} are required at the cytoplasmic face of the membrane (Lee and Shin, 1969). Schatzmann and Vincenzi (1969) showed that Ca^{2+} activates the ATPase at the cytoplasmic face only. This is analogous to the Na^+-K^+ pump, i.e., the specific ATPase and transport is activated by an ion on the face of the membrane from which the ion will be transported. Using inside-out vesicles of RBC membrane, Cha *et al.* (1971) and Weiner and Lee (1972) demonstrated ATP-dependent uptake of Ca^{2+}. It must be emphasized that the uptake monitored in such experiments is equivalent to extrusion from the intact RBC. In other words, the pump which serves to maintain low cytoplasmic Ca^{2+} turns inside-out along with the membrane as would be expected for a function based on an integral membrane protein (Steck, 1974b).

C. Ca^{2+} Transport in Plasma Membranes of Non-Erythrocytes

It appears that cells other than RBCs actively transport Ca^{2+} across the plasma membrane, although few well-studied examples are available. Active transport of Ca^{2+} from HeLa (Borle, 1969), kidney (Borle, 1972), and liver (Lamb and Lindsay, 1971) cells has been demonstrated. Na^+-Ca^{2+} exchange has also been demonstrated in some cells (Baker, 1972; Blaustein, 1976). The relative importance of these mechanistically different, but functionally equivalent phenomena has yet to be elucidated (Carafoli and Crompton, 1978). Intracellular Ca^{2+} inhibits the Na^+-K^+ pump ATPase (Dunham and Glynn, 1961; Davis and Vincenzi, 1971) and Na^+ transport (Dunn, 1974). We, therefore, believe that low intracellular Ca^{2+} is essential to allow the Na^+-K^+ pump to establish the Na^+ gradient which is necessary to drive any Na^+-Ca^{2+} exchange. We previously suggested that the Ca^{2+} pump is a primitive system present in all cells and that Na^+-Ca^{2+} exchange is a more specialized mechanism present in some excitable cells (Vincenzi and Hinds, 1976). This is similar to the suggestion by Lamb and Lindsay (1971). In apparent agreement with this suggestion, DiPolo (1978) concluded that squid axon, which has a well-documented Na^+-Ca^{2+} exchange mechanism (Baker, 1972; Blaustein, 1976), also actively transports Ca^{2+} by an ATP-driven pump (DiPolo, 1978). More recently, DiPolo and Beaugé (1979) have concluded that at normal intracellular $[Ca^{2+}]$, most Ca^{2+} efflux is related to the pump. They further concluded that exchange contributes substantially to the total Ca^{2+} efflux only when $[Ca^{2+}]$ in the cell increases above "normal physiological" levels.

While it has not been sufficiently documented, it appears likely that plasma membranes of a number of cells can form inside-out vesicles in a manner similar to RBC membranes. The uptake of Ca^{2+} into "microsomal" preparations of various tissues, for example, kidney (Moore *et al.*, 1974), uterus (Janis *et al.*, 1977), and brain (Swanson *et al.*, 1974) may represent to some extent the active transport of Ca^{2+} across the inverted plasma membranes. Unfortunately, in preparations from non-erythrocytes, contamination with organelle membranes is always possible.

D. Ca^{2+} Transport across Other Membranes

Calcium ions are actively transported by membranes other than the plasma membrane. Excellent progress has been made in the study of Ca^{2+} transport across the membranes of sarcoplasmic reticulum and mitochondria. The roles of the sarcoplasmic reticulum and mitochondria in the regulation of intracellular Ca^{2+} are considered by Carafoli and Crompton (1978). Transepithelial Ca^{2+} transport, which may involve vitamin D-dependent intracellular Ca^{2+}-binding proteins in a manner not yet elucidated (Wasserman and Feher, 1977), is also considered by Rasmussen and Bordier (1974) and Carafoli and Crompton (1978).

II. (Ca^{2+} + Mg^{2+})-ATPase

A. Introduction

The enzymatic hydrolysis of the terminal phosphate of ATP in biological preparations is defined using a variety of nomenclatures. We prefer a strict operational definition based upon the necessary and sufficient additions used to promote ATP hydrolysis. For example, we define Mg^{2+}-ATPase activity as that occurring in the presence of Mg^{2+} and ATP (and an appropriate buffer) only. Other ionic additions, if present, exert only minor nonspecific effects, if any, under such conditions. Likewise, Ca^{2+}-ATPase activity is defined as that occurring in the presence of Ca^{2+} and ATP only. ATPase activities with multiple ionic requirements should be denoted as such. Thus, the Na^+ and K^+ stimulated, Mg^{2+}-dependent ATPase is denoted as (Na^+ + K^+ + Mg^{2+})-ATPase. Because of its well-established role in Na^+ and K^+ transport, it may also be called the "Na^+–K^+ pump ATPase." The Ca^{2+}-stimulated, Mg^{2+}-dependent ATPase will be denoted as the (Ca^{2+} + Mg^{2+})-ATPase. It will be assumed that sufficient evidence is now available to accept that membrane-bound (Ca^{2+} + Mg^{2+})-ATPase is the Ca^{2+} pump ATPase. Thus, we assume that all active transport of Ca^{2+} across the plasma membrane is dependent on the activity of (Ca^{2+} + Mg^{2+})-ATPase. However, it is not proved that all the (Ca^{2+} + Mg^{2+})-ATPase activity in the plasma membrane is necessarily coupled to Ca^{2+} transport (Quist and Roufogalis, 1975b; Sarkadi *et al.*, 1977).

B. (Ca^{2+} + Mg^{2+})-ATPase of Erythrocytes

General characteristics of RBC membrane (Ca^{2+} + Mg^{2+})-ATPase have been reviewed previously (Schatzmann, 1975; Vincenzi and Hinds, 1976; Roufogalis, 1980). It is evident that there is still much to be done before a full understanding of this enzyme activity is obtained. Some characteristics are clear, however. The Ca^{2+} pump ATPase is vectorially oriented in the plasma membrane. The pump removes Ca^{2+} from the cytosol extruding it into the external milieu, at the expense of ATP which must be available at the cytoplasmic face of the membrane. The enzyme operates through a phosphorylated intermediate (Knauf *et al.*, 1974; Katz and

Blostein, 1975; Rega and Garrahan, 1975; Schatzmann and Bürgin, 1978). The phosphorylated intermediate appears to have a MW of 120,000 to 150,000 by sodium dodecyl sulfate (SDS) electrophoresis (Knauf *et al.*, 1974). The energy substrate appears to be free ATP rather than Mg^{2+}-ATP (Rega and Garrahan, 1975; Schatzmann, 1975). There may be another ATP site which is a regulatory site (Mualem and Karlish, 1979). It is thought that Ca^{2+} is necessary for phosphorylation, and Mg^{2+} is responsible for dephosphorylation (Rega and Garrahan, 1975; Schatzmann and Bürgin, 1978), notwithstanding one report (Cha and Lee, 1976). There does not appear to be a strict requirement for other ions. Activators such as CaM interact with the cytoplasmic face of the enzyme (see below).

Some essential features of Ca^{2+} pump ATPase which need emphasis at this point include a relatively high affinity for Ca^{2+} (Dunham and Glynn, 1961) and greater specific activity than the Na^+–K^+ pump ATPase in the same membranes (Farrance and Vincenzi, 1977a). The possibility that two (Ca^{2+} + Mg^{2+})-ATPases exist with different affinities for Ca^{2+} and that the Ca^{2+} pump ATPase may exhibit different affinities under different conditions has been suggested (Horton *et al.*, 1970; Quist and Roufogalis, 1975a; Hanahan *et al.*, 1978).

Given the well-established relationship between the (Ca^{2+} + Mg^{2+})-ATPase and the Ca^{2+} pump, one can predict results of transport experiments based on previously determined ATPase activities. In general, ATPase measurements are easier to perform than transport experiments, and enzymatic assays can help in analysis of mechanisms of transport and its modification by various agents. It would be useful to isolate the (Ca^{2+} + Mg^{2+})-ATPase and to reconstitute the Ca^{2+} pump. Niggli *et al.* (1979) reported isolation of the (Ca^{2+} + Mg^{2+})-ATPase using CaM affinity chromatography and Haaker and Racker (1979) recently reported reconstitution of the pump.

C. (Ca^{2+} + Mg^{2+})-ATPase of Non-Erythrocytes

Various investigators have implied the existence of a Ca^{2+} pump in nonerythrocytes based on ATPase activities from assorted preparations of tissues, including brain (Robinson, 1976; Ohashi *et al.*, 1970) and liver (Lindsay, 1976), as well as platelets (Käser-Glanzmann *et al.*, 1978). Difficulties encountered in this approach include the existence of substantial Mg^{2+}-ATPase activities and heterogeneity of the preparations. Very large Mg^{2+}-ATPase activity can obscure a small or modest (Ca^{2+} + Mg^{2+})-ATPase activity. Various Ca^{2+} pump ATPases in heterogeneous preparations cannot be dealt with in a very satisfying manner. Selective potent inhibitors of mitochondrial, sarcoplasmic reticulum, and plasma membrane Ca^{2+} pump ATPases, respectively, would be very desirable to help overcome this problem. For that matter, so would a selective inhibitor of the Mg^{2+}-ATPase. Barbiturates inhibit Mg^{2+}-ATPase (Mirčevová and Šimonová, 1973), but more selectivity is needed. Some workers have measured the Ca^{2+}-ATPase (absence of Mg^{2+}) activity of membranes. Based on the strict dependence of the RBC Ca^{2+} pump ATPase (and transport) on Mg^{2+} (Schatzmann and Vincenzi, 1969), it seems unlikely that such measurements are relevant to predictions about plasma membrane Ca^{2+} transport. The function(s) of Ca^{2+}-ATPase, if any, are unknown to us.

The possibility that the Ca^{2+} pump [or (Ca^{2+} + Mg^{2+})-ATPase] for certain cells may be linked to a hormone receptor has been raised in at least two studies. Carsten (1977) reported that oxytocin caused a decrease in ATP-dependent Ca^{2+} binding (transport?) in preparations (presumably containing some inside-out plasma membranes as well as sarcoplasmic reticulum) from pregnant and nonpregnant bovine uterus. Preparations from pregnant uteri were over 1000 times more sensitive to this effect. Åkerman and Wikstrom (1979) found that oxytocin inhibited the (Ca^{2+} + Mg^{2+})-ATPase activity of uterine cell membrane preparations. The results of these two studies are interesting, and we wish to consider them further. Let us imagine for the moment that intracellular Ca^{2+} in uterine muscle is maintained by a balance between inward "leak" and outward active transport across the plasma membrane. Intracellular Ca^{2+}, serving as a trigger for uterine contractions (probably through a calmodulin-dependent myosin light chain kinase) (Dabrowska *et al.*, 1978), could be increased either by increasing inward leak, by decreasing outward transport, or both. In the case of oxytocin, which appears to inhibit the Ca^{2+} pump ATPase (Åkerman and Wikstrom, 1979), the mechanism of decreasing outward transport may be operative. Obviously, other possible explanations exist, but the possibility that certain drugs or hormones might act as Ca^{2+} pump inhibitors needs to be explored. Specific physiological inhibition of preexisting Ca^{2+} pump sites could represent a hitherto unknown mechanism for cellular regulation.

A similar mechanism may form the basis for at least some of the effects of insulin on adipocytes. Pershadsingh and McDonald (1979) found that low concentrations of insulin inhibit the (Ca^{2+} + Mg^{2+})-ATPase of isolated adipocyte plasma membranes. Increased intracellular Ca^{2+} resulting from pump inhibition could trigger further Ca^{2+} release from intracellular stores and/or otherwise serve as the intracellular second messenger for the first messenger, insulin. To our knowledge, hormone- or drug-induced inhibition of active Ca^{2+} efflux has not been suggested previously as a physiological control mechanism. It is an area worth exploration.

III. CALMODULIN AND CA^{2+} TRANSPORT

A. Introduction

CaM is a cytoplasmic protein containing four Ca^{2+}-binding sites with $K_d = 2.38 \times 10^{-6}$ *M* (Dedman *et al.*, 1977). It is closely related to other Ca^{2+}-binding proteins and has been implicated in a variety of cellular processes (Vanaman *et al.*, 1977). Kretsinger (1977) predicted that the Ca^{2+} pump, like other high-affinity Ca^{2+} binding proteins, would contain an "EF hand." As outlined below, CaM appears to be part of the plasma membrane Ca^{2+} pump; either as an integral part of the pump mechanism, as a reversibly bound regulator of the pump, or possibly both. Evidence implicating CaM in the regulation of the plasma membrane Ca^{2+} pump may thus be in confirmation of Kretsinger's prediction. Direct proof must await isolation and reconstitution of the activity of the pump before any definitive statements can be made.

A brief review of evidence demonstrating a role of CaM in plasma membrane Ca^{2+} transport follows. Our frame of reference for this section should be stated explicitly. We conceive that CaM (and related proteins) mediates various cellular functions in response to changes in intracellular Ca^{2+} acting as a second messenger (Kretsinger, 1977). We assume that CaM is also involved in the regulation of the plasma membrane Ca^{2+} pump, probably as a Ca^{2+} recognition site. It seems likely that the level of free intracellular Ca^{2+} is the limiting factor in the operation of the pump *in vivo*. In other words, it appears that the Ca^{2+} pump tends to regulate the concentration of free Ca^{2+} at the cytoplasmic face of the membrane in the region of the "threshold" for Ca^{2+} binding to CaM (i.e., around 10^{-7} *M*). Figure 1A presents calculated binding of Ca^{2+} to CaM over a range of Ca^{2+} concentrations. In the range of free Ca^{2+} concentrations presumed to be present in the human RBC (Schatzmann, 1975; Simons, 1976b) small changes in Ca^{2+} would lead to fairly large changes in $CaM{\cdot}(Ca^{2+})_n$. This is just what would be predicted if binding of Ca^{2+} to CaM were the signal for stimulation of the pump. Figure 1A and B and various calculations will be discussed further in Section III, B.

B. Calmodulin and the Ca^{2+} Pump of Erythrocytes

Bond and Clough (1973) reported that crude hemolysates when added to isolated RBC membranes *in vitro* would stimulate ($Ca^{2+} + Mg^{2+}$)-ATPase but not (Mg^{2+})-ATPase or ($Na^+ + K^+ + Mg^{2+}$)-ATPase. The factor in hemolysates was trypsin sensitive and was shown not be be hemoglobin. This prompted Bond and Clough (1973) to suggest that a non-he-

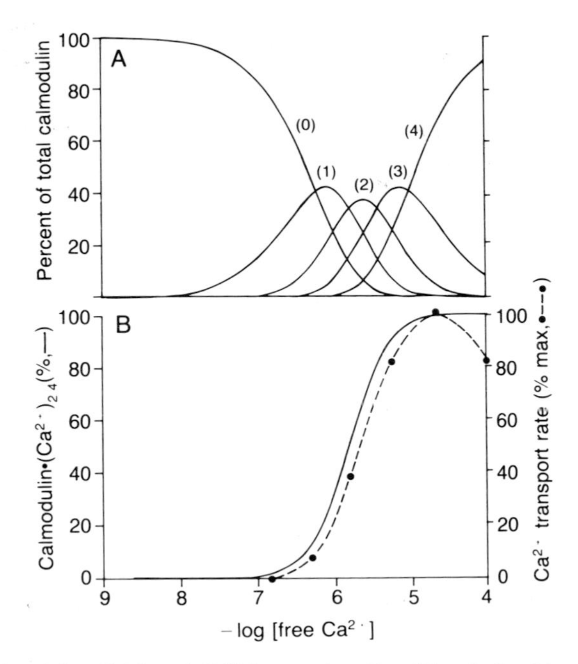

Fig. 1. Calmodulin (CaM) and Ca^{2+} interaction: Possible relationships to CaM action. In (A) the percent of total CaM existing in various forms was calculated as a function of free Ca^{2+} concentration. CaM·$(Ca^{2+})_n$ complexes are plotted separately as a percent of total CaM (assumed to be 7×10^{-6} *M*) (Jung, 1978). Thus, CaM·(Ca^{2+}) (free CaM) is labeled (0), CaM·(Ca^{2+}) is labeled (1), etc. At the K_d for Ca^{2+} of 2.38×10^{-6} *M* (Dedman *et al.*, 1977), 50% of the total binding sites are occupied, and more than 90% of the CaM molecules contain one or more bound Ca^{2+}. As shown in (B), the $[Ca^{2+}]$ dependence of Ca^{2+} transport into inside-out RBC membrane vesicles in the presence of added CaM (data adapted from Vincenzi and Larsen, 1980) closely parallels the $[Ca^{2+}]$ dependence of CaM·$(Ca^{2+})_{n\leq 2}$.

moglobin protein in the RBC cytoplasm may function as a regulator of the Ca^{2+} pump. The same phenomenon was apparent but was not noted in the earlier work of Cha *et al.* (1971). These authors were studying the ability of RBC hemolysates to convert various nucleoside triphosphates to ATP (which could then serve as a substrate for the ATPase). In the ATP "control" experiment the activity of $(Ca^{2+} + Mg^{2+})$-ATPase increased from 1.84 to 2.67 μmoles/mg protein/hr when hemolysate was added. No comment was made by the authors on this observation.

Hanahan and co-workers (Luthra *et al.*, 1976) partially purified the protein responsible for the stimulatory effect of hemolysate on $(Ca^{2+} + Mg^{2+})$-ATPase. Properties of the protein, also isolated in our laboratory (Jung, 1978), included low molecular weight, relative heat stability, sensitivity to trypsin, and acidic p*I*. Gopinath and Vincenzi (1977) compared several known Ca^{2+}-binding proteins with partially purified RBC activator protein. Neither parvalbumin, nor troponin -C was as effective as RBC activator in activation of (Ca^{2+} + Mg^{2+})-ATPase. By contrast, beef brain CaM was very potent and was equally effective with RBC activator protein (Gopinath and Vincenzi, 1977). This evidence, plus the simultaneously presented data showing comparable physicochemical properties of RBC activator and CaM (Jarrett and Penniston, 1977), prompted the suggestion that the two proteins were similar, if not identical. Subsequent isolation and characterization of CaM from human RBCs (Jarrett and Penniston, 1978; Jung, 1978) confirmed this suggestion. Hanahan *et al.* (1978) also showed that pure CaM from RBCs would activate the (Ca^{2+} + Mg^{2+})-ATPase.

Farrance and Vincenzi (1977b), Scharff (1972, 1976), and Scharff and Foder (1977) have shown that certain methods of preparing isolated RBC membranes result in membranes with high (Ca^{2+} + Mg^{2+})-ATPase activity. In studies by Farrance and Vincenzi (1977a,b), membranes of high and low (Ca^{2+} + Mg^{2+})-ATPase activity were isolated following hemolysis in 310 and 20 ideal mosm imidazole buffers, respectively (I310 and I20). It was established that the high activity membranes contained a protein activator (later shown to be CaM) (Farrance and Vincenzi, 1977b). When these membranes were washed with EGTA [ethyleneglycol bis(β-amino ethyl ether)*N,N'*-tetraacetic acid] they reverted to low activity with the loss of CaM. Low activity membranes could be activated by the addition of exogenous partially purified CaM. It was suggested that reversible Ca^{2+}-dependent binding of the "activator" might be the means by which the Ca^{2+} pump is regulated (Vincenzi, 1978). Similar results are apparent in the work of Quist and Roufogalis (1975a). These workers were studying activities of (Ca^{2+} + Mg^{2+})-ATPase, demonstrating both high and low affinity for Ca^{2+}. RBC membranes extracted with 0.1 m*M* EDTA (ethylenediaminetetraacetic acid) in low ionic strength lost the high affinity component of activity. The soluble protein fraction had no ATPase activity, but could reconstitute the high affinity ATPase. Quist and Roufogalis (1975a) concluded that an activator protein or the high affinity (Ca^{2+} + Mg^{2+})-ATPase itself is only loosely attached to the membrane. It was further suggested that the low affinity ATPase is associated with transport. This suggestion appeared to be confirmed in a later study (Quist and Roufogalis, 1977). These workers found that a concentrated, water-soluble protein fraction from RBC membranes (presumably containing CaM and other proteins), when added to inside-out RBC membrane vesicles, *did not* affect the rate of Ca^{2+} uptake, but under the same conditions *did* increase (Ca^{2+} + Mg^{2+})-ATPase activity. We have no explanation for this result. "Leaky" vesicles could account for such a finding.

In our hands, purified CaM stimulates both ATPase and transport activities of inside-out vesicles (Hinds *et al.*, 1978; Larsen and Vincenzi, 1979). Figure 2 shows the effects of CaM on both (Ca^{2+} + Mg^{2+})-ATPase and Ca^{2+} transport. It should be noted that both the ATPase and the pump display significant activities in the absence of added CaM. We have termed this activity "basal" and have usually assumed that this represents the activity of CaM-free pump ATPase (an alternative interpretation is considered below). Upon addition of purified CaM, there usually is a three- to fourfold concentration-dependent increase in ATPase or transport activity. Typical basal and activated (Ca^{2+} + Mg^{2+})-ATPase activities of isolated RBC membranes in our laboratory are approximately 12 and 48 nmoles P_i liberated per milligram of protein per minute, respectively. Depending on the [Ca^{2+}], Ca^{2+} transport typically increased from approximately 5 to 20 μmoles Ca^{2+} per milligram inside-out vesicle protein per minute upon addition of $\geq 4 \times 10^{-8}$ *M* CaM (Vincenzi and Larsen, 1979). It seems reasonable to conclude that CaM activates Ca^{2+} transport by activating (Ca^{2+} + Mg^{2+})-ATPase. The ordinates in Fig. 2 are not labeled because adequate measurements of both ATPase and transport have not been carried out simultaneously on the same preparation. Such measurements would be necessary to assess a possible influence of CaM on the stoichiometry of the Ca^{2+} pump. Even ignoring a pos-

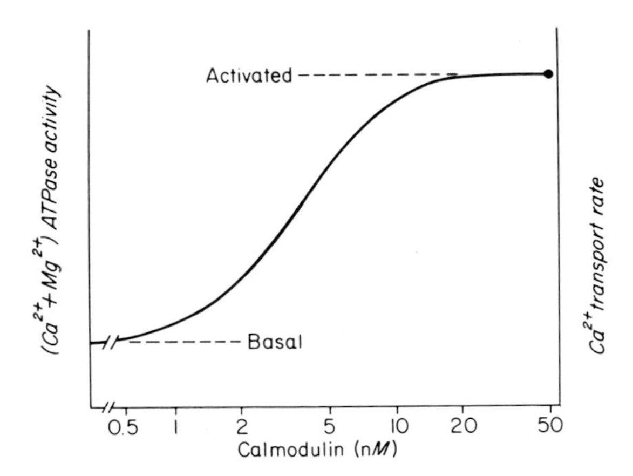

Fig. 2. Calmodulin (CaM) activation of (Ca^{2+} + Mg^{2+})-ATPase and Ca^{2+} transport. This figure is adapted from various studies in our laboratory, is intended to illustrate the approximate concentration dependence of the CaM effect and approximately relative magnitudes of basal and activated ATPase and transport values. Simultaneous quantification of ATPase and transport data in a pure inside-out plasma membrane vesicle preparation would be necessary before the ordinates could be properly labeled in a composite figure such as this. It is implied that CaM stimulates Ca^{2+} transport in a concentration-dependent fashion by stimulating the membrane-bound (Ca^{2+} + Mg^{2+})-ATPase.

sible influence of CaM, the stoichiometry of the calcium pump is not clear. It has been reported by some groups that the pump moves 2 Ca^{2+} per ATP hydrolyzed (Quist and Roufogalis, 1975b; Sarkadi *et al.*, 1977), whereas others report that only 1 Ca^{2+} ion is pumped per ATP (Schatzmann and Vincenzi, 1969; Schatzmann, 1973; Larsen *et al.*, 1978a). When the Ca^{2+} pump was reversed Rossi *et al.* (1978) found a stoichiometry of 1.55 Ca^{2+} per ATP.

To date, there have been few studies on the effect of purified CaM on the kinetics or reaction mechanism of the (Ca^{2+} + Mg^{2+})-ATPase. In our laboratory, CaM only had little effect on the K_m for ATP, increasing this only slightly from 1.6 to 6.3 × 10^{-4} *M*. Calmodulin-laden (high activity) membranes displayed a broad pH optimum around 7.1. Calmodulin-deficient (low activity) membranes displayed no pH optimum. Low activity membranes had an apparent K_d of 1.1 to 1.4 × 10^{-6} *M* for Ca^{2+}, while high activity membranes had an apparent K_d of 3.6 × 10^{-6} *M*. Low activity membranes hydrolyzed ATP at a rate of 12.7 nmoles/mg protein/min, whereas high activity membranes hydrolyzed ATP at a rate of 55.8 nmoles/mg protein/min (Farrance, 1977). Scharff (1972, 1976, 1978) has presented several kinetic studies on CaM-deficient and CaM-laden membranes.

Several laboratories reported both high and low Ca^{2+} affinities of the (Ca^{2+} + Mg^{2+})-ATPase (Horton *et al.*, 1970; Quist and Roufogalis, 1975a). We, as well as others, suggested that CaM binding might confer the high affinity. The mixture of high and low affinities often seen in isolated RBC membrane preparations could be due to some Ca^{2+} pump sites containing, and some not containing, CaM. This is still an attractive and prevalent idea, although in our hands the change in apparent Ca^{2+} affinity of neither the ATPase, nor the pump, was as striking as we had initially anticipated (Vincenzi and Larsen, 1979). We consider that the question of multiple ATPase affinities, and the possible relationships of CaM and Ca^{2+} transport to these, is not yet resolved.

A. F. Rega and P. J. Garrahan (personal communication) recently found that CaM increases the turnover of (Ca^{2+} + Mg^{2+})-ATPase. This was based on comparison of the effects of CaM on the ATPase activity and the level of phosphoenzyme. The effect of CaM on the kinetics of the ATPase is almost certainly based on direct interaction of CaM with the ATPase. Lynch and Cheung (1979) found that (Ca^{2+} + Mg^{2+})-ATPase activity in detergent-solubilized preparations or RBC membrane comigrated with CaM in the presence of Ca^{2+}. The two activities could be separated in the presence of EGTA.

Bond and Clough (1973) who first described the protein activator of (Ca^{2+} + Mg^{2+})-ATPase postulated that it would be an activator of the

Ca^{2+} pump. Hinds *et al.* (1978), using purified CaM, and MacIntyre and Green (1978), using crude RBC hemolysate, found that the rate of ATP-dependent Ca^{2+} uptake into inside-out RBC membrane vesicles was increased. As noted above, the Ca^{2+} pump exhibits asymmetry. Calmodulin is normally present in the cytoplasm and presumably interacts with the (Ca^{2+} + Mg^{2+})-ATPase (Lynch and Cheung, 1979) or some component(s) of the cytoplasmic face of the membrane in a Ca^{2+}-dependent fashion (Farrance and Vincenzi, 1977b). Calmodulin does not appear to bind to the outer surface of the RBC membrane. When intact RBCs are exposed to CaM, no activation of subsequently isolated membranes was found (Vincenzi and Farrance, 1977). In any event, inside-out vesicles took up Ca^{2+} in an ATP dependent fashion, and the rate of transport could be increased by the addition of CaM (Hinds *et al.*, 1978).

The rate of Ca^{2+} transport across RBC membranes differs in various laboratories. This is considered by Roufogalis (1979). Generally, transport rates with inside-out vesicles have been low compared with rates in intact cells or resealed ghosts. In our early experiments, initial Ca^{2+} transport rates were approximately 0.023–0.123 (basal) and 0.11–0.80 (CaM added) nmoles/mg inside-out vesicle protein/min (Hinds *et al.*, 1978). By contrast, we calculated a Ca^{2+} pump rate of 13 nmoles/mg membrane protein/min in intact RBCs from the results of Sarkadi *et al.* (1977). This estimate is based on their reported Ca^{2+} transport rate of 85 μmoles/min /liter/cells from Ca^{2+}-loaded RBCs (loaded by exposure to ionophore A23187) and on the following assumptions: (1) RBC volume $= 1 \times 10^{-13}$ liters; (2) one Dodge-type ghost (Dodge *et al.*, 1963) contains all the membrane protein of an RBC or 6.6×10^{-13} g; and (3) inside-out vesicles contain all of the membrane protein (to the extent that some membrane protein is lost during inside-out vesicle preparation this assumption will overestimate specific transport rates in inside-out vesicles).

Although it was possible to demonstrate that CaM increased the rate of Ca^{2+} transport with inside-out vesicles, both the basal and activated rates were quite low and variable (Hinds *et al.*, 1978). Another disturbing aspect of our early transport work was that no maximal stimulatory effect was obtained with CaM (Hinds *et al.*, 1978). This was true even using concentrations which gave maximal activation of (Ca^{2+} + Mg^{2+})-ATPase of isolated membranes (Jung, 1978). The difficulty, which lies in the preparation of inside-out vesicles, has now been identified, although not explained. The process of making inside-out RBC membrane vesicles consists of stimulating endocytosis of the RBC membrane and then releasing the endocytotic vesicles after rupturing the cells. Cells are ruptured by passing them forcibly through a 27-gauge needle. According to the procedure of Kant and Steck (1972) inside-out vesicles are then collected by centrifugation on a dextran density gradient. This separates broken membrane fragments from sealed vesicles, although it may not separate inside-out and right-side-out vesicles. By sampling activity at each step in the procedure, Quist and Roufogalis (1977) and Larsen (1979) found that exposure of RBC vesicles to dextran caused almost total loss of both ATPase and transport activity. The reason for this effect of dextran is not known. We have attempted to eliminate the effect by exhaustive dialysis or boiling of various dextrans. Such procedures have been unsuccessful (F. F. Vincenzi and T. R. Hinds, unpublished). Sarkadi *et al.* (1978) reported transport values of up to 10 nmoles Ca^{2+}/mg inside-out protein/min into dextran-prepared vesicles washed with EDTA. We are led to speculate that dextran contains some heavy metal in sufficient quantities to interact with the (Ca^{2+} + Mg^{2+})-ATPase pump. Until the detrimental effect of dextran is overcome or until an alternate method of purifying inside-out vesicles is developed, we have taken a compromise approach. We simply utilize the mixture of inside-out, right-side-out, and broken membranes obtained after rupturing vesicle-laden RBCs. Of the various species present, only sealed inside-out vesicles will accumulate Ca^{2+} in an ATP-dependent fashion. One can employ such a heterogeneous preparation for transport experiments, but it is obviously not well suited for measuring the stoichiometry of the Ca^{2+} pump. Transport data are expressed in terms of "inside-out vesicle protein." Quantification of the sidedness of the membrane preparation is carried out by measuring acetylcholinesterase activity in the absence and presence of low concentrations of the nonionic detergent Triton X-100, respectively (Steck, 1974a). In the absence of the detergent, only the acetylcholinesterase of broken membranes and right-side-out vesicles is accessible to substrate. In the presence of the detergent, all the acetylcholinesterase is accessible. Comparison of these activities allows one to calculate the percentage of sealed inside-out vesicles in a given preparation.

Using such a heterogeneous preparation of vesicles we obtained basal and CaM-activated rates of Ca^{2+} transport of 6.8 and 14.4 nmoles/mg inside-out protein/min, respectively (Larsen and Vincenzi, 1979), at a Ca^{2+} concentration of 150 μM. The apparent K_d (at 25°C) for CaM stimulation of (Ca^{2+} + Mg^{2+})-ATPase activity was 3.0 n*M* (Larsen, 1979). This value is in good agreement with the apparent K_d (at 37°C) for calmodulin stimulation of (Ca^{2+} + Mg^{2+})-ATPase activity (4.0 n*M*) (Larsen, 1979). These results are diagrammed in Fig. 2. It may be noted here that activation by CaM can be antagonized by phenothiazine drugs, such as chlorpromazine and trifluoperazine (Vincenzi *et al.*, 1978), or by a protein which could be called CaM-binding protein (Wang and Desai, 1977; Larsen *et al.*, 1978b) or CaM-BP_{80} (Cheung, 1980). Basal activity is not sensitive to these agents, suggesting a lack of dependence of basal activity on endogenous CaM.

The rate of active Ca^{2+} transport obtained with inside-out vesicles in the presence of added CaM is in good agreement with the values measured in more intact RBC preparations. In our judgment the rate of Ca^{2+} transport measured in intact RBCs or resealed ghosts is dependent on endogenous CaM. This assessment is based on the magnitude of the transport rates and the sensitivity of resealed ghost Ca^{2+} transport to chlorpromazine (Schatzmann, 1970).

The amount of CaM in the RBC, based on the balance sheet for its purification, was calculated to be 7×10^{-6} *M* (Jung, 1978). This level is high compared to the apparent K_d for CaM, and it is pertinent to ask why this large amount of calmodulin is present. There are a number of functional roles which CaM could have in the RBC based on its relatively high concentration. Some of these are considered below. For the moment, let us consider only the role of CaM in the regulation of the RBC Ca^{2+} pump.

The ATPase data and transport data show that CaM can stimulate the Ca^{2+} pump and its associated ATPase. Such data have usually been collected in the presence of maximal (around 10^{-5} *M*) Ca^{2+} concentrations. On the other hand, considering the low intracellular Ca^{2+} present in the RBC, the fact that the ATPase and the pump can operate without added CaM, and that EGTA-washed membranes appear to have little or no CaM, Roufogalis (1980) has suggested that the pump functions *in vivo* normally in a CaM-free state. This suggestion was based on the assumption that CaM is not an integral subunit of the Ca^{2+} pump ATPase, but can function as a regulator. We will make the same assumption for the moment and consider two alternative interpretations below.

If we assume that basal ATPase and pump activities are expressed without bound CaM and that binding of CaM increases the V_{max}, then we can calculate the apparent extent of CaM loading of the Ca^{2+} pump *in vivo*. At the outset, we emphasize that these calcultions should be viewed as semiquantitative estimates. Our justification for making these calculations is that, applied to the RBC, which lacks intracellular calcium storage, such calculations may have some validity and may help guide thinking for future work. For our calculations, a Hewlett-Packard 9830 desk top calculator was programmed to compute Ca^{2+} and CaM interaction. The K_d for Ca^{2+} binding was assumed to be 2.38×10^{-6} (Dedman *et al.*, 1977). Four equivalent, noncooperative Ca^{2+} binding sites were assumed (Dedman *et al.*, 1977). To estimate the extent of CaM binding to the Ca^{2+} pump *in vivo* we assumed that binding of Ca^{2+} to any two or more of the binding sites on CaM prompts association of $CaM{\cdot}(Ca^{2+})_n$ with the pump. This assumption is justified on the basis that the Ca^{2+} dependence of

$Ca{\cdot}(CaM^{2+})_{\geq 2}$ more closely resembles the Ca^{2+} dependence of ATPase and transport (Vincenzi and Larsen, 1980) than other possible choices [e.g., $CaM{\cdot}(Ca^{2+})_{n=1}$ or $CaM{\cdot}(Ca^{2+})_{n\geq 3}$] (see Fig. 1B). We further assumed that intracellular Ca^{2+} is 2.5×10^{-7} *M* (Simons, 1976a,b) and that total CaM = 7×10^{-6} *M* (Jung, 1978). These assumptions result in an estimate that the intact RBC $CaM{\cdot}(Ca^{2+})_{n > 2}$ constitutes about 5% of the total CaM or 3.5×10^{-7} *M*. The extent of binding of $CaM{\cdot}(Ca^{2+})_n$ to Ca^{2+} pump sites was estimated based on further assumptions: (1) one CaM binds to one pump site; (2) the number of pump sites per cell = 400 (Drickamer, 1975); (3) the volume per RBC = 1×10^{-13} liters; and (4) the *K*d of $CaM{\cdot}(Ca^{2+}(n$ for binding to pump sites = 4 n*M* (Larsen, 1979). Considering $CaM{\cdot}(Ca^{2+})_n$ affinity for pump sites and ignoring binding to other CaM binding sites, we estimate more than 98% of the Ca^{2+} pump sites will contain bound $CaM{\cdot}(Ca^{2+})_n$ under these conditions. It should be noted that the estimate of intracellular Ca^{2+} is conservative. If higher levels [e.g., 1 μM Ca^{2+} (Schatzmann, 1975; Vincenzi and Hinds, 1976)] were assumed, then a greater percentage of Ca^{2+} pump sites would contain $CaM{\cdot}(Ca^{2+})_n$.

In summary, we conclude that the RBC Ca^{2+} pump operates in a CaM-bound mode *in vivo*. One effect of the relatively high concentration of CaM in the RBC is to shift the equilibrium toward loading of the pump with CaM. This occurs even at the relatively low Ca^{2+} concentrations which are maintained by the pump which is being regulated.

The above calculations and speculations are based on the assumption that CaM is not an integral subunit of the Ca^{2+} pump ATPase. This would be analogous to the case of cyclic nucleotide phosphodiesterase of beef brain (Cheung, 1971). Calmodulin regulates phosphodiesterase by binding to it in a reversible and Ca^{2+}-dependent fashion (Cheung *et al.*, 1978; Ho *et al.*, 1977). Beef brain phosphodiesterase might be said to represent external CaM control. (Fig. 3A). Myosin light chain kinase is another example of such control (Dabrowska *et al.*, 1978). On the other hand, some enzymes [such as cyclic nucleotide phosphodiesterase from rabbit lung (Sharma and Wirch, 1979)] may contain CaM as an integral subunit which does not readily dissociate. These might be said to represent internal CaM control (Fig. 3B). A third possibility which has been brought forth by Cohen and co-workers is that both modes of control may apply to a given enzyme (Fig. 3C). Thus, skeletal muscle phosphorylase kinase contains CaM as an integral subunit (Cohen *et al.*, 1978). This subunit appears to be important to the basal activity of the enzyme (internal CaM control). In addition, the enzyme responds with an increased V_{max} to exogenous CaM (P. Cohen, personal communication). We will avoid the temptation of suggesting that this latter represents a "dual overhead CaM model," but

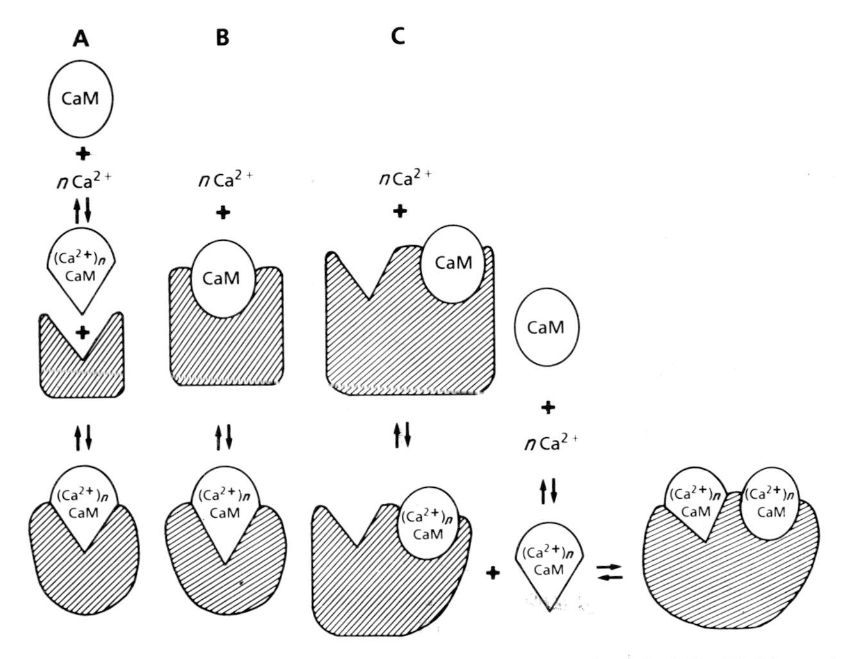

Fig. 3. Schematic models of three hypothetical modes of calmodulin (CaM) regulation. (A) External calmodulin control. In this case, CaM is readily dissociable from the enzyme which it regulates, binding only in the presence of Ca^{2+}. Experimentally, the enzyme would appear to be activated by Ca^{2+} and CaM, depending on the circumstances (e.g., beef brain cyclic nucleotide phosphodiesterase). (B) Internal CaM control. In this case, CaM is not dissociable from the enzyme; it is a subunit of the enzyme. The enzyme will be activated by Ca^{2+} (e.g., rabbit lung phosphodiesterase). (C) Dual CaM control. Both types of control. The enzyme will be activated by both Ca^{2+} and by Ca^{2+} and CaM, depending on the experimental conditions (e.g., skeletal muscle phosphorylase kinase). It remains to be determined which of these models most closely represents CaM control of the plasma membrane $(Ca^{2+} + Mg^{2+})$-ATPase. In these models no specific values for n are implied.

hope that the idea will be clear without a catchy name. Which of these models of CaM regulation apply to the Ca^{2+} pump ATPase is unknown at this time. None of these possibilities can be excluded, although internal CaM control is the least likely because of the ease with which CaM can be removed (Farrance and Vincenzi, 1977a,b). More work is needed on this question.

What are the roles of CaM (if any) in addition to its presumed regulation of the Ca^{2+} pump? Some of the other roles seem inescapable, while others can only be speculated upon at this time. We wish to make explicit our opinion that the high concentration of CaM in the RBC is not a chance event. It is not simply a vestigial genetic expression of a Ca^{2+}-binding protein that is important in other cells or which was needed only by immature red cells. Indeed, regulation of intracellular Ca^{2+} in the RBC, which we presume to be dependent on CaM, is critical to thc life of the RBC and in turn to the like of the organism.

An almost inescapable conclusion is that CaM in the RBC serves as a Ca^{2+} sink or buffer. The same assumptions for CaM content and binding constants were made as outlined above. The total amount of calcium in the RBC was estimated to be 1.6×10^{-6} *M*. This is one-tenth of the total calcium content of RBCs (1.6×10^{-5} *M*) measured by Harrison and Long (1968). These workers found that nonpenetrating chelating agents removed over 90% of the calcium associated with RBCs. At the assumed total cellular calcium of 1.6×10^{-6} *M*, and assuming no Ca^{2+} binding to other species, 80% of the CaM molecules would still be calcium-free (Fig. 4). Taking into account the multiple sites present on each CaM molecule, over 93% of the total Ca^{2+} binding sites would be free at this concentration of calcium. Thus, CaM can serve as a potential sink of Ca^{2+} in the cytoplasm.

Another way of looking at CaM as a Ca^{2+} sink is shown in Fig.5. Within limits, an increment in total cellular calcium will result in a smaller increment in free Ca^{2+} because of the presence of CaM. As shown in Fig. 5,

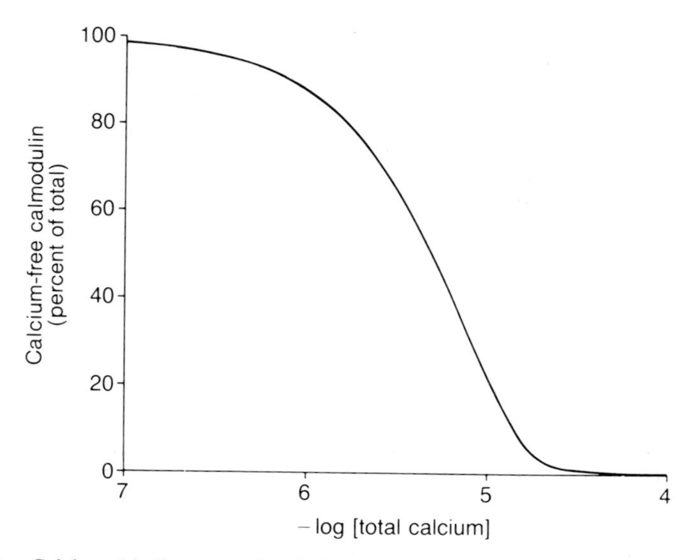

Fig. 4. Calcium binding to calmodulin (CaM). Calcium ion binding to CaM was calculated as in Fig. 1. As a function of total calcium the percentage of Ca^{2+}-free CaM molecules decreased. At the assumed total calcium present in the normal human red blood cell (1.6×10^{-6} *M*) approximately 80% of the CaM molecules would be Ca^{2+}-free. At this same calcium concentration, over 93% of the total Ca^{2+}-binding sites would be free. Thus, CaM can act as a Ca^{2+} sink when total calcium in the cell increases.

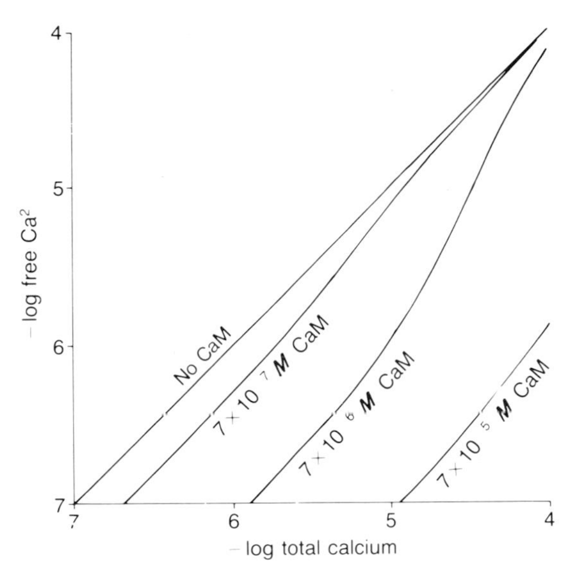

Fig. 5. Calmodulin binding of Ca^{2+} in a model system. Free Ca^{2+} was calculated as a function of total calcium concentration as in Fig. 1, but in the presence of various assumed concentrations of calmodulin (CaM), as indicated. Under conditions presumed to be applicable to the cytoplasm of the normal human red blood cell (7×10^{-6} *M* CaM, 1.6×10^{-6} *M* total calcium) free Ca^{2+} would be 1.3×10^{-7} *M*. In other words, about 92% of the cytoplasmic calcium would be bound to CaM, assuming no binding to other components. Because saturation of CaM is not reached under these assumptions, CaM would serve as a sink, and not as a typical buffer.

calculated free $[Ca^{2+}]$s are below total [calcium] because of binding to CaM. The effect is, of course, dependent on the amount of CaM. At the assumed total RBC calcium of 1.6×10^{-6} *M* and with CaM at 7×10^{-6} *M*, free Ca^{2+} would be approximately 1.3×10^{-7} *M*. This value is in good agreement with a totally independent estimate of Ca^{2+} in the RBC (2.5×10^{-7} *M*) (Simons, 1976b). This does not prove, but is compatible with, the interpretation that most RBC cytoplasmic calcium is bound to CaM.

Changes in cellular CaM content are, of course, possible. Figure 5 plots data for assumed CaM ranging from 7×10^{-7} to 7×10^{-5} *M*. If it were assumed that CaM concentration were double or one-half the apparent cellular content[9] of 7×10^{-6} *M* and at the constant total calcium of 1.6×10^{-6} *M*, Ca^{2+} would range from approximately 6.6×10^{-8} to 2.5×10^{-7} *M*, respectively. We suspect that the cells regulate Ca^{2+} or $CaM{\cdot}(Ca^{2+})_n$ and that total CaM and total calcium may vary considerably in certain cells in response to various stimuli and stresses.

We believe that these semiquantitative data allow some qualitative interpretations of the normal human RBC. It seems clear that while there is sufficient CaM·$(Ca^{2+})_n$ to keep pump sites occupied (see above), most of the CaM is free and, thus, can serve as a sink for intracellular Ca^{2+}. This is a result of (1) the high Ca^{2+} affinity of CaM, (2) the relatively high CaM concentration in the cell, and (3) the low total amount of calcium in the cell. The amount of Ca^{2+} present in the cytoplasm of other cells is probably comparable to that in RBCs (Carafoli and Crompton, 1978). Thus, CaM, which appears to be ubiquitous and present in high concentrations in many cells, presumably functions as a Ca^{2+} sink in those cells as well.

Another predicted effect which would result from the presence of CaM in cells (including the RBC) has been pointed out to us by Professor James B. Bassingthwaighte (personal communication). In the presence of fixed Ca^{2+}-binding sites, such as present on cell membranes, diffusion of Ca^{2+} is somewhat limited when compared to free Ca^{2+} in solution. In the presence of fixed binding sites inclusion of a mobile Ca^{2+}-binding sites (such as CaM) with a similar K_d will increase the apparent diffusion constant of Ca^{2+}, at least when $[Ca^{2+}]$ is of the same order as the K_d of the binding sites. The phenomenon may serve to facilitate the distribution of Ca^{2+} throughout the cytoplasm. In other words, CaM may aid in rapid movement of Ca^{2+} in those cases where diffusion would have been limiting in the absence of the mobile Ca^{2+}-binding site. It should be noted that binding of Ca^{2+} to CaM *in vitro* would decrease the diffusion constant. However, in the presence of a significant number of (nonsaturated) fixed binding sites, the presence of mobile binding sites will serve to increase the apparent diffusion constant. The signifigance of this phenomenon in living cells has yet to be demonstrated, but seems obvious.

A number of responses occur in RBCs in which Ca^{2+} is elevated. One of these phenomena is the so-called "Gárdos effect" (Gárdos, 1958). This is an increase in K^+ conductance of the membrane promoted by increased intracellular Ca^{2+} (Lew and Ferriera, 1978). The Gárdos effect is so sensitive to Ca^{2+} that is has been used to estimate Ca^{2+} in RBC ghosts (Simons, 1976b). The Gárdos effect is, thus, a prime candidate for a "physiological" response in the RBC that is mediated by CaM. Whether this is so remains to be determined. The fact that Ca^{2+} induced no K^+ leak when added to the outside of (CaM-free) inside-out vesicles (Grinstein and Rothstein, 1978) is consistent with, but does not prove, this interpretation. A Gárdos-like effect has been documented in a variety of cells, including nerve (Meech, 1978).

Most workers imply that CaM mediates only "physiological" responses to cells. Indeed, that may be so, but needs to be investigated. Most of the responses of RBC membranes to increased intracellular Ca^{2+} are probably "pathological" and are probably not mediated by CaM.

These changes include increased Na^+ conductance and impaired Na^+-K^+ transport (Dunn, 1974), decreased membrane deformability (LaCelle *et al.*, 1973), changes in shape (Weed and Chailley, 1973; LaCelle *et al.*, 1973), and loss of membrane area associated with altered phospholipid metabolism (Allan and Michell, 1978). A molecular mechanism of membrane damage induced by intracellular Ca^{2+} has been elucidated by Lorand and co-workers (Lorand *et al.*, 1978, 1979). These workers showed extensive cross-linking of RBC membrane proteins promoted by a cytoplasmic transglutaminase. The transglutaminase is activated by Ca^{2+} in the millimolar range. Like most of these effects, transglutaminase activation occurs only when the CaM sink and the plasma membrane Ca^{2+} pump have failed to keep up with inward "leak." As an extension of the work of Lorand *et al.* (1978), we have measured the (Ca^{2+} + Mg^{2+})-ATPase of membranes isolated from RBCs which were preloaded with high levels of Ca^{2+}. As expected, membrane proteins were cross-linked in these cells. In addition, the (Ca^{2+} + Mg^{2+})-ATPase activity, and its responsiveness to CaM was impaired (Vincenzi and Raess, 1979). This could suggest that prolonged elevations of Ca^{2+} in the cell can damage the Ca^{2+} pump; Ca^{2+}-induced impairment of the Ca^{2+} pump could lead to a viscious cycle, but no evidence for such a cycle has been put forth.

C. Calmodulin in the Regulation of the Plasma Membranes of Non-Erythrocytes

We are unaware of any published reports unambiguously demonstrating involvement of CaM in the plasma membrane Ca^{2+} transport of cells other than RBCs. We anticipate that CaM will eventually be shown to be involved in plasma (and possibly other) membrane transport. Phenothiazine-induced inhibition of Ca^{2+} efflux (Schatzmann, 1970) from other cells might be a means of exploring this possibility. It must be noted that, in contrast to our anticipation, Leslie and Borowitz (1975) concluded that the cells of the adrenal medulla, but not those of the adrenal cortex do not have a plasma membrane Ca^{2+} pump. Greengard and his colleagues have demonstrated that CaM promotes Ca^{2+}-dependent protein phosphorylation in synaptosomal membrane fractions (Greengard, 1979). Whether this is associated with Ca^{2+} transport or some other CaM-regulated cellular activity has not been determined. It appears that the plasma membranes of plant cells will invert and take up Ca^{2+} in an ATP-dependent fashion (Gross and Marme, 1978). Although CaM is known to be present in plants, to our knowledge, its possible effects on Ca^{2+} transport are not known.

D. Calmodulin and Ca^{2+} Transport of Non-Plasma Membranes

It is too early to conclude whether CaM is involved directly in the regulation of Ca^{2+} transport of non-plasma membranes. Indirect effects on all Ca^{2+}-dependent processes seem inescapable (e.g., Fig. 5). It seems unlikely that CaM is directly involved in mitochondrial Ca^{2+} transport (E. Carafoli, personal communication). At least one report suggests that CaM increases the rate of Ca^{2+} transport into a preparation enriched in cardiac sarcoplasmic reticulum (Katz and Remtulla, 1978). The effect was found to be independent of, and additive with, the well-known cyclic AMP-dependent increase in sarcoplasmic reticular Ca^{2+} transport (Tada and Kirchberger, 1978). A difficulty in interpreting the report of Katz and Remtulla (1978) is that the effect could be due to plasma membrane contamination of the sarcoplasmic reticulum preparation. Other workers (B. Van Winkle and M. Entman, personal communication) failed to find an effect of CaM on cardiac sarcoplasmic reticular Ca^{2+} transport.

E. Pharmacologic Antagonism of Calmodulin

Whatever the precise mechanism(s) by which CaM regulates (Ca^{2+} + Mg^{2+})-ATPase and Ca^{2+} transport, it is clear that CaM may be a target for drug action. Levin and Weiss (1977) reported that phenothiazines bind to CaM in a Ca^{2+}-dependent fashion. Thereby phenothiazines antagonize Ca^{2+}-dependent activation of cyclic nucleotide phosphodiesterase. With the growing number of established roles for CaM (e.g., Vanaman *et al.*, 1977) it seems pertinent to ask whether some of the therapeutic and/or side effects of phenothiazines, or other drugs, are dependent on modification of CaM or its effectors. Weiss and Levin (1978), for example, implied that the neuroleptic effects of phenothiazines and butyrophenones are related to binding of these drugs to CaM. We found that trifluoperazine antagonized Ca^{2+} transport (Raess and Hinds, 1979) and (Ca^{2+} + Mg^{2+})-ATPase (Vincenzi *et al.*, 1978), but that neither haloperidol nor butaclamol antagonized CaM in the ATPase assay (B. U. Raess and F. F. Vincenzi, unpublished). We interpreted these data to suggest that neuroleptic activity does not correlate with generalized antagonism of CaM. It seems reasonable to suggest that the multiple side effects of phenothiazines are to a large extent related to antagonism of CaM. It also seems reasonable to assume that specific neuroleptic activity is based upon antagonism of central dopamine receptors (Seeman, 1977). In any event, the notion of "anti-CaM" drug activity as a bona fide pharmacological mechanism may evolve from the work of Levin and Weiss (1977), and we believe it will be a useful idea upon which to base further experimental work. Hidaka and his co-workers (Hidaka *et al.*, 1978, 1979) have independently developed a series of anti-CaM drugs which are napthalene sulfonamides. We think it likely that these compounds will eventually be shown to be more specific anti-CaM compounds than phenothiazines. It is, of course, too early to predict what therapeutic uses might be made of selective anti-CaM compounds. Even if no therapeutic applications were to arise, such compounds could help to elucidate more fully the multiple role(s) and mechanism(s) of this ubiquitous protein. A major question is whether it will be possible to interfere selectively with various CaM effectors in the cell (thereby reducing the effectiveness of intracellular Ca^{2+}) or to interfere selectively with the plasma membrane Ca^{2+} pump (thereby increasing the level and, possibly, the effectiveness of intracellular Ca^{2+}). A recent review of the role of Ca^{2+} antagonism in the therapeutic action of drugs (Rahwan *et al.*, 1979) considers various ways in which drugs may alter cellular responses to Ca^{2+}. Although CaM and anti-CaM drugs were not considered explicitly, the review forms an excellent conceptual framework. We consider it likely that some drugs with generalized "Ca^{2+} antagonist" properties will eventually be shown to have at least some anti-CaM activity and that newer and more specific anti-CaM drugs will be developed in the future.

IV. REGULATION OF THE ERYTHROCYTE PLASMA MEMBRANE Ca^{2+} PUMP

A. Introduction

There are striking parallels between the Ca^{2+} pump and the Na^+–K^+ pump (Schatzmann and Vincenzi, 1969). Thus, it was natural to assume that regulation of the Ca^{2+} pump occurred simply as a consequence of the amount of Ca^{2+} available at the inner membrane surface. It is now apparent that multiple influences on the Ca^{2+} pump are possible, and that CaM is, perhaps, only one of a number of regulators of plasma membrane Ca^{2+} transport. The interrelationships between these various regulators and their relative importance in health and disease remain to be determined. Influences on both Ca^{2+} "leak" and "pump" are possible. Some agents may affect one or the other process, or both.

A caveat is in order. Various experimental manipulations may be shown to modify a process in a certain preparation *in vitro*. However, whether this manipulation is relevant to the *in vivo* situation is often not determined. For example, CaM-binding protein (Wang and Desai, 1977) is a protein isolated from bovine brain. Calmodulin-binding protein was shown to antagonize both the activation of (Ca^{2+} + Mg^{2+})-ATPase and Ca^{2+} transport induced by CaM (Larsen *et al.*, 1978b). Activity resembling CaM-binding protein was recently reported in pig RBCs by Au (1978). What remains to be determined is whether CaM in RBCs (or other cells) is regulated by CaM-binding protein or similar proteins and, if so, under what conditions. In all likelihood, mechanisms for both acute and chronic regulation of the plasma membrane Ca^{2+} pump exist, but these have not been examined. Some of the potential regulatory mechanisms will be mentioned.

B. Monovalent Cations

Schatzmann and Vincenzi (1969) found no specific monovalent cation sensitivity of the (Ca^{2+} + Mg^{2+})-ATPase by substituting Na^+ for K^+ and could not demonstrate the need for a transported counterion. The possibility that at least some of the Ca^{2+} pump ATPase was dependent upon a monovalent cation (e.g., Na^+ or K^+) was raised by Schatzmann and Rossi (1971). When choline was substituted for Na^+ and K^+ replacement of either Na^+ or K^+ increased the (Ca^{2+} + Mg^{2+})-ATPase activity of isolated membranes. On the other hand, when Ca^{2+}-loaded resealed RBC ghosts (presumably sealed with K^+, although not stated) were suspended in media without Na^+ and K^+, no decrease in the rate of Ca^{2+} extrusion was noted as compared to extrusion from ghosts suspended in Na^{2+} or K^+. Asymmetry of the RBC membrane may account for these apparently contradictory data. Thus, Sarkadi *et al.* (1978) found that Na^+ or K^+ increased the Ca^{2+} influx rate by approximately fourfold into inside-out vesicles of RBC membranes. The monovalent ions were essentially equipotent and equally effective. Presumably no CaM was present (see below). The most parsimonious explanation of all of these findings seems to be that a rather nonspecific monovalent cation sensitive site is present on the cytoplasmic face of the RBC membrane. Binding at this monovalent cation site can increase the rate of (Ca^{2+} + Mg^{2+})-ATPase, and transport. It seems unlikely that this is a very important regulatory mechanism *in vivo* because Na^+ and K^+ are always present at the cytoplasmic face in rather substantial concentrations. If, as Roufogalis (1979) suggests, the Ca^{2+} pump operates in a CaM-free manner *in vivo*, then based on the results of Scharff (1978; see below), perhaps small changes in intracellular Na^+ would be relatively more effective.

Whether CaM has any influence on the possible monovalent cation sensitivity of the Ca^{2+} pump ATPase is not known. Suggestive results in this area were recently presented by Scharff (1978). He found that RBC membranes prepared by lysis in a buffer containing Ca^{2+} (CaM-laden, "B state") were more sensitive to activation by K^+ than Na^+, and that the degree of activation was more than in membranes prepared by lysis in a buffer containing EGTA (CaM-free, "A state"). It was suggested that K^+, Na^+, or both contribute to the regulation of the Ca^{2+} pump, possibly by enhancing the role of Ca^{2+}. As noted, these effects are almost certainly mediated at the cytoplasmic surface of the membrane.

C. Phospholipids

Among the potential regulators of the plasma membrane Ca^{2+} pump are the phospholipids which surround the ATPase. Ronner *et al.* (1977) found that phospholipase A_2 treatment resulted in loss of (Ca^{2+} + Mg^{2+})-ATPase activity of RBC membranes. The activity could be restored by oleate and several negatively charged phospholipids, including phosphatidyl serine. Thus, the lipid requirement for the (Ca^{2+} + Mg^{2+})-ATPase appears to be similar to that of (Na^+ + K^+ + Mg^{2+})-ATPase. Roelofsen and Schatzmann (1977), using pure phospholipases, found somewhat different results. They concluded that outer leaflet phospholipids could be degraded completely (intact cells) without affecting (Ca^{2+} + Mg^{2+})-ATPase activity. On the other hand, exposure of isolated membranes to phospholipases which caused a loss of glycerophospholipids (phosphatidylcholine, phosphatidylethanolamine, and phosphatidylserine) resulted in complete loss of activity. Activity could be restored by any of the glycerphospholipids or lysolecithin, but not by fatty acids. Whether physiologically relevant regulation of the Ca^{2+} pump takes place by influence on phospholipids is not clear at this time. An intriguing paper which raises this possibility appeared recently (Strittmatter *et al.*, 1979). It was found that enzymatic methylation of membrane phospholipids using *S*-adenosyl-L-methionine as methyl donor increased the activity of the (Ca^{2+} + Mg^{2+})-ATPase, while there was no effect on Mg^{2+}-ATPase. The stimulation of (Ca^{2+} + Mg^{2+})-ATPase was related to an increase in the V_{max}, rather than a change in the apparent affinities for Ca^{2+} or ATP. Likewise, inclusion of *S*-adenosyl-L-methionine in $^{45}Ca^{2+}$ and ATP-loaded ghosts resulted in an increased efflux rate of $^{45}Ca^{2+}$. *S*-Adenosyl-L-methionine did not change the passive flux of $^{45}Ca^{2+}$. It was suggested that increased membrane fluidity produced by phospholipid methylation may be the mechanism of the effect (Strittmatter *et al.*, 1979).

D. Overdesign of the Pump?

An important question remains in the general area of regulation of the Ca^{2+} pump. Why is there such a tremendous capacity of the plasma membrane Ca^{2+} pump in the RBC? Various workers have found that the passive Ca^{2+} permeability of the RBC membrane is very low. In reviewing this, Lew and Ferriera (1978) noted that passive influx of Ca^{2+} into fresh RBCs is between 1 and 10 μmoles/liter cells/hr, whereas pumping rates are in the range of 5–40 mmoles/liter cells/hr. Roufogalis (1980) noted variations in pump rates in different laboratories. In spite of these variations, the general conclusion must be that pump capacity is in the range of three orders of magnitude greater than the passive permeability. Several ideas have been expressed in attempts to account for the apparent "overdesign" of the RBC membrane Ca^{2+} pump. Such overdesign would, of course, help to ensure that intracellular Ca^{2+} is maintained at very low levels. With a pump of a given affinity, overdesign would mainly affect the rate at which given low levels of Ca^{2+} could be attained. Is there any reason to believe that the apparent overdesign of the pump is necessary for such purposes? Alternatively, should the apparent overdesign of the RBC Ca^{2+} pump be regarded as an unnecessary process which is needed in most cells but which has not yet been eliminated in the evolution of the RBC? That is certainly possible. We think it rather more likely that the capacity of the Ca^{2+} pump is appropriate to the RBC and that it may be utilized from time to time under physiological circumstances, not only pathological conditions. The pitfall which may have led to the apparent paradox may have been to measure Ca^{2+} permeability of RBCs at "rest" *in vitro*. Thus, there may be both chemical and mechanical influences which increase the passive flux of Ca^{2+} into RBCs *in vivo*.

There is no compelling evidence for coupling of either Ca^{2+} channels or pumps in RBCs to hormone or other cell surface receptors. However, the results of Rasmussen and his co-workers could be interpreted to suggest that catecholamines (Allen and Rasmussen, 1971) and/or prostaglandins (Rasmussen and Lake, 1977) bind to human RBCs and promote the accumulation of Ca^{2+}. This probably happens without the intermediacy of cyclic AMP. Such an idea needs further study. It could drastically alter current thinking about how responses to catecholamines are mediated. It could also alter our ideas about the Ca^{2+} permeability of the RBC. RBCs exposed to such chemicals, possibly in high concentrations in the microcirculation, may take on a considerable obligatory load of Ca^{2+}. Other mechanisms by which intracellular Ca^{2+} may increase in RBCs while in the microcirculation include deoxygenation (Chau-Wong and Seeman, 1971). We would like to suggest, in addition, that mechanically induced

Ca^{2+} leak may be an important phenomenon in the microcirculation. We imagine that deformation of RBCs in the microcirculation causes increased influx of Ca^{2+}, which is then pumped out when the cells "recover" in the venous blood. To our knowledge, such an idea has never been investigated.

E. The Plasma Membrane Ca^{2+} Pump in Disease

Many, and perhaps seemingly unrelated, diseases may be related to variations in plasma membrane passive permeability to Ca^{2+}, the membrane Ca^{2+} pump, its ATPase, and their regulation. This seems a promising area and many laboratories are investigating relationships between various diseases and alterations in Ca^{2+} regulation. None of the suggested relationships can be considered proven. However, some ideas are attractive and well formulated, if not yet established. In order to illustrate the concepts, and the potential impact on Ca^{2+} binding proteins, a few examples will be mentioned. Other references to disease are contained in an earlier review (Vincenzi and Hinds, 1976).

Cystic fibrosis (CF) is a disease which affects mainly exocrine secretory function, including abnormal pancreatic function and altered fat absorption. Because lipid metabolism may be grossly abnormal in CF, it seems reasonable to suggest that some of the reported effects of the disease on membrane ATPase may be related indirectly to changes in membrane fluidity. For example, Horton *et al.* (1970) reported that the (Ca^{2+} + Mg^{2+})-ATPase activity or RBC membranes was decreased in patients with CF. The decrease in activity was correlated with the clinical severity of the disease. In apparent agreement with Horton *et al.* (1970), Katz (1978a) also reported a specific decrease in the (Ca^{2+} + Mg^{2+})-ATPase activity or RBC membranes of cystic fibrosis patients. The V_{max} was reduced by about 50%, whereas the activities of the Mg^{2+}-ATPase and (Na^{+} + K^{+} + Mg^{2+})-ATPase were not different from control values. The implication is that there is not a generalized membrane defect, but that some difference in Ca^{2+} transport may exist in RBCs as well as in secretory cells. Extending this finding, Katz (1978b) also found decreased (Ca^{2+} + Mg^{2+})-ATPase activity in membranes prepared from cystic fibrosis-derived fibroblasts grown in tissue culture. This latter result shows that genetically determined changes in cellular function may be demonstrated in CF cells. In our laboratory, measurements of CF RBC membranes were performed over the past four years. Our results show considerable variation which we cannot ascribe to uncontrolled factors in measurement techniques. The variation could arise from patient variability, in the preparation of membranes, in variability in the amount of Ca^{2+} in the ATPase assay (before EGTA was used as a Ca^{2+} buffer in our system) and possibly in the amount of CaM which remained bound to the membranes. This latter source of possible variability would have been a factor before sensitivity to CaM was a part of our routine ATPase assay method. Comparing RBC membranes from normals and CF homozygotes, we saw no difference in the Mg^{2+}-ATPase, (Na^+ + K^+ + Mg^{2+})-ATPase, or (Ca^{2+} + Mg^{2+})-ATPase in the absence of added CaM. However, in the presence of added autologous hemolysate (presumably containing CaM) CF membranes exhibited statistically significant lower activity than normals (F. F. Vincenzi, unpublished). The significance of this observation is not known. The relationship of our results and the differences from those of Katz (1978a) may be due to methodology.

There are a number of results which are compatible with the idea that Ca^{2+} influx or Ca^{2+} levels may be altered in cells of patients with various dystrophies (Duncan, 1978). Ruitenbeek (1979) found that Ca^{2+}-stimulated *p*-nitrophenylphosphatase activity of RBC membranes from patients with muscular dystrophy was twice that of normals. Likewise (Ca^{2+} + Mg^{2+})-ATPase was increased in RBC membranes from patients with Duchenne and congenital myotonic muscular dystrophy. It was suggested that the increased (Ca^{2+} + Mg^{2+})-ATPase activity may represent a compensatory response for some change in calcium metabolism of dystrophic red cells (Ruitenbeek, 1979). A possible change in the monovalent cation sensitivity (Schatzmann and Rossi, 1971) of the (Ca^{2+} + Mg^{2+})-ATPase was demonstrated but preliminary results showed no major change in the response (Ruitenbeek, 1979). Luthra *et al.* (1979) recently reported that the (Ca^{2+} + Mg^{2+})-ATPase activities of membranes isolated from RBCs of Duchenne and myotonic dystrophy patients were comparable to controls. In the presence of added hemolysate (containing CaM) the activity of Duchenne dystrophy RBCs was higher and the activity of myotonic dystrophy RBCs was lower than controls. Likewise, in saponin-lysed RBCs (endogenous CaM present), there were slight differences in the group means as compared to controls. The significance, if any, of these findings remains to be determined, since there was great variation in the control values.

Although the molecular basis of sickle cell anemia is known, the mechanism of sickling, especially the formation of irreversibly sickled cells (ISCs) is not known. Calcium ion is involved in irreversible sickling. Eaton *et al.* (1973) found approximately eight times as much calcium in RBCs of sickle cell patients as in normals. ISCs were found to contain twice as much calcium as reversibly sickled cells with normal morphology. Many functional changes in the sickle RBCs could be traced to calcium accumulation. Artificial calcium loading of sickle RBCs produced cells with characteristics like ISCs (Eaton *et al.*, 1975). Intracellular accumulation of calcium was suggested as the cause of most of the defects in ISCs. Just why calcium accumulation occurs in sickle cells is not clear. Our laboratory reported that the (Ca^{2+} + Mg^{2+})-ATPase of sickle cell membranes is less responsive to CaM than normal. The CaM-induced increase in activity was less, while the apparent affinity for CaM was unchanged (Vincenzi and Gopinath, 1977). It was suggested that a decreased maximal activity of the plasma membrane Ca^{2+} pump may be important in the sickling phenomenon.

A relationship may exist between the (Ca^{2+} + Mg^{2+})-ATPase activity of RBCs and bone marrow activity. This suggestion was made by Scharff and Foder (1975) who found that the (Ca^{2+} + Mg^{2+})-ATPase activity of patients with polycythemia vera was 67% of the control value. The degree of reduction of ATPase activity was related to the degree of leukocytosis. RBCs from patients with secondary polycythemia showed no such effect. Scharff and Foder (1975) prepared membranes which were presumably CaM-associated. In light of the findings of Vincenzi and Gopinath (1977) referred to above, it would be of interest to know the responsiveness of the CaM-free polycythemia RBC membranes to added CaM.

V. CONCLUSIONS

The evidence reviewed in this chapter supports the notion that the cytoplasmic Ca^{2+}-binding protein, calmodulin (CaM), is involved in the regulation of the plasma membrane Ca^{2+} pump of the erythrocyte. Calmodulin appears to significantly increase the V_{max} of the Ca^{2+} pump ATPase and Ca^{2+} transport, with little or no change in the K_m (ATP) or K_d (Ca^{2+}). The molecular mechanism of CaM in the regulation of the Ca^{2+} pump has not been elucidated. The presence of a substantial amount of CaM in the RBC, coupled with fairly low total cellular calcium, leads to the conclusions that most intracellular calcium is bound to CaM, most intracellular CaM is calcium-free, and most Ca^{2+} pump sites are occupied by CaM *in vivo*. Variations in the regulation of the plasma membrane Ca^{2+} pump may occur in or in response to certain diseases or drugs.

ACKNOWLEDGMENTS

We are indebted to the following individuals who kindly provided prepublication information for us: J. Bassingthwaighte, E. Carafoli, P. Cohen, M. Entman, B. Roufogalis, J. Penniston, and W. B. Van Winkle. Mr. Robert Hartshorne of our department programmed the

Hewlett-Packard desk top calculator to compute Ca^{2+} and calmoulin interactions. We gratefully acknowledge his help and advice.

REFERENCES

Åkerman, K. E., and Wikstrom, M. K. F. (1979). (Ca^{2+} + Mg^{2+})-stimulated ATPase activity of rabbit myometrium plasma membrane is blocked by oxytocin. *FEBS Lett.* **97,** 283–287.

Allan, D., and Michell, R. H. (1978). A calcium-activated polyphosphoinositide phosphodiesterase in the plasma membrane of human and rabbit erythrocytes. *Biochim. Biophys. Acta* **508,** 277–286.

Allen, J. E., and Rasmussen, H. (1971). Human red blood cells: Prostaglandin E_2, epinephrine, and isoproterenol alter deformability. *Science* **174,** 512–514.

Au, K. S. (1978). An endogenous inhibitor of erythrocyte membrane (Ca^{2+} + Mg^{2+})-ATPase. *Int. J. Biochem.* **9,** 477–480.

Baker, P. F. (1972). Transport and metabolism of calcium ions in nerve. *Prog. Biophys. Mol. Biol.* **24,** 179–223.

Blaustein, M. P. (1976). The ins and outs of calcium transport in squid axons: Internal and external ion activation of calcium efflux. *Fed. Proc., Fed. Am. Soc. Exp. Biol.* **35,** 2574–2578.

Bodemann, H., and Passow, H. (1972). Factors controlling the resealing of the membrane of human erythrocyte ghosts after hypotonic hemolysis. *J. Membr. Biol.* **8,** 1–26.

Bond, G. H., and Clough, D. L. (1973). A soluble protein activator of (Mg^{2+} + Ca^{2+})-dependent ATPase in human red cell membranes. *Biochim. Biophys. Acta* **323,** 592–599.

Borle, A. B. (1969). Kinetic analyses of calcium movements in HeLa cell cultures. II. Calcium efflux. *J. Gen. Physiol.* **53,** 57–69.

Borle, A. B. (1972). Kinetic analysis of calcium movements in cell culture. V. Intracellular calcium distribution in kidney cells. *J. Membr. Biol.* **10,** 45–66.

Brostrom, C. O., Huang, Y. -C., Breckenridge, B. McL., and Wolff, D. O. (1975). Identification of a calcium-binding protein as a calcium-dependent regulator of brain adenylate cyclase. *Proc. Natl. Acad. Sci. U.S.A.* **72,** 64–68.

Carafoli, E., and Crompton, M. (1978). The regulation of intracellular calcium. *Curr. Top. Membr. Transp.* **10,** 151–216.

Carsten, M. E. (1977). Can hormones regulate myometrial calcium transport? *In* "The Biochemistry of Smooth Muscle" (N. L. Stephens, ed.), pp. 617–639. Univ. Park Press, Baltimore, Maryland.

Cha, Y. N., and Lee, K. S. (1976). Phosphorylation of the red blood cell membrane during the active transport of Ca^{++}. *J. Gen. Physiol.* **67,** 251–261.

Cha, Y. N., Shin, B. C., and Lee, K. S. (1971). Active uptake of Ca^{++} and Ca^{++}-activated Mg^{++} ATPase in red cell membrane fragments. *J. Gen. Physiol.* **57,** 202–215.

Chau-Wong, M., and Seeman, P. (1971). The control of membrane-bound Ca^{2+} by ATP. *Biochim. Biophys. Acta* **241,** 473–482.

Cheung, W. Y. (1971). Cyclic 3′,5′-nucleotide phosphodiesterase. *J. Biol. Chem.* **246,** 2859–2869.

Cheung, W. Y. (1980). Calmodulin plays a pivotal role in cellular regulation. *Science* **207,** 19–27.

Cheung, W. Y., Lynch, T. J., and Wallace, R. W. (1978). An endogenous Ca^{2+}-dependent activator protein of brain adenylate cyclase and cyclic nucleotide phosphodiesterase. *Adv. Cyclic Nucleotide Res.* **9,** 233–251.

Cohen, P., Burchell, A., Foulkes, J. G., Cohen, P. T. W., Vanaman, T. C., and Nairn, A. C. (1978). Identification of the Ca^{2+}-dependent modulator protein as the fourth subunit of rabbit skeletal muscle phosphorylase kinase. *FEBS Lett.* **92,** 287–293.

Dabrowska, R., Sherry, J. M. F., Aromatorio, D. K., and Hartshorne, D. J. (1978). Modulator protein as a component of the myosin light chain kinase from chicken gizzard. *Biochemistry* **17,** 253–258.

Danon, D., and Marikovsky, Y. (1964). Determination of density distribution of red cell population. *J. Lab. Clin. Med.* **64,** 668–674.

Davis, P. W., and Vincenzi, F. F. (1971). Ca-ATPase activation and NaK-ATPase inhibition as a function of calcium concentration in human red cell membranes. *Life Sci.* **10,** 401–406.

Dedman, J. R., Potter, J. D., Jackson, R. L., Johnson, J. D., and Means, A. R. (1977). Physicochemical properties of rat testis Ca^{2+}-dependent regulator protein of cyclic nucleotide phosphodiesterase. Relationship of Ca^{2+} binding, conformational changes, and phosphodiesterase activity. *J. Biol. Chem.* **252,** 8415–8422.

DeLoach, J., and Ihler, G. (1977). A dialysis procedure for loading erythrocytes with enzymes and lipids. *Biochim. Biophys. Acta* **496,** 136–145.

DiPolo, R. (1978). Ca pump driven by ATP in squid axons. *Nature (London)* **274,** 390–392.

DiPolo, R., and Beaugé, L. (1979). Physiological role of ATP-driven calcium pump in squid axon. *Nature (London)* **278,** 271–273.

Dodge, J. T., Mitchell, C., and Hanahan, D. J. (1963). The preparation of chemical characteristics of hemoglobin-free ghosts of human erythrocytes. *Biochim. Biophys. Acta* **100,** 119–130.

Drickamer, L. K. (1975). The red cell membrane contains three different adenosine triphosphatases. *J. Biol. Chem.* **250,** 1925–1954.

Duncan, C. J. (1978). Role of intracellular calcium in promoting muscle damage: A strategy for controlling the dystrophic condition. *Experientia* **34,** 1531–1672.

Dunham, E. T., and Glynn, I. M. (1961). Adenosinetriphosphatase activity and the active movements of alkali metal ions. *J. Physiol. (London)* **156,** 274–293.

Dunn, M. J. (1974). Red blood cell calcium and magnesium: Effects upon sodium and potassium transport and cellular morphology. *Biochim. Biophys. Acta* **352,** 97–116.

Eaton, J. W., Skelton, T. D., Swofford, H. S., Kolpin, C. E., and Jacob, H. S. (1973). Elevated erythrocyte calcium in sickle cell disease. *Nature (London)* **246,** 105–106.

Eaton, J. W., Berger, E., White, J. G., and Jacob, H. S. (1975). Effects of calcium on hemoglobin SS erythrocytes: Evidence that calcium accumulation underlies formation of irreversibly sickled red cells. *In* "Proceedings of the Symposium on Molecular and Cellular Aspects of Sickle Cell Disease" (J. I. Hercules, G. L. Cottam, M. R. Watterman, and A. N. Schechter, eds.), Publ. No. (NIH) 76-1007. pp. 327–345. U. S. Department of Health, Education, and Welfare, Washington, DC.

Farrance, M. L. (1977). Mechanism of enhancement of (Ca-Mg) ATPase activity of membranes from human erythrocytes hemolyzed in isosmotic imidazole buffer. Doctoral Dissertation, University of Washington, Seattle.

Farrance, M. L., and Vincenzi, F. F. (1977a). Enhancement of (Ca^{2+} + Mg^{2+})-ATPase activity of human erythrocyte membranes by hemolysis in isosmotic imidazole buffer. I. General properties of variously prepared membranes and the mechanism of the isosmotic imizadole effect. *Biochim. Biophys. Acta* **471,** 49–58.

Farrance, M. L., and Vincenzi, F. F. (1977b). Enhancement of (Ca^{2+} + Mg^{2+})-ATPase activity of human erythrocyte membranes of hemolysis in isosmotic imidazole buffer. II. Dependence on calcium and a cytoplasmic activator. *Biochim. Biophys. Acta* **471,** 59–66.

Gárdos, G. (1958). The role of calcium in the potassium permeability of human erythrocytes. *Acta Physiol. Acad. Sci. Hung.* **15,** 121–125.

Gopinath, R. M., and Vincenzi, F. F. (1977). Phosphodiesterase protein activator mimics red blood cell cytoplasmic activator of (Ca^{2+} + Mg^{2+})-ATPase. *Biochem. Biophys. Res. Commun.* **17,** 1203–1209.

Greengard, P. (1979). Cyclic nucleotides, phosphorylated proteins, and the nervous system. *Fed. Proc., Fed. Am. Soc. Exp. Biol.* **38,** 2217.

Grinstein, S., and Rothstein, A. (1978). Chemically-induced cation permeability in red cell membrane vesicles. The sidedness of the response and the proteins involved. *Biochim. Biophys. Acta* **508,** 236–245.

Gross, J., and Marme, D. (1978). ATP-dependent Ca^{2+} uptake into plant membrane vesicles. *Proc. Natl. Acad. Sci. U.S.A.* **75,** 1232–1236.

Haaker, H., and Racker, E. (1979). Purification and reconstitution of Ca^{2+}-ATPase from plasma membranes of pig erythrocytes. *J. Biol. Chem.* **254,** 6598–6602.

Hanahan, D. J., Taverna, R. D., Flynn, D. D., and Ekholm, J. E. (1978). The interaction of Ca^{2+}/Mg^{2+} ATPase activator protein and Ca^{2+} with human erythrocyte membranes. *Biochem. Biophys. Res. Commun.* **84,** 1009–1015.

Harrison, D. G., and Long, C. (1968). The calcium content of human erythrocytes. *J. Physiol.* (*London*) **199,** 367–381.

Hidaka, H., Asano, M., Iwadare, S., Matsumoto, I., Totsuka, T., and Aoki, N. (1978). A novel vascular relaxing agent, *N*-(6-aminohexyl)-5-chloro-1-naphthalenesulfonamide which affects vascular smooth muscle actomyosin. *J. Pharmacol. Exp. Ther.* **207,** 8–15.

Hidaka, H., Yamaki, T., Totsuka, T., and Asano, M. (1979). Selective inhibitors of Ca^{2+}-binding modulator of phosphodiesterase produce vascular relaxation and inhibit actin-myosin interaction. *Mol. Pharmacol.* **15,** 49–59.

Hinds, T. R., Larsen, F. L., and Vincenzi, F. F. (1978). Plasma membrane Ca^{2+} transport: Stimulation by soluble proteins. *Biochem. Biophys. Res. Commun.* **81,** 455–461.

Ho, H. C., Wirch, E., Stevens, F. C., and Wang, J. H. (1977). Purification of a Ca^{2+}-activatable cyclic nucleotide phosphodiesterase from bovine heart by specific interaction with its Ca^{2+}-dependent modulator protein. *J. Biol. Chem.* **252,** 43–50.

Horton, C. R., Cole, W. Q., and Bader, H. (1970). Depressed (Ca^{++})-transport ATPase in cystic fibrosis erythrocytes. *Biochem. Biophys. Res. Commun.* **40,** 505–509.

Janis, R. A., Crankshaw, D. J., and Daniel, E. E. (1977). Control of intracellular Ca^{2+} activity in rat myometrium. *Am. J. Physiol.* **232,** C50–C58.

Jarrett, H. W. and Penniston, J. T. (1977). Partial purification of the Ca^{2+} + Mg^{2+} ATPase activator from human erythrocytes: its similarity to the activator of 3′:5′-cyclic nucleotide phosphodiesterase. *Biochem. Biophys. Res. Commun.* **77,** 1210–1216.

Jarrett, H. W., and Penniston, J. T. (1978). Purification of the Ca^{2+}-stimulated ATPase activator from human erythrocytes. *J. Biol. Chem.* **253,** 4676–4682.

Jay, A. W. L., and Burton, A. C. (1969). Direct measurement of potential difference across the human red blood cell membrane. *Biophys. J.* **9,** 115–121.

Jung, N. S. G. T. (1978). Purification and characterization of the cytoplasmic (Ca^{2+} + Mg^{2+})-ATPase activator protein of human red blood cells. Master's Thesis, University of Washington, Seattle.

Kant, J. A., and Steck, T. L. (1972). Cation-impermeable inside-out and right-side-out vesicles from human erythrocyte membranes. *Nature* (*London*), *New Biol.* **240,** 26–28.

Käser-Glanzmann, R., Jakábová, M., George J. N., and Lüscher, E. F. (1978). Further characterization of calcium-accumulating vesicles from human blood platelets. *Biochim. Biophys. Acta* **512,** 1–12.

Katz, S. (1978a). Calcium and sodium transport processes in patients with cystic fibrosis. I. A specific decrease in Mg^{2+}-dependent, Ca^{2+}-adenosine triphosphatase activity in erythrocyte membranes from cystic fibrosis patients. *Pediatr. Res.* **12,** 1033–1038.

Katz, S. (1978b). Calcium and sodium transport processes in patients with cystic fibrosis. 2. Mg^{2+}-dependent, Ca^{2+}-ATPase activity in fibroblast membrane preparations from cystic fibrosis in patients and control. *Res. Commun. Chem. Pathol. Pharmacol.* **19,** 491–503.

Katz, S., and Blostein, R. (1975). Ca^{2+}-stimulated membrane phosphorylation and ATPase activity of the human erythrocyte. *Biochim. Biophys. Acta* **389,** 314–324.

Katz, S., and Remtulla, M. A. (1978). Phosphodiesterase protein activator stimulates calcium transport in cardiac microsomal preparations enriched in sarcoplasmic reticulum. *Biochem. Biophys. Res. Commun.* **83,** 1373–1379.

Knauf, P. A., Proverbio, F., and Hoffman, J. F. (1974). Electrophoretic separation of different phosphoproteins associated with Ca-ATPase and Na,K-ATPase in human red cell ghosts. *J. Gen. Physiol.* **63,** 324–336.

Kretsinger, R. H. (1976). Calcium binding proteins. *Annu. Rev. Biochem.* **45,** 239–266.

Kretsinger, R. H. (1977). Evolution of the information role of calcium in eukaryotes. *In* "Calcium Binding Proteins and Calcium Function" (R. H. Wasserman, R. A. Corradino, E. Carafoli, R. H. Kretsinger, D. H. MacLennon, and F. L. Siegel, eds.), pp. 63–72. Am. Elsevier, New York.

LaCelle, P. L., Kirkpatrick, F. H., Udkow, M. P., and Arkin, B. (1973). Membrane fragmentation and Ca^{++}-membrane interaction: Potential mechanisms of shape change in the senescent red cell. *In* "Red Cell Shape" (M. Bessis, R. I. Weed, and P. F. Leblond, eds.), pp. 69–78, Springer-Verlag, Berlin and New York.

Lamb, J. F., and Lindsay, R. (1971). Effect of Na, metabolic inhibitors, and ATP on Ca movements in L cells. *J. Physiol. (London)* **218,** 691–708.

Larsen, F. L. (1979). Studies on the plasma membrane Ca^{2+} pump and (Ca^{2+} + Mg^{2+})-ATPase of human red blood cells. Doctoral Dissertation, University of Washington, Seattle.

Larsen, F. L., and Vincenzi, F. F. (1979). Calcium transport across the plasma membrane: Stimulation by calmodulin. *Science* **204,** 306–309.

Larsen, F. L., Hinds, T. R., and Vincenzi, F. F. (1978a). On the red blood cell Ca^{2+}-pump: An estimate of stoichiometry. *J. Membr. Biol.* **41,** 361–376.

Larsen, F. L., Raess, B. U., Hinds, T. R., and Vincenzi, F. F. (1978b). Modulator binding protein antagonizes activation of (Ca^{2+} + Mg^{2+})-ATPase and Ca^{2+} transport of red blood cell membranes. *J. Supramol. Struct.* **9,** 269–274.

Lee, K. S., and Shin, B. C. (1969). Studies on the active transport of calcium in human red cells. *J. Gen. Physiol.* **54,** 713–729.

Leslie, S. W., and Borowitz, J. L. (1975). Evidence for a plasma membrane calcium pump in bovine adrenal medulla but not adrenal cortex. *Biochim. Biophys. Acta* **394,** 227–238.

Levin, R. M., and Weiss, B. (1977). Binding of trifluoperazine to the calcium-dependent activator of cyclic nucleotide phosphodiesterase. *Mol. Pharmacol.* **13,** 690–697.

Lew, V. L., and Ferreira, H. G. (1978). Calcium transport and the properties of a calcium-activated potassium channel in red cell membranes. *Curr. Top. Membr. Transp.* **10,** 217–277.

Lindsay, R. (1976). Properties of the calcium-activated adenosine triphosphatase from L-cell membranes. *Q. J. Exp. Physiol. Cogn. Med. Sci.* **61,** 95–104.

Lorand, L., Siefring, G. E., and Lowe-Krentz, L. (1978). Formation of γ-glutamyl-ϵ-lysine bridges between membrane proteins by a Ca^{2+}-regulated enzyme in intact erythrocytes. *J. Supramol. Struct.* **9,** 427–440.

Lorand, L., Siefring, G. E., Jr., and Lowe-Krenetz, L. (1979). Enzymatic basis of membrane stiffening in human erythrocytes. *Semin. Hematol.* **16,** 65–73.

Luthra, M. G., Hildenbrandt, G. R., and Hanahan, D. J. (1976). Studies on an activator of the (Ca^{2+} + Mg^{2+})-ATPase of human erythrocyte membranes. *Biochim. Biophys. Acta* **419,** 164–179.

Luthra, M. G., Stern, L. Z., and Kim, K. D. (1979). (Ca^{++} + Mg^{++})-ATPase of red cells in Duchenne and myotonic dystrophy: Effect of soluble cytoplasmic activator. *Neurology* **29,** 835–841.

Lynch, T. J., and Cheung, W. Y. (1979). Human erythrocyte Ca^{2+} – Mg^{2+}-ATPase: Mechanism of stimulation by Ca^{2+}. *Arch. Biochem. Biophys.* **194,** 165–170.

MacIntyre, J. D., and Green, J. W. (1978). Stimulation of calcium transport in inside-out vesicles of human erythrocyte membranes by a soluble cytoplasmic activator. *Biochim. Biophys. Acta* **510,** 373–377.

Meech, R. W. (1978). Calcium-dependent potassium activation in nervous tissues. *Annu. Rev. Biophys. Bioeng.* **7,** 1–18.

Mırčevová, L., and Šimonová, A. (1973). Effect of barbiturate on Mg^{++}-dependent ATPase in human erythrocytes. *Experientia* **29,** 660.

Moore, L., Fitzpatrick, D. F., Chen, T. S., and Landon, E. J. (1974). Calcium pump activity of the renal plasma membrane and renal microsomes. *Biochim. Biophys. Acta* **345,** 405–418.

Mualem, S., and Karlish, S. J. D. (1979). Is the red cell calcium pump regulated by ATP? *Nature (London)* **277,** 238–240.

Niggli, V., Penniston, J. T., and Carafoli, E. (1979). Purification of the Ca^{2+} – Mg^{2+} ATPase from human erythrocyte membranes using a calmodulin affinity column. *J. Biol. Chem.* **254,** 9955–9958.

Ohashi, T., Uchida, S., Nagai, K., and Yoshida, H. (1970). Studies on phosphate hydrolyzing activities in the synaptic membrane. *J. Biochem. (Tokyo)* **67,** 635–641.

Olson, E. J., and Cazort, R. J. (1969). Active calcium and strontium transport in human erythrocyte ghosts. *J. Gen. Physiol.* **53,** 311–322.

Penzias, A. A. (1979). The origin of the elements. *Science* **205,** 549–554.

Pershadsingh, H. A., and McDonald, J. M. (1979). Direct addition of insulin inhibits a high affinity Ca^{2+}-ATPase in isolated adipocyte plasma membranes. *Nature (London)* **281,** 495–497.

Quist, E. E., and Roufogalis, B. D. (1975a). Calcium transport in human erythrocytes. Separation and reconstitution of high and low Ca affinity (Mg + Ca)-ATPase activities in membranes prepared at low ionic strength. *Arch. Biochem. Biophys.* **168,** 240–251.

Quist, E. E., and Roufogalis, B. D. (1975b). Determination of the stoichiometry of the calcium pump in human erythrocytes using lanthanum as a selective inhibitor. *FEBS Lett.* **50,** 135–139.

Quist, E. E., and Roufogalis, B. D. (1977). Association of (Ca + Mg)-ATPase activity with ATP-dependent Ca uptake in vesicles prepared from human erythrocytes. *J. Supramol. Struct.* **6,** 375–381.

Raess, B. U., and Hinds, T. R. (1979). Plasma membrane Ca^{2+} transport: Inhibition by trifluoperazine. *Pharmacologist* **21,** 277.

Rahwan, R. G., Piascik, M. F., and Witiak, D. T. (1979). The role of calcium antagonism in the therapeutic action of drugs. *Can. J. Physiol. Pharmacol.* **57,** 443–460.

Rasmussen, H., and Bordier, P. (1974). *"The Physiological and Cellular Basis of Metabolic Bond Disease."* Williams & Wilkins, Baltimore, Maryland.

Rasmussen, H., and Lake, W. (1977). Prostaglandins and the mammalian erythrocyte. *In* "Prostaglandins in Hematology" (M. J. Silver and J. B. Smith, eds.), pp. 187–202. Spectrum Publ., New York.

Rega, A. F., and Garrahan, P. J. (1975). Calcium ion-dependent phosphorylation of human erythrocyte membranes. *J. Membr. Biol.* **22,** 313–327.

Robinson, J. D. (1976). (Ca + Mg)-stimulated ATPase activity of a rat brain microsomal preparation. *Arch. Biochem. Biophys.* **176,** 366–374.

Roelofsen, B., and Schatzmann, H. J. (1977). The lipid requirement of the (Ca^{2+} + Mg^{2+})-ATPase in the human erythrocyte membrane, as studied by various highly purified phospholipases. *Biochim. Biophys. Acta* **464,** 17–36.

Ronner, P., Gazzotti, P., and Carafoli, E. (1977). A lipid requirements for the (Ca^{2+} + Mg^{2+})-activated ATPase of erythrocyte membranes. *Arch. Biochem. Biophys.* **179,** 578–583.

Rossi, J. P. F. C., Garrahan, P. J., and Rega, A. F. (1978). Reversal of the calcium pump in human red cells. *J. Membr. Biol.* **44,** 37–46.

Roufogalis, B. D. (1979). Regulation of calcium translocation across the red blood cell membrane. *Can. J. Physiol. Pharmacol.* **57,** 1331–1349.

Ruitenbeek, W. (1979). Membrane-bound enzymes of erythrocytes in human muscular dystrophy. *J. Neurol. Sci.* **41,** 71–80.

Sarkadi, B., Szász, I., and Gárdos, G. (1976). The use of ionophores for rapid loading of human red cells with radioactive cations for cation-pump studies. *J. Membr. Biol.* **26,** 357–370.

Sarkadi, B., Szász, I., Gerlóczy, A., and Gárdos, G. (1977). Transport parameters and stoichiometry of active calcium ion extrusion in intact human red cells. *Biochim. Biophys. Acta* **464,** 93–107.

Sarkadi, B., MacIntyre, J. D., and Gárdos, G. (1978). Kinetics of active calcium transport in inside-out red cell membrane vesicles. *FEBS Lett.* **89,** 78–82.

Scharff, O. (1972). The influence of calcium ions on the preparation of the (Ca^{2+} + Mg^{2+})-activated membrane ATPase in human red cells. *Scand. J. Clin. Lab. Invest.* **30,** 313–320.

Scharff, O. (1976). Ca^{2+} activation of membrane-bound (Ca^{2+} + Mg^{2+})-dependent ATPase from human erythrocytes prepared in the presence or absence of Ca^{2+}. *Biochim. Biophys. Acta* **443,** 206–218.

Scharff, O. (1978). Stimulating effects of monovalent cations on activator-dissociated and activator-associated states of Ca^{2+}-ATPase in human erythrocytes. *Biochim. Biophys. Acta* **512,** 309–317.

Scharff, O., and Foder, B. (1975). Decreased (Ca^{2+} + Mg^{2+})-stimulated ATPase activity in erythrocyte membranes from polycythemia vera patients. *Scand. J. Clin. Lab. Invest.* **35,** 583–389.

Scharff, O., and Foder, B. (1977). Low Ca^{2+} concentrations controlling two kinetic states of Ca^{2+}-ATPase from human erythrocytes. *Biochim. Biophys. Acta* **483,** 416–424.

Schatzmann, H. J. (1966). ATP-dependent Ca^{++} extrusion from human red cells. *Experientia* **22,** 364–365.

Schatzmann, H. J. (1970). Transmembrane calcium movements in resealed human red cells. *In* "Calcium and Cellular Function" (A. W. Cuthbert, ed.), pp. 85–95. St. Martin's Press, New York.

Schatzmann, H. J. (1973). Dependence on calcium concentration and stoichiometry of the calcium pump in human red cells. *J. Physiol.* (*London*) **235,** 551–569.

Schatzmann, H. J. (1975). Active calcium transport and Ca^{2+}-activated ATPase in human red cells. *Curr. Top. Membr. Transp.* **6,** 125–168.

Schatzmann, H. J., and Bürgin, H. (1978). Calcium in human red blood cells. *Ann. N. Y. Acad. Sci.* **307,** 125–147.

Schatzmann, H. J., and Rossi, G. L. (1971). (Ca^{2+} + Mg^{2+})-Activated membrane ATPase in human red cells and their possible relations to cation transport. *Biochim. Biophys. Acta* **241,** 379–392.

Schatzmann, H. J., and Vincenzi, F. F. (1969). Calcium movements across the membrane of human red cells. *J. Physiol.* (*London*) **201,** 369–395.

Seeman, P. (1977). Anti-schizophrenic drugs—Membrane receptor sites of action. *Biochem. Pharmacol.* **26,** 1741–1748.

Sharma, R. K., and Wirch, E. (1979). Ca^{2+}-dependent cyclic nucleotide phosphodiesterase from rabbit lung. *Biochem. Biophys. Res. Commun.* **91,** 338–344.

Simons, T. J. B. (1976a). The preparation of human red cell ghosts containing calcium buffers. *J. Physiol.* (*London*) **256,** 209–225.

Simons, T. J. B. (1976b). Calcium-dependent potassium exchange in human red cell ghosts. *J. Physiol.* (*London*) **256,** 227–244.

Steck, T. L. (1974a). The organization of proteins in the human red blood cell membrane. *J. Cell Biol.* **62,** 1–19.

Steck, T. L. (1974b). Preparation of impermeable inside-out and right-side-out vesicles from erythrocyte membranes. *Methods Membr. Biol.* **2,** 245–281.

Strittmatter, W. J., Hirata, F., and Axelrod, J. (1979). Increased Ca^{2+}-ATPase activity associated with methylation of phospholipids in human erythrocytes. *Biochem. Biophys. Res. Commun.* **88,** 147–153.

Swanson, P. D., Anderson, L., and Stahl, W. L. (1974). *Biohim. Biophys. Acta* **356,** 174–183.

Tada, M., and Kirchberger, M. A. (1978). Significance of the membrane protein phospholamban in cyclic AMP-mediated regulation of calcium transport by sarcoplasmic reticulum. *In* "Recent Advances on Cardiac Structure and Metabolism" (T. Kobayashi, T. Sano, and N. S. Dhalla, eds.), Vol. II, pp.265–272. Univ. Park Press, Baltimore, Maryland.

Vanaman, T. C., Sharief, F., and Watterson, D. M. (1977). Structural homology between brain modulator protein and muscle TnCs. *In* "Calcium-Binding Proteins and Cell Function" (R. H. Wasserman, R. A. Corradino, E. Carafoli, R. H. Kretsinger, D. H. MacLennon, and F. L. Siegel, eds.), pp. 107–116. Am. Elsevier, New York.

Vincenzi, F. F. (1971). A calcium pump in red cell membranes. *In* "Cellular Mechanism for Calcium Transport and Homeostasis" (G. Nichols, Jr. and R. H. Wasserman, eds.), pp. 135–148. Academic Press, New York.

Vincenzi, F. F. (1978). Regulation of a plasma membrane calcium pump: A speculative model. *Ann. N. Y. Acad. Sci.* **307,** 229–231.

Vincenzi, F. F., and Farrance, M. L. (1977). Interaction between cytoplasmic (Ca^{2+} + Mg^{2+})-ATPase activator and the erythrocyte membrane. *J. Supramol. Struct.* **7,** 301–306.

Vincenzi, F. F., and Gopinath, R. M. (1977). (Ca^{2+} + Mg^{2+})-ATPase activator of red blood cells: Decreased response in sickle cell membranes. *In* "Calcium Binding Proteins and Calcium Function" (R. H. Wasserman, R. A. Corradino, E. Carafoli, R. H. Kretsinger, D. H. MacLennon, and F. L. Siegel, eds.), pp. 507–509. Am. Elsevier, New York.

Vincenzi, F. F., and Hinds, T. R. (1976). Plasma membrane calcium transport and membrane-bound enzymes. *Enzymes Bio. Membr.* **2,** 261–281.

Vincenzi, F. F., and Larsen, F. L. (1980). The plasma membrane calcium pump: Regulation by a soluble Ca^{2+} binding protein. *Fed. Proc., Fed. Am. Soc. Exp. Biol.* **39,** 2427–2431.

Vincenzi, F. F., and Raess, B. U. (1979). Red blood cell calcium pump ATPase: Reversible and irreversible changes caused by intracellular Ca^{2+}. *Fed. Proc., Fed. Am. Soc. Exp. Biol.* **38,** 304.

Vincenzi, F. F., Raess, B. U., Larsen, F. L., Jung, N. S. G. T., and Hinds, T. R. (1978). The plasma membrane calcium pump: A potential target for drug action. *Pharmacologist* **20,** 195.

Wang, J. H., and Desai, R. (1977). The modulator binding protein—A bovine brain protein exhibiting Ca^{2+}-dependent association with the protein modulator of cyclic nucleotide phosphodiesterase. *J. Biol. Chem.* **252,** 4175–4184.

Wasserman, R. H., and Feher, J. J. (1977). Vitamin D-dependent calcium-binding proteins. *In* "Calcium Binding Proteins and Calcium Function" (R. H. Wasserman, R. A. Corradino, E. Carafoli, R. H. Kretsinger, D. H. MacLennan, and F. L. Siegel, eds.), pp. 292–302. Am. Elsevier, New York.

Weed, R. I., and Chailley, B. (1973). Calcium–pH interactions in the production of shape changes in erythrocytes. *In* "Red Cell Shape" (M. Bessis, R. I. Weed, and P. F. Leblond, eds.), pp. 55–68, Springer-Verlag, Berlin and New York.

Weiner, M. L., and Lee, K. S. (1972). Active calcium ion uptake by inside-out and right-side-out vesicles of red blood cell membranes. *J. Gen. Physiol.* **59,** 462–475.

Weiss, B., and Levin, R. M. (1978). Mechanisms for selectively inhibiting the activation of cyclic nucleotide phosphodiesterase and adenylate cyclase by antipsychotic agents. *Adv. Cyclic Nucleotide Res.* **9,** 285–303.

Chapter 8

Smooth Muscle Myosin Light Chain Kinase

ROBERT S. ADELSTEIN
CLAUDE B. KLEE

I. INTRODUCTION

The contractile proteins actin, myosin and tropomyosin (Korn, 1978), like the calcium-binding protein, calmodulin (Cheung, 1980; Klee, *et al.*, 1980; Wang and Waisman, 1979) are widely distributed throughout nature. Since Ca^{2+} is known to initiate contractile activity in a wide variety of cells, it is not surprising that calcium-receptor proteins (Kretsinger, 1980) have been found to be intimately involved in the contractile process. Two examples of these calcium-binding proteins in muscle are calmodulin, which is present in all types of muscle, as well as in non-muscle cells, and troponin C, which has only been found in skeletal and cardiac muscle (see Section I,B). The purpose of this chapter is to review what is known about the calmodulin-dependent enzyme, smooth muscle myosin

CALCIUM AND CELL FUNCTION, VOL. I

ISBN 0-12-171401-2

light chain kinase, and to understand the role played by this enzyme in regulating smooth muscle contraction. In reviewing the properties of the myosin kinase isolated from smooth muscle, we shall compare this enzyme to the other myosin kinases isolated from skeletal muscle (Yagi *et al.*, 1978; Nairn and Perry, 1979a), cardiac muscle (Walsh *et al.*, 1979; Rappaport and Adelstein, 1980), and non-muscle cells, such as platelets (Dabrowska and Hartshorne, 1978; Hathaway and Adelstein, 1979), BHK cells (Yerna *et al.*, 1979), and brain cells (Dabrowska and Hartshorne, 1978; Hathaway *et al.*, 1980a). For a comprehensive review on smooth muscle see Small and Sobieszek (1980).

A. Smooth Muscle Contractile Proteins

The major contractile proteins found in smooth muscle are actin, myosin, and tropomyosin (see Fig. 1). Actin is a globular protein with a monomeric molecular weight of 43,000. At physiological ionic strength it is polymerized in the form of a double helical filament. In addition to its structural role in muscle cells, actin has the unique property of activating the rate at which myosin hydrolyzes ATP. This *in vitro* enzyme activation (the actin-activated ATPase activity of myosin) is thought to correlate with the *in vivo* hydrolysis of ATP by myosin that accompanies muscle contraction.

Tropomyosin, a fibrous protein which has a helical content of more than 90%, lies in the groove created by the two actin filaments. It has a molecular weight of 66,000, and in smooth muscle it is composed of two different subunits. The exact function of tropomyosin in smooth muscle is

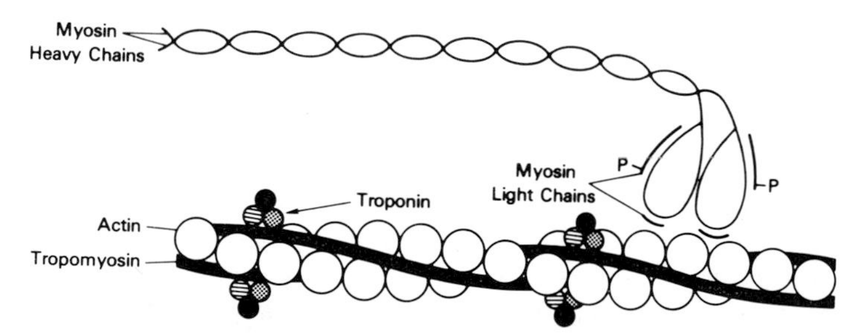

Fig. 1. Diagram of the muscle contractile proteins. The phosphorylated light chains are indicated (P), but the exact location of the myosin light chains with respect to the heavy chains is not known. Skeletal, cardiac, and smooth muscle contain actin, myosin, and tropomyosin but troponin has only been positively identified in skeletal and cardiac muscle. (Modified from Adelstein and Eisenberg, 1980, by permission.)

uncertain, but it is known to amplify the actin-activation of myosin ATPase activity (Sobieszek and Small, 1977; Chacko *et al.*, 1977).

Smooth muscle myosin, similar to most myosins isolated from muscle and non-muscle cells, is a hexamer. It is composed of two heavy chains (MW 200,000) and two pairs of light chains (MW 20,000 and 15,000). The 20,000 dalton light chain can undergo a covalent, reversible, phosphorylation (Fig. 1).

B. Regulation of the Actin–Myosin Interaction

Muscle contraction consists of the cyclic attachment and detachment of the globular portion of the myosin molecule to the actin filament (see Fig. 1). The attachment is followed by a change in the angle of myosin–actin orientation in a manner that causes the myosin and actin filaments to slide past each other (Huxley and Niedergerke, 1954; Huxley and Hanson, 1954). The energy for this physical process is provided by ATP and is released by the interaction of actin with myosin, which activates the myosin ATPase activity.

Actin activation of smooth muscle myosin ATPase activity differs from actin activation of myosin from skeletal and cardiac muscle in one significant manner: actin activation of smooth muscle myosin ATPase activity requires prior enzymatic phosphorylation of the 20,000 dalton light chain of myosin. Dephosphorylation of the 20,000 dalton light chain by a phosphatase restores myosin to a form that cannot be activated by actin (for a review, see Adelstein and Eisenberg, 1980). Evidence that phosphorylation of smooth muscle myosin plays a regulatory role in actin activation of myosin ATPase activity has been reported for a number of smooth muscles including: fowl gizzard (Sobieszek and Small, 1977; Gorecka *et al.*, 1976), pig stomach (Small and Sobieszek, 1977), guinea pig vas deferens (Chacko *et al.*, 1977), ewe uterus (Lebowitz and Cooke, 1979), bovine aorta (DiSalvo *et al.*, 1978), and human placenta (Huszar and Bailey, 1979). In addition to these *in vitro* studies, smooth muscle myosin phosphorylation has been shown to correlate with an increase in tension in preparations using skinned muscle fibers (Hoar *et al.*, 1979) and strips of smooth muscle (Aksoy and Murphy, 1979; de Lanerolle and Stull, 1980).

Myosin light chain phosphorylation has also been documented as a major regulatory mechanism in non-muscle cells such as platelets (Adelstein and Conti, 1975), proliferative myoblasts (Scordilis and Adelstein, 1977), macrophages (Trotter and Adelstein, 1979), and fibroblasts (Yerna *et al.*, 1979).

Whereas the reversible phosphorylation of myosin appears to be the major regulatory system in smooth muscle and non-muscle cells, it does not appear to play a major role in regulating skeletal and cardiac muscle contractile proteins. Nevertheless, these muscles also contain complete myosin phosphorylating systems, including calmodulin-dependent myosin kinases. In skeletal and cardiac muscle there is evidence that phosphorylation of myosin, while not required for actin activation of myosin ATPase activity, may act to modulate muscle contraction (Stull and High, 1977; Bárány *et al.*, 1979; Kopp and Bárány, 1979). For example, in an experiment with intact rat skeletal muscle, phosphorylation of myosin was found to be correlated with post-tetanic potentiation of peak twitch tension (Manning and Stull, 1979).

In skeletal and cardiac muscles, the major regulatory system consists of tropomyosin and a complex of proteins made up of troponin T (for tropomyosin binding), troponin I (for inhibitory), and troponin C (for Ca^{2+} binding). At Ca^{2+} concentrations below 10^{-7} *M*, actin activation of the myosin ATPase activity is inhibited. This inhibition is relieved when Ca^{2+} levels rises to 10^{-5} *M* which results in Ca^{2+} binding to troponin C (for reviews, see Weber and Murray, 1973; Adelstein and Eisenberg, 1980). There is a marked similarity in the primary structure of the two Ca^{2+}-binding proteins, calmodulin and troponin C. This possibility was initially suggested by Wang *et al.* (1975) and confirmed by the primary structure studies of Watterson *et al.* (1980).

To date no troponin complex has been isolated from smooth muscle [but see Marston *et al.* (1980) for preliminary evidence on the possible existence of such a complex]. Another Ca^{2+}-binding protein, leiotonin C has been isolated from smooth muscle (Mikawa *et al.*, 1978), and the possible existence of an alternate regulatory system cannot be ruled out (Ebashi *et al.*, 1978). As noted above, however, the major regulatory system controlling actin–myosin interaction in smooth muscle appears to consist of the Ca^{2+}-binding protein calmodulin and the enzyme activated by calmodulin, myosin light chain kinase.

II. SMOOTH MUSCLE MYOSIN LIGHT CHAIN KINASE (MYOSIN KINASE)

The active kinase isolated from smooth muscle is made up of two proteins: a large protein which confers the marked specificity of the enzyme and contains at least part, if not all, of the catalytic site. The second protein is calmodulin. This important observation was first reported for smooth muscle myosin kinase by Dabrowska *et al.* (1978) and for skeletal muscle myosin kinase by Yagi *et al.* (1978). The amino acid sequence of calmodulin isolated from bovine uterus smooth muscle (Grand and Perry, 1978) is essentially the same as the one isolated from bovine brain (Watterson *et al.,* 1980). Turkey gizzard myosin kinase comprises about 0.1% of the total muscle protein. It has a molecular weight of 134,000 in sodium dodecyl sulfate (SDS)-polyacrylamide gel electrophoresis and 124,000 by sedimentation equilibrium centrifugation under nondenaturing conditions. The value for $s_{20,w}$ is 4.45 without calmodulin bound and 5.05 with calmodulin bound (Adelstein and Klee, 1980). Chicken gizzard myosin kinase has a molecular weight of 105,000 by SDS-polyacrylamide gel electrophoresis (Dabrowska *et al.,* 1977).

Table I shows the absolute dependence of turkey gizzard smooth muscle myosin kinase activity on Ca^{2+} and calmodulin. Unlike other calmodulin-dependent enzymes (e.g., cyclic nucleotide phosphodiesterase) smooth muscle myosin kinase, as well as myosin kinases isolated from other sources, have no basal activity in the absence of calmodulin. Table II compares the properties of turkey gizzard smooth muscle myosin kinase to myosin kinases purified from skeletal muscle and a non-muscle source.

A. Preparation of Smooth Muscle Myosin Kinase

Two important features in the preparation of this enzyme are the use of proteolytic inhibitors to prevent degradation of the kinase and the use of a calmodulin affinity column in the final purification step (Adelstein *et al.,* 1978).

TABLE I

Dependence of Smooth Muscle Myosin Light Chain Kinase on Calcium and Calmodulin

Myosin kinase assay[a]	dpm Incorporated into light chain/min	Kinase specific activity (μmole/mg/min)
+ Ca^{2+} + calmodulin	278,000	3.0
+ EGTA + calmodulin	3,000	—
+ Ca^{2+} only	3,900	—
+ EGTA only	2,900	—

[a] Myosin light chain kinase activity was assayed in 0.1 ml, 20 m*M* Tris-HCl, pH 7.3, 10 m*M* $MgCl_2$, 0.1 m*M* [γ-^{32}P]ATP (0.5 Ci/mmole), 10^{-8} *M* myosin kinase, and 0.2 mg/ml smooth muscle myosin light chain at 24°C. Either 0.2 m*M* $CaCl_2$ (in excess over EGTA) or 2 m*M* EGTA was present in the assay, and 10^{-6} *M* calmodulin was added to the assay as indicated above. Background level of dpm for the Millipore assay performed under the above conditions is 3000–4000 dpm. Assays in the absence of calmodulin or calcium showed no increase in dpm with time over the background. (From Conti and Adelstein 1980.)

TABLE II

Properties of Myosin Kinases

Property	Smooth[a] (turkey gizzard)	Platelet[b] (human)	Skeletal[c] (rabbit)
MW (SDS-PAGE)	134,000	105,000	80,000
MW (sedimentation equilibrium)	124,000		
Specific activity (μmole P_i/mg/min)	30	8	30
K_m ATP (μM)	20	121	
K_m light chains (μM)	6	18	40–50
Can undergo phosphorylation	Yes	Yes	?
Effect of phosphorylation	Decreases activity	Decreases activity	
App K_m calmodulin	10^{-9}	10^{-8}	10^{-9}[d]

[a] Adelstein and Klee, 1980.
[b] Hathaway *et al.*, 1980b.
[c] Nairn and Perry, 1979a.
[d] Stull *et al.*, 1980.

With respect to proteolysis, Ca^{2+} and Mg^{2+}-activated proteases are inhibited by including 2–10 m*M* EDTA and EGTA throughout most of the preparation. In addition, the following proteolytic inhibitors are included in the extracting mixture: leupeptin, pepstatin, diisopropyl fluorophosphate, L-1-tosylamido-2-phenylethyl chloromethyl ketone, benzylarginylmethyl ester, soyabean trypsin inhibitor, and phenylmethylsulfonyl fluoride. The latter was then included in all the buffers, with the exception of the final step of purification (Adelstein *et al.*, 1978).

Following gel filtration and ion exchange chromatography, the partially purified kinase is applied, in the presence of Ca^{2+}, to a column of calmodulin covalently linked to Sepharose 4B (Klee and Krinks, 1978). The kinase which binds to calmodulin under these conditions (pH 7.8, 0.05–0.2 *M* NaCl) is eluted by replacing Ca^{2+} with 2 m*M* EGTA.

A number of myosin kinases from non-muscle cells were originally isolated as Ca^{2+}-independent kinases (Daniel and Adelstein, 1976; Adelstein, 1978). However, the myosin kinase isolated from platelets, when prepared in the presence of the proteolytic inhibitors as outlined above consists of a single form of the kinase that is entirely dependent on Ca^{2+}-calmodulin for activity (Hathaway and Adelstein, 1979). Omission of the proteolytic inhibitors results in the isolation of two forms of the kinase in the same preparation; a Ca^{2+}-calmodulin-independent form which does not bind to the calmodulin affinity column and a Ca^{2+}-calmodulin-dependent form which does bind. This suggests that limited proteolysis of the Ca^{2+}-calmodulin-dependent kinase during the preparative procedure removes the calmodulin-binding site, and produces an active kinase which no longer requires Ca^{2+} for activity (D. R. Hathaway and R. S. Adelstein, unpublished).

Recent experiments, with the smooth muscle kinase, appear to confirm this hypothesis. Thus, brief proteolysis of the smooth muscle myosin kinase with trypsin first liberates a 20,000 dalton peptide containing the site of phosphorylation (see Section II,C) on the kinase (Adelstein *et al.*, 1978). This has no major effect on the kinase activity nor on its dependence on Ca^{2+}-calmodulin. Further digestion with trypsin generates an enzyme which retains about one-half of its original activity, but is no longer dependent on either Ca^{2+} or calmodulin for activity. This modified enzyme does not bind to a calmodulin affinity column. Thus, the calmodulin-binding site can be removed by proteolysis, resulting in an active kinase that is Ca^{2+} and calmodulin independent (Conti and Adelstein, unpublished). As noted in Table I, the intact enzyme is completely inactive in the absence of calmodulin. This ability to generate a Ca^{2+}-calmodulin-independent enzyme by proteolysis is similar to the results with cyclic nucleotide phosphodiesterase (Cheung, 1971).

B. Properties of Smooth Muscle Myosin Kinase

Purified smooth muscle myosin kinase has marked specificity for the phosphorylatable light chain of myosin. It does not phosphorylate mixed histones, casein, phosphorylase *b*, or phosphorylase kinase. The amino acid sequence around the phosphorylatable serine has been determined for the 20,000 dalton light chain of myosin isolated from chicken gizzard smooth muscle. For comparison, the sequence around the phosphorylated residue of chicken skeletal muscle myosin is also given:

chicken (smooth): -Arg-Ala-Thr-Ser*-Asn-Val-Phe-Ala-Met-

chicken (skeletal): -Glu-Gly-Ser-Ser*-Asn-Val-Phe-Ser-Met-

In each case the serine residue that is phosphorylated is located near the blocked amino-terminal end, at residue 13.

As noted above, all myosin kinases isolated to date (in the Ca^{2+}-dependent form), appear to be regulated by Ca^{2+} through the binding of Ca^{2+} to calmodulin. Since the free concentration of Ca^{2+} in resting smooth muscle

* Phosphorylated residue: chicken smooth, Jakes *et al.* (1976); chicken skeletal, Matsuda *et al.* (1977).

is approximately 10^{-7} *M* and since myosin kinase is inactive at this concentration of Ca^{2+} (Sobieszek, 1977), it is reasonable to assume that calmodulin is dissociated from myosin kinase in resting muscle. A rise in Ca^{2+} concentration to 10^{-5} *M* is sufficient to activate myosin kinase *in vitro* (Sobieszek, 1977). This activation most likely occurs by Ca^{2+} first binding to calmodulin and the Ca^{2+}-calmodulin complex then binding to the inactive form of myosin kinase (see Fig. 2, lower left). A two-step mechanism of this type was first demonstrated for the activation of cyclic nucleotide phosphodiesterase (Teo and Wang, 1973; Teshima and Kakiuchi, 1974; Lin *et al.*, 1975; for a review, see Wolff and Brostrom, 1979). Evidence that this sequence might occur for Ca^{2+} activation of skeletal muscle myosin kinase has been published by Stull *et al.* (1980).

Calmodulin cannot bind to smooth muscle myosin kinase in the absence of Ca^{2+} (Adelstein and Klee, 1980). ^{14}C-Labeled (guanidinated) calmodulin and myosin kinase were sedimented in the presence of 10^{-4} *M* Ca^{2+} or 2 mM EGTA. In the presence of Ca^{2+} the [^{14}C]calmodulin cosedimented with myosin kinase, whereas they sedimented as two separate peaks in the presence of EGTA.

The stoichiometry of calmodulin binding to the 130,000 dalton myosin kinase appears to be 1 : 1, in the presence of 10^{-4} *M* Ca^{2+}. This was determined by three different techniques. First, [^{14}C]calmodulin was cross-linked with the 130,000 dalton myosin kinase using dimethyl suberimidate. The molecular weight of the complex thus formed (145,000) was found to correspond to a 1 : 1 complex of calmodulin (16,000) : myosin kinase (130,000), even when an excess of [^{14}C]calmodulin was employed. The second method involved titration of the enzyme with calmodulin in the presence of 10^{-4} *M* Ca^{2+}. The concentration of myosin kinase was 10^{-6} *M*, well above the binding constant of calmodulin for the kinase, which is approximately 10^{9} M^{-1} in the presence of 10^{-4} *M* Ca^{2+}. Kinase activity was found to increase linearly with the addition of calmodulin, leveling out when the amount of calmodulin reached 10^{-6} *M* (i.e., 1 : 1). Third, the kinase was sedimented in the absence and presence of Ca^{2+}. At equilibrium the molecular weight of the complex (140,000), in the presence of Ca^{2+} corresponded to a 1 : 1 complex of calmodulin : kinase. In the presence of EGTA the molecular weight was that of the kinase alone (Adelstein and Klee, 1980).

C. Phosphorylation of Myosin Kinase by cAMP-Dependent Protein Kinase

Smooth muscle myosin kinase can be phosphorylated by the catalytic subunit of cyclic AMP-dependent protein kinase (Adelstein *et al.*, 1978).

Approximately 1 mole of phosphate can be incorporated per mole of myosin kinase. Phosphorylation of a specific amino acid residue was suggested by finding that tryptic digestion of ^{32}P-labeled myosin kinase resulted in the release of a 20,000 dalton peptide which coelectrophoresed in an SDS-polyacrylamide gel with all the incorporated radioactivity (Adelstein *et al.*, 1978). However, since there is undoubtedly more than one potentially phosphorylatable residue in this fragment, and since in certain preparations more than 1 mole of phosphate can be incorporated into myosin kinase, the possibility of a second site for phosphorylation has not been excluded.

Phosphorylation of smooth muscle myosin kinase decreases the rate at which the enzyme transfers phosphate to the 20,000 dalton light chain of myosin (Adelstein *et al.*, 1978). The decreased activity of myosin kinase after phosphorylation appears to be related to a decrease in the affinity of the phosphorylated kinase for calmodulin (Conti and Adelstein, 1980). This was determined by studies of the calmodulin activation of low concentrations of phosphorylated and unphosphorylated kinase (10^{-9} *M*). When reciprocal plots of these activation curves are analyzed, they show that phosphorylation decreases the ability of myosin kinase to bind calmodulin by 20-fold. Under similar circumstances the value for V_{max} is decreased only 1.5-fold.

Myosin light chain kinase purified from human platelets can also be phosphorylated by the catalytic subunit of cyclic AMP-dependent protein kinase. Similar to the findings with smooth muscle myosin kinase, phosphorylation of the platelet enzyme results in a decrease in activity (Hathaway *et al.*, 1980b).

Recently, rapid stoichiometric autophosphorylation of canine cardiac muscle myosin light chain kinase has been reported (Rappaport and Adelstein, 1980). This autophosphorylation, which appears to be intramolecular, requires both calcium and calmodulin. At present the effect of this phosphorylation on cardiac kinase activity is not known. A form of cardiac myosin kinase that cannot be phosphorylated was purified from dog heart and characterized by Walsh *et al.* (1979).

D. Dephosphorylation of Myosin Kinase

Smooth muscle myosin as well as myosin light chain kinase can be dephosphorylated by two different phosphatases isolated from turkey gizzard smooth muscle. Both of these phosphatases have been purified using an affinity column of thiophosphorylated 20,000 dalton smooth muscle light chains bound to Sepharose 4B. The two phosphatases differ in molecular weight as determined by SDS-polyacrylmide gel electrophoresis.

Phosphatase I is comprised of equimolar amounts of three different polypeptide chains of 60,000, 55,000, and 38,000 daltons. Phosphatase II was found to be comprised of a single polypeptide chain of 43,000 daltons. The latter enzyme also differed from the former in requiring Mg^{2+} for activity (Pato and Adelstein, 1980).

Neither phosphatase exhibited the marked substrate specificity shown by myosin light chain kinase and skeletal muscle myosin phosphatase (Morgan *et al.*, 1976), but phosphatase I was fivefold more active in dephosphorylating smooth muscle myosin, the isolated myosin light chain, and the enzyme myosin light chain kinase than histone, casein, or a variety of other substrates. Phosphatase II also showed a marked preference for myosin light chains as a substrate, but not for myosin light chain kinase (Pato and Adelstein, 1980; Adelstein *et al.*, 1980).

III. SUMMARY: REGULATORY OF SMOOTH MUSCLE CONTRACTION

In a previous section (II,B) we discussed how Ca^{2+} binding to calmodulin might serve to initiate smooth muscle contraction. The mechanism by which this might occur is depicted in Fig. 2 (lower left).

The active complex, Ca^{2+}-calmodulin-myosin kinase catalyzes the phosphorylation of myosin, converting smooth muscle myosin from a form that cannot interact with actin to one that can undergo actin activation. A phosphatase, which is not dependent on Ca^{2+} for activity, restores myosin to the unphosphorylated form.

Figure 2 also depicts how epinephrine, which binds to the β-receptor (upper right), might act to decrease smooth muscle tension. This modulation of smooth muscle contractility is mediated through phosphorylation of myosin light chain kinase. Specifically, the interaction of epinephrine with the β-receptor results in an increase in cyclic AMP, which is synthesized from ATP. The increase in cAMP activates protein kinase by binding to the regulatory subunit (R), liberating the catalytic subunit (C) of cAMP-dependent protein kinase (for a recent review on cAMP-dependent protein kinase, see Krebs and Beavo, 1979). The catalytic subunit catalyzes phosphorylation of myosin light chain kinase.

Phosphorylation of the myosin kinase results in a decreased ability of the kinase to bind calmodulin (indicated qualitatively by the size of the vertical arrows on the right side of Fig. 2) and thus leads to an inactive form of the kinase. A decrease in myosin kinase activity results in an increase in the amount of unphosphorylated myosin, a species that cannot interact with actin, and smooth muscle tension would, therefore, decrease.

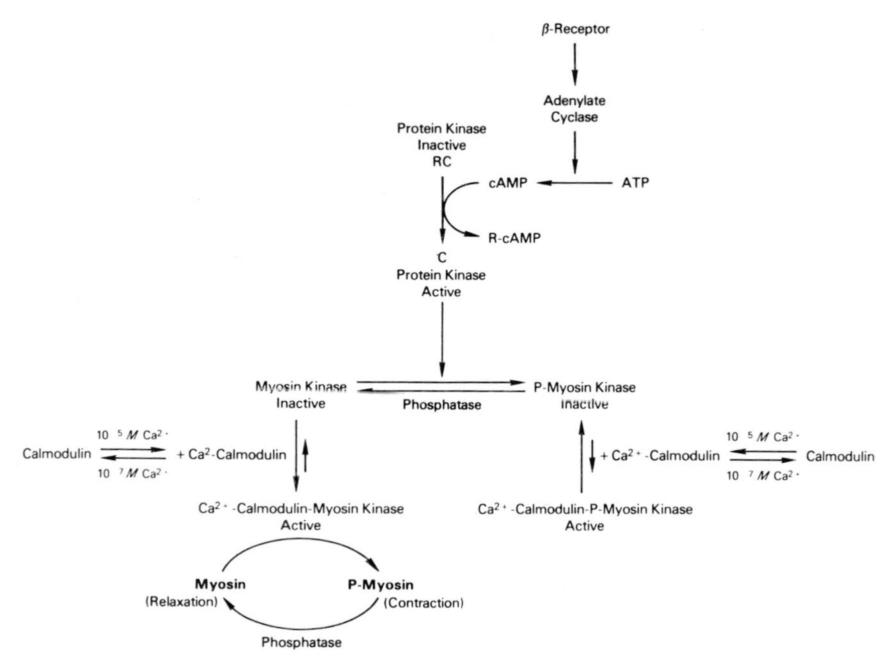

Fig. 2. Regulation of smooth muscle contraction by Ca^{2+}-calmodulin and cyclic AMP. The scheme is discussed in detail in the text. It has been published previously. (From Conti and Adelstein, 1980, by permission.)

Recent experiments with a crude preparation of native actomyosin containing myosin kinase from vascular smooth muscle are consistent with this model in that increased levels of cyclic AMP resulted in a decrease in the actin-activated ATPase activity of smooth muscle myosin (Silver and DiSalvo, 1979).

The mechanism presented in Fig. 2 could explain how epinephrine acts to relax certain smooth muscles. For example, epinephrine is routinely given to relax bronchial smooth muscle in caring for an attack of acute asthma.

The model shown in Fig. 2 emphasizes the importance of the concentration and distribution of the various reactants in regulating smooth muscle contraction: Ca^{2+}, calmodulin, myosin kinase, and myosin. In speculating as to which of these regulates smooth muscle contraction it should be recalled that myosin is present as a filament in muscle and not as a soluble protein, and the distribution of myosin kinase with respect to myosin has not been determined.

Most of the data at present indicate that changes in the concentration of Ca^{2+} regulate the binding of calmodulin to myosin kinase. Both calmodulin and myosin kinase appear to be present in smooth muscle in micromolar concentrations, and since the binding of calmodulin to myosin kinase is very tight ($10^9\ M^{-1}$ at $10^{-4}\ M\ Ca^{2+}$), it is likely that a rise in Ca^{2+} from 10^{-7} to $10^{-5}\ M$ can rapidly activate the phosphorylating system. Even in the presence of other calmodulin-binding proteins, it is likely that the concentration of free calmodulin is high enough so that Ca^{2+} will be the determining factor in regulating calmodulin–myosin kinase interaction.

A second related problem, concerns the presence of four binding sites for Ca^{2+} on calmodulin (Klee *et al.*, 1980) and the affinity of each of these species, [calmodulin $(Ca)_1$, calmodulin $(Ca)_2$, etc.] for myosin kinase. Recently, Stull *et al.* (1980) have obtained information on the binding of these various species to skeletal muscle myosin kinase.

Two other factors bearing on myosin kinase activity must be considered. Nairn and Perry (1979b) reported that the affinity of skeletal muscle myosin kinase for the P-light chain (i.e., phosphorylatable) is increased in the presence of calmodulin. This implies that the concentration of this light chain in muscle would have an effect on the binding of calmodulin to myosin kinase. Finally, it is likely that contraction of smooth muscle may not require the phosphorylation of every myosin molecule, and, hence, only partial activation of the enzyme myosin kinase will be necessary to initiate contraction. These are only a few of the relevant problems that must be considered in understanding how Ca^{2+}, calmodulin, and myosin kinase might interact to regulate smooth muscle contraction.

Confirmation of the mechanism proposed in Fig. 2 for modulation of the actin–myosin interaction by cyclic AMP will require relevant physiological data. Thus, it will be of interest to learn whether a rise in cyclic AMP concentration *in vivo* can be correlated with an increase in phosphorylation of myosin kinase and whether phosphorylation of myosin kinase can in turn be correlated with a decrease in tension in the case of smooth muscle and a decrease in a contractile activity in non-muscle cells. Finally, it will be of interest to learn, in definitive molecular terms, why the interaction of epinephrine with the β-receptor causes a decrease in tension in certain smooth muscles and yet leads to an increase in tension in the case of cardiac muscle.

ACKNOWLEDGMENTS

The authors gratefully acknowledge all those investigators who provided them with preprints of their work. They also wish to thank Mrs. Beverly Minnich for her editorial assistance.

REFERENCES

Adelstein, R. S. (1978). Myosin phosphorylation, cell motility and smooth muscle contraction. *Trends Biochem. Sci.* **3,** 27–30.

Adelstein, R. S., and Conti, M. A. (1975). Phosphorylation of platelet myosin increases actin-activation of myosin ATPase activity. *Nature* (*London*) **256,** 597–598.

Adelstein, R. S., and Eisenberg, E. (1980). Regulation and kinetics of the actin-myosin-ATP interaction. *Annu. Rev. Biochem.* **49,** 921–956.

Adelstein, R. S., and Klee, C. B. (1980). In preparation.

Adelstein, R. S., Conti, M. A., Hathaway, D. R., and Klee, C. B. (1978). Phosphorylation of smooth muscle myosin light chain kinase by the catalytic subunit of adenosine 3′:5′-monophosphate-dependent protein kinase. *J. Biol. Chem.* **253,** 8347–8350.

Adelstein, R. S., Pato, M. D., Rappaport, L., Conti, M. A., Hathaway, D. R., and Eaton, C. R. (1980). "Regulation of Contractile Proteins by Kinases and Phosphatases." Karger, Basel (in press).

Aksoy, M. O., and Murphy, R. A. (1979). Temporal relationship between phosphorylation of the 20,000 dalton myosin light chain (LC) and force generation in arterial smooth muscle. *Physiologist* **22,** 2abs.

Bárány, K., Bárány, M., Gillis, J. M., and Kushmerick, M. J. (1979). Phosphorylation–dephosphorylation of the 18,000-dalton light chain of myosin during the contraction –relaxation cycle of frog muscle. *J. Biol. Chem.* **254,** 3617–3623.

Chacko, S., Conti, M. A., and Adelstein, R. S. (1977). Effect of phosphorylation of smooth muscle myosin on actin-activation and on Ca^{++} regulation. *Proc. Natl. Acad. Sci. U.S.A.* **74,** 129–133.

Cheung, W. Y. (1971). Cyclic 3′,5′-nucleotide phosphodiesterase: Evidence for and properties of a protein activator. *J. Biol. Chem.* **246,** 2859–2869.

Cheung, W. Y. (1980). Calmodulin plays a pivotal role in cellular regulation. *Science* **207,** 19–27.

Conti, M. A., and Adelstein, R. S. (1980). Phosphorylation by cyclic 3′:5′-monophosphate-dependent protein kinase regulates myosin light chain kinase. *Fed. Proc. Fed. Am. Soc. Exp. Biol.* **39,** 1569–1573.

Dabrowska, R., and Hartshorne, D. J. (1978). A Ca^{2+} and modulator-dependent myosin light chain kinase from non-muscle cells. *Biochem. Biophys. Res. Commun.* **85,** 1352–1359.

Dabrowska, R., Aromatorio, D., Sherry, J. M. F., and Hartshorne, D. J. (1977). Composition of the myosin light chain kinase from chicken gizzard. *Biochem. Biophys. Res. Commun.* **78,** 1263–1272.

Dabrowska, R., Aromatorio, D., Sherry, J. M. F., and Hartshorne, D. J. (1978). Modulator protein as a component of the myosin light chain kinase from chicken gizzard. *Biochemistry* **17,** 253–258.

Daniel, J. L., and Adelstein, R. S. (1976). Isolation and properties of platelet myosin light chain kinase. *Biochemistry* **15,** 2370–2377.

de Lanerolle, P., and Stull, J. T. (1980). *J. Biol. Chem.* (in press).

DiSalvo, J., Gruenstein, E., and Silver, P. (1978). Ca^{2+} dependent phosphorylation of bovine aortic actomyosin. *Proc. Soc. Exp. Biol. Med.* **158,** 410–414.

Ebashi, S., Mikawa, T., Hirata, M., and Nonomura, Y. (1978). The regulatory role of calcium in muscle. *Ann. N.Y. Acad. Sci.* **307,** 451–461.

Gorecka, A., Aksoy, M. O., and Hartshorne, D. J. (1976). The effect of phosphorylation of gizzard myosin on actin activation. *Biochem. Biophys. Res. Commun.* **71,** 325–331.

Grand, R. J. A., and Perry, S. V. (1978). The amino acid sequence of the troponin C-like protein (modulator protein) from bovine uterus. *FEBS Lett.* **92,** 137–142.

Hathaway, D. R. Eaton, C. R., and Adelstein, R. S. (1980b). Regulation of human platelet requires the calcium-binding protein calmodulin for activity. *Proc. Natl. Acad. Sci. U.S.A.* **76,** 1653–1657.

Hathaway, D. R., Adelstein, R. S., and Klee, C. B. (1980a). In preparation.

Hathaway, D. R., Eaton, C. R., and Adelstein, R. S. (1980b). Regulation of human platelet myosin kinase by calcium-calmodulin and cyclic AMP. *In* "The Regulation of Coagulation" (K. G. Mann, F. B. Taylor, eds.) pp. 271–276. Elsevier, New York.

Hoar, P. E., Kerrick, W. G. L., and Cassidy, P. (1979). Chicken gizzard: Relation between calcium-activated phosphorylation and contraction. *Science* **204,** 503–506.

Huszar, G., and Bailey, P. (1979). Relationship between actin–myosin interaction and myosin light chain phosphorylation in human placental smooth muscle. *Am. J. Obstet. Gynecol.* **135,** 718–726.

Huxley, A. F., and Niedergerke, R. (1954). Structural changes in muscle during contraction. *Nature (London)* **173,** 971–973.

Huxley, H. E., and Hanson, J. (1954). Changes in the cross-striation of muscle during contraction and stretch and their structural interpretation. *Nature (London)* **173,** 973–976.

Jakes, R., Northrup, F., and Kendrick-Jones, J. (1976). Calcium binding regions of myosin regulatory light chains. *FEBS Lett.* **70,** 229–234.

Klee, C. B., and Krinks, M. H. (1978). Purification of cyclic 3′,5′-nucleotide phosphodiesterase inhibitory protein by affinity chromatography on activator protein coupled to Sepharose. *Biochemistry* **17,** 120–126.

Klee, C. B., Crouch, T. H., and Richman, P. G. (1980). Calmodulin *Annu. Rev. Biochem.* **49,** 489–515.

Kopp, S. J., and Bárány, M. (1979). Phosphorylation of the 19,000-dalton light chain of myosin in perfused rat heart under the influence of negative and positive inotropic agents. *J. Biol. Chem.* **254,** 12007–12012.

Korn, E. D. (1978). Biochemistry of actomyosin-dependent cell motility (a review). *Proc. Natl. Acad. Sci. U.S.A.* **75,** 588–599.

Krebs, E. G., and Beavo, J. A. (1979). Phosphorylation–dephosphorylation of enzymes. *Annu. Rev. Biochem.* **48,** 923–959.

Kretsinger, R. H. (1980). Structure and evolution of calcium modulated proteins. *CRC Crit. Rev. Biochem.* **8,** 119–174.

Lebowitz, E., and Cooke, R. (1979). Phosphorylation of uterine smooth muscle myosin permits actin activation. *J. Biochem. (Tokyo)* **85,** 1489–1494.

Lin, Y. M., Liu, Y. P., and Cheung, W. Y. (1975). Cyclic 3′,5′ nucleotide phosphodiesterase. Ca^{2+} dependent formation of bovine brain enzyme–activator complex. *FEBS Lett.* **49,** 356–360.

Manning, D. R., and Stull, J. T. (1979). Myosin light chain phosphorylation and phosphorylase *a* activity in rat extensor digitorum longus muscle. *Biochem. Biophys. Res. Commun.* **90,** 164–170.

Marston, S. B., Trevett, R. M., and Walters, M. (1980). Calcium regulated thin filaments from vascular smooth muscle. *Biochem. J.* **185,** 355–365.

Matsuda, G., Suzuyuma, Y., Maita, T., and Umegane, T. (1977). The L-2 light chain of chicken skeletal muscle myosin. *FEBS Lett.* **84,** 53–56.

Mikawa, T., Nonomura, Y., Hirata, M., Ebashi, S., and Kakiuchi, S. (1978). Involvement of an acidic protein in regulation of smooth muscle contraction by the tropomyosin leiotonin system. *J. Biochem. (Tokyo)* **84,** 1633–1636.

Morgan, M., Perry, S. V., and Ottaway, J. (1976). Myosin light chain phosphatase. *Biochem. J.* **157,** 687–697.

Nairn, A. C., and Perry, S. V. (1979a). Calmodulin and myosin light chain kinase of rabbit fast skeletal muscle. *Biochem. J.* **179,** 89–97.

Nairn, A. C., and Perry, S. V. (1979b). The role of calmodulin in the myosin light chain kinase system. *Biochem. Soc. Trans.* **7,** 966–967.

Pato, M. D., and Adelstein, R. S. (1980). Dephosphorylation of the 20,000 dalton light chain of myosin by two different phosphatases from smooth muscle. *J. Biol. Chem.* **255,** 6535–6538.

Rappaport, L., and Adelstein, R. S. (1980). Phosphorylation of cardiac myosin light chain kinase. *Fed. Proc., Fed. Am. Soc. Exp. Biol.* 2117 (Abstr.).

Scordilis, S. P., and Adelstein, R. S. (1977). Myoblast myosin phosphorylation is a prerequisite for actin-activation. *Nature (London)* **268,** 558–560.

Silver, P. J., and DiSalvo, J. (1979). Adenosine 3′:5′-monophosphate-mediated inhibition of myosin light chain phosphorylation in bovine aortic actomyosin. *J. Biol. Chem.* **254,** 9951–9954.

Small, J. V., and Sobieszek, A. (1977). Ca-Regulation of mammalian smooth muscle actomyosin via a kinase phosphatase dependent phosphorylation and dephosphorylation of the 20,000 M_r light chain of myosin. *Eur. J. Biochem.* **76,** 521–530.

Small, J. V., and Sobieszek, A. (1980). The contractile apparatus of smooth muscle. *Int. Rev. Cytol.* **64,** 241–306.

Sobieszek, A. (1977). Vertebrate smooth muscle myosin: Enzymatic and structural properites. *In* "The Biochemistry of Smooth Muscle" (N. L. Stephens, ed.), pp 413–443. Univ. Park Press, Baltimore, Maryland.

Sobieszek, A., and Small, J. V. (1977). Regulation of the actin–myosin interaction in vertebrate smooth muscle: Activation via a myosin light chain kinase and the effect of tropomyosin. *J. Mol. Biol.* **112,** 559–576.

Stull, J. T., and High, C. W. (1977). Phosphorylation of skeletal muscle contractile proteins *in vivo*. *Biochem. Biophys. Res. Commun.* **77,** 1078–1083.

Stull, J. T., Manning, D. R., High, C. W., and Blumenthal, D. K. (1980). Phosphorylation of contractile proteins in heart and skeletal muscle. *Fed. Proc., Fed. Am. Soc. Exp. Biol.* **39,** 1552–1557.

Teo, T. S., and Wang, J. H. (1973). Mechanism of activation of a cyclic adenosine 3′,5′-monophosphate phosphodiesterase from bovine heart by calcium ions: Identification of the protein activator as a Ca^{2+}-binding protein. *J. Biol. Chem.* **248,** 5950–5955.

Teshima, Y., and Kakiuchi, S. (1974). Mechanism of stimulation of Ca^{2+} plus Mg^{2+}-dependent phosphodiesterase from rat cerebral cortex by the modulator protein and Ca^{2+}. *Biochem. Biophys. Res. Commun.* **56,** 489–495.

Trotter, J. A., and Adelstein, R. S. (1979). Macrophage myosin: Regulation of actin-activated ATPase activity by phosphorylation of the 20,000 dalton light chain. *J. Biol. Chem.* **254,** 8781–8785.

Walsh, M. P., Vallet, B., Autric, F., and Demaille, J. G. (1979). Purification and characterization of bovine cardiac calmodulin-dependent myosin light chain kinase. *J. Biol. Chem.* **254,** 12136–12144.

Wang, J. H., and Waisman, D. M. (1979). Calmodulin and its role in the second messenger system. *Curr. Top. Cell. Regul.* **15,** 47–107.

Wang, J. H., Teo, T. S., Ho, H. C., and Stevens, F. C. (1975). Bovine heart protein activator of cyclic nucleotide phosphodiesterase. *Adv. Cyclic Nucleotide Res.* **5,** 179–194.

Watterson, D. M., Sharief, F., and Vanaman, T. C. (1980). The complete amino acid sequence of the Ca^{2+}-dependent modulator protein (calmodulin) of bovine brain. *J. Biol. Chem.* **255,** 962–975.

Weber, A., and Murray, J. M. (1973). Molecular control mechanisms in muscle contraction. *Physiol. Rev.* **53,** 612–673.

Wolff, D., and Brostrom, C. W. (1979). Properties and functions of the calcium-dependent regulator protein. *Adv. Cyclic Nucleotide Res.* **11,** 27–88.

Yagi, K., Yazawa, M., Kakiuchi, S., Ohshima, M., and Uenishi, K. (1978). Identification of an activator protein for myosin light chain kinase as the Ca^{2+}-dependent modulator protein. *J. Biol. Chem.* **253,** 1338–1340.

Yerna, M. J., Dabrowska, R., Hartshorne, D. J., and Goldman, R. D. (1979). Calcium sensitive regulation of actin–myosin interaction in baby hamster kidney cells. *Proc. Natl. Acad. Sci. U.S.A.* **76,** 184–188.

Chapter 9

The Role of Calmodulin and Troponin in the Regulation of Phosphorylase Kinase from Mammalian Skeletal Muscle

PHILIP COHEN

CALCIUM AND CELL FUNCTION, VOL. I

ISBN 0-12-171401-2

I. INTRODUCTION

Phosphorylase kinase was the first protein kinase to be discovered (Krebs and Fischer, 1956), and its identification therefore represents an important landmark in our understanding of cellular control mechanisms. The enzyme catalyzes the phosphorylation of a single serine residue on glycogen phosphorylase (Titani *et al.*, 1975) converting it from a *b* form, which is dependent on the allosteric activator 5′-AMP, to an *a* form, which is almost fully active in the absence of 5′-AMP. The formation of phosphorylase *a* is therefore a mechanism for activating phosphorylase and mobilising glycogen, and it is well established that an increase in the proportion of phosphorylase *a* takes place *in vivo* during muscle contraction or in response to epinephrine (Danforth and Lyon, 1964; Helmreich and Cori, 1966; Posner *et al.*, 1965; Drummond *et al.*, 1969).

Phosphorylase kinase is of central importance in mediating both the neural and hormonal control of glycogenolysis. Its activity is completely dependent on calcium ions (Osawa *et al.*, 1967; Heilmeyer *et al.*, 1970; Brostrom *et al.*, 1971), and it is also stimulated 50-fold or more by phosphorylation catalysed by cyclic AMP-dependent protein kinase (Walsh *et al.*, 1971; Cohen, 1973).

Although information on the structure and regulation of phosphorylase kinase up to the beginning of 1978 has been reviewed in detail (Nimmo and Cohen, 1977; Cohen, 1978), the subsequent discovery that calmodulin is a subunit of the enzyme (Cohen *et al.*, 1978) has caused a resurgence of interest in this enzyme. It has also become clear over the past year that phosphorylase kinase catalyzes a rapid phosphorylation of glycogen synthase (Roach *et al.*, 1978, 1979; Cohen *et al.*, 1979a; Embi *et al.*, 1979a,b) so that it may exert control over the synthesis as well as the breakdown of glycogen. This chapter presents these recent developments which have placed our understanding of the control of glycogen metabolism on a much firmer molecular basis.

II. REGULATION OF PHOSPHORYLASE KINASE BY CALCIUM IONS AND IDENTIFICATION OF CALMODULIN AS A SUBUNIT

It has been established for 10 years that the activity of phosphorylase kinase is almost completely dependent on calcium ions, and that the activation occurs in the range 0.1–1.0 μM (Osawa *et al.*, 1967; Heilmeyer *et al.*, 1970; Brostrom *et al.*, 1971). Furthermore, the conversion of phosphorylase *b* to *a in vitro* can be blocked instantaneously by the addition of

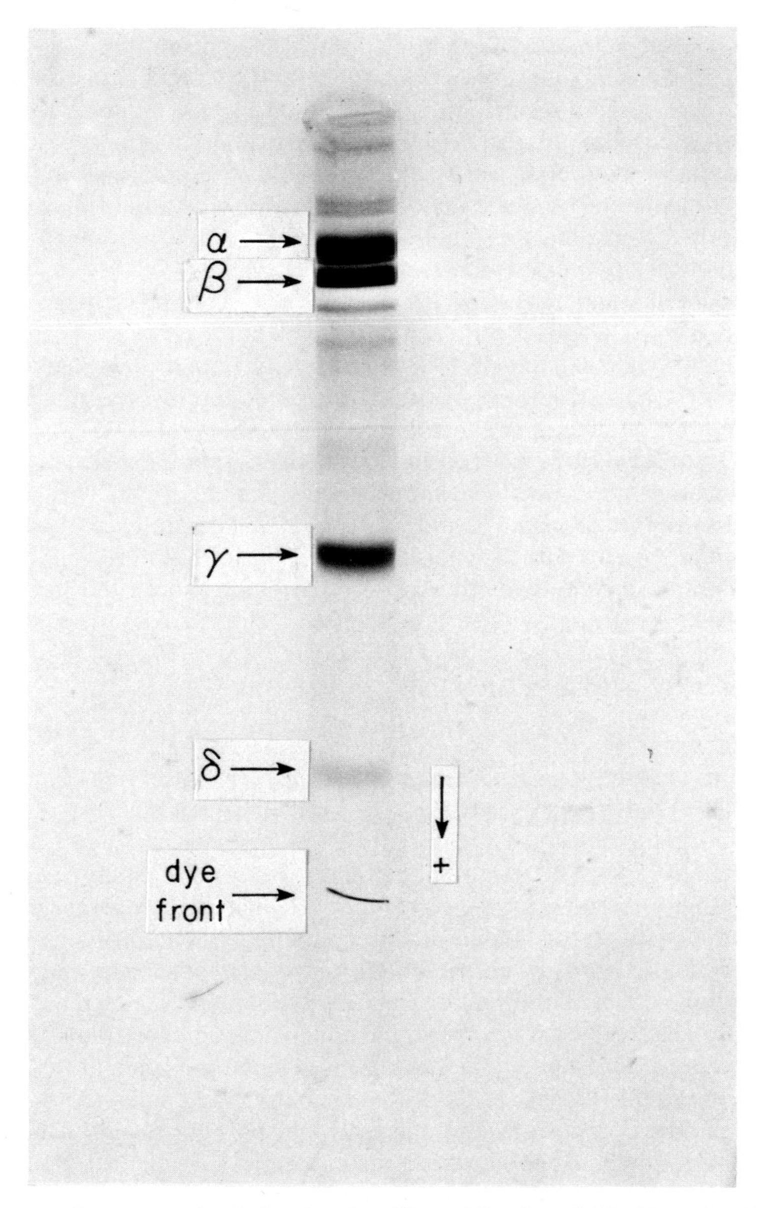

Fig. 1. Electrophoresis of phosphorylase kinase (10 μg) on 7.5% polyacrylamide gels in the presence of sodium dodecyl sulfate. The gels were stained with Coomassie blue and the migration is from top to bottom. At these high loadings which are required to visualize the δ subunit clearly, some faint minor bands are also visible.

a sarcoplasmic reticulum membrane preparation and restarted by the readdition of calcium ions (Brostrom *et al.*, 1971). This demonstrates that the naturally occurring calcium chelator in muscle can compete successfully for the calcium ions that are bound to phosphorylase kinase, indicating that the process could be freely reversible *in vivo* (Brostrom *et al.*, 1971). These findings suggest that the reversible activation of phosphorylase kinase by calcium ions represents a mechanism for synchronizing glycogenolysis and muscle contraction.

Several years ago phosphorylase kinase was reported to possess the structure $(\alpha\beta\gamma)_4$, where the molecular weights of the α, β, and γ subunits were 145,000, 128,000, and 45,000, respectively (Cohen, 1973; Hayakawa *et al.*, 1973). This structure appeared to exclude the involvement of calmodulin (MW 17,000) in the regulation of phosphorylase kinase activity. However, in July, 1978, we reported that phosphorylase kinase contains a fourth component termed the δ subunit (Cohen *et al.*, 1978). This subunit is much smaller than the α, β, and γ subunits, and due to its acidic nature stains rather poorly with Coomassie blue. It was, therefore, missed for several years, since it migrated as a faintly staining band with the bromophenol blue dye-front on the 5% polyacrylamide gels used previously to resolve the α, β, and γ subunits. This component is, however, readily detectable on 10% polyacrylamide gels as shown in Fig. 1.

III. IDENTIFICATION OF THE δ SUBUNIT OF PHOSPHORYLASE KINASE AS CALMODULIN

The identity of the δ subunit with calmodulin was originally suggested by its stability to boiling, its electrophoretic mobility on polyacrylamide gels in the presence of sodium dodecyl sulfate, its elution behavior on DEAE-cellulose, and its characteristic ultraviolet absorption spectrum (Cohen *et al.*, 1978). Confirmation that the δ subunit was indeed calmodulin was obtained in two ways. First, the amino acid composition was identical to calmodulin from bovine brain including the presence of the single residue of trimethyllysine; second, the δ subunit could activate two other calmodulin-dependent enzymes, namely, myosin light chain kinase and cyclic nucleotide phosphodiesterase, in an identical manner to calmodulin isolated from bovine brain (Cohen *et al.*, 1978).

The amino acid sequence of the δ subunit is currently being determined by Dr. Roger Grand at the University of Birmingham, England, using material supplied from this laboratory. Currently, 70% of the sequence has been determined, and no differences from the bovine uterine protein have so far been detected. This work constitutes direct proof that the δ subunit is identical to calmodulin.

IV. MOLAR PROPORTION OF THE δ SUBUNIT AND ITS INTERACTION WITH PHOSPHORYLASE KINASE

The δ subunit can be released quantitatively from phosphorylase kinase by heating the enzyme at 100°C for several minutes, since the α, β, and γ subunits are precipitated completely by this treatment. The amount of calmodulin bound to phosphorylase kinase can therefore be determined very simply by measuring the protein concentration of the starting material, and the protein concentration of the supernatant after the heat treatment. This procedure showed that the molar ratio $\delta/\alpha\beta\gamma\delta$ was 0.94 (SD ±0.05 for seven different preparations). The δ subunit is therefore stoichiometric with the other subunits, and phosphorylase kinase has the structure $(\alpha\beta\gamma\delta)_4$, MW 1,300,000 (Shenolikar *et al.*, 1979).

Many calmodulin-dependent enzymes are resolved from calmodulin during their isolation, and it is therefore of interest that calmodulin remains associated with phosphorylase kinase throughout the purification of the enzyme. Since 2.0 m*M* EDTA is present throughout, this implies that calmodulin interacts with phosphorylase kinase in the absence of calcium ions. Even when the 2.0 m*M* EDTA is replaced by 20 m*M* EGTA at the final two steps of the purification (ammonium sulfate precipitation and gel filtration on Sepharose 4B), there is no decrease in the amount of calmodulin bound to the enzyme (Cohen *et al.*, 1978).

Although the δ subunit is firmly associated with phosphorylase kinase in the absence of calcium ions, its interaction with phosphorylase kinase is even tighter in the presence of calcium ions. Thus, the δ subunit remains associated with the α, β, and γ subunits even in the presence of 8 *M* urea, provided that calcium ions are present (Shenolikar *et al.*, 1979). In this respect the situation resembles the interaction of troponin C and troponin I, which also form complexes which are stable to 8 *M* urea in the presence of calcium ions (Head and Perry, 1974). It is hoped that this property will facilitate the identification of the subunit with which calmodulin interacts.

V. EVIDENCE THAT THE δ SUBUNIT IS THE COMPONENT WHICH CONFERS CALCIUM SENSITIVITY TO THE PHOSPHORYLASE KINASE REACTION

Kilimann and Heilmeyer (1977) measured the binding of calcium ions to phosphorylase kinase in 1.0 m*M* sodium glycerophosphate buffer, pH 6.8. Under these conditions each molecule of enzyme (MW 1,300,000) bound 12 calcium ions with an association constant of 6×10^{-7} *M*, and a further 4 calcium ions with an association constant of 1.7×10^{-6} *M*. These measurements, which were made before calmodulin was identified as a subunit, strongly suggests that all the high-affinity calcium-binding sites on phosphorylase kinase are located on the calmodulin subunit. This in turn suggests that calmodulin is the component which confers Ca^{2+} sensitivity to the phosphorylase kinase reaction. However, Wolff *et al.* (1977) reported that in 20 m*M* Tris/HCl pH 7.4 calmodulin binds three Ca^{2+} per mole with an affinity of 2×10^{-7} *M* and one Ca^{2+} with an affinity of 1×10^{-6} *M*. The interaction of the δ subunit with phosphorylase kinase may therefore after the Ca^{2+}-binding properties of calmodulin.

Recently I have found that phosphorylase kinase is inhibited by antibody prepared to rat testis calmodulin. This antibody, which was purified by affinity chromatography on protein A-Sepharose and calmodulin-Sepharose, was provided by Dr. John Dedman. In the standard assay at pH 8.2 (Shenolikar *et al.*, 1979), which contains 0.15 μg/ml phosphorylase kinase (7 ng/ml calmodulin), the activity was decreased fourfold in the presence of 60 μg/ml antibody. The inhibition could be completely prevented by the addition of calmodulin either before or after the addition of antibody. These experiments are consistent with the idea that calmodulin is important in controlling the activity of the enzyme.

VI. PHOSPHORYLASE KINASE CONTAINS A SECOND CALMODULIN-BINDING SITE WHICH ACTIVATES THE ENZYME

Although all preparations of phosphorylase kinase contain stoichiometric quantities of calmodulin, the activity of the enzyme is increased by the addition of further calmodulin to the assay. Freshly prepared preparations of phosphorylase kinase are activated approximately fivefold in the standard assay at pH 8.2 (Shenolikar *et al.*, 1979). The concentration of calmodulin required for half-maximal activation is 0.15 ± 0.05 μg/ml (0.9×10^{-8} *M*) corresponding to a molar ratio calmodulin/phosphorylase kinase of 20:1 (Fig. 2).

Several lines of evidence indicate that this additional activation is caused by the binding of a second molecule of calmodulin to phosphorylase kinase. The calmodulin-stimulated activity can be prevented by the addition of either the antipsychotic drug trifluoperazine (Fig. 3) or by the calmodulin-binding protein isolated ($CaM\text{-}BP_{80}$) from brain, previously referred to as the inhibitory protein since it inhibits the calmodulin-stimulated cyclic nucleotide phosphodiesterase (Shenolikar *et al.*, 1979; Klee and Krinks, 1978; Cohen *et al.*, 1979b). In contrast, these compounds have almost no effect on the calcium-dependent activity in the absence of cal-

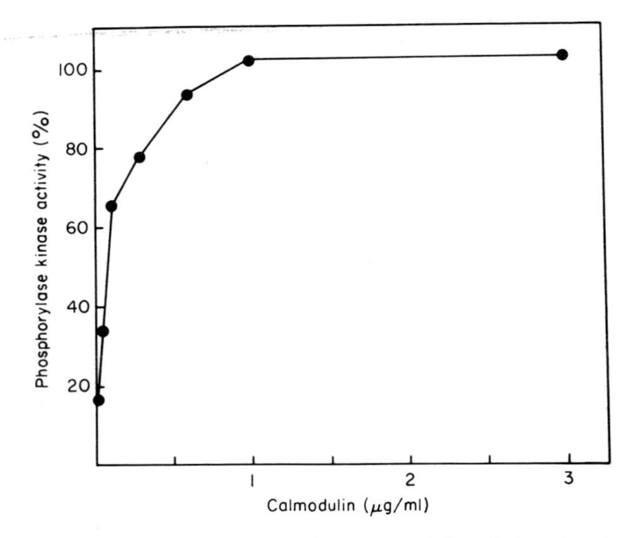

Fig. 2. Effect of rabbit muscle calmodulin on the activity of phosphorylase kinase. The assays were carried out at pH 8.2 in the presence of Ca^{2+} using 0.14 μg/ml phosphorylase kinase. (From Shenolikar *et al.*, 1979.)

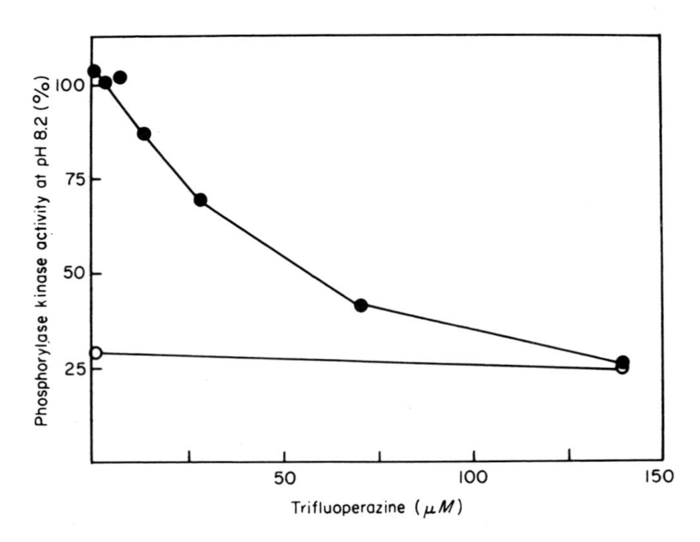

Fig 3. Effect of trifluoperazine on the activity of phosphorylase kinase at pH 8.2. The assays were carried out in the presence of Ca^{2+}, 0.14 μg/ml phosphorylase kinase, and in the presence (closed circles) and absence (open circles) of calmodulin (6 μg/ml). In the absence of Ca^{2+} there was essentially no activity in the presence or absence of calmodulin. 100% activity corresponds to the activity in the presence of saturating concentrations of calmodulin. (From Shenolikar *et al.*, 1979.)

modulin (Fig. 3). It would appear that trifluoperazine and the inhibitory protein can differentiate between the tightly bound molecule of calmodulin, the δ subunit, which is an integral component of the enzyme, and a second more weakly bound molecule of calmodulin, which stimulates the activity.

Direct evidence establishing the presence of a second calmodulin-binding site on phosphorylase kinase has been obtained in two ways. First, phosphorylase kinase binds to calmodulin-Sepharose in the presence of calcium ions and can be eluted from the column by excluding calcium ions from the buffer (Shenolikar *et al.*, 1979). Second, glycerol density gradient centrifugation in the presence of ^{14}C-labeled-guanidylated calmodulin has demonstrated that one additional molecule of calmodulin can bind to phosphorylase kinase per $\alpha\beta\gamma\delta$ unit, in the presence but not in the absence of calcium ions. These latter experiments have also emphasised how tightly the δ subunit is attached to the enzyme, since $<10\%$ exchange with ^{14}C-labeled guanidylated calmodulin takes place in 7 days in the absence of calcium ions (C. B. Klee and P. Cohen, unpublished).

VII. TROPONIN C AND THE TROPONIN COMPLEX—SUBSTITUTES FOR THE SECOND MOLECULE OF CALMODULIN IN THE ACTIVATION OF PHOSPHORYLASE KINASE

Since calmodulin is very similar in amino acid sequence to troponin C, and troponin C is present at very high concentrations in skeletal muscle, it seemed important to test whether troponin C could substitute for calmodulin in the activation of phosphorylase kinase. The results of this experiment are illustrated in Fig. 4. It can be seen that troponin C activates phosphorylase kinase to the same extent as calmodulin but a 200-fold higher concentration is required. Half-maximal activation is attained at 2.0 μM troponin C (Cohen *et al.*, 1979b).

Troponin C does not exist as such in skeletal muscle but as a complex with the troponin I and troponin T components, and the effect of the troponin complex on the activity of phosphorylase kinase is also illustrated in Fig. 4. The troponin complex can substitute for calmodulin in the activation of phosphorylase kinase and on a molar basis it is twofold more effective than troponin C. Half-maximal activation occurs at 1.0 μM troponin complex (Cohen *et al.*, 1979b). The finding that the troponin complex is slightly more effective as an activator could be related to the method used to purify troponin C, which involves disruption of the tro-

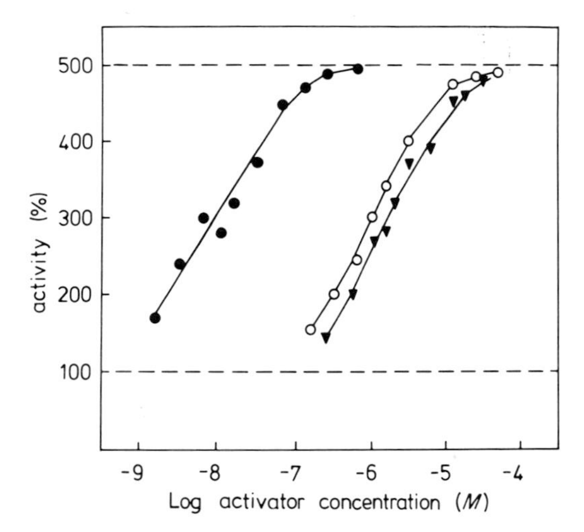

Fig. 4. Influence of calmodulin and troponin on the activity of phosphorylase kinase. The dashed lines show the level of activity in the presence (500) and absence (100) of saturating amounts of calmodulin. Calmodulin, ●; troponin C, ▼; troponin complex, ○. (From Cohen *et al.*, 1979b.)

ponin complex in 8 *M* urea followed by chromatography on DEAE-cellulose (Perry and Cole, 1974).

Since 100 to 200-fold higher concentrations of troponin C and the troponin complex are required for the activation of phosphorylase kinase, this raises the possibility that the activation by these proteins is really caused by trace contamination with calmodulin. This possibility has been excluded by making use of the calmodulin-binding protein (CaM-Bp_{80}) from brain and the results are illustrated in Fig. 5. It has already been mentioned that CaM-BP_{80} completely prevents the activation of phosphorylase kinase by the second molecule of calmodulin without affecting the calcium-dependent activity in the absence of calmodulin (channels 1–4). The results in Fig. 5 also show that CaM-BP_{80} does not prevent the activation of phosphorylase kinase by either troponin C or the troponin complex (channels 5 and 6), nor do troponin C and the troponin complex prevent CaM-BP_{80} from inhibiting the activation of phosphorylase kinase by calmodulin (channels 7 and 8). These experiments prove that the activation of phosphorylase kinase by troponin C and the troponin complex is not due to trace contamination with calmodulin, and the methodology illustrated in Fig. 5 would appear to represent a simple general procedure for detecting traces of calmodulin in troponin C or other proteins (Cohen *et al.*, 1979b). It should also be mentioned that the effects of troponin C and

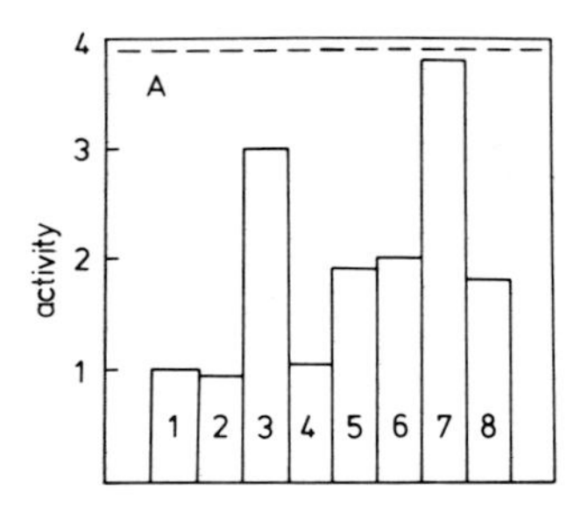

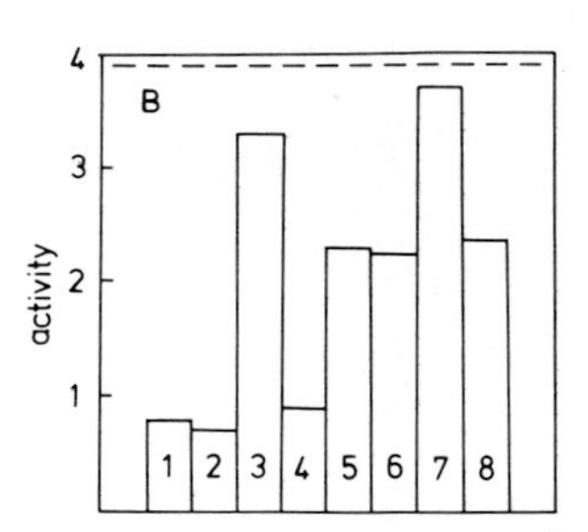

Fig. 5. Influence of the inhibitory protein on the activation of phosphorylase kinase by troponin C (A) and the troponin complex (B). The ordinates show the activity in arbitrary units and the broken line the activity in the presence of saturating concentrations of calmodulin. Channel 1, no additions; channel 2, activity in the presence of 0.3 μM CaM-BP_{80}; channel 3, activity in the presence of 0.06 μM calmodulin; channel 4, activity in the presence of 0.06 μM calmodulin + 0.3 μM CaM-BP_{80}; channel 5, activity in the presence of 1.2 μM troponin C (A) or 1.2 μM troponin complex (B); channel 6, activity in the presence of 0.3 μM CaM-BP_{80} and either 1.2 μM troponin C (A) or 1.2 μM troponin complex (B); channel 7, activity in the presence of 0.06 μM calmodulin and either 1.2 μM troponin C (A) or 1.2 μM troponin complex (B); channel 8, activity in the presence of 0.06 μM calmodulin, 0.3 μM CaM-BP_{80} and either 1.2 μM troponin C (A) or 1.2 μM troponin complex (B). Different preparations of phosphorylase kinase were used in the experiments with troponin C (A) and the troponin complex (B), and the concentration of phosphorylase kinase in the assays was 0.45 nM (0.15 μg/ml.) (From Cohen *et al.*, 1979b).

calmodulin on the activation of phosphorylase kinase are not additive. The two proteins activate phosphorylase kinase to the same extent when added alone or in combination (Cohen *et al.*, 1979b). The question of whether the second molecule calmodulin or troponin is the physiological activator of phosphorylase kinase will be considered in Section X.

VIII. INFLUENCE OF THE SECOND MOLECULE OF CALMODULIN AND TROPONIN ON THE ACTIVITY OF PHOSPHORYLASE KINASE

The results illustrated in Fig. 2 and 4 show that the second molecule of calmodulin and troponin can activate phosphorylase kinase fivefold at the pH optimum (pH 8.2). At physiological pH (6.8) the activation observed is even greater and depends on the concentration of calcium ions in the assay. Half-maximal activation by the troponin complex occurs at 4×10^{-6} M calcium ions, whereas activation by calmodulin occurs at 2×10^{-5} M calcium ions. At 4 μM calcium ions, the troponin complex activates phorphorylase kinase 20-fold, whereas little activation by calmodulin is observed under these conditions. At 20 μM calcium ions, activation by calmodulin is seven- to tenfold (Philip Cohen, unpublished experiments).

Maximal phosphorylation of phosphorylase kinase by cyclic AMP-dependent protein kinase leads to the phosphorylation of one site on the α subunit and one site on the β subunit of the enzyme (Cohen, 1973; Hayakawa *et al.*, 1973; Yeaman *et al.*, 1975). At 4 and 20 μM calcium ions, phosphorylation increases the activity 150- and 30-fold respectively at pH 6.8. The phosphorylated enzyme can only be activated slightly 1.3- to 1.4-fold) by either calmodulin or troponin.

IX. PROPORTION OF THE CALMODULIN IN SKELETAL MUSCLE THAT IS BOUND TO PHOSPHORYLASE KINASE

The concentration of calmodulin in dilute EDTA extracts of rabbit skeletal muscle is estimated in this laboratory to be 50 mg/kg of muscle by gel electrophoresis in the presence of 8 *M* urea (Shenolikar *et al.*, 1979). Since the concentration of phosphorylase kinase in rabbit skeletal muscle is 350–400 mg per 1000 g of muscle (Cohen, 1978), and calmodulin represents 5% of the phosphorylase kinase molecule by weight, it can be calculated that 35–40% of the calmodulin in rabbit skeletal muscle should be bound to phosphorylase kinase as the δ subunit.

This result has been confirmed by a second method which makes use of a strain of mice (ICR/IAn) which has <0.2% of normal phosphorylase kinase activity in their skeletal muscles (Cohen *et al.*, 1976). It has been established previously in this laboratory that the α, β, and γ subunits are completely absent in ICR/IAn mice, and that the mice are likely to be defective in a gene required for the induction of the α, β, and γ subunits (Cohen *et al.*, 1976). When muscle extracts from rabbits or C3H/He-mg mice (which have normal phosphorylase kinase activity) are fractionated from 0 to 35% ammonium sulfate, the 0–35% ammonium sulfate precipitates which contains all the phosphorylase kinase activity, possess 35% of the calmodulin present in the muscle extract. However, when muscle extracts from ICR/IAn mice are fractionated in an identical manner, the 35% ammonium sulfate precipitate contains 10–20 times less calmodulin. This experiment shows that 90–95% of the calmodulin which precipitates at 35% ammonium sulfate is bound to phosphorylase kinase, and again demonstrates that the δ subunit represents about 35% of the calmodulin in rabbit and mouse skeletal muscle (Shenolikar *et al.*, 1979).

These experiments also showed that the 35% ammonium sulfate supernatants from ICR/IAn mice and C3H/He-mg mice contain identical amounts of calmodulin (Shenolikar *et al.*, 1979). ICR/IAn mice therefore contain only 60–65% of the calmodulin that is present in normal mice, and the calmodulin is reduced by exactly the amount that is bound to phosphorylase kinase. This suggests that calmodulin is synthesized in sufficient amounts to saturate calmodulin-dependent enzymes and that there exists a mechanism for preventing further accummulation of calmodulin beyond this concentration.

ICR/IAn mice contain the same calmodulin-dependent myosin light chain kinase activity found in C3H/He-mg mice (Shenolikar *et al.,* 1979). There is therefore not a generalized defect in calmodulin-dependent enzymes in ICR/IAn mice. A review of the molecular basis of phosphorylase kinase deficiency in ICR/IAn mice can be found in Cohen (1978)

X. DOES TROPONIN C OR THE SECOND MOLECULE OF CALMODULIN ACTIVATE PHOSPHORYLASE KINASE *In Vivo?*

The finding that troponin C can substitute for the second molecule of calmodulin in the activation of phosphorylase kinase (Fig. 4), raises the question of which of these calcium-binding proteins is the activator of phosphorylase kinase *in vivo.* Several lines of evidence indicate that troponin C *cannot* be excluded as the activator of phosphorylase kinase.

1. The troponin complex, the form in which troponin C exists *in vivo,* is at least as effective as troponin C as an activator of phosphorylase kinase. Furthermore, artificial thin filaments made by mixing actin, tropomyosin and the troponin complex at a 7:1:1 molar ratio can also maximally activate phosphorylase kinase (Philip Cohen, unpublished experiments).
2. Although a 100-fold higher concentration of troponin (1.0 μM) is required for the activation of phosphorylase kinase compared to calmodulin, troponin is present at very high concentrations in skeletal muscle. The average concentration is 100 μM (Perry, 1974), which is clearly sufficient to maximally activate phosphorylase kinase *in vitro* in the presence of calcium ions (Fig. 4).
3. Troponin is a much more effective activator than calmodulin at concentrations of calcium ions believed to exist in contracting muscle (1–5 μM) (see Section VIII).
4. The δ subunit of phosphorylase kinase accounts for 35–40% of the calmodulin in skeletal muscle. Activation by the second molecule of calmodulin would therefore require most of the remaining calmodulin in the tissue, and much of this protein is likely to be complexed to other calmodulin-dependent proteins such as myosin light chain kinase (Yagi *et al.,* 1978).

TABLE I

Effect of EDTA and Calcium Chloride on the Extraction of the Enzymes of Glycogen Metabolism from Rabbit Skeletal Muscle[a]

Activity measurement	Relative activities	
	EDTA extraction	$CaCl_2$ extraction
Protein (mg/ml)	16.5	14.8
Lactate dehydrogenase	100	104
Phosphorylase kinase	100	42
Phosphorylase phosphatase	100	54
Glycogen phosphorylase	100	57
Glycogen synthase	100	50

[a] Minced rabbit muscle (500 *g*) was divided into two portions. One-half was homogenized in 600 ml of 4.0 m*M* EDTA–0.1% (v/v) mercaptoethanol, pH 7.0, and the other half in 600 ml of 2.0 m*M* $CaCl_2$–0.1% (v/v) mercaptoethanol. The homogenizations were carried out for 30 sec at low speed in a Waring blender at 0°C. The homogenates were centrifuged at 10,000 *g* for 10 min.; and the supernatants decanted through glass wool. The extract prepared by homogenization in $CaCl_2$ was immediately made 5 m*M* in EDTA. The pH 6.8/8.2 activity ratio of phosphorylase kinase was identical in the $CaCl_2$ and EDTA extracts (0.07–0.08) showing that no significant proteolytic activation of the enzyme had taken place during the 15 min exposure to high concentrations of $CaCl_2$ (From Cohen *et al.*, 1979b.)

In order to obtain further information about this important question the experiment described in Table I was carried out. Minced rabbit skeletal muscle was divided into two parts. One-half was homogenized in the presence of 4.0 m*M* EDTA and the other half with 1.0 m*M* $CaCl_2$, and the myofibrillar proteins were removed by centrifugation. It can be seen that the activities of the enzymes of glycogen metabolism were much lower in the calcium chloride extracts, and the activity of phosphorylase kinase was 2.5-fold lower (Table I). In contrast, the lactate dehydrogenase activities in both extracts were indistinguishable, and the protein concentration in the extract prepared in the presence of calcium ions was only 10% lower. This experiment indicates that a considerable proportion of the enzymes of glycogen metabolism remain associated with the myofibrils in the presence of calcium ions. The similar protein concentrations of the two supernatants (Table I) indicate that this association is relatively selective.

The enzymes of glycogen metabolism appear to be linked together *in vivo* on protein–glycogen particles (Cohen, 1978; Meyer *et al.*, 1970), which are known to be localized in skeletal muscle at the level of the thin filaments (Sigel and Pette, 1969). Since phosphorylase kinase is the only enzyme involved in glycogen metabolism which is known to interact with a myofibrillar component (troponin C) in a calcium-dependent manner, it might be thought that this interaction is responsible for the relatively selective calcium-dependent association of the enzymes of glycogen metabolism with the myofibrils. However, this does not appear to be the case, since a calcium-dependent association of phosphorylase, glycogen synthase, and phosphorylase phosphatase is still observed in skeletal muscle of ICR/IAn mice that lack phosphorylase kinase (P. Cohen, unpublished). The molecular basis for this observation is therefore unclear.

XI. PHOSPHORYLATION OF GLYCOGEN SYNTHASE BY PHOSPHORYLASE KINASE

Several years ago a glycogen synthase kinase distinct from cyclic AMP-dependent protein kinase was identified in this laboratory, which was present as a trace contaminant in purified glycogen synthase preparations (Nimmo and Cohen, 1974). This enzyme, termed glycogen synthase kinase-2, was unaffected by calcium ions and therefore appeared to be distinct from phosphorylase kinase (Nimmo and Cohen, 1974). However, subsequently, glycogen synthase kinase-2 was shown to be activated by calmodulin in the presence of calcium ions (Rylatt and Cohen, 1979; Rylatt *et al.*, 1979).

In 1978, Roach *et al,* (1978) reported that purified phosphorylase kinase catalyzes the phosphorylation of glycogen synthase, and this raised the question of the relationship between glycogen synthase kinase-2 and phosphorylase kinase. As a result of further experimentation, it has now become completely clear that glycogen synthase kinase-2 is merely a modified form of phosphorylase kinase, which has lost its ability to be regulated by calcium ions, as a result of the procedure used to purify glycogen synthase. Presumably the subunit interactions in the enzyme have become modified in such a way that the activity can no longer be regulated by the δ subunit. On the other hand, the modified enzyme still retains the activation by the second molecule of calmodulin. The evidence which demonstrates the identity of glycogen synthase kinase-2 and phosphorylase kinase is detailed in Embi *et al.* (1979a,b).

Glycogen synthase is an extremely good substrate for phosphorylase kinase and its rate of phosphorylation is only twofold slower than for phosphorylase when identical concentrations of the two substrates are used (Embi *et al.*, 1979b). At physiological concentrations of phosphorylase (8 mg/ml) and glycogen synthase (0.3 mg/ml), the time required for half-maximal phosphorylation of either enzyme does not differ more than twofold. Phosphorylation of glycogen synthase by phosphorylase kinase decreases the activity of the enzyme in the absence of glucose-6P (Embi *et al.*, 1979b).

ICR/IAn mice which completely lack phosphorylase kinase activity do not contain any calcium-dependent glycogen synthase kinase activity in skeletal muscle (Embi *et al.*, 1979a,b). This confirms the identity of glycogen synthase kinase-2 and phosphorylase kinase and also demonstrates that phosphorylase kinase is the only calcium-dependent glycogen synthase kinase in skeletal muscle. Myosin light chain kinase does not phosphorylate glycogen synthase (Embi *et al.*, 1979b; Srivastava *et al.*, 1979).

XII. SUMMARY

Although it has been believed for a number of years that the activation of phosphorylase kinase by calcium ions is likely to be the mechanism by which glycogenolysis and muscle contraction are synchronized (Heilmeyer *et al.*, 1970; Brostrom *et al.*, 1971), recent work has started to place this concept on a much firmer molecular basis, and our current ideas are summarized in Fig. 6.

Electrical excitation of muscle causes the release of calcium ions from the sarcoplasmic reticulum. The calcium ions bind to troponin C, activating actomyosin ATPase and initiating muscle contraction. The calcium ions also bind to calmodulin, a protein which has a similar structure and similar calcium binding properties to troponin C, and which is an integral

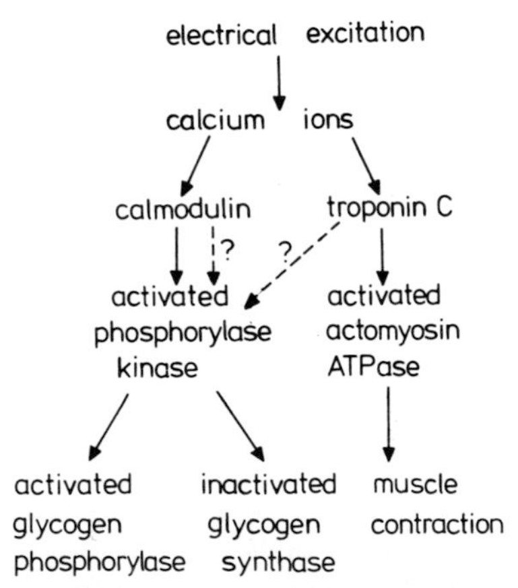

Fig. 6. The role of calmodulin and troponin C in the synchronous control of glycogen metabolism and muscle contraction.

component of the structure of phosphorylase kinase. The interaction of calcium with calmodulin leads to the activation of phosphorylase kinase, but, in addition, phosphorylase kinase can be further activated by either a second molecule of calmodulin or by troponin C. It is not yet clear which of these two molecules activates phosphorylase kinase *in vivo* (see Section X).

It is well known that when isolated frog muscles are made to contract by stimulating them electrically, that phosphorylase *b* is converted to phosphorylase *a*. However, the steady state level of phosphorylase *a* that is attained increases, and the time taken to reach the steady state decreases as a function of the frequency of stimulation (Helmreich and Cori, 1966; Danforth *et al.,* 1962). It is possible that the regulation of phosphorylase kinase activity by two molecules of calmodulin (or by calmodulin and troponin) enables the activity of phosphorylase kinase to respond to a much greater range of electrical stimulation of the tissue, allowing more sophisticated coordination of the processes of contraction and glycogenolysis. In addition, phosphorylase kinase not only phosphorylates and activates phosphorylase kinase but also phosphorylates and inactivates glycogen synthase, allowing coordinated control of the two opposing pathways of glycogen breakdown and synthesis.

ACKNOWLEDGMENTS

This work was supported by grants from the Medical Research Council, London and the British Diabetic Association.

REFERENCES

Brostrom, C. O., Hunkeler, F. L., and Krebs, E. G. (1971). *J. Biol. Chem.* **246,** 1961–1967.

Cohen, P. (1973). *Eur. J. Biochem.* **34,** 1–14.

Cohen, P. (1974). *Biochem. Soc. Symp.* **39,** 51–73.

Cohen, P. (1978). *Curr. Top. Cell. Reg.* **14,** 117–196.

Cohen, P. T. W., Burchell, A., and Cohen, P. (1976). *Eur. J. Biochem.* **66,** 347–356.

Cohen, P., Burchell, A., Foulkes, J. G., Cohen, P. T. W., Vanaman, T. C., and Nairn, A. C. (1978). *FEBS Lett.* **92,** 287–293.

Cohen, P., Embi, N., Foulkes, J. G., Hardie, D. G., Nimmo, G. A., Rylatt, D. B., and Shenolikar, S. (1979a). *Miami Winter Symp.* **16,** 463–481.

Cohen, P., Picton, C., and Klee, C. B. (1979b). *FEBS Lett.* **104,** 25–30.

Danforth, W. H., and Lyon, J. B. (1964). *J. Biol. Chem.* **239,** 4047–4050.

Danforth, W. H., Helmreich, E., and Cori, C. F. (1962). *Proc. Natl. Acad. Sci. U.S.A.* **48,** 1191–1199.

Dedman, J. R., Potter, J. D., Jackson, R. L., Johnson, J. D., and Means, A. R. (1977). *J. Biol. Chem.* **252,** 8415–8423.

DePaoli-Roach, A. A., Roach, P. J., and Larner, J. (1979). *J. Biol. Chem.* **254,** 12062–12068.
Drummond, G. I., Harwood, J. P., and Powell, C. A. (1969) *J. Biol. Chem.* **214,** 4235–4340.
Embi, N., Rylatt, D. B., and Cohen, P. (1979a). *ICN-UCLA Symp.* **13,** 257–271.
Embi, N., Rylatt, D. B., and Cohen, P. (1979b). *Eur. J. Biochem.* **100,** 339–347.
Hayakawa, T., Perkins, J. P., and Krebs, E. G. (1973a). *Biochemistry* **12,** 574–580.
Hayakawa, T., Perkins, J. P., and Krebs, E. G. (1973b). *Biochemistry* **12,** 567–573.
Head, J. F., and Perry, S. V. (1974). *Biochem. J.* **137,** 145–154.
Heilmeyer, L. M. G., Meyer, F., Haschke, R. H., and Fischer, E. H. (1970). *J. Biol. Chem.* **245,** 6649–6656.
Helmreich, E., and Cori, C. F. (1966). *Adv. Enzyme Reg.* **3,** 91–107.
Kilimann, M., and Heilmeyer, L. M. G. (1977). *Eur. J. Biochem.* **73,** 191–197.
Klee, C. B., and Krinks, M. H. (1978). *Biochemistry* **17,** 120–126.
Krebs, E. G., and Fischer, E. H. (1956). *Biochim. Biophys. Acta* **20,** 150–157.
Meyer, F., Heilmeyer, L. M. G., Haschke, R. H., and Fischer, E. H. (1970). *J. Biol. Chem.* **245,** 6642–6648.
Nimmo, H. G., and Cohen, P. (1974). *FEBS Lett.* **47,** 162–167.
Nimmo, H. G., and Cohen, P. (1977). *Adv. Cyc. Nuc. Res.* **8,** 145–266.
Osawa, E., Hosoi, K, and Ebashi, S. (1967). *J. Biochem.* **61,** 531–533.
Perry, S. V. (1974). *Biochem. Soc. Symp.* **39,** 115–132.
Perry, S. V., and Cole, H. A. (1974). *Biochem. J.* **141,** 733–743.
Posner, J. R., Stern, R., and Krebs, E. G. (1965). *J. Biol. Chem.* **240,** 982–985.
Roach, P. J., Roach, A. A., and Larner, J. (1978). *J. Cyc. Nuc. Res.* **4,** 245–257.
Rylatt, D. B., and Cohen, P. (1979). *FEBS Lett.* **98,** 71–75.
Rylatt, D. B., Embi, N., and Cohen, P. (1979). *FEBS Lett.* **98,** 76–80.
Shenolikar, S., Cohen, P. T. W., Cohen, P., Nairn, A. C., and Perry, S. V. (1979). *Eur. J. Biochem.* **100,** 329–337.
Sigel, P., and Pette, D. (1969). *J. Histochem. Cytochem.* **17,** 225–237.
Srivastava, A. K., Waisman, D. M., Brostrom, C. O., and Soderling, T. R. (1979). *J. Biol. Chem.* **254,** 583–586.
Titani, K., Cohen, P., Walsh, K. A., and Neurath, H. (1975). *FEBS Lett.* **55,** 120–123.
Walsh, D. A., Perkins, J. P., Brostrom, C. O., Ho, E. S., and Krebs, E. G. (1971). *J. Biol. Chem.* **246,** 1968–1976.
Wolff, D. J., Poirier, P. G., Brostrom, C. O., and Brostrom, M. A. (1977). *J. Biol. Chem.* **252,** 4108–4117.
Yagi, K., Yazawa, M., Kakiuchi, S., Oshimo, M., and Uenishi, K. (1978). *J. Biol. Chem.* **253,** 1338–1340.
Yeaman, S. J., Cohen, P., Watson, D. C., and Dixon, G. H. (1975). *Biochem. J.* **162,** 411–421.

NOTE ADDED IN PROOF

It has recently been found that the δ subunit is complexed to the γ subunit of phosphorylase kinase, while the δ'-subunits interacts, with both the α- and β-subunits (Picton, C., Klee, C. B., and Cohen, P., *Eur. J. Biochem,* in press).

Chapter 10

Plant and Fungal Calmodulin and the Regulation of Plant NAD Kinase

MILTON J. CORMIER
JAMES M. ANDERSON
HARRY CHARBONNEAU
HAROLD P. JONES
RICHARD O. McCANN

I. INTRODUCTION

Diverse sources of animal tissues ranging from coelenterates to mammals have been found to contain a low molecular weight Ca^{2+}-binding protein known as calmodulin (Teo *et al.*, 1973; Teo and Wang, 1973; Lin *et al.*, 1974; Wolff and Brostrom, 1974; Childers and Siegel, 1975; Dedman *et al.*, 1977b; Yagi *et al.*, 1978; Dabrowska *et al.*, 1978; Waisman *et al.*, 1978a; Jones *et al.*, 1979; Head *et al.*, 1979). A comparison of the physical

CALCIUM AND CELL FUNCTION, VOL. I

ISBN 0-12-171401-2

and biological properties of calmodulin isolated from these sources suggested that a high degree of structural conservation exists in calmodulin isolated from widely diverse species (Jones *et al.*, 1979). This high degree of structural conservation has been shown recently by a comparison of the amino acid sequences of coelenterate and mammalian calmodulin (Sharief, *et al.*, 1980).

The observed structural conservation among animal calmodulin suggests one or more fundamental roles for this protein in Ca^{2+}-dependent regulatory processes. *In vitro* studies have indeed shown that calmodulin will activate a number of enzymes in a Ca^{2+}-dependent manner and that it is involved in a number of Ca^{2+}-dependent cellular processes (see Cheung, 1980 for a review).

Calmodulin-like activity has been observed in tissue extracts of plants and fungi (Anderson and Cormier, 1978; Waisman *et al.*, 1978a). Since these original observations we have obtained homogeneous preparations of plant and fungal calmodulin (Anderson and Cormier, 1979). We provide evidence here that the chemical, physical, and biological properties of plant and fungal calmodulin are similar to calmodulin isolated from animal sources and that calmodulin, isolated from either plant, fungal, or animal sources, will activate plant NAD kinase in a Ca^{2+}-dependent manner.

II. ISOLATION OF CALMODULIN BY FLUPHENAZINE-SEPHAROSE CHROMATOGRAPHY

The isolation of plant and fungal calmodulin proved to be difficult when using procedures developed in various laboratories for the isolation of calmodulin from animal tissues. To overcome this problem we developed an affinity chromatography procedure which simplifies the purification of calmodulin regardless of tissue source (Charbonneau and Cormier, 1979b). A similar procedure has also been developed by Jamieson and Vanaman (1979). The procedure is based on the observations by Weiss and Levin (1978) that the antipsychotic phenothiazine drugs, such as trifluoperazine, will bind to calmodulin in the presence of Ca^{2+} with a K_d of 1 μM; in the absence of Ca^{2+} the affinity of calmodulin for trifluoperazine is greatly reduced.

Fluphenazine-Sepharose Chromatography of Calmodulin

The phenothiazine drug, fluphenazine, was immobilized on Sepharose 4B as previously described (Charbonneau and Cormier, 1979b). This fluphenazine-Sepharose matrix was successfully used in the isolation of calmodulin from various tissue sources including mammals (porcine brain), coelenterates (*Renilla reniformis*), higher plants (peanuts and pea seedlings), and a fungus (mushroom). A typical elution profile from this affinity column is shown in Fig. 1. Tissue extracts (crude or partially purified) are prepared in neutral buffers containing 1 m*M* Ca^{2+} and loaded on the column equilibrated in the same buffer. Under these conditions most of the protein in the extract is eluted while all of the calmodulin binds to the column. Addition of NaCl (0.5 *M* final concentration) to the elution buffer removes additional protein absorbed to the column in a non-specific manner. Finally, calmodulin is eluted from the column with a pulse of 10 m*M* Tris and 5 m*M* EGTA, pH 8.0. Alkaline discontinuous and SDS polyacrylamide gel analysis of the protein eluted with EGTA showed that plant

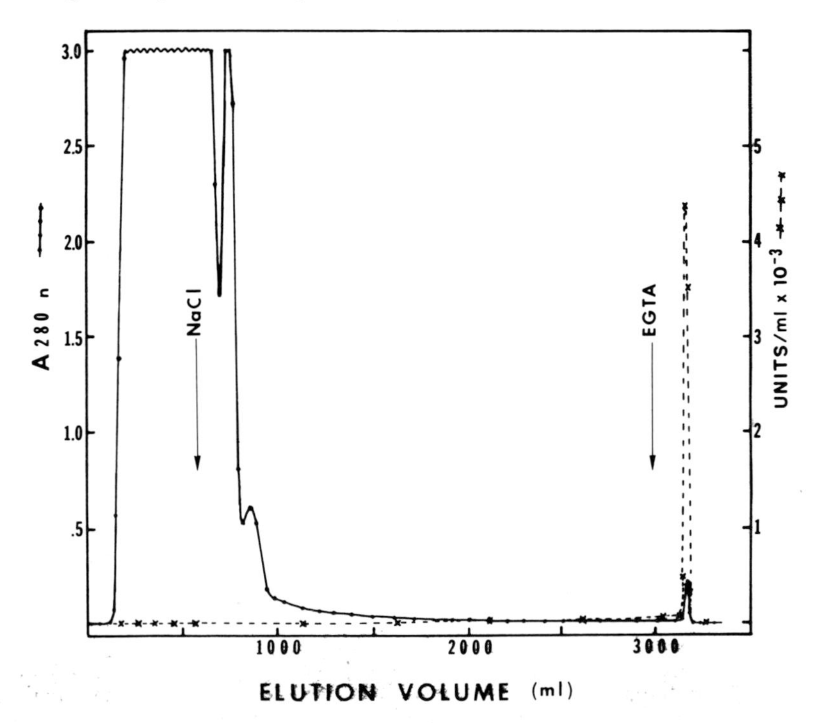

Fig. 1. Typical elution profile for the chromatography of calmodulin on fluphenazine-Sepharose. Crude or partially purified tissue extracts are prepared in neutral buffers containing 1 m*M* Ca^{2+} and loaded on the column equilibrated in the same buffer. Column capacity was 3 mg of porcine brain calmodulin per milliliter of packed gel. NaCl (0.5 *M* final concentration) was added to the elution buffer at the point indicated by the arrow. Calmodulin was eluted from the column with a pulse 10 m*M* Tris and 5 m*M* EGTA, pH 8.0, as indicated by the arrow. Both absorbance at 280 nm and calmodulin activity were monitored. Calmodulin activity was measured by the bioluminescence assay of Matthews and Cormier (1978).

and porcine brain calmodulin preparations were homogeneous; however, calmodulin isolated from coelenterate and fungal sources was contaminated with a low molecular weight protein (<10%) which could be removed by ion exchange chromatography. Furthermore, SDS gel analyses showed that all of the protein species loaded on the column could be accounted for in the fractions washed through the column and eluted with EGTA. Therefore, in extracts of higher plants and porcine brain (100,000 *g* supernatant) calmodulin alone binds to fluphenazine-Sepharose in a Ca^{2+}-dependent manner.

These observations support the suggestion of Weiss and Levin (1978) that some of the diverse pharmacological effects of the phenothiazine drugs may be explained by their ability to inhibit calmodulin function. Because of the high selectivity of binding of the phenothiazine drugs to calmodulin, it may be possible to utilize these drugs as specific inhibitors of calmodulin-dependent processes. Although further work will be necessary to assess the use of the phenothiazine drugs as functional probes for the role(s) of calmodulin in various Ca^{2+}-dependent cellular events, this approach appears to be promising.

III. PROPERTIES OF PLANT AND FUNGAL CALMODULIN

A. Electrophoretic Data

Alkaline discontinuous polyacrylamide gel electrophoresis of plant and fungal calmodulin showed that these proteins are highly acidic and exhibit migration patterns similar to that observed for bovine brain calmodulin. Both proteins produced single bands on SDS-polyacrylamide slab gel electrophoresis as shown in Fig. 2. The apparent molecular weights observed for plant and fungal calmodulin were similar but not identical to bovine brain calmodulin. Sedimentation equilibrium studies gave molecular weights of 14,600 and 16,000 for plant and fungal calmodulin, respectively (see Table III).

An additional similarity between plant, fungal and bovine brain calmodulin is their ability to form hybrid complexes with troponin I. It is well known that mammalian calmodulin will form a hybrid protein–protein complex with troponin I in the presence of Ca^{2+} (Amphlett *et al.*, 1976; Dedman *et al.*, 1977a). Such complexes are easily observed by using polyacrylamide gel electrophoretic techniques (Amphlett *et al.*, 1976). Using such procedures we have found that plant, fungal, and bovine brain calmodulin all form Ca^{2+}-dependent calmodulin–troponin I complexes that are electrophoretically similar.

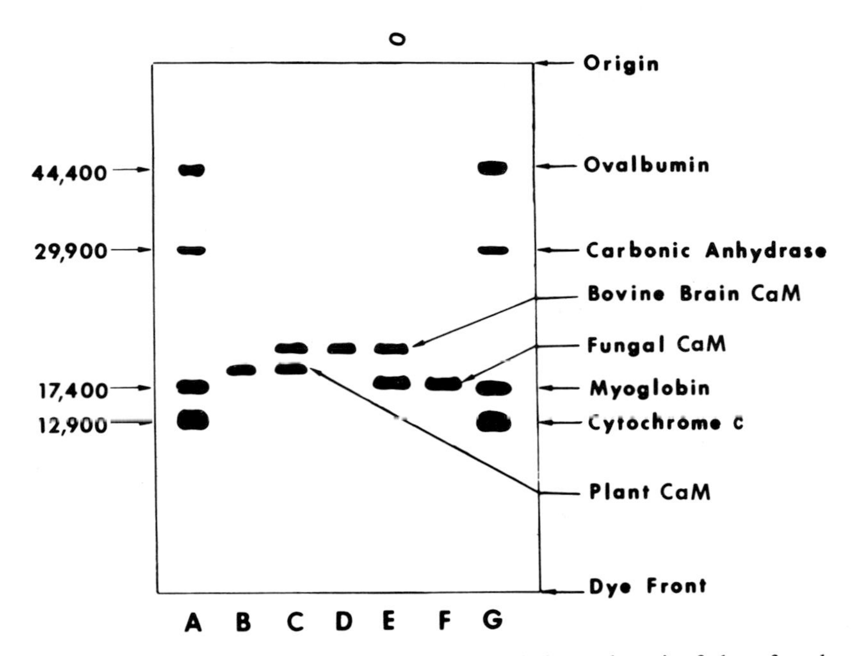

Fig. 2. Sodium dodecyl sulfate polyacrylamide gel electrophoresis of plant, fungal and bovine brain calmodulin (CaM). Proteins used as molecular weight standards are also shown in the figure (A and G). Wells were loaded with 10 μg of each protein. B, Plant CaM; C, bovine brain + plant CaM; D, bovine brain CaM; E, bovine brain + fungal CaM; F, fungal CaM.

B. Amino Acid Composition and Spectral Data

As shown in Table I the amino acid compositions of plant and fungal calmodulin are remarkably similar to that found for bovine brain calmodulin. The similarities include the presence of trimethyllysine, a relative preponderance of negatively charged amino acids, two proline residues, and the lack of tryptophan. Of interest is the presence of one residue of cysteine in plant calmodulin.

A subtle but interesting difference in plant calmodulin is the high phenylalanine to tyrosine ratio as shown in Table I. Plant calmodulin contains one tyrosyl residue while fungal and bovine brain calmodulin each contain two such residues; all three proteins contain eight phenylalanine residues. These differences in the phenylalanine to tyrosine ratios are reflected in the absorption spectra of the respective calmodulins.

As seen in Fig. 3 plant and fungal calmodulin exhibit multiple absorption maxima at 276–279, 268, 265, 258, and 253 nm which are characteristic of calmodulin isolated from mammalian sources. However, a relatively low absorbance at 279 nm is observed for plant calmodulin when com-

TABLE I

Amino Acid Composition of Calmodulins

	Plant (peanut) (mole/16,700 g)	Fungal (mushroom) (mole/16,700 g)	Bovine brain[a]
Lysine	8	10	7
Histidine	1	1	1
Trimethyllysine	1–2	1	1
Arginine	4	4	6
Aspartic acid	27	25	23
Threonine	9	10	12
Serine	5	10	4
Glutamic acid	27	27	27
Proline	2	2	2
Glycine	11	12	11
Alanine	10	10	11
Half-cystine	1[b]	0	0
Valine	6	5	7
Methionine	7	9	9
Isoleucine	5	7	8
Leucine	11	11	9
Tyrosine	1	2	2
Phenylalanine	8	8	8
Tryptophan	0	0	0

[a] By sequence from Vanaman *et al.*, 1977.

[b] One residue of cysteine was observed by Dr. T. C. Vanaman during the course of sequence work (personal communication). We have also noted the presence of cysteine by performic acid oxidation.

pared to porcine brain or fungal calmodulin. The presence of a single tyrosyl residue in calmodulin isolated from the marine coelenterate, *Renilla reniformis,* has recently been shown by amino acid sequence data (Sharief *et al.,* 1980). It appears that plant calmodulin represents a second such example.

In contrast, bovine brain calmodulin contains two tyrosyl residues located at positions 99 and 138 in the polypeptide chain (Vanaman *et al.,* 1977). Upon binding of Ca^{2+} to calmodulin, or upon changes in ionic strength, spectral perturbations of the tyrosyl residues are observed (Wang, 1977; Liu and Cheung, 1976; Klee, 1977; Wolff *et al.,* 1977; Dedman *et al.,* 1977b). Interpretations of recent data suggest that Ca^{2+}-binding to calmodulin selectively produces large changes in the microenvironment of tyrosyl residue 138 in contrast to negligible effects on tyrosyl residue 99 (Richman and Klee, 1979).

Since both *Renilla* and plant calmodulin contain a single tyrosyl resi-

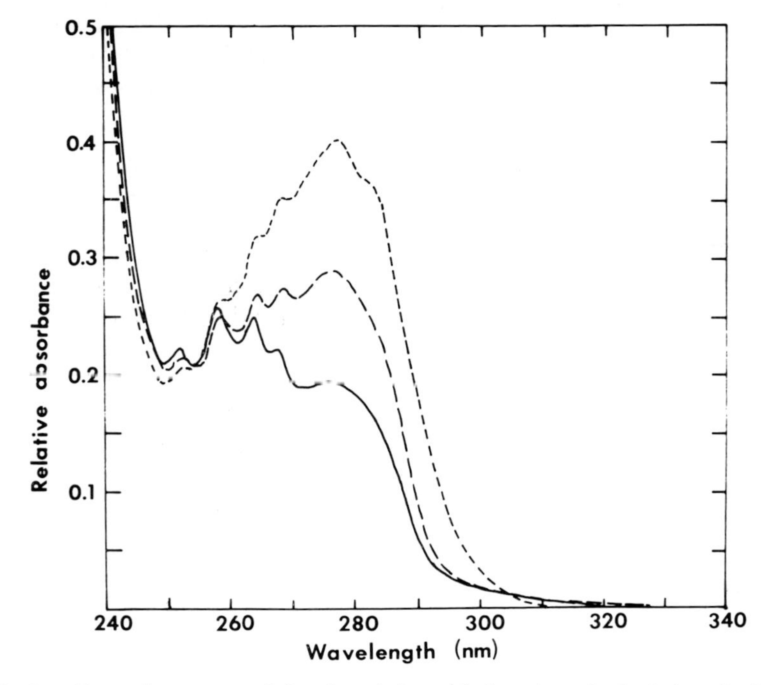

Fig. 3. Absorption spectra of plant (——), fungal (– –), and porcine brain (- - - -) calmodulin. Spectra were normalized at 255 nm.

TABLE II

Effects of Ca^{2+} and Trifluoperazine on the Activation of Porcine Brain Phosphodiesterase by Calmodulins Isolated from Bovine Brain, Plant (Peanuts), and Fungus (Mushroom)[a]

	Relative phosphodiesterase activity		
Calmodulin source	$+Ca^{2+}$	+EGTA	$+Ca^{2+}$ plus trifluoperazine
None	15	10	14
Bovine brain	100	12	15
Plant	87	9	13
Fungal	106	29	18

[a] Calmodulin-dependent phosphodiesterase activity was measured by the bioluminescence assay of Matthews and Cormier, 1978.

Final concentrations of Ca^{2+}, EGTA, and trifluoperazine were 0.1 m*M*. Estimated coefficient of variation was ±20%. Final concentration of calmodulin was 1 mg/ml which assured saturation of phosphodiesterase.

due, and since Ca^{2+} induced spectral perturbations of this tyrosyl residue have been observed by us (see Table III), a careful study of the effects of Ca^{2+}-binding and ionic strength on the spectral perturbation of this single tyrosyl residue may be informative.

C. Biological Activity of Plant and Fungal Calmodulin

As shown in Table II, plant and fungal calmodulin, as well as bovine brain calmodulin, will activate porcine brain cyclic nucleotide phosphodiesterase. A comparative study of the concentration dependence of plant and bovine brain calmodulin on the activation of this phosphodiesterase showed that the amount of calmodulin required for 50% maximal activation of phosphodiesterase was the same. All three calmodulins exhibit a dependency on Ca^{2+} for activation of phosphodiesterase.

It is well known that the Ca^{2+}-dependent binding of trifluoperazine to bovine brain calmodulin will inhibit its ability to activate a number of enzymes such as calmodulin-dependent phosphodiesterase (Levin and Weiss, 1977; Weiss and Levin, 1978; Brostrom *et al.*, 1978). This is also true for plant and fungal calmodulin (Table II).

Thus the Ca^{2+}-dependent biological activities of plant and fungal calmodulin are indistinguishable from those observed with bovine brain calmodulin. A comparison of the physical and biological properties of these calmodulins is summarized in Table III.

Taken together these data suggest that calmodulin is a Ca^{2+}-dependent regulatory protein associated with eukaryotic organisms and that its structure has been highly conserved throughout evolution.

IV. ACTIVATION OF PLANT NAD KINASE BY CALMODULIN

NAD kinase has been implicated as an important regulatory protein in higher plants since many key enzymes involved in the biosynthesis of plant sugars utilize NADP while catabolic enzymes utilize NAD. Muto and Miyachi (1977) showed that partially purified plant NAD kinase is activated by a protein activator which is heat stable and highly acidic. Anderson and Cormier (1978) noted that there were similarities in the properties of this plant activator protein and bovine brain calmodulin. Furthermore, they recognized similarities in the activation of plant NAD kinase by the plant activator protein and calmodulin-dependent enzyme regulation in animal systems.

At this point the question of immediate interest was whether or not the

TABLE III

Properties of Calmodulin Isolated from Plants, Fungal, and Mammalian Sources

Property	Source of calmodulin		
	Plant	Fungal	Mammalian
Stokes radius (Å)	22	21	23[a]
$s_{20,w}$	1.9	1.8	1.9[a]
Molecular weight:			
From SDS-gel electrophoresis	17,300	15,800	18,000[b]
From gel filtration (K_D versus M_r)	23,300	21,300	28,200[a]
From sedimentation equilibrium	14,600	16,000	17,800[a]
Extinction coefficients 1% ε276	0.9		1.8[b]
Phenylalanine/tyrosine ratio	8	4	4[b]
Absorption maxima (nm)	279, 268, 265, 259, 253	276, 269, 265, 258, 253	277, 269, 265, 259, 253[c]
Ca^{2+}-dependent enhancement of tyrosine fluorescence	+		+[a]
Ca^{2+}-dependent interaction with troponin I	+	+	+[d]
Amount required for 50% activation of phosphodiesterase (ng)[e]	9		8
Stelazine inhibition of phosphodiesterase activity	+	+	+[f]
Activation of plant NAD kinase	+	+	+

[a] Dedman *et al.*, 1977b.
[b] Watterson *et al.*, 1976.
[c] Cheung *et al.*, 1978.
[d] Amphlett *et al.*, 1976.
[e] This phosphodiesterase preparation catalyzed the conversion of 32 pmoles of cAMP min^{-1} in the absence of calmodulin.
[f] Levin and Weiss, 1977.

activation of plant NAD kinase was Ca^{2+} dependent. Since Muto and Miyachi (1977) had not used metal chelators in their assay buffers for NAD kinase, this question remained unanswered. When partially purified plant NAD kinase from pea seedlings was assayed in the presence and absence of EGTA we found that the activation of plant NAD kinase by the protein activator was indeed Ca^{2+} dependent (Anderson and Cormier, 1978). The enzyme activity was stimulated about fourfold in the presence of saturating amounts of the activator and Ca^{2+}, but little or no stimulation was observed in the absence of Ca^{2+}.

The plant activator protein was then partially purified by chromatography on DEAE-cellulose. Column fractions were tested for their ability to activate plant NAD kinase and for their ability to activate calmodulin-dependent porcinc brain cyclic nucleotide phosphodiesterase. We found that the activator for plant NAD kinase also activated porcine brain phosphodiesterase (Anderson and Cormier, 1978). For both the NAD kinase and phosphodiesterase assays activation by the plant activator protein was Ca^{2+} dependent. Furthermore, activation of both NAD kinase and phosphodiesterase was not inhibited by prior heat treatment of the plant activator. In both cases a single protein peak exhibiting both activitites was noted at a salt concentration of 330 m*M* NaCl indicating that the activator protein is highly acidic. This is approximately the same elution profile observed for the activator of plant NAD kinase by Muto and Miyachi (1977).

Additional observations suggested that the plant NAD kinase activator and calmodulin were related (Anderson and Cormier, 1978). For example, both the plant activator and bovine brain calmodulin were found to have similar elution volumes on Sephadex G-75 superfine indicating similar Stokes' radii. Furthermore, the plant NAD kinase activator, like calmodulin, interacts with troponin I in a Ca^{2+}-dependent manner. In addition, bovine brain calmodulin will replace the plant activator protein in the Ca^{2+}-dependent activation of plant NAD kinase.

We subsequently purified the plant NAD kinase activator to homogeneity by procedures outlined in Section II. Both pea seedlings and peanuts were used as source material. Table IV shows that the activator protein isolated from plants can be replaced by bovine brain and fungal calmodulin in the Ca^{2+}-dependent activation of plant NAD kinase. Furthermore

TABLE IV

Ca^{2+}-Dependent Stimulation of Plant NAD Kinase Activity by Calmodulin

Calmodulin source	Relative NAD kinase activity	
	$-Ca^{2+}$	$+Ca^{2+}$
None	23	45[a]
Bovine brain	29	100
Plant	26	112
Fungal	19	121

[a] This NAD kinase preparation was contaminated with small amounts of plant calmodulin which accounts for the stimulation by Ca^{2+} in the absence of added calmodulin. NAD kinase activity was measured by the method of Muto and Miyachi, 1977. Calcium was added to 1.0 m*M* for $+Ca^{2+}$ and EGTA was added to 1.0 m*M* for $-Ca^{2+}$. Final concentration of calmodulin was 30 μg/ml in the assay.

the Ca^{2+} concentration dependency of this activation is approximately the same as that observed for the calmodulin-dependent activation of phosphodiesterase (Fig. 4). That is, the activator of plant NAD kinase, like calmodulin, responds to micromolar concentrations of free Ca^{2+}.

The homogeneous preparation of the NAD kinase activator was then characterized as outlined in Section III. The results listed in Section III demonstrate that the plant NAD kinase activator (termed plant calmodulin in Section III) is indeed calmodulin whose physical, chemical, and biological properties are similar to those observed for calmodulin isolated from animal sources.

Of immediate interest is the isolation of plant NAD kinase and a study of its Ca^{2+}-dependent interaction with calmodulin. It will also be of interest to examine the possible calmodulin-dependent activation of NAD kinase in animal systems.

V. SIGNIFICANCE OF PLANT AND FUNGAL CALMODULIN

Crystal structure studies on parvalbumin have shown that Ca^{2+} is bound in a site possessing a distinctive conformation which provides the six ligands necessary for the coordination of Ca^{2+} (Moews and Kret-

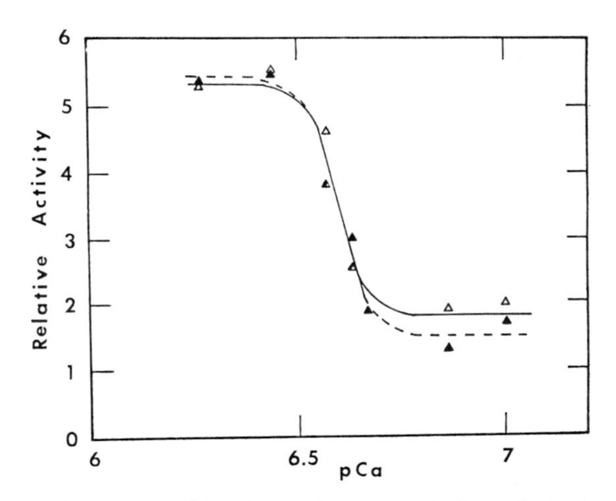

Fig. 4. Calcium dependence of the plant activator protein-dependent stimulation of NAD kinase (△–△) and calmodulin-dependent porcine brain phosphodiesterase (▲–▲). The assays were done in 25 m*M* Tris-HCl, pH 8.0 at 25°C, 10 m*M* $MgCl_2$, 2.0 m*M* EGTA. The values for NAD kinase activity relate to 0.01 absorbance units lost at 600 nm per 200 sec. Free Ca^{2+} in the Ca^{2+}-EGTA buffer system was determined using a modification of the program of Perrin and Sayce (1967).

singer, 1975); this conformation has been termed the EF-hand. Other investigators have suggested that homologous EF-hands are present in several other Ca^{2+}-binding proteins based on amino acid sequence homologies with parvalbumin. These include troponin C (Collins *et al.*, 1973; Van Eerd and Takahashi, 1975), alkali-extractable myosin light chains (Tufty and Kretsinger, 1975; Weeds and McLachlan, 1974), vitamin D-dependent Ca^{2+}-binding protein (Huang *et al.*, 1975), and calmodulin (Watterson *et al.*, 1976). The two Ca^{2+}-binding proteins of bioluminescence, i.e., aequorin and luciferin binding protein, are also among the postulated EF hand proteins (Kretsinger, 1977; Charbonneau and Cormier, 1979a).

The postulated EF hand conformation is believed to be responsible for two important properties of all of the above mentioned Ca^{2+}-binding proteins. These are (1) the ability to bind Ca^{2+} with dissociation constants (K_d) in the range 0.1–1 μM and (2) to undergo significant conformational changes upon Ca^{2+}-binding. For example, calmodulin possesses four putative EF hands, or Ca^{2+}-binding loci, whose K_d for Ca^{2+}-binding is approximately 1 μM (Vanaman *et al.*, 1977; Dedman *et al.*, 1977b). Furthermore, significant changes in conformation occur upon binding of Ca^{2+} to calmodulin (Wang, 1977; Liu and Cheung, 1976; Klee, 1977; Wolff *et al.*, 1977; Dedman *et al.*, 1977b). Presumably, it is this change in conformation, brought about by Ca^{2+}-binding, which allows the Ca–calmodulin complex to bind to calmodulin-dependent enzymes thus turning them "on."

The probable physiological significance of the observed K_d values for Ca^{2+}-binding to calmodulin can be appreciated by an examination of the changes in free Ca^{2+} concentration within the cytosol of cells during stimulus–response coupling events. The most reliable estimates of these changes have been obtained by the use of aequorin, a Ca^{2+}-dependent bioluminescent protein (Shimomura *et al.*, 1962) whose K_d for Ca^{2+}-binding is 0.14 μM (Allen *et al.*, 1977). As shown in Fig. 5 aequorin is viewed as an oxygenase which is most unique since in the absence of Ca^{2+}, it exists as an oxygen-containing, enzyme–substrate intermediate which is remarkably stable (see Cormier, 1978, and Blinks *et al.*, 1976, for reviews). In the presence of Ca^{2+} a conformational change presumably occurs resulting in the interaction of the substrate (luciferin) and the oxygenated species. This interaction generates an electronic excited state of the protein–product complex resulting in the production of blue light.

Since aequorin will produce visible light in the presence of low concentrations of free Ca^{2+}, this protein has been used by physiologists to monitor changes in cytosolic free Ca^{2+} concentration during stimulus–response coupling. Aequorin has been pressure injected into the cytosol of a variety of cell types including muscle, nerve, photoreceptor, protozoan,

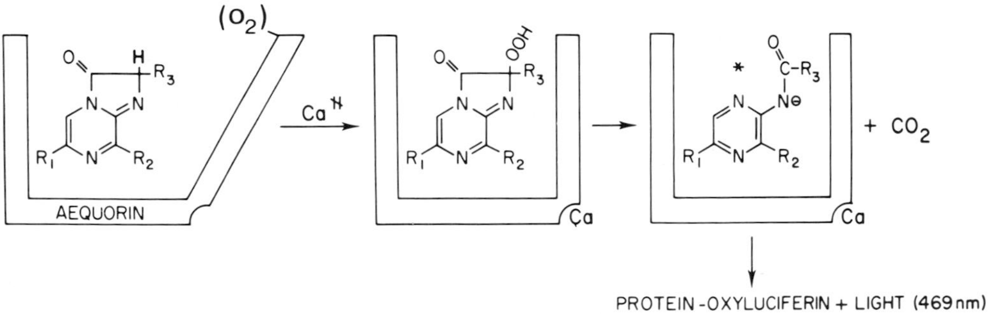

Fig. 5. A model for Ca^{2+}-dependent light production from the photoprotein, aequorin. The structure shown is the substrate (luciferin) where R_1, *p*-hydroxyphenyl; R_2, benzyl; and R_3, *p*-hydroxybenzyl. See text for details.

salivary gland, and egg cells (see Blinks *et al.*, 1976, for a review). Luminescence output measurements from these experiments provide an estimate of the cytosolic free Ca^{2+} concentration of approximately 10^{-8} to 10^{-7} *M* in the unstimulated cell. Upon receiving the appropriate stimulus this value has been found to rise to approximately 10^{-5} to 10^{-4} *M* and is relatively independent of the cell type.

Thus the role of calmodulin in stimulus–response coupling in animal systems may be viewed as illustrated in Fig. 6. In this model Ca^{2+} acts as a second messenger through calmodulin. In the unstimulated cell, the free

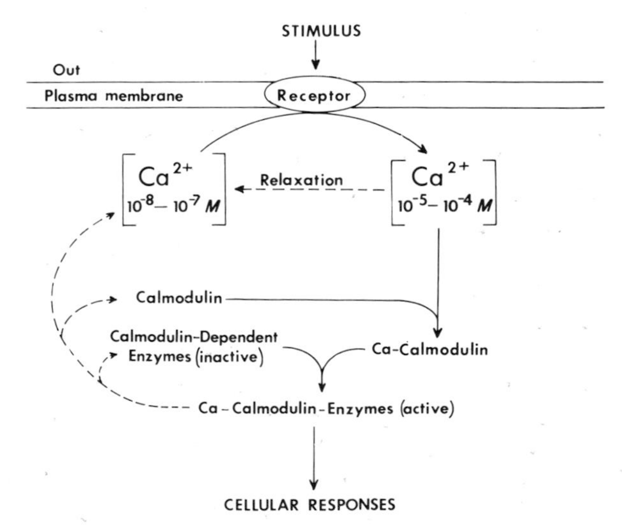

Fig. 6. A model for calmodulin-dependent regulation during stimulus–response coupling. See text for details.

Ca^{2+} concentration is too low to bind to calmodulin since the concentration is considerably below the K_d value of 1 μM. In this state the calmodulin-dependent enzymes are turned "off." Upon stimulation, the cytosolic free Ca^{2+} concentration rises within milliseconds to a level considerably above the K_d value for Ca^{2+}-binding to calmodulin. In this state Ca^{2+} binds to calmodulin and the Ca–calmodulin complex in turn binds to a series of calmodulin-dependent enzymes turning them "on." *In vitro* evidence for this mechanism of calmodulin-dependent regulation has been obtained in several laboratories (Watterson and Vanaman, 1976; Lynch *et al.*, 1976; Ho *et al.*, 1977; Cheung *et al.*, 1978; Lynch and Cheung, 1979; Nairn and Perry, 1979). During relaxation Ca^{2+} is presumably pumped out of the cell or rapidly resequestered. This results in dissociation of the calmodulin–enzyme complexes thus turning them "off."

The model illustrated in Fig. 6 may also be used to explain many physiological events in plants and fungi. Support for this concept is provided by the presence of calmodulin in these organisms and by the observation that plant NAD kinase is activated by calmodulin in a Ca^{2+}-dependent manner. However, in order to explain how a stimulus such as light can orient the direction of growth in plants, as in phototropism, the recent work on stimulus-induced ion currents in plant cells must be taken into account.

Using eggs of the brown algae, *Fucus pelvetia,* it has been observed that unilateral light establishes an ion current which passes through the cell and which originates from the point of stimulus (see Jaffe, 1979, for a review). One of the major components of this ion current is Ca^{2+}, and it is interesting to note that among the divalent cations in the cell, Ca^{2+} has the lowest effective mobility (Hodgkin and Keynes, 1957; Kushmerick and Podolsky, 1969). Thus a stimulus-induced Ca^{2+} current through a cell should result in the formation of a concentration gradient across the cytosol. That such a Ca^{2+} concentration gradient can form in response to light has been demonstrated in experiments employing $^{45}Ca^{2+}$ and low-temperature autoradiography (Jaffe, 1979). In these experiments Ca^{2+} was found to be concentrated about threefold within the cytosol of the growing tip relative to opposite regions of the cell.

This Ca^{2+} gradient, coupled with the ability of Ca^{2+} to activate enzymes via calmodulin, could provide the mechanism for orienting cellular growth. With the stimulus removed the ion current would cease followed by the elimination of the Ca^{2+} gradient and directional growth. Pollen tube growth also appears to be directed by a Ca^{2+} gradient generated by an ion current (Weisenseel and Jaffe, 1976; Herth, 1978; Reiss and Herth, 1979). If this gradient is disrupted by treatment of the pollen tube with the Ca^{2+} ionophore A23187 then tip growth ceases. The Ca^{2+} chelator EDTA also prevents pollen tube growth. Thus not only is Ca^{2+} required for pollen tube growth but Ca^{2+} must specifically enter the tip of the growing tube.

Thus the model illustrated in Fig. 6 can be used to envision a mechanism for Ca^{2+}-mediated physiological events in both animals and plants. Calmodulin is one of the logical targets for Ca^{2+} in either case. As Kretsinger (1977) has pointed out, there are no known cellular components other than Ca^{2+}-binding proteins which can bind Ca^{2+} in the 10^{-5} *M* concentration range. It is thus possible that Ca^{2+}, acting through calmodulin, may be involved in regulatory events that can be either rapid, as one would expect from a Ca^{2+} transient, or long-lived, as one would expect from a sustained intracellular Ca^{2+} concentration gradient. In the latter case, the result could be directional growth.

Calcium ion and its acceptor calmodulin could therefore be involved in such diverse processes as phototropism, photoperiodism, geotropism, and hormonal regulation of growth. Consistent with this model is the observation that NAD is converted to NADP upon illumination of plant leaves (Krogmann and Ogren, 1965).

An experimental approach for testing the above model is currently being developed in our laboratory. It involves aequorin injected cells and on-line minicomputer controlled image intensification analysis for monitoring changes in free Ca^{2+} within the cytosol of various plant and fungal cell types.

ACKNOWLEDGMENTS

This work was supported by National Science Foundation Grants No. PCM79-05043 and PCM 76-84593. We wish to thank Dr. T. C. Vanaman for the bovine brain calmodulin used in these experiments and Dr. Mary Anne Conti for supplying us with synthetic trimethyllysine.

REFERENCES

Allen, D. G., Blinks, J. R., and Prendergast, F. G. (1977). Aequorin luminescence: Relation of light emission to calcium concentration—A calcium-independent component. *Science* **195,** 996–998.

Amphlett, G. W., Vanaman, T. C., and Perry, S. V. (1976). Effect of the troponin C-like protein from bovine brain (brain modulator protein) on the Mg^{2+}-stimulated ATPase of skeletal muscle actomyosin. *FEBS Lett.* **72,** 163–167.

Anderson, J. M., and Cormier, M. J. (1978). Calcium-dependent regulator of NAD kinase in higher plants. *Biochem. Biophys. Res. Commun.* **84,** 595–602.

Anderson, J. M., and Cormier, M. J. (1979). Isolation of calcium-dependent modulator protein from higher plants and fungi. *Fed. Proc., Fed. Am. Soc. Exp. Biol.* **38,** 1310.

Blinks, J. R., Prendergast, F. G., and Allen, D. G. (1976). Photoproteins as biological calcium indicators. *Pharmacol. Rev.* **28,** 1–93.

Brostrom, C. O., Huang, Y. C., Breckenridge, B. McL., and Wolff, D. J. (1975). Identification of a calcium-binding protein as a calcium-dependent regulator of brain adenylate cyclase. *Proc. Natl. Acad. Sci. U.S.A.* **72,** 64–68.

Brostrom, M. A., Brostrom, C. O., Breckenridge, B. McL., and Wolff, D. J. (1978). Calcium-dependent regulation of brain adenylate cyclase. *Adv. Cyclic Nucleotide Res.* **9,** 85–99.

Charbonneau, H., and Cormier, M. J. (1979a). Ca^{2+}-Induced bioluminescence in *Renilla reniformis*. Purification and characterization of a calcium-triggered luciferin-binding protein. *J. Biol. Chem.* **254,** 769–780.

Charbonneau, H., and Cormier, M. J. (1979b). Purification of plant calmodulin by fluphenazine-sepharose affinity chromatography. *Biochem. Biophys. Res. Commun.* **90,** 1039–1047.

Cheung, W. Y. (1980). Calmodulin plays a pivotal role in cellular regulation. *Science* **207,** 19–27.

Cheung, W. Y., Lynch, T. J., and Wallace, R. W. (1978). An endogenous Ca^{2+}-dependent activator protein of brain adenylate cyclase and cyclic nucleotide phosphodiesterase. *Adv. Cyclic Nucleotide Res.* **9,** 233–251.

Childers, S. R., and Siegel, F. L. (1975). Isolation and purification of a calcium-binding protein from electroplax of *Electrophorus Electricus*. *Biochim. Biophys. Acta* **405,** 99–108.

Cohen, P., Burchell, A., Foulkes, J. G., Cohen, P. T. W., Vanaman, T. C., and Nairn, A. C. (1978). Identification of the Ca^{2+}-dependent modulator protein as the fourth subunit of rabbit skeletal muscle phosphorylase kinase. *FEBS Lett.* **92,** 287–293.

Collins, J. H., Potter, J. D., Horn, M. j., Wilshire, G., and Jackman, N. (1973). The amino acid sequence of rabbit skeletal muscle troponin C: gene replication and homology with calcium-binding proteins from carp and hake muscle. *FEBS Lett.* **36,** 268–272.

Cormier, M. J. (1978). Comparative biochemistry of animal systems. *In* "Bioluminescence in Action" (P. J. Herring, ed.), pp 75–108. Academic Press, New York.

Dabrowska, R., Sherry, J. M. F., Aromatorio, D. K., and Hartshorne, D. J. (1978). Modulator protein as a component of the myosin light chain kinase from chicken gizzard. *Biochemistry* **17,** 253–258.

Dedman, J. R., Potter, J. D., and Means, A. R. (1977a). Biological cross-reactivity of rat testis phosphodiesterase activator protein and rabbit skeletal muscle troponin-C. *J. Biol. Chem.* **252,** 2437–2440.

Dedman, J. R., Potter, J. D., Jackson, R. L., Johnson, J. D., and Means, A. R. (1977b). Physicochemical properties of rat testis Ca^{2+}-dependent regulator protein of cyclic nucleotide phosphodiesterase. *J. Biol. Chem.* **252,** 8415–8422.

Head, J. F., Mader, S., and Kaminer, B. (1979). Calcium-binding modulator protein from the unfertilized egg of the sea urchin *Arbacia Punctulata*. *J. Cell. Biol.* **80,** 211–218.

Herth, W. (1978). Ionophore A23187 stops tip growth, but not cytoplasmic streaming, in pollen tubes of *Lilium longiflorum*. *Protoplasma* **96,** 275–282.

Ho, H. C., Wirch, E., Stevens, F. C., and Wang, J. H. (1977). Purification of a Ca^{2+}-activatable cyclic nucleotide phosphodiesterase from bovine heart by specific interaction with its Ca^{2+}-dependent modulator protein. *J. Biol. Chem.* **252,** 43–50.

Hodgkin, A. L., and Keynes, R. D. (1957). Movements of labelled calcium in squid giant axons. *J. Physiol. (London)* **138,** 253–281.

Huang, W. Y., Cohn, D. V., Fullmer, C., Wasserman, R. H., and Hamilton, J. W. (1975). Calcium-binding protein of bovine intestine. The complete amino acid sequence. *J. Biol. Chem.* **250,** 7647–7655.

Jaffe, L. F. (1979). Control of development by ionic currents. *In* "Membrane Transduction Mechanisms" (R. A. Cone, and J. E. Dowling, eds.), pp. 199–231. Raven, New York.

Jamieson, G. A., and Vanaman, T. C. (1979). Calcium-dependent affinity chromatography of calmodulin on an immobilized phenothiazine. *Biochem. Biophys. Res. Commun.* **90,** 1048–1056.

Jones, H. P., Matthews, J. C., and Cormier, M. J. (1979). Isolation and characterization of Ca^{2+}-dependent modulator protein from the marine invertebrate *Renilla reniformis*. *Biochemistry* **18,** 55–60.

Klee, C. B. (1977). Conformational transition accompanying the binding of Ca^{2+} to the protein activator of 3′,5′-cyclic adenosine monophosphate phosphodiesterase. *Biochemistry* **16,** 1017–1024.

Krctsingei, R. H. (1977). Evaluation ot the informational role of calcium in eukaryotes. *In* "Calcium-Binding Proteins and Calcium Function" (R. H. Wasserman *et al.*, eds.), pp. 63–72. Am. Elsevier, New York.

Krogmann, D. W., and Ogren, W. L. (1965). Studies on pyridine nucleotides in photosynthetic tissue. Concentrations, interconversions, and distribution. *J. Biol. Chem.* **240,** 4603–4608.

Kushmerick, M. J., and Podalsky, R. J. (1969). Ionic mobilities in muscle cells. *Science* **166,** 1297–1298.

Levin, R. M., and Weiss, B. (1977). Binding of trifluoperazine to the calcium-dependent activator of cyclic nucleotide phosphodiesterase. *Mol. Pharmacol.* **13,** 690–697.

Lin, Y. M., Liu, Y. P., and Cheung, W. Y. (1974). Cyclic 3′,5′-nucleotide phosphodiesterase. Purification, characterization and active form of the protein activator from bovine brain. *J. Biol. Chem.* **249,** 4943–4954.

Liu, Y. P., and Cheung, W. Y. (1976). Cyclic 3′,5′-nucleotide phosphodiesterase. Ca^{2+} confers more helical conformation to the protein activator. *J. Biol. Chem.* **251,** 4193–4198.

Lynch, T. J., and Cheung, W. Y. (1979). Human erythrocyte $Ca^{2+} - Mg^{2+}$-ATPase: Mechanism of stimulation by Ca^{2+}. *Arch. Biochem. Biophys.* **194,** 165–170.

Lynch, T. J., Tallant, A., and Cheung, W. Y. (1976). Ca^{2+}-Dependent formation of brain adenylate cyclase–protein activator complex. *Biochem. Biophys. Res. Commun.* **68,** 616–625.

Matthews, J. C., and Cormier, M. J. (1978). Rapid microassay for the calcium-dependent protein modulator of cyclic nucleotide phosphodiesterase. *In* "Methods in Enzymology" (M. A. DeLuca, ed.), Vol. 57, pp. 107–112. Academic Press, New York.

Moews, P. C., and Kretsinger, R. H. (1975). Refinement of the structure of carp muscle calcium-binding parvalbumin by model building and difference. Fourier analysis. *J. Mol. Biol.* **91,** 201–228.

Muto, S., and Miyachi, S. (1977). Properties of a protein activator of NAD kinase from plants. *Plant Physiol.* **59,** 55–60.

Nairn, A. C., and Perry, S. V. (1979). Calmodulin and myosin light-chain kinase of rabbit fast skeletal muscle. *Biochem. J.* **179,** 89–97.

Perrin, D. D., and Sayce, I. G. (1967). Computer calculation of equilibrium concentrations in mixtures of metal ions and complexing species. *Talanta* **14,** 833–842.

Reiss, H.-D., and Herth, W. (1979). Calcium ionophore A23187 affects localized wall secretion in the tip region of pollen tubes of *Lilium longiflorum*. *Planta* **145,** 225–232.

Richman, P. G., and Klee, C. B. (1979). Specific perturbation by Ca^{2+} of tyrosyl residue 138 of calmodulin. *J. Biol. Chem.* **254,** 5372–5376.

Sharief, F., Jones, H. P., Cormier, M. J. and Vanaman, T. C.(1980). In preparation.

Shimomura, O., Johnson, F. H., and Saiga, Y. (1962). Extraction, purification and properties of aequorin, a bioluminescent protein from the luminous hydromedusan, *Aequorea. J. Cell. Comp. Physiol.* **59,** 223–240.

Teo, T. S., and Wang, J. H. (1973). Mechanism of activation of a cAMP phosphodiesterase from bovine heart by Ca^{2+}. *J. Biol. Chem.* **248,** 5950–5955.

Teo, T. S., Wang, T. H., and Wang, J. H. (1973). Purification and properties of the protein activator of bovine heart cAMP phosphodiesterase. *J. Biol. Chem.* **248,** 588–595.

Tufty, R. M., and Kretsinger, R. H. (1975). Troponin and parvalbumin calcium binding regions predicted in myosin light chain and T_4 lysozyme. *Science* **187,** 167–169.

Vanaman, T. C., Sharief, F., and Watterson, D. M. (1977). Structural homology between brain modulator protein and muscle TnC. *In* "Calcium-Binding Proteins and Calcium Function" (R. H. Wasserman *et al.*, eds.), pp. 107–116. Am. Elsevier, New York.

Van Eerd, J. P., and Takahashi, K. (1975). The amino acid sequence of bovine cardiac troponin-C. Comparison with rabbit skeletal troponin C. *Biochem. Biophys. Res. Commun.* **64,** 122–127.

Waisman, D. M., Stevens, F. C., and Wang, J. H. (1978a). Purification and characterization of a Ca^{2+}-binding protein in *Lumbricus terrestris. J. Biol. Chem.* **253,** 1106–1113.

Wang, J. H. (1977). Calcium-regulated protein modulator in cyclic nucleotide systems. *In* "Cyclic 3',5'-Nucleotides. Mechanism of Action" (H. Cramer and J. Schultz, eds.), pp. 37–56. Wiley, New York.

Watterson, D. M., and Vanaman, T. C. (1976). Affinity chromatography purification of a cyclic nucleotide phosphodiesterase using immobilized modulator protein, a troponin C-like protein from brain. *Biochem. Biophys. Res. Commun.* **73,** 40–46.

Watterson, D. M., Harrelson, W. G., Keller, P. M., Sharief, F., and Vanaman, T. C. (1976). Structural similarities between the Ca^{2+}-dependent regulatory proteins of 3',5'-cyclic nucleotide phosphodiesterase and actomyosin ATPase. *J. Biol. Chem.* **251,** 4501–4513.

Weeds, A. G., and McLachlan, A. D. (1974). Structural homology of myosin alkali light chains, troponin C and carp calcium binding protein. *Nature* (*London*) **252,** 646–649.

Weisenseel, M. H., and Jaffe, L. F. (1976). The major growth current through lily pollen tubes enters as K^+ and leaves as H^+. *Planta* **133,** 1–7.

Weiss, B., and Levin, R. M. (1978). Mechanism for selectively inhibiting the activation of cyclic nucleotide phosphodiesterase and adenylate cyclase by antipsychotic agents. *Adv. Cyclic Nucleotide Res.* **9,** 285–303.

Wolff, D. J., and Brostrom, C. O. (1974). Calcium-binding phosphoprotein from pig brain: Identification as a calcium-dependent regulator of brain cyclic nucleotide phosphodiesterase. *Arch. Biochem. Biophys.* **163,** 349–358.

Wolff, D. J., Poirier, P. G., Brostrom, C. O., and Brostrom, M. A. (1977). Divalent cation binding properties of bovine brain Ca^{2+}-dependent regulator protein. *J. Biol. Chem.* **252,** 4108–4117.

Yagi, K., Yazawa, M., Kakiuchi, S., Ohshima, M., and Uenishi, K. (1978). Identification of an activator protein for myosin light chain kinase as the Ca^{2+}-dependent modulator protein. *J. Biol. Chem.* **253,** 1338–1340.

Chapter 11

Calcium-Dependent Protein Phosphorylation in Mammalian Brain and Other Tissues

HOWARD SCHULMAN
WIELAND B. HUTTNER
PAUL GREENGARD

CALCIUM AND CELL FUNCTION, VOL. I

ISBN 0-12-171401-2

I. INTRODUCTION

Early work on nerve and muscle physiology implicated calcium ion as an important intracellular messenger (Harvey and MacIntosh, 1940; Heilbrunn, 1952; Luttgau and Niedergerke, 1958). Subsequent biochemical studies demonstrated that the effect of calcium on actomyosin ATPase is a critical feature of the molecular basis of action of this cation in skeletal muscle (Huxley, 1964). There was also strong evidence for a role of calcium in excitation–secretion coupling including the stimulus–coupled release of neurotransmitters and peptide hormones (Katz and Miledi, 1967, 1969; Douglas, 1968). The molecular basis for calcium action in these and other nonmuscle systems remains to be determined.

Interest in calcium action waned following the elegant studies of Sutherland and his colleagues demonstrating the involvement of cAMP in the action of many hormones. These studies led to their hypothesis postulating that cAMP is a ubiquitous "second messenger," the link between many extracellular messengers and control of intracellular events (Sutherland *et al.,* 1968; Robison *et al.,* 1971). In 1968 Krebs and his associates discovered a cAMP-dependent protein kinase in muscle and presented evidence that it mediated the breakdown of muscle glycogen (Walsh *et al.,* 1968). Subsequently, Kuo and Greengard (1969) found that cAMP-dependent protein kinase activity was widespread in nature. In addition, they proposed that the diverse biological effects of cAMP in various tissues might be mediated by regulating protein kinase activity. The appealing feature of this protein kinase hypothesis was that it provided a mechanism by which a single effector molecule, cAMP, could elicit such a diversity of physiological and biochemical effects. According to this hypothesis cAMP need only interact with one intracellular receptor, the regulatory subunit of the protein kinase; the specificity of the action of cAMP would be determined by the nature of the substrates for cAMP-dependent protein kinase present in the various tissues.

II. CALCIUM-REGULATED PROTEIN PHOSPHORYLATION

In recent years there has been renewed appreciation of the importance of calcium in various modes of cell activation. In many systems, it is apparent that the role of calcium as a second messenger is no less important than the role of cyclic nucleotides and that their actions are often interrelated. A major apparent difference between the actions of calcium and cAMP is that, unlike cAMP, whose action in eukaryotic cells seems to be mediated exclusively by protein phosphorylation, calcium's effects appear to be mediated by several classes of enzymes. This chapter considers the possibility, however, that although there is not a single universal mechanism for calcium action, certain of the physiological effects of calcium may also be mediated by protein phosphorylation.

It has been known for many years that a variety of biological systems contain protein kinases which are not responsive to fluctuating levels of cAMP. These kinases have often been referred to as cAMP-independent protein kinases (Rubin and Rosen, 1975). Recent studies in our laboratory were designed to test the possibility that phosphorylated proteins may be involved in the actions of many classes of regulatory agents. The results indicate that many regulatory agents affect the phosphorylation of specific proteins in their target cells. In addition to regulatory agents that utilize cAMP as a second messenger, these agents include cGMP, steroid hormones, insulin, and calcium. Moreover, studies in other laboratories indicate that phosphorylated proteins are also involved in certain of the actions of interferon, thyroid hormone, hemin, vaccinia virus, and transforming viruses (Greengard, 1978a; Collett *et al.*, 1979). It would appear that phosphorylation of specific proteins is a final common pathway for some of the actions of a diverse array of stimuli. This concept is discussed more fully in two recent reviews and will not be further elaborated here (Greengard, 1978a,b). In this review we will summarize our recent studies which indicate that some of the biological effects of calcium in brain as well as in other tissues may be mediated or modulated by calcium-dependent protein phosphorylation. Several soluble calcium-dependent protein kinases, such as phosphorylase kinase and myosin light-chain kinase have been identified and characterized. This class of protein kinase has recently been reviewed (Schulman, 1980). To avoid redundancy, we have limited our discussion only to the membrane-bound protein kinases that we have described, since other chapters in this volume contain authoritative reviews of the soluble calcium-dependent protein kinases.

A. Calcium-Dependent Protein Phosphorylation in Intact Synaptosomes

The important role of calcium in the physiology of the nervous system is well documented (Baker, 1972; Rubin, 1972). Stimulation of nerve cells leads to a wave of depolarization traveling down the axon which, at the nerve terminal, causes an influx of calcium and the subsequent release of neurotransmitters (Baker *et al.*, 1971). There is also evidence that calcium regulates the activity of neurotransmitter-synthesizing enzymes (e.g., Patrick and Barchas, 1974) and the electrical excitability of nerve cells (Frankenhaeuser, 1957; Kelly *et al.*, 1969). The biochemical mechanisms underlying these and other physiological effects of calcium have not yet been elucidated. It seemed possible that one or more of the effects of calcium in the presynaptic nerve terminal might be mediated or modulated through the phosphorylation of specific proteins. For this reason, Krueger *et al.* (1976, 1977) studied the effect of calcium influx on protein phosphorylation in intact synaptosomes. These particles, produced by homogenization of brain tissue, retain many morphological and functional characteristics of the nerve endings from which they are derived (Gray and Whittaker, 1962; Rodriguez de Lores Arnaiz *et al.*, 1967). Synaptosomes are metabolically active. They can synthesize ATP through oxidative phosphorylation, and they exhibit uptake, synthesis, and release of neurotransmitters. Such preparations have therefore been used extensively to study the phenomena of neurotransmitter synthesis and release.

For the study of protein phosphorylation in intact synaptosomes, a synaptosomal fraction (P_2) from rat cerebral cortex was preincubated with $^{32}P_i$ of high specific activity for 30 min in order to label the intrasynaptosomal pool of ATP. The synaptosomes were then incubated for 30 sec in the absence or in the presence of veratridine or high potassium, each either with or without calcium present in the medium. Both veratridine and high potassium cause an influx of calcium from the medium into the nerve terminals by depolarizing the synaptosomal membrane and thereby opening voltage-sensitive calcium channels (Blaustein, 1975). The incubation was then terminated by addition of sodium dodecyl sulfate (SDS), and the state of phosphorylation of the synaptosomal proteins analyzed by SDS-polyacrylamide gel electrophoresis and autoradiography. An autoradiograph illustrating the incorporation of ^{32}P into proteins of a crude synaptosomal fraction is shown in Fig. 1. The control lane indicates that during the 30 min of preincubation about 20 polypeptide bands incorporated ^{32}P. Following the 30 sec treatment with either veratridine (100 μM) or high K^+ (60 mM) in the presence of 1 mM calcium in the incubation medium, there was a change in the state of phosphorylation of several polypeptides. The major change seen was a marked increase of ^{32}P incorporation into two polypeptides with apparent molecular weights of 86,000 and 80,000. These polypeptides, designated Ia and Ib, are also substrates for a membrane-bound cAMP-dependent protein kinase in brain and are collectively referred to as Protein I (see Section II,B) (Ueda and Greengard, 1977). There was also a significant increase in incorporation of ^{32}P into three polypeptide bands in the 50,000 to 60,000 dalton range. Two other bands of approximately 90,000 daltons exhibited a decreased amount of ^{32}P incorporation in the presence of either veratridine or high potassium. Similar changes were seen if calcium was present only during the 30 sec period of treatment with veratridine or high potassium and not during the

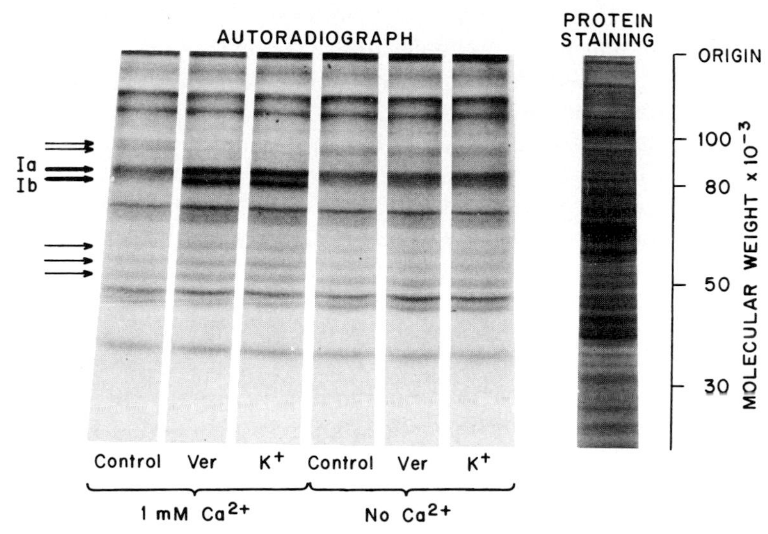

Fig. 1. Effect of veratridine and of high potassium, in the absence and presence of Ca^{2+}, on the phosphorylation of endogenous proteins in crude synaptosomal preparation from rat cerebral cortex. The synaptosome fraction was preincubated with $^{32}P_i$ for 30 min in the absence of Ca^{2+}. Aliquots of this suspension were then incubated for 30 sec in the absence (control) or presence of 100 μM veratridine (Ver) or 60 mM K^+ (K^+); 1 mM Ca^{2+} was present where indicated. The incubation was terminated by the addition of sodium dodecyl sulfate, and an aliquot of each sample subjected to sodium dodecyl sulfate-polyacrylamide gel electrophoresis and autoradiography. The molecular weight scale was generated by determining the position of marker proteins of known molecular weight. The positions of Proteins Ia and Ib are indicated by bold arrows to the left of the autoradiograph. Light arrows indicate the positions of other protein bands whose phosphorylation was inhibited (upper two arrows) or stimulated (lower three arrows) by veratridine or high K^+. (From Krueger *et al.*, 1977.)

preincubation. When calcium was omitted from the incubation medium neither veratridine nor high potassium had any effect on phosphorylation of the seven polypeptide bands. These results suggested that the observed changes in protein phosphorylation were due to the entry of calcium into the synaptosomes.

The structures involved in the calcium-dependent phosphorylation were shown to be intact synaptosomes by several criteria. The preparation of crude synaptosomes (P_2) was further fractionated on a discontinuous Ficoll/sucrose flotation gradient to yield subfractions enriched in myelin, synaptosomes, and mitochondria. In the subfraction enriched in synaptosomes, veratridine produced effects on protein phosphorylation similar to those seen in the initial P_2 fraction. Furthermore, the phosphorylation, both basal and stimulated, was inhibited by metabolic poisons and could not be accomplished by addition of [γ-^{32}P]ATP to the medium, indicating that intact organelles that synthesize [^{32}P]ATP from $^{32}P_i$ were involved. In addition, protein phosphorylation has recently been studied in P_2 fractions prepared from brain regions which had been pretreated by injection of kainic acid, a substance which causes selective destruction of neuronal but not glial cells (Sieghart *et al.,* 1980). In the P_2 fractions prepared from the kainic acid treated animals there was a marked decrease in both basal and veratridine stimulated ^{32}P incorporation into Protein I.

The two stimuli used, depolarization with veratridine and depolarization with high potassium, have been shown to increase the uptake of calcium by synaptosomes (Blaustein *et al.,* 1972; Blaustein, 1975). The effects of these agents on ^{32}P incorporation into Protein I were therefore compared with the effects of these agents on $^{45}Ca^{2+}$ uptake, both in the absence and in the presence of tetrodotoxin, a substance that has been shown to selectively inhibit the depolarizing effects of veratridine (Ulbricht, 1969). The results are shown in Table I. Veratridine and 60 m*M* K^+ each caused a 100% increase in the phosphorylation of Protein I. In agree-

TABLE I

Effect of Veratridine, High K^+, and A23187 in Absence and Presence of Tetrodotoxin on Protein I Phosphorylation and Ca^{2+} Uptake in Crude Synaptosomal Preparation[a]

Conditions	TTX	Protein I phosphorylation (arbitrary units)	Calcium uptake (nmoles/mg)
A. Control	−	4.6	3.0
	+	5.2	3.4
Veratridine	−	9.7	7.6
	+	5.3	3.6
High K^+	−	9.8	12.3
	+	10.0	13.4
B. Control	−	4.7	7.3
	+	5.0	7.1
A23187	−	7.4	10.4
	+	7.3	10.2

[a] Aliquots of a crude synaptosome preparation were incubated under the indicated conditions. The final concentrations of the additions were veratridine, 100 μM; high KCl, 60 m*M*; tetrodotoxin (TTX), 5 μM; A23187, 40 μM. Values for protein I phosphorylation are the means of duplicate determinations which varied by less than 10%. Values for Ca^{2+} uptake are the means of quadruplicate determinations with standard errors between 4 and 7%.

From Krueger *et al.*, 1977.

ment with earlier reports, veratridine stimulated calcium uptake by 150% while 60 mM K^+ increased calcium uptake by 300%. The effects of veratridine both on Protein I phosphorylation and on calcium uptake were almost abolished by tetrodotoxin, but the effects of 60 mM K^+ were resistant to tetrodotoxin. The effects of the calcium ionophore A23187 (Pressman, 1973) on Protein I phosphorylation and on $^{45}Ca^{2+}$ uptake are also shown in Table I. A23187 stimulated both Protein I phosphorylation and

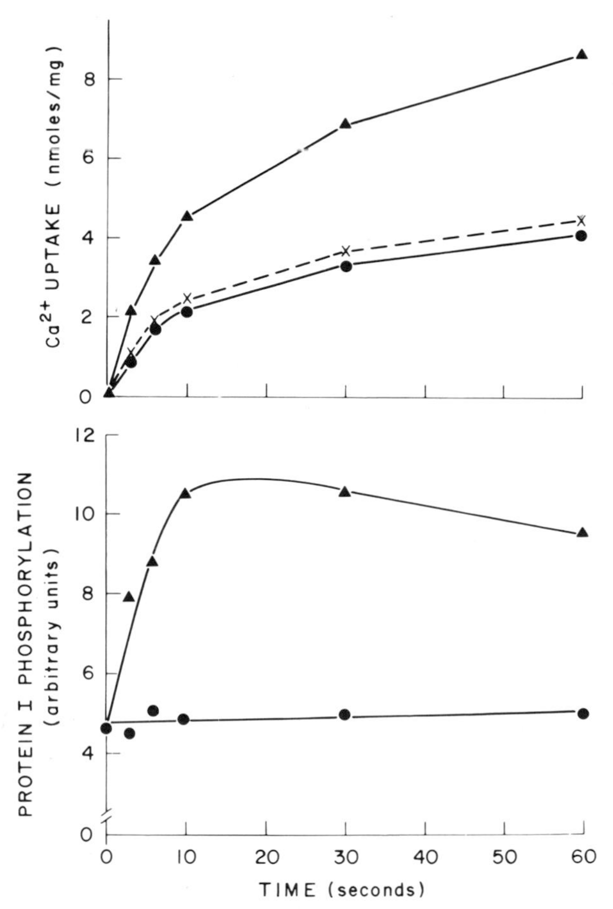

Fig. 2. Time course of $^{45}Ca^{2+}$ uptake (top) and Protein I phosphorylation (bottom) in a crude synaptosome preparation. Incubations were carried out in the absence (●) or presence (▲) of 100 μM veratridine, which was added at zero time. The tracer amount of $^{45}Ca^{2+}$ was added together with the veratridine for the measurements of Ca^{2+} uptake. The accumulation of $^{45}Ca^{2+}$ in the absence of veratridine is attributable to an exchange of labeled for unlabeled Ca^{2+}, and not to a net accumulation of Ca^{2+}. The increment in Ca^{2+} uptake caused by veratridine is shown by the broken curve (×—×) (From Krueger *et al.*, 1977.)

calcium uptake by about 50% and these effects were not inhibited by tetrodotoxin. The stimulation of Protein I phosphorylation by A23187 was abolished when calcium was omitted from the incubation buffer.

The time course of $^{45}Ca^{2+}$ uptake into synaptosomes and of Protein I phosphorylation, in the absence and in the presence of veratridine, is shown in Fig. 2. The veratridine-independent component of $^{45}Ca^{2+}$ uptake is due to an exchange of labeled for unlabeled calcium rather than to a net accumulation of calcium by the synaptosomes. The veratridine-dependent net accumulation of calcium is indicated by the dashed curve (Fig. 2, top). This component of uptake reaches its half-maximal level within 10 sec after addition of veratridine.

Increased ^{32}P incorporation into Protein I (Fig. 2, bottom) may represent either increased turnover of protein-bound phosphate or an actual increase in the state of phosphorylation of Protein I. In other experiments, when EGTA was added to the incubation mixture after a 30 sec treatment of synaptosomes with veratridine there was a rapid dephosphorylation of Protein I (Krueger *et al.*, 1977). Incorporation of ^{32}P into Protein I was reduced back to the control level within 2 min after addition of EGTA, indicating complete reversibility of the calcium-dependent phosphorylation of Protein I. This reversal indicates that calcium had stimulated the net incorporation of phosphate into Protein I rather than simply stimulated the exchange of unlabeled phosphate with $^{32}PO_4$ (Rudolph *et al.*, 1978).

Despite the fact that Protein I is also a substrate for the cAMP-dependent protein kinase (see below), the calcium-stimulated phosphorylation of Protein I in intact synaptosomes does not appear to be a consequence of some indirect effect of calcium on cyclic nucleotide levels. Neither veratridine nor 60 m*M* K^+, under conditions which stimulated phosphorylation of Protein I, had any significant effect on the level of cAMP in the preparation, either in the absence or in the presence of 3-isobutyl-l-methylxanthine (IBMX), a potent inhibitor of cyclic nucleotide phosphodiesterase. Further evidence against the possibility that cAMP mediates the effects of calcium on Protein I phosphorylation was obtained from recent studies on the effects of cyclic nucleotides on protein phosphorylation in intact synaptosomes. Thus, although 8-bromo-cAMP (but not 8-bromo-cGMP) was also found to stimulate phosphorylation of Protein I, it did so with different site specificities than were observed with veratridine (see Section IV) (Huttner and Greengard, 1979).

Protein I and the other proteins whose states of phosphorylation are regulated by calcium may be involved in calcium-dependent physiological processes at nerve terminals. Of particular interest among such physiological processes are neurotransmitter synthesis and release, which, like the protein phosphorylation system just described, are stimulated by depolarizing conditions in a calcium-dependent manner.

B. Properties of Protein I

1. *Chemical Properties of Protein I*

Earlier work from this laboratory focused on the most prominent endogenous substrates for the cAMP-dependent protein kinase in synaptic membrane preparations from brain (Greengard, 1978b). The two most prominent polypeptides, designated Ia (86,000 daltons) and Ib (80,000 daltons), and collectively referred to as Protein I, are structurally related (Ueda and Greengard, 1977). Protein I has been purified to apparent homogeneity from bovine cerebral cortex. It is a very basic protein comprised of Ia and Ib in a molar ratio of 1 to 2. It is a highly elongated molecule, and is enriched in glycine and proline. Treatment of Protein I with collagenase leads to its partial degradation and yields a globular collagenase-resistant fragment which contains the cAMP-regulated phosphorylation site.

2. *Biological Properties of Protein I*

Protein I has several unique biological characteristics. Not only does Protein I appear to be confined to the nervous system (Ueda *et al.*, 1973; Ueda and Greengard, 1977; DeCamilli *et al.*, 1979), but within the nervous system Protein I appears to be present only in neurons (Sieghart *et al.*, 1978a). Proteins Ia and Ib are present in highest concentrations in subcellular fractions enriched in synaptic vesicles and postsynaptic densities (Ueda *et al.*, 1979). In agreement with the results of subcellular fractionation studies, immunocytochemical studies on rat brain also indicate that Protein I is associated with synaptic structures and in particular with synaptic vesicles and postsynaptic densities (Bloom *et al.*, 1979). A good correlation between the time course of synapse formation and the appearance of Proteins Ia and Ib in the brains of developing rats and guinea pigs has provided further evidence that Proteins Ia and Ib are associated with synaptic structures (Lohmann *et al.*, 1978).

C. Regulation of Protein I Phosphorylation in Intact Brain Tissue

Regulation of the state of phosphorylation of Protein I in whole brains of mice *in vivo* has recently been demonstrated (Strombom *et al.*, 1979). Several central nervous system depressants, including phenobarbital, chloral hydrate, and urethane, administered to mice intraperitoneally, decreased the state of phosphorylation of Protein I. Conversely, the convulsant drugs pentylenetetrazole and picrotoxin each increased the state of phosphorylation of Protein I.

Protein I phosphorylation has also been shown to be altered by depolarizing agents and by cyclic nucleotides in intact slices of rat cerebral cortex incubated in Krebs–Ringer buffer (Forn and Greengard, 1978). In resting slices Protein I appeared to exist almost entirely in the dephosphorylated form. Upon addition of veratridine, high potassium, 8-Br-cAMP, or IBMX, there was a rapid and marked increase in the level of phosphorylation of Proteins Ia and Ib in the slice preparation.

The studies carried out thus far indicate that phosphorylation of Protein I can be regulated in synaptosomes, in brain slices, and *in vivo* by agents that mimic neuronal activity, supporting the possibility that it may play an important role in the physiology of the synapse.

III. CALCIUM- AND CALMODULIN-DEPENDENT PROTEIN KINASE

A. Calcium-Dependent Protein Phosphorylation in Isolated Membranes

The finding of calcium-stimulated changes in protein phosphorylation in intact synaptosomes incubated with $^{32}P_i$ and the absence of detectable increases in cyclic nucleotide levels accompanying these changes suggested the presence of a cyclic nucleotide-independent protein phosphorylation system in brain. Evidence concerning the molecular mechanism by which calcium entry might stimulate phosphorylation of synaptosomal proteins has come recently from experiments using lysed synaptosomes (Schulman and Greengard, 1978a). In these experiments, synaptosomes obtained from rat cerebral cortex were lysed by hypoosmotic shock, and incorporation of ^{32}P from [γ-^{32}P]ATP into protein was analyzed by SDS-polyacrylamide gel electrophoresis and autoradiography. The effect of calcium on endogenous protein phosphorylation in a synaptosomal lysate is shown in Fig. 3 (lanes 1 and 2). Calcium markedly stimulated ^{32}P incorporation into many polypeptides in the lysed synaptosome preparation that contained membranes (synaptic membranes plus synaptic vesicles) and synaptosomal cytoplasm. Proteins Ia and Ib were among the protein bands that showed calcium-dependent phosphorylation. However, the relative amount of radioactive phosphate incorporated into the various polypeptides differed from that seen with intact synaptosomes incubated with $^{32}P_i$ (compare Figs. 1 and 3). In the experiment with lysed synaptosomes and [γ-^{32}P]ATP, the largest net increase in ^{32}P incorporation was seen in polypeptides of about 51,000 and 62,000 daltons, although an increase in ^{32}P incorporation into Protein I was also observed. Similar patterns of stimulation by calcium have been described by DeLorenzo and Freedman (1977). These series of studies suggest that the calcium-stimulated phosphorylation seen in intact synaptosomes may be a direct consequence of the activation of a calcium-sensitive protein kinase by calcium influx.

The calcium-dependent phosphorylation observed in lysed synaptosomes was lost on preparation of cytoplasm-free membranes (Fig. 3, lanes 3 and 4). This phenomenon was not due to inactivation of the calcium-dependent protein kinase activity during experimental manipulations as it could be regained by addition of an amount of boiled (Fig. 3, lanes 5 and 6) or unboiled synaptosomal cytoplasm commensurate with the amount of membrane protein present. This sample of boiled cytosol which conferred calcium-dependent phosphorylation on cytosol-free membranes did not exhibit any endogenous phosphorylation when incubated in the absence of membranes (Fig. 3, lane 7), nor did it show any protein kinase activity using histone, protamine, or casein as substrate. Thus, the calcium-dependent phosphorylation in synaptic membranes required a soluble heat-stable factor present in synaptosomal cytosol.

The biochemical mechanisms underlying the physiological effects of calcium are not well understood. A common feature of those systems that have been investigated at the molecular level, however, is the involvement of a calcium-binding protein (Kretsinger, 1976). Thus, the trigger for contraction of skeletal muscle is the interaction of calcium with troponin C (a calcium-binding protein) (Ebashi, 1974). A similar but distinct calcium-binding protein (calmodulin, calcium-dependent regulator or CDR)* regulates the activity of cyclic nucleotide phosphodiesterase (Cheung, 1970; Kakiuchi and Yamazaki, 1970) and of a detergent solubilized preparation of adenylate cyclase (Brostrom *et al.*, 1975; Cheung *et al.*, 1975) from mammalian brain. It, therefore, seemed possible that a similar function might be served by the factor in the synaptosomal cytosol. Indeed, the heat stability of the stimulating factor in the cytosol suggested a possible relationship to calmodulin. Therefore, calmodulin was purified to homogeneity by the method of Teo *et al.* (1973) in order to test its effect in the calcium-dependent protein phosphorylation system. Addition of this protein to washed synaptosomal membrane fractions restored

* In this chapter we refer to the phosphodiesterase activator as calmodulin in the text and by the abbreviation CDR (calcium-dependent regulator) in the figures.

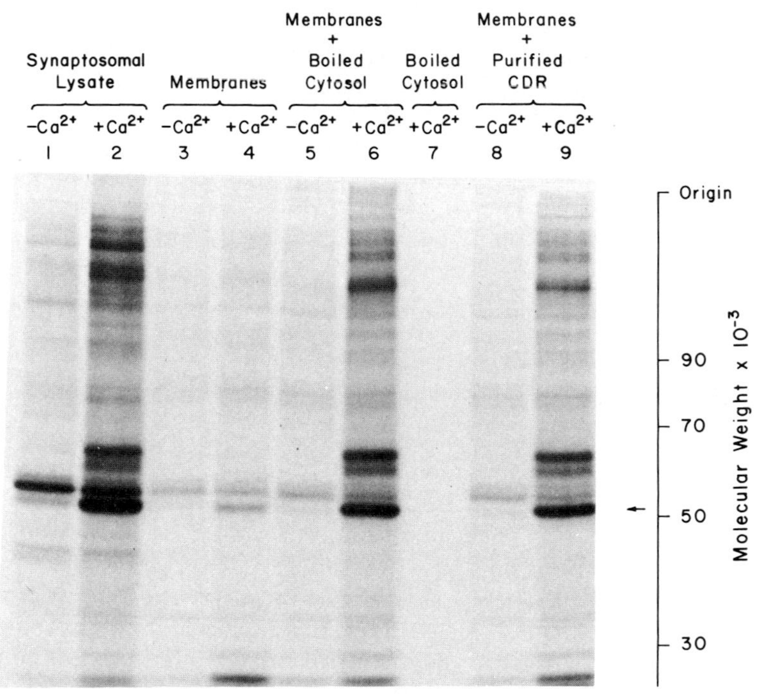

Fig. 3. Effect of Ca^{2+}, synaptosomal cytoplasm, and purified calmodulin on endogenous protein phosphorylation of brain membranes. Synaptosomes obtained from rat cerebral cortex were subjected to osmotic shock by homogenization in 10 volumes of ice-cold water (15 ml) and kept on ice for 30 min to optimize lysis. Total membranes were obtained from lysed synaptosomes by centrifugation at 150,000 *g* for 30 min. The released synaptosomal cytosol was removed, heated at 95° to 100°C for 5 min, and used as the source of activator (boiled cytosol). The pellet was resuspended in 15 ml of 5m*M* Tris-HCl (pH 7.0), centrifuged as above, and again resuspended in 15 ml of 5 m*M* Tris-HCl (pH 7.0). This resuspended pellet served as the source of activator-deficient membranes. The reaction mixture for assay of endogenous protein phosphorylation (final volume, 100 μ1) contained 50 m*M* PIPES buffer (pH 7.0), 10 m*M* $MgCl_2$, 1m*M* dithioerythritol, 0.2 m*M* EGTA (minus calcium) or 0.2 m*M* EGTA + 0.5 m*M* $CaCl_2$ (plus calcium), 5 μM [γ-^{32}P]ATP (5.2 × 10^4 cpm/pmole), and, where indicated, lysed synaptosomes (50 μg protein), membranes (34 μg protein), boiled cytosol (14 μg protein), or purified calmodulin (0.25 μg protein). Incubation was carried out for 10 sec at 30°C, the reaction terminated by the addition of sodium dodecyl sulfate, and an aliquot of the sample was analyzed for protein phosphorylation by sodium dodecyl sulfate-polyacrylamide gel electrophoresis and autoradiography. The arrow indicates the 51,000-dalton phosphoprotein used for quantitative measurements of stimulation of ^{32}P incorporation into membrane protein. (From Schulman and Greengard, 1978a.)

calcium-dependent protein phosphorylation (Fig. 3, lanes 8 and 9). The pattern of phosphorylation obtained on addition of boiled synaptosomal cytoplasm was indistinguishable from that obtained on addition of calmodulin (compare Fig. 3, lanes 6 and 9), suggesting that the two factors are functionally equivalent.

Stimulation of endogenous protein phosphorylation in synaptic membrane fractions was dependent on the presence of both calcium and either the cytosol factor or calmodulin. Thus, calcium in the absence of boiled cytosol and boiled cytosol in the absence of calcium were ineffective, whereas addition of both calcium and boiled cytosol restored phosphorylation (Fig. 3, compare lanes 3, 4, 5, and 6). Analogous results were obtained with calmodulin (Fig. 3, compare lanes 3, 4, 8, and 9). This endogenous kinase activator, like calmodulin, was extremely heat stable, nondialyzable, resistant to DNase and RNase, and sensitive to trypsin (Schulman and Greengard, 1978a).

B. Identification of Protein Kinase Activator as Calmodulin

Since calmodulin is present in synaptosomal cytosol of bovine and rat cerebral cortex, it should account for some of the activation seen by fortifying membranes with synaptosomal cytosol. However, this does not exclude the possibility that other factors also present in the cytosol may be physiologically important regulators of the calcium-dependent protein phosphorylation observed in synaptosomal lysates. Thus, it was necessary to purify the endogenous kinase activator and determine its relationship to calmodulin. We therefore fractionated a cytosol preparation from bovine cerebral cortex to determine the number and nature of protein kinase activators present, assaying column fractions for the ability to stimulate calcium-dependent protein phosphorylation in membrane preparations from rat brain (Schulman and Greengard, 1978b). Throughout the purification, only a single peak of kinase activator was found, suggesting that only one protein in the cytosol is able to reconstitute calcium-dependent protein kinase activity. When the same column fractions were examined for the presence of calmodulin by assaying their ability to activate calmodulin-depleted phosphodiesterase, the presence of calmodulin coincided with the kinase activator peak. The yields of the two activities were nearly identical at each step of the purification, and both copurified with an 18,000-dalton polypeptide as revealed by SDS-polyacrylamide gel electrophoresis, thus suggesting that the endogenous activator of the calcium-dependent protein phosphorylation system is calmodulin. Further studies indicated that the protein kinase activator is indistinguishable from authentic calmodulin based on a number of criteria.

1. Effect on Endogenous Protein Phosphorylation

As shown in Fig. 4, stimulation of protein phosphorylation by calmodulin and by kinase activator is identical. Synaptic membrane fractions washed free of cytosol were nearly devoid of calcium-dependent protein phosphorylation (Fig. 4, lanes 1 and 2). Upon addition of authentic calmodulin, prepared from bovine cerebral cortex by the method of Teo *et al.* (1973) (Fig. 4, lanes 5 and 6) or upon addition of the purified kinase activator (Fig. 4, lanes 7 and 8), the same pattern of phosphorylation was obtained. The pattern with calmodulin and with kinase activator was qualitatively and quantitatively similar (Fig. 4, compare lanes 6 and 8).

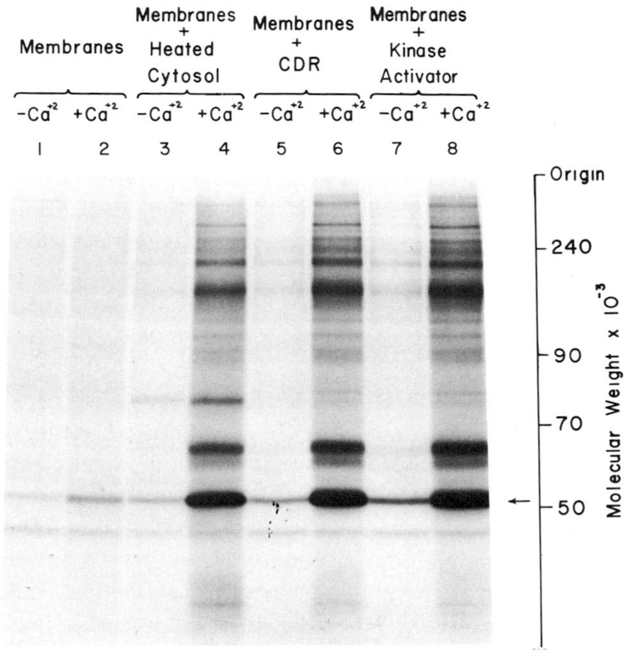

Fig. 4. Autoradiograph illustrating endogenous calcium-dependent protein phosphorylation of brain membranes, and activation by heated cytosol, purified calmodulin (CDR), or purified kinase activator. Cytosol-free membranes (34 μg of protein) were incubated in the standard reaction mixture containing 0.2 m*M* EGTA (minus calcium) or 0.2 m*M* EGTA + 0.5 m*M* $CaCl_2$ (plus calcium), 5 μM [γ-^{32}P]ATP (5×10^4 cpm/pmole), and, where indicated, heated cytosol (14 μg protein), purified calmodulin (0.25 μg of protein), or purified kinase activator (0.25 μg of protein). The arrow indicates the 51,000-dalton phosphoprotein used for quantitative measurements of stimulation of ^{32}P incorporation into membrane protein. (From Schulman and Greengard, 1978b.)

2. *Analysis by Nondenaturing Polyacrylamide Gel Electrophoresis*

A comparison of calmodulin and protein kinase activator by nondenaturing polyacrylamide gel electrophoresis is presented in Fig. 5 (top). Each preparation showed only a single band detected by protein staining, and these two bands migrated at identical rates. A duplicate gel of the kinase activator was frozen after completion of electrophoresis, sliced, and assayed for the presence of factors activating the calcium-dependent protein kinase and phosphodiesterase. As shown in Fig. 5 (bottom), only a single peak of either protein kinase activator or phosphodiesterase activator was found in the preparation, its mobility corresponding to the band of protein staining. When a duplicate gel, containing calmodulin was treated and analyzed as above, it also showed a single component stimulating both calmodulin-depleted phosphodiesterase and kinase activator-depleted membranes.

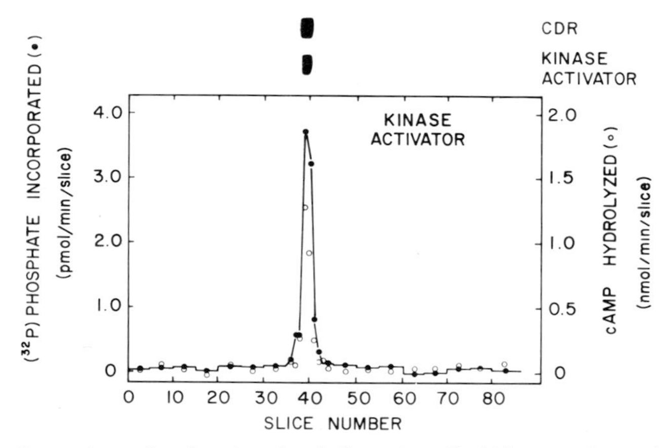

Fig. 5. Comparison of authentic calmodulin and purified kinase activator by nondenaturing polyacrylamide gel electrophoresis. (Top) Samples containing 30 μg of purified calmodulin (upper gel) or 20 μg of purified kinase activator (lower gel) were subjected to nondenaturing polyacrylamide gel electrophoresis in long gels (9.8 cm) containing 15% acrylamide, pH 8.9. (Bottom) A duplicate sample of kinase activator (10 μg) was electrophoresed simultaneously with the above samples, frozen, and sliced into 1.1-mm segments. Protein was eluted from each slice by incubation at 4°C for 48 hr in 200 μl of 50 mM PIPES buffer, pH 7.0. Aliquots (8 μl) of each fraction, or a pool of five consecutive fractions (40 μl total), were assayed for activation of Ca^{2+}-dependent protein phosphorylation (●). Aliquots (2 μl) of each fraction, or a pool of five consecutive fractions (10 μl total), were also assayed for activation of calmodulin-depleted phosphodiesterase (○). Distances along the two gels are directly related to distances in the profiles of activity. The solid line is drawn to indicate the profile of protein kinase activator. (From Schulman and Greengard, 1978b.)

3. Physicochemical Characterization

On SDS-polyacrylamide gel electrophoresis both calmodulin and the protein kinase activator migrated as homogeneous polypeptides. The two proteins comigrated and had an apparent molecular weight of 18,000 daltons. A more complete biochemical characterization using criteria such as ultraviolet spectrum, isoelectric focusing, and peptide mapping after partial proteolysis indicated that the two proteins, designated as authentic calmodulin and protein kinase activator, are, in fact, identical (H. Schulman, P. Freiman, and P. Greengard, unpublished data). Indeed, analysis of the amino acid composition of the protein kinase activator (Table II) shows that, within experimental error, it is identical to the composition of calmodulin reported by Watterson *et al.* (1976).

C. Characteristics of Protein Kinase Activity

For quantitative determinations of protein kinase activity, the radioac-

TABLE II

Amino Acid Composition of Kinase Activator and of Authentic Calmodulin

	Single column analysis	
Amino acid	Kinase activator (mole/18,000 g)	Calmodulin[a] (mole/18,000 g)
Aspartic acid	22.8	24.0
Threonine	11.9	12.2
Serine	4.4	4.7
Glutamic acid	31	31
Proline	2.1	1.7
Glycine	12.0	12.0
Alanine	11.8	11.9
Half-cystine	0	0
Valine	7.7	7.8
Methionine	8.6	8.4
Isoleucine	8.2	7.8
Leucine	9.3	9.5
Tyrosine	1.8	2.2
Histidine	1.1	1.1
Lysine	7.4	7.6
Arginine	6.6	6.4
Phenylalanine	8.9	8.7
Tryptophan	0	0
ε-*N*-Trimethyllysine	1.0	1.0

[a] Taken from Watterson *et al.*, 1976.

tivity in the band of 51,000 daltons (see arrow, Figs. 3 and 4) was routinely measured. Several other protein bands gave qualitatively similar results. The radioactivity was stable in 0.5*N* NaOH at 4°C and stable in 5% trichloracetic acid at 100°C indicating that it is incorporated into protein by phosphomonoester linkage. These characteristics as well as the presence of a large number of polypeptides that incorporate ^{32}P suggest that the ^{32}P-labeled polypeptides are not intermediates of a calcium-dependent ATPase but, rather, are substrates for a calcium-dependent protein kinase.

Calmodulin activates calcium-dependent protein phosphorylation in a dose-dependent manner. Our preparations of phosphodiesterase activator and of protein kinase activator stimulated ^{32}P incorporation into the 51,000 dalton polypeptide with equal efficacy (Fig. 6). Each caused about a sixfold stimulation, with half-maximal stimulation at about 0.25 μg of protein. Incorporation of ^{32}P into several other polypeptides was measured quantitatively in the same way and showed a similar response to the two factors.

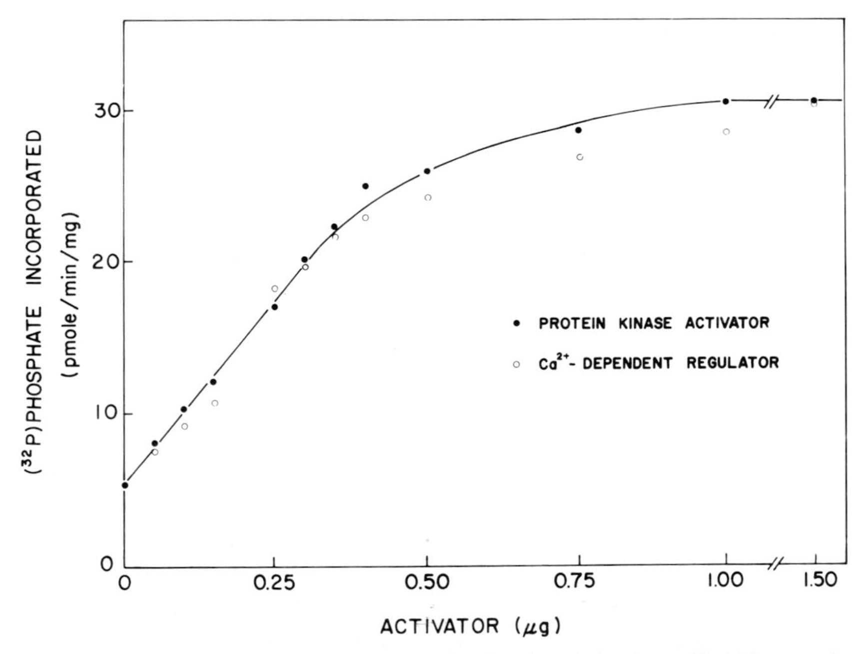

Fig. 6. Activation of Ca^{2+}-dependent protein phosphorylation by purified kinase activator and by authentic calmodulin. Cytosol-depleted membranes were incubated with various amounts of purified kinase activator (●) or calmodulin (○) in the absence or presence of Ca^{2+} under standard conditions, and assayed for Ca^{2+}-dependent protein phosphorylation of the 51,000 dalton substrate protein. (From Schulman and Greengard, 1978b.)

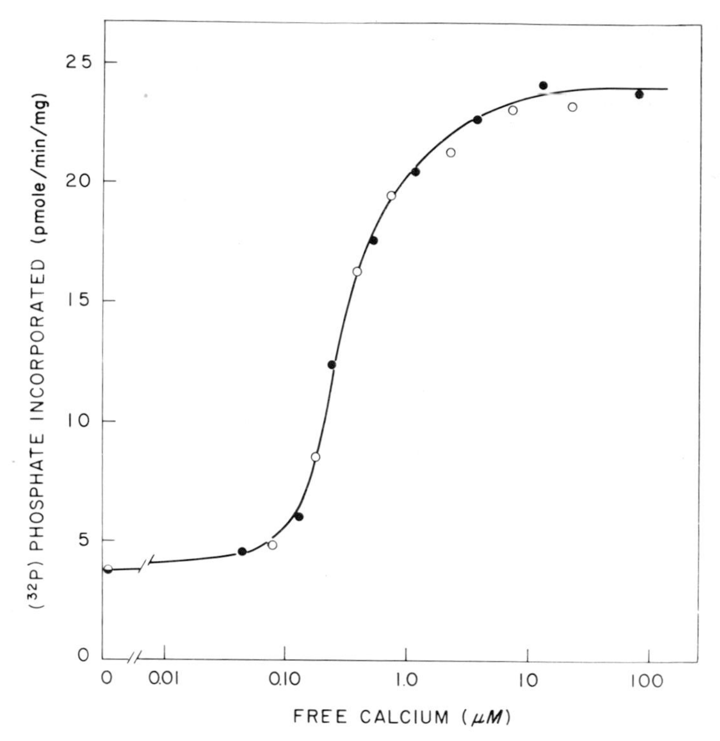

Fig. 7. Activation by Ca^{2+} of protein phosphorylation, with either purified kinase activator or authentic calmodulin as the source of activator. Activator-depleted membranes were incubated with either 1 μg of kinase activator (●) or 1 μg of calmodulin (○) under standard conditions at pH 7.1. Free Ca^{2+} concentration was varied by use of Ca^{2+}/EGTA buffer containing 0.4 mM EGTA. Free Ca^{2+} concentration was determined by using an apparent binding constant for Ca^{2+}-EGTA of $7.61 \times 10^6 M^{-1}$ (Portzehl *et al.*, 1964). (From Schulman and Greengard, 1978b.)

A physiological role for this calcium-dependent phosphorylation system would dictate that the kinase be sensitive to low concentrations of calcium. It has been estimated that the basal cytoplasmic concentration of free Ca^{2+} in eukaryotic cells is 0.01 to 1 μM. Following cell stimulation free Ca^{2+} concentration may be elevated to between 1 and 10 μM (Kretsinger, 1976). Indeed, as with other calmodulin-mediated processes, endogenous protein phosphorylation is sensitive to physiologically relevant levels of calcium. The effect of various concentrations of calcium on protein phosphorylation is shown in Fig. 7. With either preparation of activator, half-maximal stimulation occurred at a free Ca^{2+} concentration of about 0.3 μM. This is consistent with the findings that, under various conditions, half-maximal activation of calmodulin-dependent phosphodiesterase and binding of calcium to calmodulin in binding studies occur in the range of 0.2 to 3 μM free Ca^{2+} (Teo and Wang, 1973; Wolff *et al.*, 1977). *In vivo* this phosphorylation system would then be expected to undergo dynamic changes in activity cycling between a largely "off" and a largely "on" state.

D. Function of Calcium-Dependent Protein Phosphorylation

Depolarization of intact synaptosomes with either veratridine or high potassium has been shown to stimulate a calcium-dependent release of neurotransmitters (Blaustein *et al.*, 1972; Blaustein, 1975), to regulate neurotransmitter synthesis (Patrick and Barchas, 1974), and to stimulate phosphorylation of Protein I and other specific proteins (Fig. 1). It is possible that phosphorylation of Protein I, a vesicle-associated protein, and of other synaptosomal proteins, mediates or modulates some of the presynaptic effects of calcium, such as neurotransmitter synthesis and release. The calcium- and calmodulin-dependent protein kinase may have a role in such calcium-dependent processes. The presence of multiple phosphorylation sites in Protein I, which are differentially regulated by cAMP- and calcium-dependent protein kinases (see below), suggests a possible molecular basis for physiological interactions between these two second messengers in certain synaptic functions.

The findings that depolarization-induced calcium influx regulates protein phosphorylation in intact synaptosomes (Krueger *et al.*, 1976, 1977) and that synaptosomal membranes contain a calmodulin and calcium-dependent protein kinase (Schulman and Greengard, 1978a,b) have recently been confirmed (DeLorenzo *et al.*, 1979). However, the claim of the latter investigators that this calcium-dependent phosphorylation system mediates neurotransmitter release may be premature. The "release" of neurotransmitter which they measured was dependent on calcium and calmodulin but occurred in a vesicle preparation apparently in the absence of presynaptic plasma membrane (DeLorenzo and Freedman, 1978), a necessary component of the physiological release process. Their observations may represent an effect of calcium, calmodulin, magnesium, and ATP on vesicle integrity or permeability rather than on a specific parameter of neurotransmitter release. Similar objections to the conclusions of these investigators have recently been raised by Wolff and Brostrom (1979).

Recently, the possibility that release processes may in general be associated with calcium-stimulated changes in protein phosphorylation was investigated using rat peritoneal mast cells (Sieghart *et al.*, 1978b). In experiments analogous to those with intact synaptosomes, the mast cells were preincubated with $^{32}P_i$ and then exposed to the classic mast cell secretagogue 48/80 or to the calcium ionophore A23187. A rapid phosphorylation of several polypeptides with molecular weights of 68,000, 59,000, and 42,000 daltons accompanied secretion of granules by these cells. With 48/80, both degranulation and phosphorylation were stimulated by external calcium. In addition, at a later stage in the course of the mast cell response to 48/80, as secretion terminated, phosphorylation of a 78,000 dalton polypeptide was seen to increase. Interestingly, the phosphorylation of this 78,000 dalton protein was selectively stimulated by the antiallergic drug cromolyn sodium (cromoglycate), which simultaneously inhibited histamine release in the mast cell preparation (Theoharides *et al.*, 1979). The effect of cromolyn in stimulating the phosphorylation of this peptide paralleled its effect in inhibiting histamine release under a variety of experimental conditions. The results suggested a possible association between calcium-dependent protein phosphorylation and the initiation and termination of the secretory response in mast cells.

IV. COMPARISON OF cAMP-REGULATED AND CALCIUM-REGULATED PHOSPHORYLATION OF PROTEIN I

The designation "Protein Ia and Protein Ib" originally referred to two polypeptides in synaptic membrane preparations, with apparent molecular weights of 86,000 and 80,000 daltons, which were prominent substrates for an endogenous cAMP-dependent protein kinase. Two polypeptides with mobilities on SDS polyacrylamide gel electrophoresis similar to those of Proteins Ia and Ib were found to be major targets for calcium-regulated protein phosphorylation in intact synaptosomes. Recent studies have indicated that these two calcium-regulated phosphoproteins are, in fact, identical to Proteins Ia and Ib (Sieghart *et al.*, 1979). Both sets of proteins were extracted from membranes under similar acid-extraction conditions, had similar molecular weights in several SDS-polyacrylamide gel electrophoresis systems, and had similar isoelectric points. Like Protein I phosphorylated by a cAMP-dependent protein kinase in synaptic membrane preparations, the protein phosphorylated in response to calcium influx in intact synaptosomes was sensitive to digestion with collagenase, and was converted to the same phosphopeptide intermediates and end products at the same rate. Upon limited proteolysis with either *Staphylococcus aureus* V8 protease (SAP), chymotrypsin, or papain, both sets of proteins yielded phosphopeptides of similar if not identical molecular weights. It thus appears that phosphorylation of the same neuronal protein, Protein I, can be regulated by two different second messengers, cAMP, and calcium.

A. Phosphorylation of Protein I by cAMP and Calcium in Intact and Lysed Synaptosomes

The relationship of cAMP and calcium-regulated phosphorylation of Protein I was recently investigated in intact as well as in lysed synaptosome preparations from rat brain (Huttner and Greengard, 1979). As indicated above, in intact synaptosomes incubated in the presence of calcium, addition of the depolarizing agent veratridine led to a pronounced increase in the phosphorylation of Protein I (Fig. 1), whereas an effect of 8-Br-cAMP could not be detected (Krueger *et al.*, 1977; Huttner and Greengard, 1979) (however, see Sections IV, B and IV, C). In lysed synaptosomes, the addition of either calcium or 8-Br-cAMP resulted in stimulation of Protein I phosphorylation as well as increased phosphorylation of a number of other protein bands.

In order to compare the phosphorylation site(s) that were regulated by cAMP and calcium, phosphorylated Protein I was analyzed by peptide mapping techniques. Protein I was purified after its phosphorylation in intact and lysed synaptosomes. Purification was performed under condi-

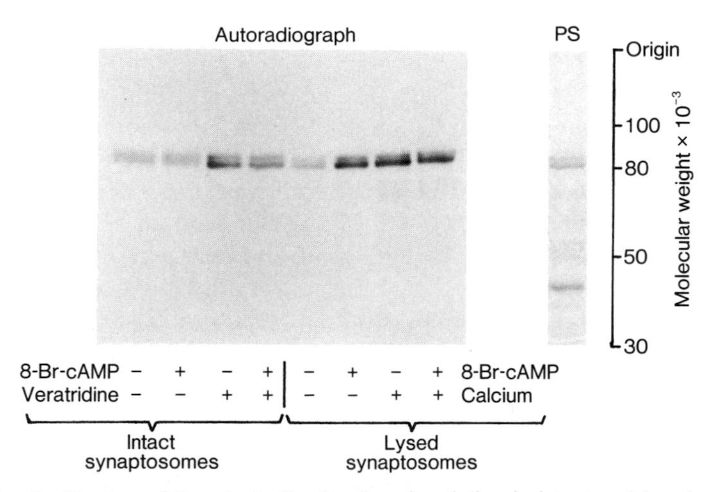

Fig. 8. Purification of Protein I after its phosphorylation in intact and lysed synaptosomes. After phosphorylation, samples were subjected to nonequilibrium pH gradient electrophoresis on a polyacrylamide gel. The Protein I region of the gel was cut out and subjected to SDS-polyacrylamide gel electrophoresis, followed by protein staining (PS) and autoradiography of the latter gel. (From Huttner and Greengard, 1979.)

tions which did not alter the state of phosphorylation of Protein I. For this purpose, the synaptosomal proteins were denatured and subjected to a two-dimensional separation technique for basic proteins (O'Farrell *et al.*, 1977) involving nonequilibrium pH gradient electrophoresis (NEPHGE) and SDS-polyacrylamide gel electrophoresis in slab gels. Because of the high isoelectric points of Proteins Ia and Ib, both proteins migrated ahead of the bulk of the other proteins in the NEPHGE gel. When the region of the NEPHGE gel which contained Proteins Ia and Ib was subjected to a second electrophoresis step on SDS-polyacrylamide gels, Proteins Ia and Ib appeared as major proteins by Coomassie blue staining, and as the only ^{32}P-labeled proteins (Fig. 8).

Protein I phosphorylation in intact synaptosomes appeared to be stimulated only by veratridine plus calcium (Fig. 8), but not by 8-Br-cAMP, when Protein I that had been purified by NEPHGE gel electrophoresis plus SDS-polyacrylamide gel electrophoresis was examined. (However, see the results of peptide mapping described below.) In contrast, both calcium and 8-Br-cAMP affected the total amount of phosphate incorporated into Protein I in lysed synaptosomes, when purified Protein I was used for analysis (Fig. 8).

B. Peptide Mapping after Limited Proteolysis of Phosphorylated Protein I

The phosphopeptide patterns of purified Protein Ia and Protein Ib, obtained after limited proteolysis with SAP, were qualitatively similar to one another (W. B. Huttner, L. J. De Gennaro, and P. Greengard, unpublished data). When Protein Ia and Protein Ib were combined prior to limited proteolysis with SAP, the phosphopeptide pattern shown in Fig. 9 was obtained. Autoradiography of the gel revealed some undigested Proteins Ia and Ib at the top of the gel, together with an upper peptide fragment of approximately 35,000 daltons and a lower peptide fragment of approximately 10,000 daltons. In intact as well as in lysed synaptosomes, 8-Br-cAMP selectively stimulated the phosphorylation of the lower fragment without increasing the phosphorylation of the upper fragment. Veratridine-induced calcium influx into intact synaptosomes, or the addition of calcium to lysed synaptosomes, stimulated the phosphorylation of both the upper and lower fragments. In intact synaptosomes, 8-Br-cAMP reduced the amount of radioactive phosphate incorporated into the upper fragment, both in the absence and in the presence of veratridine. The inhibition of phosphate incorporation into one fragment of Protein I simultaneously with the stimulation of phosphate incorporation into another fragment by 8-Br-cAMP in intact synaptosomes may explain the inability to

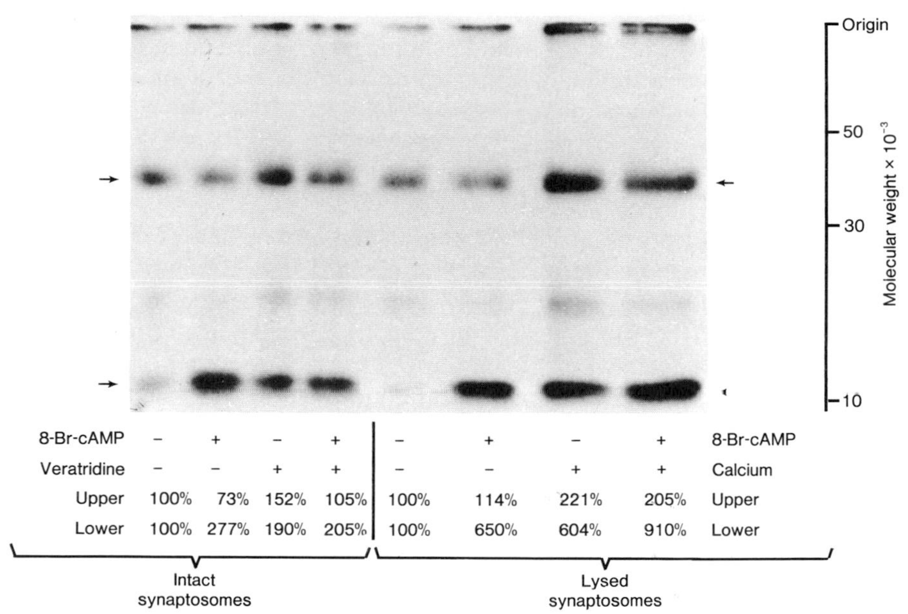

Fig. 9. Autoradiograph showing the phosphopeptide pattern obtained upon digestion with *Staphylococcus aureus* V_8 protease (SAP) of purified phosphorylated Protein I. Protein I that had been phosphorylated, in intact or lysed synaptosomes, was purified as described in the legend to Fig. 8. Proteins Ia and Ib were cut out of the gel together and subjected to limited proteolysis during SDS-polyacrylamide gel electrophoresis as described by Cleveland *et al.* (1977). The arrows indicate the upper fragment (approximately 35,000 daltons) and the lower fragment (approximately 10,000 daltons). The amount of phosphate incorporated into the upper and lower fragments under various incubation conditions is given for both intact and lysed synaptosomes as percent of control. (From Huttner and Greengard, 1979.)

detect significant effects of 8-Br-cAMP on Protein I phosphorylation by SDS-polyacrylamide gel electrophoresis of undigested Protein I (Krueger *et al.*, 1977; Huttner and Greengard, 1979). The same differential effects of cAMP and calcium on the two fragments were obtained when phosphorylated Protein I was purified by precipitation by a highly specific antibody rather than by separation on the basis of its high isoelectric point, prior to peptide mapping (W. B. Huttner and P. Greengard, unpublished data).

In experiments with intact synaptosomes, the stimulation by 8-Br-cAMP (4 m*M*) of the phosphorylation of the lower fragment could be mimicked by monobutyryl cAMP (4 m*M*) or IBMX (1 m*M*), but not by 8-Br-cGMP (4 m*M*). Omission of calcium from the incubation medium did not affect the stimulation of phosphorylation of the lower fragment by any of these agents. The effect of veratridine on the phosphorylation of both the upper and lower fragments could be mimicked by high potassium. The actions of both veratridine and potassium were dependent on the presence of external calcium ions.

C. Tryptic Fingerprints of Phosphorylated Protein I

Tryptic fingerprints of purified Protein Ia and Protein Ib were qualitatively similar to one another (W. B. Huttner, L. J. De Gennaro, and P. Greengard, unpublished data). The two-dimensional separation of phosphopeptides derived from purified Protein Ib after exhaustive digestion with trypsin is shown in Fig. 10. Five phosphopeptides, designated peptides 1–5, were observed both in intact and in lysed synaptosomes. 8-Br-cAMP selectively stimulated the phosphorylation of peptide 1, both in intact and in lysed synaptosomes. Veratridine-induced calcium influx into intact synaptosomes or the addition of calcium to lysed synaptosomes stimulated phosphorylation of peptide 1, as did 8-Br-cAMP, but in addition stimulated the phosphorylation of peptides 2–5. In intact synaptosomes, 8-Br-cAMP reduced the amount of radioactive phosphate incorporated into peptides 2–5, both in the absence and in the presence of veratridine.

In order to relate the phosphopeptide pattern obtained after limited proteolysis with SAP to peptides 1–5 of the tryptic fingerprint, the upper and lower fragments obtained upon limited proteolysis with SAP (Fig. 9) were separately subjected to exhaustive digestion with trypsin, followed by two-dimensional separation of phosphopeptides. The SAP-generated lower fragment yielded only peptide 1, whereas the SAP-generated upper fragment yielded only peptides 2–5.

These studies indicate that there are multiple phosphorylation sites in Protein I which show differential regulation by cAMP and by calcium. The data are consistent with the possibility that cAMP stimulates the phosphorylation of only one site in Protein I (represented by the lower

Fig. 10. Autoradiographs showing tryptic fingerprints of purified phosphorylated Protein Ib. Protein I that had been phosphorylated, in intact or lysed synaptosomes, was purified as described in the legend to Fig. 8. Proteins Ia and Ib were separately cut out of the gel and subjected to exhaustive digestion by trypsin. Phosphopeptides were spotted on cellulose plates (O designates the origin) and were separated in two dimensions, first by electrophoresis in the horizontal dimension (negative pole, left; positive pole, right) and then by ascending chromatography in the vertical dimension. In each autoradiograph, the five arrows indicate peptides 1–5, as designated in (A). The amount of phosphate incorporation into peptide 1 and into peptides 2 through 5 under various incubation conditions is given for both intact and lysed synaptosomes as percent of control. (From Huttner and Greengard, 1979.)

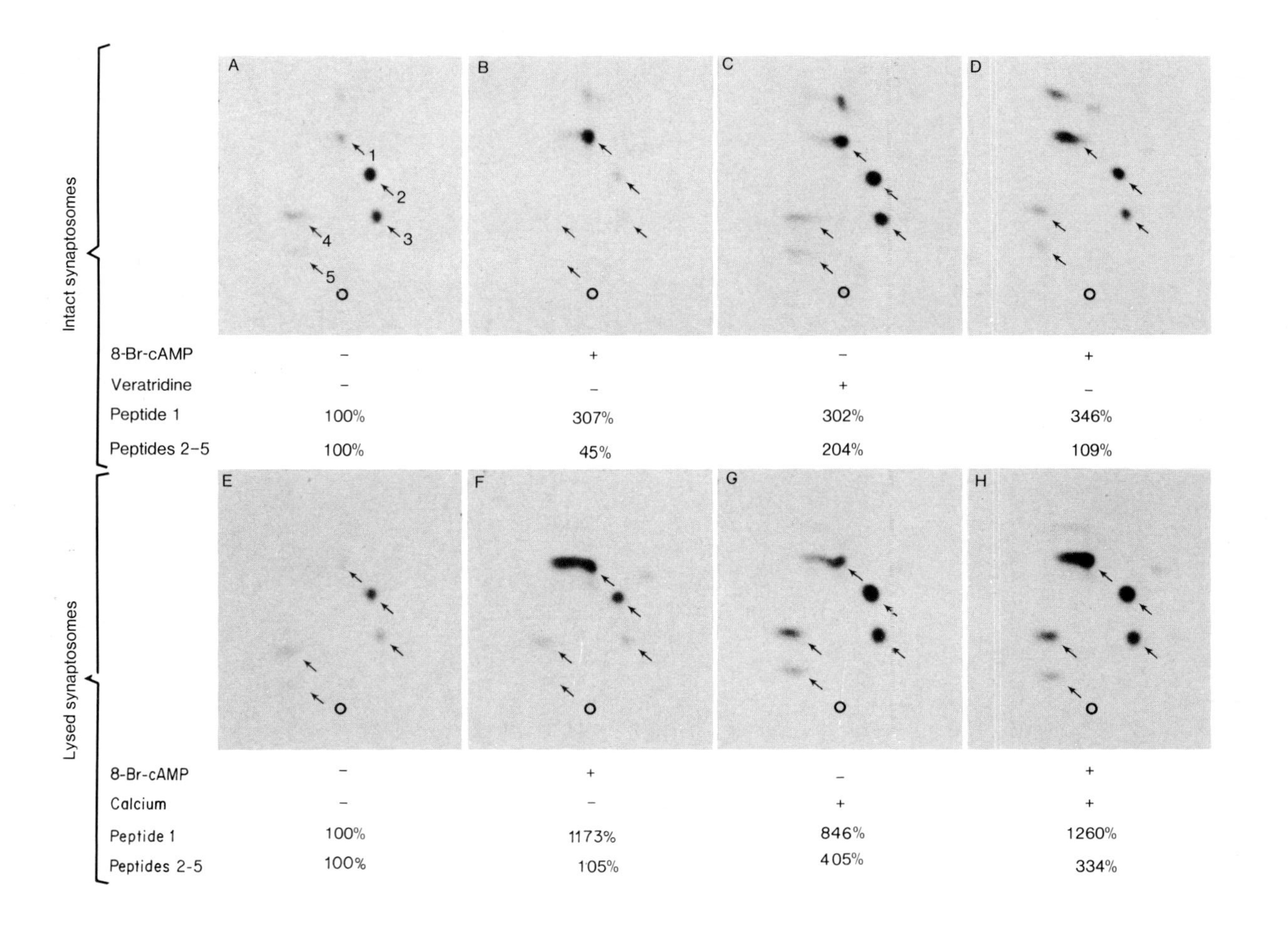
Intact synaptosomes
A
B
C
D
1
2
3
4
5
8-Br-cAMP – + – +
Veratridine – – + –
Peptide 1 100% 307% 302% 346%
Peptides 2–5 100% 45% 204% 109%
Lysed synaptosomes
E
F
G
H
8-Br-cAMP – + – +
Calcium – – + +
Peptide 1 100% 1173% 846% 1260%
Peptides 2-5 100% 105% 405% 334%

SAP fragment and by peptide I), whereas calcium stimulates the phosphorylation not only of this site, but also of additional site(s) (represented by the upper SAP fragment and by peptides 2–5). The finding that in intact organelles 8-Br-cAMP modified the effect of veratridine on the phosphorylation of Protein I suggests that the action of 8-Br-cAMP on the phosphorylation of Protein I, like the action of depolarizing agents, occurred in intact isolated axon terminals. Consistent with this suggestion, recent studies involving subcellular fractionation (Ueda *et al.*, 1979) and immunocytochemical techniques (Bloom *et al.*, 1979; DeCamilli *et al.*, 1979) have provided evidence that Protein I is localized primarily in axon terminals at certain types of synapses, where it appears to be associated primarily with synaptic vesicles.

The finding that cAMP and calcium cause differential regulation of multiple-site phosphorylation of the same neuronal protein, Protein I, is of interest. Interaction between cAMP and calcium at the level of phosphorylation of Protein I may provide a molecular basis for physiological interactions between these two intracellular second messengers in those synaptic terminals containing this protein.

Phosphorylation of phospholamban, a membrane protein of cardiac sarcoplasmic reticulum, has recently been found to be regulated both by cAMP and by calcium (Le Peuch *et al.*, 1979). In that study, evidence for different sites of phosphate incorporation into phospholamban catalyzed by cAMP-dependent and calcium- calmodulin-dependent protein kinases was obtained. Multiple-site phosphorylation, catalyzed by cAMP-dependent and calcium–calmodulin-dependent protein kinases, has also been found in certain enzymes involved in the regulation of glycogen metabolism (Cohen, 1978; Roach *et al.*, 1978; Rylatt *et al.*, 1979; Srivastava *et al.*, 1979) as well as in an integral protein present in the plasma membrane of turkey erythrocytes (Alper *et al.*, 1980).

V. WIDESPREAD OCCURRENCE OF CALCIUM- AND CALMODULIN-DEPENDENT PROTEIN PHOSPHORYLATION

Calcium ion has been implicated in important regulatory functions in a variety of tissues (Rasmussen and Goodman, 1977). For most of these functions, as for the neuronal functions discussed earlier, little is known about the biochemical basis for the action of calcium in these tissues. In light of the finding of calcium- and calmodulin-dependent protein phosphorylation in brain membranes, it was of interest to determine whether the action of calcium on membrane function in other tissues might also be

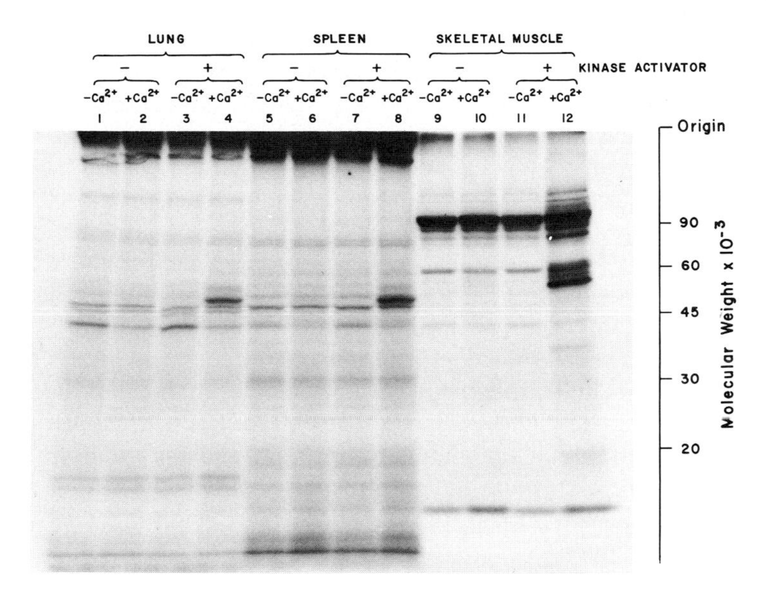

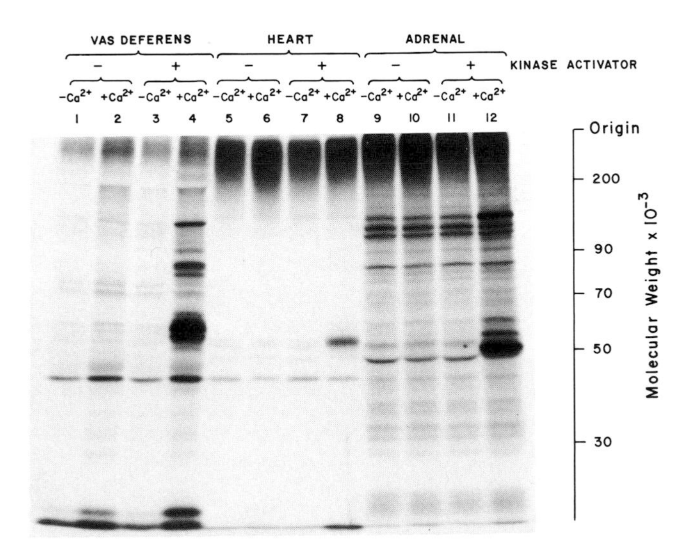

Fig. 11. Autoradiograph illustrating endogenous calcium- and calmodulin-dependent phosphorylation of membrane proteins from various rat tissues. Total membrane fractions, washed free of calmodulin, were assayed for endogenous calcium-dependent phosphorylation in either the absence or the presence of 1 μg of calmodulin (kinase activator purified from bovine brain). The tissues studied, and the amounts of membrane protein applied to each lane, were lung, 7.7 μg; spleen, 11.4 μg; skeletal muscle, 5.0 μg; vas deferens, 5.4 μg; heart, 5.3 μg; and adrenal, 6.8 μg. (From Schulman and Greengard, 1978b.)

mediated by such a mechanism (Schulman and Greengard, 1978b). Analysis of endogenous protein phosphorylation in total membrane fractions from six representative rat tissues (lung, spleen, skeletal muscle, vas deferens, heart, and adrenal) is shown in Fig. 11. A total membrane fraction, washed free of cytosol and calmodulin, was prepared from each of the tissues and then assayed for endogenous calcium-dependent phosphorylation in either the absence or the presence of kinase activator purified from bovine brain. It can be seen that membranes from each of the six tissues contained a calcium- and calmodulin-dependent protein phosphorylation system. For example, in membranes from vas deferens, neither calcium alone nor protein kinase activator alone was effective, whereas addition of both calcium and protein kinase activator stimulated protein phosphorylation (Fig. 11 right, compare lanes 1–4). Each tissue displayed a different pattern of calcium-stimulated protein phosphorylation that was distinct from the cAMP-dependent phosphorylation seen in the same tissue (data not shown). One of the low molecular weight phosphoproteins (20,000 daltons) seen in the particulate fraction from vas deferens may be residual myosin light chain, which has been shown to be phosphorylated by myosin light chain kinase, an enzyme requiring both calmodulin and calcium for activity (Dabrowska *et al.,* 1978; Yagi *et al.,* 1978).

The conditions used to assay calcium-dependent protein phosphorylation in various tissues in this study were those that had been standardized for brain membranes and were not optimized for any other tissue. It is, therefore, possible that substrates for the calcium-dependent kinase, in addition to those detected here, exist in those membranes. The analysis does demonstrate, however, a widespread distribution of calcium- and calmodulin-dependent protein phosphorylation in membranes from various tissues. Each tissue differs in the number and molecular weight of the phosphoprotein substrates for this enzyme as well as in the proportion of proteins whose phosphorylation is regulated by calcium. For the vast majority of proteins, calcium-dependent phosphorylation also required the presence of calmodulin.

VI. CONCLUDING COMMENTS

Publication of this volume attests to the renewed interest by scientists from diverse fields in the molecular mechanisms underlying the physiological effects elicited by calcium. It is also evident that tremendous progress has been made towards our understanding of calcium action in numerous systems since the discovery of calmodulin.

The active form of Ca^{2+} for a variety of reactions appears to be the Ca^{2+}·calmodulin complex. It has already been demonstrated that a large number of physiological effects known to be regulated by calcium are mediated via interactions of a Ca^{2+}·calmodulin complex with a variety of biochemical systems. Microtubule polymerization-depolymerization (Marcum *et al.*, 1978), Ca^{2+} transport in erythrocytes (Gopinath and Vincenzi, 1977; Jarret and Penniston, 1977), and the activities of phosphorylase kinase (Cohen *et al.*, 1978; Shenolikar *et al.*, 1979), and myosin light chain kinase (Dabrowska *et al.*, 1978; Yagi *et al.*, 1978) are regulated by a direct interaction of the Ca^{2+}·calmodulin complex with the respective enzymes. cAMP-dependent protein kinases would appear to be regulated indirectly by the effect of the Ca^{2+}·calmodulin complex on cAMP levels, that is, by its regulation of adenylate cyclase and cyclic nucleotide phosphodiesterase activities. Synthesis and degradation of cGMP may also be regulated by the Ca^{2+}·calmodulin complex, thereby indirectly affecting the activity of cGMP-dependent protein kinase. Thus, the Ca^{2+}·calmodulin complex may indirectly control numerous biochemical processes.

The effect of the Ca^{2+}·calmodulin complex on the newly discovered class of protein kinase, the calcium- and calmodulin-dependent protein kinase, provides a mechanism for calcium to regulate diverse biochemical processes directly. The finding of calcium- and calmodulin-dependent protein phosphorylation in membranes from a variety of tissues suggests that such a phosphorylation system may be of general importance as a mechanism by which some of the actions of calcium are mediated. In addition to calcium- and calmodulin-dependent protein kinase, membranes from these tissues also contain a tissue-specific array of endogenous substrates for the kinase. These findings offer a promising approach for examining the biochemical basis for the action of calcium in a variety of tissues. Our findings also suggest an interface for certain of the interactions of cyclic nucleotide- and calcium-mediated events. Since calcium, as well as the cyclic nucleotides, are able to regulate protein kinases, these second messengers may interact at the level of protein phosphorylation. In any system, depending on the protein or proteins that these classes of kinases regulate, their individual effects may be independent, antagonistic, additive, or synergistic. As discussed earlier, it has been found that both cAMP and calcium stimulate the phosphorylation of Proteins Ia and Ib in synaptosomal preparations. Different specificities with regard to the sites of phosphorylation on Protein I are displayed by the two protein kinase systems. It will be interesting to determine the extent to which multiple-site phosphorylation of other substrate proteins occurs, and whether phosphorylation of the different sites elicits different changes in the activity of the substrate proteins.

ACKNOWLEDGMENTS

The authors wish to thank Mrs. Karen Benight for typing this manuscript.

REFERENCES

Alper, S. L., Palfrey, H. C., DeRiemer, S. A., and Greengard, P. (1980). Hormonal control of protein phosphorylation in turkey erythrocytes. Phosphorylation by cAMP-dependent and Ca^{2+}-dependent protein kinases of distinct sites in goblin, a high molecular weight protein of the plasma membrane. *J. Biol. Chem* (in press).

Baker, P. F. (1972). Transport and metabolism of calcium ions in nerve. *Prog. Biophys. Mol. Biol.* **24,** 177–223.

Baker, P. F., Hodgkin, A. L., and Ridgeway, E. B. (1971). Depolarization and calcium entry in squid giant axons. *J. Physiol. (London)* **218,** 709–755.

Blaustein, M. P. (1975). Effects of potassium, veratridine and scorpion venom on calcium accumulation and transmitter release by nerve terminals *in vitro*. *J. Physiol. (London)* **247,** 617–655.

Blaustein, M. P., Johnson, E. M., Jr., and Needleman, P. (1972). Calcium-dependent norepinephrine release from presynaptic nerve endings *in vitro*. *Proc. Natl. Acad. Sci. U.S.A.* **69,** 2237–2240.

Bloom, F. E., Ueda, T., Battenberg, E., and Greengard, P. (1979). Immunocytochemical localization in synapses of Protein I, an endogenous substrate for protein kinases in mammalian brain. *Proc. Natl. Acad. Sci. U.S.A.* **76,** 5982–5986.

Brostrom, C. O., Huang, Y.-C., Breckenridge, B. McL., and Wolff, D. J. (1975). Identification of a calcium-binding protein as a calcium-dependent regulator of brain adenylate cyclase. *Proc. Natl. Sci. U.S.A.* **72,** 64–68.

Cheung, W. Y. (1970). Cyclic 3′,5′-nucleotide phosphodiesterase. Demonstration of an activator. *Biochem. Biophys. Res. Commun.* **38,** 533–538.

Cheung, W. Y., Bradham, L. S., Lynch, T. J., Lin, Y. M., and Tallant, E. A. (1975). Protein activator of cyclic 3′:5′-nucleotide phosphodiesterase of bovine or rat brain also activates its adenylate cyclase. *Biochem. Biophys. Res. Commun.* **66,** 1055–1062.

Cleveland, D. W., Fischer, S. G., Kirschner, M. W., and Laemmli, U. K. (1977). Peptide mapping by limited proteolysis in sodium dodecyl sulfate and analysis by gel electrophoresis. *J. Biol. Chem.* **252,** 1102–1106.

Cohen, P. (1978). The role of cyclic-AMP-dependent protein kinase in the regulation of glycogen metabolism in mammalian skeletal muscle. *Curr. Top. Cell. Regul.* **14,** 117–196.

Cohen, P., Burchell, A., Foulkes, J. G., Cohen, P. T. W., Vanaman, T. C., and Nairn, A. C. (1978). Identification of the Ca^{2+}-dependent modulator protein as the fourth subunit of rabbit skeletal muscle phosphorylase kinase. *FEBS Lett.* **92,** 287–293.

Collett, M. S., Brugge, J. S., Erikson, R. L., Lau, A. F., Krzyzek, R. A., and Faras, A. J. (1979). The *src* gene product of transformed and morphologically reverted ASV-infected mammalian cells. *Nature (London)* **205,** 195–198.

Dabrowska, R., Sherry, J. M. F., Aromatorio, D. K., and Hartshorne, D. J. (1978). Modulator protein as a component of the myosin light chain kinase from chicken gizzard. *Biochemistry* **17,** 253–258.

DeCamilli, P., Ueda, T., Bloom, F. E., and Greengard, P. (1979). Widespread distribution of Protein I in the central and peripheral nervous system. *Proc. Natl. Acad. Sci. U.S.A.* **76,** 5977–5981.

DeLorenzo, R. J., and Freedman, S. D. (1977). Calcium-dependent phosphorylation of synapitc vesicle proteins and its possible role in mediating neurotransmitter release and vesicle function. *Biochem. Biophys. Res. Commun.* **77,** 1036–1043.

DeLorenzo, R. J., and Freedman, S. D. (1978). Calcium-dependent neurotransmitter release and protein phosphorylation in synaptic vesicles. *Biochem. Biophys. Res. Commun.* **80,** 183–192.

DeLorenzo, R. J., Freedman, S. D., Yohe, W. B., and Maurer, S. C. (1979). Stimulation of Ca^{2+}-dependent neurotransmitter release and presynaptic nerve terminal protein phosphorylation by calmodulin and a calmodulin-like protein isolated from synaptic vesicles. *Proc. Natl. Acad. Sci. U.S.A.* **76,** 1838–1842.

Douglas, W. W. (1968). Stimulus-secretion coupling: The concept and clues from chromaffin and other cells. *Br. J. Pharmacol.* **34,** 451–474.

Ebashi, S. (1974). Regulatory mechanism of muscle contraction with special reference to the Ca-troponin-tropomyosin system. *Essays Biochem.* **10,** 1–36.

Forn, J., and Greengard, P. (1978). Depolarizing agents and cyclic nucleotides regulate the phosphorylation of specific neuronal proteins in rat cerebral cortex slices. *Proc. Natl. Acad. Sci. U.S.A.* **75,** 5195–5199.

Frankenhaeuser, B. (1957). The effect of calcium on the myelinated nerve fibre. *J. Physiol.* (*London*) **137,** 245–260.

Gopinath, R. M., and Vincenzi, F. F. (1977). Phosphodiesterase protein activator mimics red blood cell cytoplasmic activator (Ca^{2+}-Mg^{2+}) ATPase. *Biochem. Biophys. Res. Commun.* **77,** 1203–1209.

Gray, E. G., and Whittaker, V. P. (1962). The isolation of nerve endings from brain: An electron microscopic study of cell fragments derived by homogenization and centri fugation. *J. Anat.* **96,** 79–88.

Greengard, P. (1978a). Phosphorylated proteins as physiological effectors. *Science* **199,** 146–152.

Greengard, P. (1978b). "Cyclic Nucleotides, Phosphorylated Proteins, and Neuronal Function," Raven Press, New York.

Harvey, A. M., and MacIntosh, F. C. (1940). Calcium and synaptic transmission in a sympathetic ganglion. *J. Physiol.* (*London*) **97,** 408–416.

Heilbrunn, L. V. (1952). "An Outline of General Physiology," 3rd ed., Saunders, Philadelphia, Pennsylvania.

Huttner, W. B., and Greengard, P. (1979). Multiple phosphorylation sites in Protein I and their differential regulation by cyclic AMP and calcium. *Proc. Natl. Acad. Sci. U.S.A.* **76,** 5402–5406.

Huxley, A. F. (1964). Muscle. *Annu. Rev. Physiol.* **26,,** 131–152.

Jarrett, H. W., and Penniston, J. T. (1977). Partial purification of the Ca^{2+}-Mg^{2+} ATPase activator from human erythrocytes: Its similarity to the activator of 3′:5′-cyclic nucleotide phosphodiesterase. *Biochem. Biophys. Res. Commun.* **77,** 1210–1216.

Kakiuchi, S., and Yamazaki, R. (1970). Calcium-dependent phosphodiesterase activity and its activating factor (PAF) from brain: Studies on cyclic 3′:5′-nucleotide phosphodiesterase (III). *Biochem. Biophys. Res. Commun.* **41,** 1104–1110.

Katz, B., and Miledi, R. (1969). Tetrodotoxin-resistant electrical activity in presynaptic terimpulses. *J. Physiol.* (*London*) **192,** 407–436.

Katz, B., and Miledi, R. (1969). Tetrodotoxin-resistant electrical activity in presynapitc terminals. *J. Physiol.* (*London*) **203,** 459–487.

Kelly, J. S., Krnjevic, K., and Somjen, G. J. (1969). Divalent cations and electrical properties of cortical cells. *J. Neurobiol.* **1,** 197–208.

Kretsinger, R. H. (1976). Calcium-binding proteins. *Annu. Rev. Biochem.* **45,** 239–266.

Krueger, B. K., Forn, J., and Greengard, P. (1976). Calcium-dependent protein phosphorylation in rat brain synaptosomes. *Neurosci. Abstr.* **2,** 1007.

Krueger, B. K., Forn, J., and Greengard, P. (1977). Depolarization-induced phosphorylation of specific proteins, mediated by calcium ion influx, in rat brain synaptosomes. *J. Biol. Chem.* **252,** 2764–2773.

Kuo, J. F., and Greengard, P. (1969). Cyclic nucleotide-dependent protein kinases. IV. Widespread occurrence of adenosine 3′,5′-monophosphate-dependent protein kinase in various tissues and phyla of the animal kingdom. *Proc. Natl. Acad. Sci. U.S.A.* **64,** 1349–1355.

Le Peuch, C. J., Haiech, J., and DeMaille, J. G. (1979). The concerted regulation of cardiac sarcoplasmic reticulum calcium transport by cAMP-dependent and calcium-dependent phosphorylations. *Biochemistry* **18,** 5150–5157.

Lohmann, S. M., Ueda, T., and Greengard, P. (1978). Ontogeny of synaptic phosphoproteins in brain. *Proc. Natl. Acad. Sci. U.S.A.* **75,** 4037–4041.

Luttgau, H. C., and Neidergerke, R. (1958). The antagonism between Ca and Na ions on the frog's heart. *J. Physiol.* (*London*) **143,** 486–505.

Marcum, J. M., Dedman, J. R., Brinkley, B. R., and Means, A. R. (1978). Control of microtubule assembly-disassembly by calcium-dependent regulator protein. *Proc. Natl. Acad. Sci. U.S.A.* **75,** 3771–3775.

O'Farrell, P. Z., Goodman, H. W., and O'Farrell, P. H. (1977). High resolution two-dimensional electrophoresis of basic as well as acidic proteins. *Cell* **12,** 1133–1142.

Patrick, R. L., and Barchas, J. D. (1974). Stimulation of synaptosomal dopamine synthesis by veratridine. *Nature* (*London*) **250,** 737–739.

Portzehl, H., Caldwell, P. C., and Reugg, J. C. (1964). The dependence of contraction and relaxation of muscle fibres from the crab *maia squinado* on the internal concentration of free calcium ions. *Biochim. Biophys. Acta* **79,** 581–591.

Pressman, B. C. (1973). Properties of ionophores with broad range cation selectivity. *Fed. Proc., Fed. Am. Soc. Exp. Biol.* **32,** 1698–1703.

Rasmussen, H., and Goodman, D. B. P. (1977). Relationships between calcium and cyclic nucleotides in cell activation. *Physiol. Rev.* **57,** 421–509.

Roach, P. J., DePaoli-Roach, A. A., and Larner, J. (1978). Ca^{2+}-stimulated phosphorylation of muscle glycogen synthase by phosphorylase b kinase. *J. Cycl. Nucl. Res.* **4,** 245–257.

Robison, G. A., Butcher, R. W., and Sutherland, E. W. (1971). "Cyclic AMP." Academic Press, New York.

Rodriguez de Lores Arnaiz, G., Alberici, M., and DeRobertis, E. (1967). Ultrastructural and enzymatic studies of cholinergic and noncholinergic synaptic membranes isolated from brain cortex. *J. Neurochem.* **14,** 215–225.

Rubin, C. S., and Rosen, O. M. (1975). Protein phosphorylation. *Annu. Rev. Biochem.* **44,** 831–887.

Rubin, R. P. (1972). The role of calcium in the release of neurotransmitter substances and hormones. *Pharmacol. Rev.* **22,** 389–428.

Rudolph, S. A., Beam, K. G., and Greengard, P. (1978). Studies of protein phosphorylation in relation to hormonal control of ion transport in intact cells. "Coupled Transport Phenomena in Cells and Tissues" (J. Hoffman, ed.) pp.107–123. Raven Press, New York.

Rylatt, D. B., Embi, N., and Cohen, P. (1979). Glycogen synthase kinase-2 from rabbit skeletal muscle is activated by the calcium-dependent regulator protein. *FEBS Lett.* **98,** 76–80.

Schulman, H. (1980). Calcium-dependent protein phosphorylation. *In* "Handbuch der Experimentellen Pharmakologie" (J. A. Nathanson, and J. W. Kebabian, eds.). Springer-Verlag, Berlin and New York (in press).

Schulman, H., and Greengard, P. (1978a). Stimulation of brain membrane protein phosphorylation by calcium and an endogenous heat-stable protein. *Nature (London)* **271,** 478–479.

Schulman, H., and Greengard, P. (1978b). Ca^{2+}-dependent protein phosphorylation system in membranes from various tissues, and its activation by "calcium-dependent regulator." *Proc. Natl. Acad. Sci. U.S.A.* **75,** 5432–5436.

Shenolikar, S., Cohen, P. T. W., Cohen, P., Nairn, A. C., and Perry, S. V. (1979). The role of calmodulin in the structure and regulation of phosphorylase kinase from rabbit skeletal muscle. *Eur. J. Biochem.* **100,** 329–337.

Sieghart, W., Forn, J., Schwarz, R., Coyle, J. T., and Greengard, P. (1978a). Neuronal localization of specific brain phosphoproteins. *Brain Res.* **156,** 345–350.

Seighart, W., Theoharides, T. C., Alper, S. L., Douglas, W. W., and Greengard, P. (1978b). Calcium-dependent protein phosphorylation during secretion by exocytosis in the mast cell. *Nature (London)* **275,** 329–331.

Sieghart, W., Forn, J., and Greengard, P. (1979). Ca^{2+} and cyclic AMP regulate phosphorylation of same two membrane-associated proteins specific to nerve tissue. *Proc. Natl. Acad. Sci. U.S.A.* **76,** 2475–2479.

Sieghart, W., Schulman, H., and Greengard, P. (1980). Neuronal localization of Ca^{2+}-dependent protein phosphorylation in brain. *J. Neurochem.* **34,** 548–553.

Srivastava, A. K., Waisman, D. M., Brostrom, C. O., and Soderling, T. R. (1979). Stimulation of glycogen synthase phosphorylation by calcium-dependent regulator protein. *J. Biol. Chem.* **254,** 583–586.

Strombom, U., Forn, J., Dolphin, A. C., and Greengard, P. (1979). Regulation of the state of phosphorylation of specific neuronal proteins in mouse brain by *in vivo* administration of anesthetic and convulsant agents. *Proc. Natl. Acad. Sci. U.S.A.* **76,** 4687–4690.

Sutherland, E. W., Robison, G. A., and Butcher, R. W. (1968). Some aspects of the biological role of adenosine 3′,5′-monophosphate (cyclic AMP). *Circulation* **37,** 279–306.

Teo, T. S., and Wang, J. H. (1973). Mechanism of activation of a cyclic adenosine 3′:5′-monophosphate phosphodiesterase from bovine heart by calcium ions. Identification of the protein activator as a Ca^{2+}-binding protein. *J. Biol. Chem.* **248,** 5950–5955.

Teo, T. S., Wang, H., and Wang, J. H. (1973). Purification and properties of the protein activator of bovine heart cyclic adenosine 3′:5′-monophosphate phosphodiesterase. *J. Biol. Chem.* **248,** 588–595.

Theoharides, T. C., Sieghart, W., Greengard, P., and Douglas, W. W. (1980). The antiallergic drug cromolyn may inhibit histamine secretion by regulating phosphorylation of a mast cell protein. *Science* **207,** 80–82.

Ueda, T., and Greengard, P. (1977). Adenosine 3′:5′-monophosphate-regulated phosphoprotein system of neuronal membranes. I. Solubilization, purification and some properties of an endogenous phosphoprotein. *J. Biol. Chem.* **252,** 5155–5163.

Ueda, T., Maeno, H., and Greengard, P. (1973). Regulation of endogenous phosphorylation of specific proteins in synaptic membrane fractions from rat brain by adenosine 3′:5′-monophosphate. *J. Biol. Chem.* **248,** 8295–8305.

Ueda, T., Greengard, P., Berzins, K., Cohen, R. S., Blomberg, F., Grab, D. J., and Siekevitz, P. (1979). Subcellular distribution in cerebral cortex of two proteins phosphorylated by a cAMP-dependent protein kinase. *J. Cell. Biol.* **83,** 308–319.

Ulbricht, W. (1969). The effect of veratridine on excitable membranes of nerve and muscle. *Ergeb. Physiol., Biol. Chem. Exp. Pharmakol.* **61,** 18–71.

Walsh, D. A., Perkins, J. P., and Krebs, E. G. (1968). An adenosine 3′,5′-monophosphate dependent protein kinase from rabbit skeletal muscle. *J. Biol. Chem.* **243,** 3763–3765.

Watterson, D. M., Harrelson, W. G., Jr., Keller, P. M., Sharief, F., and Vanaman, T. C. (1976). Structural similarities between the Ca^{2+}-dependent regulatory proteins of 3′:5′-cyclic nucleotide phosphodiesterase and actomyosin ATPase. *J. Biol. Chem.* **251,** 4501–4513.

Wolff, D. J., and Brostrom, C. O. (1979). Properties and functions of the calcium-dependent regulator protein. *Adv. Cyclic Nucleotide Res.* **11,** 27–88.

Wolff, D. J., Poirier, P. G., Brostrom, C. O., and Brostrom, M. A. (1977). Divalent cation binding properties of bovine brain Ca^{2+}-dependent regulator protein. *J. Biol. Chem.* **252,** 4108–4117.

Yagi, K., Yazawa, M., Kakiuchi, S., Ohshima, M., and Uenishi, K. (1978). Identification of an activator protein for myosin light chain kinase as the Ca^{2+}-dependent modulator protein. *J. Biol. Chem.* **253,** 1338–1340.

Chapter 12

Role of Calmodulin In Dopaminergic Transmission

I. HANBAUER
E. COSTA

I. DOPAMINE RECEPTORS AS SUPRAMOLECULAR ENTITIES

Dopamine receptors, like many other postsynaptic receptors, function as supramolecular entities, that is, that receptor activity is generated by interactions among a number of membrane proteins. In brief, when the receptor recognition site is occupied by its agonists, it triggers specific molecular interactions among a characteristic group of membrane proteins. In the case of dopamine receptors, these molecular interactions often lead to the activation of adenylate cyclase (Kebabian *et al.*, 1972; Brown and Makman, 1972; Kebabian and Calne, 1979) which, in turn, mediates the internalization of the stimulus into the postsynaptic cells via the

CALCIUM AND CELL FUNCTION, VOL. I
ISBN 0-12-171401-2

activation of a specific protein kinase. However, this is not the only type of molecular interaction which occurs during dopamine receptor stimulation. In fact, there are physiological responses elicited by dopamine, which do not depend on the stimulation of adenylate cyclase (MacLeod, 1976). Thus, in certain neurons, the occupancy of dopamine receptors might also trigger a change in the ionic environment of the postsynaptic cell following the opening of specific ion channels located in the membrane.

According to the fluid mosaic model of cell membranes proposed by Singer and Nicolson (1972), proteins are imbedded in the membrane lipids where they possess a great deal of mobility, which can be regulated by various factors. For instance, when the recognition site of the postsynaptic receptor is occupied by the agonist, protein interactions are accelerated, therefore collision coupling becomes prominent in promoting the activation of specific enzymes (Lefkowitz *et al.*, 1976). In striatum, the occupancy of dopamine recognition sites activates adenylate cyclase, and the increment of cAMP formed stimulates the cAMP-dependent protein kinase, which triggers a number of metabolic responses. It was first shown (Weiss and Costa, 1967, 1968) in pineal extracts, during β-adrenergic receptor stimulation that the activation of adenylate cyclase was coupled to the stimulation of postsynaptic receptors. Moreover, adenylate cyclase activation by norepinephrine was increased during receptor supersensitivity following ganglionectomy (Weiss and Costa, 1967, 1968). The coupling of adenylate cyclase to postsynaptic receptors was extended to striatal dopamine receptors by Kebabian and colleagues (1972). Since the activation of striatal adenylate cyclase by dopamine was inhibited by neuroleptics (Clement-Cormier *et al.*, 1974), it was proposed that these drugs ameliorated the symptoms of psychoses by blocking dopamine activation of striatal adenylate cyclase. However, the ranking order of a number of neuroleptics as blockers of dopamine-dependent adenylate cyclase and as antipsychotics shows important differences, suggesting that an inhibition of the dopamine-dependent adenylate cyclase may not be a relevant element to explain the therapeutic activity of these compounds (Creese *et al.*, 1976). Later it was reported that striatum contains two populations of high-affinity binding sites for dopamine (Kebabian and Calne, 1979), but only one of the two is coupled to adenylate cyclase (Schwarcz *et al.*, 1978). From these results one can infer that the antipsychotic action of the neuroleptics relates to their ability to block those dopamine receptors which are not coupled to adenylate cyclase. While the enzymatic system which expresses this second type of dopamine receptor is still unknown, the proteins that couple dopamine recognition sites to adenylate cyclase gradually are being identified. A specific GTP-binding protein which is essential in the regulation of the adenylate cyclase coupled to β-adrenergic receptors (Lefkowitz *et al.*, 1976) appears to be operative in the coupling of adenylate cyclase to the recognition sites of dopamine receptors (Blume, 1978; Creese *et al.*, 1978). Also other proteins, such as calmodulin, have been implicated in the dopamine receptor function (Costa *et al.*, 1977). Calmodulin is a thermostable protein which mediates many of the actions of Ca^{2+} (for a review, see Wolff and Brostrom, 1979). Membrane-bound calmodulin regulates the activity of adenylate cyclase (Brostrom *et al.*, 1975) and of a Ca^{2+}-dependent protein kinase (Waisman *et al.*, 1978), while cytosolic calmodulin modulates phosphodiesterase (Cheung, 1970; Hanbauer *et al.*, 1979a,b) and other enzymes. Since adenylate cyclase and phosphodiesterase are involved in the regulation of the second messenger cAMP, the interaction of these enzymes with calmodulin appears to be important in the function of striatal neurons. In fact, it was shown that in synaptosomal membranes prepared from striatum, calmodulin regulates adenylate cyclase (Gnegy *et al.*, 1976), and, in turn, cAMP modulates calmodulin translocation from the cell membrane into the cytosol (Gnegy *et al.*, 1977a). Membrane-bound calmodulin may regulate its own translocation by a mechanism involving activation of Ca^{2+}-dependent protein kinases and/or activation of a cAMP-dependent protein kinase. It was shown that the activation of phosphodiesterase by calmodulin and Ca^{2+} "down-regulates" the cytosolic concentration of cAMP (Hanbauer *et al.*, 1979b). This mechanism may be operative in the subsensitivity of dopamine receptors elicited by their persistent stimulation.

II. CALMODULIN MODULATION OF ADENYLATE CYCLASE

Brostrom *et al.* (1975) first showed that calmodulin regulates solubilized adenylate cyclase. This finding was soon confirmed by Cheung *et al.* (1975). Ion exchange chromatography of solubilized adenylate cyclase preparations resulted in a loss of adenylate cyclase activity, which could be restored by the addition of both Ca^{2+} and purified calmodulin (Brostrom *et al.*, 1975). Studies were undertaken to ascertain whether calmodulin while embedded in the postsynaptic membrane could modulate adenylate cyclase activity and whether this regulation could be of physiological relevance. Work from various laboratories indicates that several distinct forms of adenylate cyclase are present in particulate fractions prepared from brain homogenates (Brostrom *et al.*, 1978; Neer, 1978). Some of these forms are calmodulin dependent (Gnegy *et al.*, 1976), while others are calmodulin independent (Brostrom *et al.*, 1978). Both types of adenylate cyclase can be activated by F^-, although the F^- concentrations that elicit maximal activation differ (Brostrom *et al.*, 1977). Moreover, the F^- activation of calmodulin-dependent adenylate cyclase was completely reversed by EGTA, while that of calmodulin-independent enzymes was not (Brostrom *et al.*, 1978).

III. RELEASE OF CALMODULIN BY MEMBRANE PHOSPHORYLATION

Since the amount of calmodulin bound to synaptosomal membranes decreases following either the *in vitro* phosphorylation of membrane proteins or a sustained stimulation of dopamine receptors, studies were undertaken to determine whether in striatal membranes the ability of dopamine to stimulate adenylate cyclase is decreased as a result of membrane protein phosphorylation (Gnegy *et al.*, 1977a).

The data shown in Fig. 1 suggest that calmodulin (indicated in the figure as PDEA) can be released spontaneously from a membrane suspension. This spontaneous release of calmodulin can be inhibited by the addition of 2 m*M* EGTA, suggesting that a Ca^{2+}-activated process (phosphorylation) may be operative. The data in Fig. 1 show that the release rate of calmodulin is enhanced by the addition of ATP, protein kinase, and cAMP, whereas in the absence of these components the spontaneous release of calmodulin is only about 33% of the maximal release. The regulatory role of Ca^{2+} in the spontaneous release of calmodulin and the possiblity that membrane protein phosphorylation could accelerate this release are currently being investigated by Dr. Gnegy. The concept that calmodulin translocation is the rate-limiting step in Ca^{2+} regulation of cyclic nucleotide metabolic enzymes has been recently criticized on theoretical and experimental grounds (Wolff and Brostrom, 1979). The criticism was mainly based on the fact that a nonpurified protein kinase preparation was used, and, therefore, other membrane processes might have been modified by this addition. Since a crude membrane preparation was used, it was important to demonstrate that the release of calmodulin caused by the phosphorylation depended on the presence of ATP and cAMP. This dependence is demonstrated in Fig. 1 and in our view supports the participation of phosphorylation in the calmodulin release. To substantiate the involvement of protein phosphorylation in the calmodulin release, β- and γ-methylene ATP were studied as substitutes for ATP in eliciting the release of calmodulin from membrane, using the same protein kinase preparation (Uzunov *et al.*, 1976). The results obtained clearly showed that these ATP

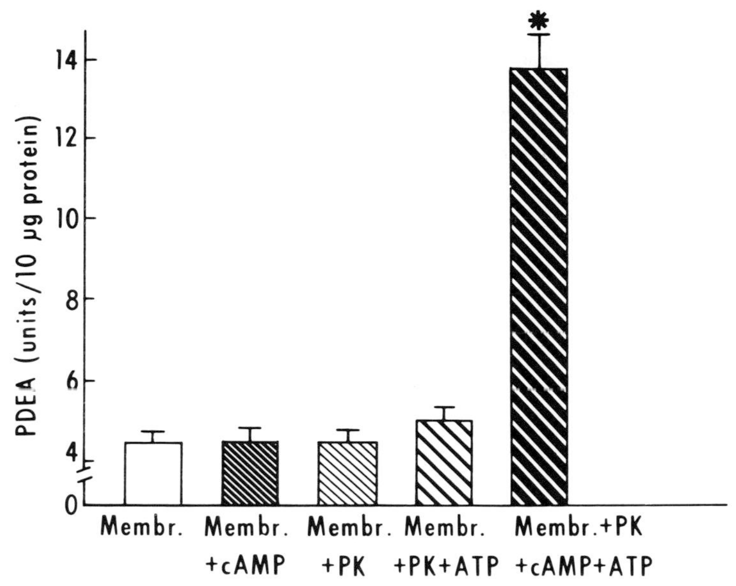

Fig. 1. A synaptic membrane enriched fraction was prepared from rat brain tissue homogenized with 0.32 *M* sucrose containing 150 m*M* KCl. A pellet prepared by centrifuging at 10^5 *g* for 1 hr was washed several times and rehomogenized in 5 ml of 32 m*M* Tris-HCl (pH 7.5) containing 1 m*M* $MgSO_4$, 20 *M* $CaCl_2$, and 0.6 m*M* dithiothreitol. This homogenate was then dialyzed for 3 hr against 200 volumes of the same buffer used for homogenization. A cyclic AMP-dependent protein kinase was prepared according to Miyamato and co-workers (1969) and added to the membrane preparation described above. Phosphodiesterase activator (PDEA) is now called calmodulin. * $p < 0.01$, $n = 6$.

analogues failed to substitute for ATP in the calmodulin release. Further support of the role of protein phosphorylation comes from experiments shown in Table I. Incubation of membrane preparations with various neurotransmitters for a protracted time period releases calmodulin from its membrane binding site only if the medium contains ATP and a cAMP-dependent protein kinase. Furthermore, the membrane was phosphorylated only by the addition of the protein kinase. Calmodulin which was released as a result of the membrane phosphorylation from its binding sites was not phosphorylated (Gnegy *et al.*, 1976). In addition, it was found that a membrane protein was phosphorylated which on gel filtration appears to have a molecular weight smaller than calmodulin. Whether the protein that functions as a binding site for calmodulin on synaptic membranes is the endogenous inhibitor described by Klee and Krinks (1978) cannot be assessed from these experiments and requires further investigations.

TABLE I

Release of Calmodulin from Synaptic Membranes by Various Neurotransmitters in the Absence and in the Presence of Protein Kinase and ATP[a]

	Calmodulin release (U/20 μg protein)	
Neurotransmitter[b] (10^{-6} *M*)	−ATP and −PK	+ATP and +PK
None	2.6 ± 0.4	2.6 ± 0.5
NE	3.0 ± 0.7	13 ± 2
DA	2.7 ± 0.8	11 ± 2
5HT	2.8 ± 0.4	2.4 ± 0.2

[a] Incubation at 30°C in 32 m*M* Tris-HCl buffer, pH 7.5, 1 m*M* $MgSO_4$ 20 μM $CaCl_2$, and 0.6 m*M* dithiothreitol. Purified cAMP-dependent protein kinase (PK) (160 μg) (Kuo and Greengard, 1972) and 25 μM ATP were added as indicated.

[b] NE, norepinephrine; DA, dopamine; 5HT, 5-hydroxytryptamine.

IV. CALMODULIN AS A POSSIBLE MODULATOR OF DOPAMINE RECEPTORS

The foregoing results suggest that persistent stimulation of the dopamine receptors elicits an accumulation of cAMP which causes the activation of cAMP-dependent protein kinase. This enzyme phosphorylates the membrane of postsynaptic cells, and, in turn, this phosphorylation triggers the release into the cytosol of some of the calmodulin bound to synaptic membranes. In various brain areas, the potency of the various putative transmitters as activators of adenylate cyclase appears to correlate with their ability to release calmodulin. It can be inferred that the release of calmodulin reflects the degree of protein kinase activation caused by the cAMP produced. In line with this postulate are the experiments showing that, in striatum, dopamine was tenfold more active than norepinephrine in the stimulation of adenylate cyclase and in the release of calmodulin from membrane (Revuelta *et al.*, 1976). In contrast, norepinephrine was more active than dopamine in membrane preparations from cerebellum (Revuelta *et al.*, 1976). The basal activity of adenylate cyclase remained unchanged when synaptic membranes prepared from striatum were phosphorylated in the presence of ATP, cAMP, and protein kinase (Table II). However, if phosphorylation of striatal membranes preceded the stimulation of adenylate cyclase by dopamine and by F^- the stimulatory action of the latter components was reduced (Table II). Although many uncontrolled variables still remain in these experiments, it is possible to suggest that membrane-bound calmodulin participates in the stimulation of adenylate cyclase by dopamine. M. Gnegy (personal communi-

TABLE II

Adenylate Cyclase Activity in the Presence and in the Absence of F^- in Membranes Depleted of Calmodulin by Phosphorylation[a]

Membrane phosphorylation	Calmodulin in medium (U/mg protein)	Adenylate cyclase activity (pmole cAMP/mg protein/min)		
		Basal	+ 10m*M* NaF	+10 μ*M* Dopamine
No	88 ± 11	96 + 10	261 ± 29	191 ± 24
Yes	953 ± 80[b]	91 ± 14	136 ± 29[c]	117 ± 9[c]

[a] Phosphorylation media contained 160 μg protein kinase (PK) purified from beef heart according to Kuo and Greengard (1972), ATP, and 1 μ*M* cyclic AMP. Nonphosphorylation media contained PK and 1 μ*M* cyclic AMP but no ATP. Other conditions as described in Table I. Results are expressed as the mean ± SEM of ten experiments.

[b] $p < 0.01$.

[c] $p < 0.05$.

cation) conducted experiments on the action of calmodulin on agonist binding to dopamine recognition sites and concluded that the binding of dopamine was not influenced by the amount of calmodulin stored in postsynaptic membranes. Further experiments are necessary to reveal the mechanism by which calmodulin facilitates the coupling of dopamine recognition sites to adenylate cyclase.

Kebabian *et al.* (1972) suggested that the stimulation of adenylate cyclase is an important step during stimulus internalization following dopamine receptor activation. However, as mentioned above, the activation of adenylate cyclase may not be the only transducer mechanism operating in the regulation of dopamine receptors. Kebabian and Calne (1979) dis-

TABLE III

Effect of Apomorphine on the Calmodulin Content of Striatal Slices[a]

Homogenizing condition	Calmodulin (μg/mg protein) ± SEM	
	Control	Apomorphine (10^{-7} *M*)
0.32 *M* Sucrose	0.50 ± 0.03	0.90 ± 0.06 +
0.32 *M* Sucrose + 0.1 m*M* EGTA	0.46 ± 0.03	0.91 ± 0.04[b]

[a] Striatal slices were preincubated in Krebs–Ringer solution for 45 min (95% O_2 + 5% CO_2). The incubation was continued for 30 min in presence of 10^{-7} *M* apomorphine. At the end of the incubation the slices were homogenized in 0.32 *M* sucrose in absence or presence of 0.1 m*M* EGTA and centrifuged in 10^5 g_{av} for 60 min. The calmodulin content was assayed in the supernatant fraction by an enzyme-linked immunosorbent assay.

[b] $p < 0.05$ ($n = 5$).

cussed the existence of two types of dopamine receptors which differ not only in the characteristics of the recognition sites but also in the transducer system that is coupled to them. These receptors have been termed D-1 and D-2. The D-1 receptor is coupled to adenylate cyclase, whereas the D-2 receptor fails to be linked to this enzyme. The present view on the molecular architecture of the D-1 receptor is derived from a number of experimental findings, showing that adenylate cyclase and calmodulin are located on postsynaptic membranes presumably in the vicinity of the inner part of the membrane facing the cytosol. *In vivo* persistent stimulation of the dopamine receptors in the caudate nucleus by (+)-amphetamine caused a transient increase of cAMP content (Carenzi *et al.,* 1975) due to activation of adenylate cyclase. While the sterotype behavior elicited by (+)-amphetamine is dose dependent (Randrup and Munkvad, 1970), the duration of the increase in the cAMP content of caudate nucleus fails to be dose dependent and is shorter lasting than the behavioral response (Carenzi *et al.,* 1975). This discrepancy suggests that involvement of other factors in the regulation of the cyclase which are not linked to the activation of dopamine receptors by (+)-amphetamine. The possibility that a transient increase in intracellular Ca^{2+} content could be operative in regulating the cAMP content is frequently discussed. However, in cell cultures, which could be used as a model of striatal postsynaptic cells, the increase in intracellular Ca^{2+} content fails to change the cAMP content (Wolff and Brostrom, 1979). This lack of response favors the possibility that the translocation of calmodulin rather than a change in the Ca^{2+} content might be rate limiting in the "down-regulation" of cAMP.

To assess the validity of such a hypothesis we studied the redistribution of calmodulin during the stimulation of dopamine receptors in homogenates of striatal slices. These experiments revealed that the calmodulin content in the supernatant fraction was increased following persistent activation of the dopamine receptors. The data in Table III show that this increase in calmodulin content is also detectable when striatal slices are homogenized in the presence of 0.1 m*M* EGTA. Since *in vivo* the duration of the increase in cAMP content following receptor stimulation appears to be related to a redistribution of calmodulin, studies on the regulation of calmodulin synthesis, storage, and metabolism in postsynaptic cells will be of great interest.

V. RECEPTOR/MEDIATED MODULATION OF PHOSPHODIESTERASE: ROLE OF CALMODULIN TRANSLOCATION

The cyclic nucleotide phosphodiesterase is the only known enzyme which catalyzes the hydrolytic degradation of cAMP and cGMP (Sutherland and Rall, 1957; Cheung, 1970). In brain there are various molecular forms of this enzyme which differ in regard to the affinity for the substrate and in their ability to be regulated by calmodulin in the presence of Ca^{2+} (Uzunov and Weiss, 1972). Calmodulin regulates only phosphodiesterase isozymes which have a high K_m for cAMP (Teo and Wang, 1973; Uzunov

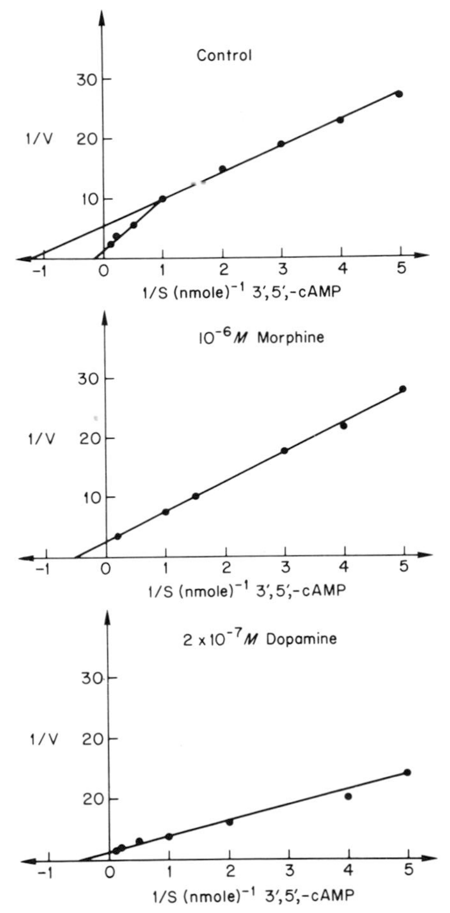

Fig. 2. Double reciprocal plot of the initial velocity of phosphodiesterase in the supernatant fraction versus cAMP concentration. (Top) Striatal slices were incubated in Krebs–Ringer bicarbonate buffer, pH 7.4 (95% O_2 + 5% CO_2) for 90 min. The slices were then homogenized in 0.32 *M* sucrose and centrifuged at 10^5 *g* for 60 min. (Center) Striatal slices were preincubated for 60 min as described above: Thereafter 10^{-6} *M* morphine sulfate was added and the incubation continued for 30 min. (Bottom) Striatal slices were preincubated for 60 min as described above. Thereafter 2 × 10^{-7} *M* dopamine was added and the incubation continued for 30 min. The apparent K_m values were determined from Lineweaver–Burk plots of five such experiments. (Reproduced from Hanbauer *et al.* 1979c.)

et al., 1975) and requires Ca^{2+} (Thompson and Appleman, 1971; Campbell and Oliver, 1972; Uzunov and Weiss, 1972). On a short-term basis phosphodiesterase modulation occurs by changing the kinetic properties of the enzyme via formation of an enzyme–calmodulin–Ca^{2+} complex (Lin *et al.*, 1975; Wang and Desai, 1976). In striatal slices such changes in the kinetic properties of phosphodiesterase occur during persistent activation of dopamine receptors. As shown in Fig. 2 (top), the biphasic double reciprocal plot of phosphodiesterase activity versus cAMP concentrations denotes the presence of a heterogenous population of enzyme molecules (possible various molecular forms) in the supernatant of nonstimulated striatal slices. In contrast, when striatal slices are incubated with dopamine the double reciprocal plot becomes monophasic, and the apparent K_m for cAMP of the enzyme is lowered (Fig. 2, bottom). Preincubation of striatal slices with dopamine receptor blockers prevents the enzyme modification elicited by dopamine (Hanbauer *et al.*, 1979b). The change in the properties of the phosphodiesterase first proposed by Cheung (1970) may be instrumental in the regulation of the amount of cAMP that can accumulate intracellularly following persistent stimulation of dopamine receptors. While the increase of cAMP itself elicited by short-term membrane depolarization cannot evoke calmodulin translocation, it was shown that in the adrenal medulla a persistent activation of the cAMP-dependent protein kinase is associated with a redistribution of calmodulin (Uzunov *et al.*, 1975). Thus, the functional implication of the calmodulin translocation in the modulation of phosphodiesterase activity can be also extended to the persistent stimulation of nicotinic receptors in adrenal medulla. Since extensive biochemical studies show that stimulation of various postsynaptic receptors increases the intracellular Ca^{2+} uptake (for a review, see Kretsinger, 1979), it is of interest to examine whether an increase in the Ca^{2+} availability without a redistribution of calmodulin determines the change in the kinetic profile of the phosphodiesterase isozymes. In the cytosol the amount of calmodulin present appears to be in excess of the amount of calmodulin-dependent phosphodiesterase, though no accurate measurements of this enzyme form are available. The assumption that Ca^{2+} is the sole regulatory factor for calmodulin-dependent phosphodiesterase implies that the Ca^{2+} influx is the rate-limiting step in this regulation. However, recent experiments from our laboratory using Ca^{2+} ionophores show that these considerations do not apply to the regulation of striatal phosphodiesterase (Hanbauer *et al.*, 1979b). However, in striatal slices persistent stimulation of dopamine receptors elicits a redistribution of calmodulin. By gel filtration it was possible to show that after dopamine receptor stimulation, the amount of calmodulin which

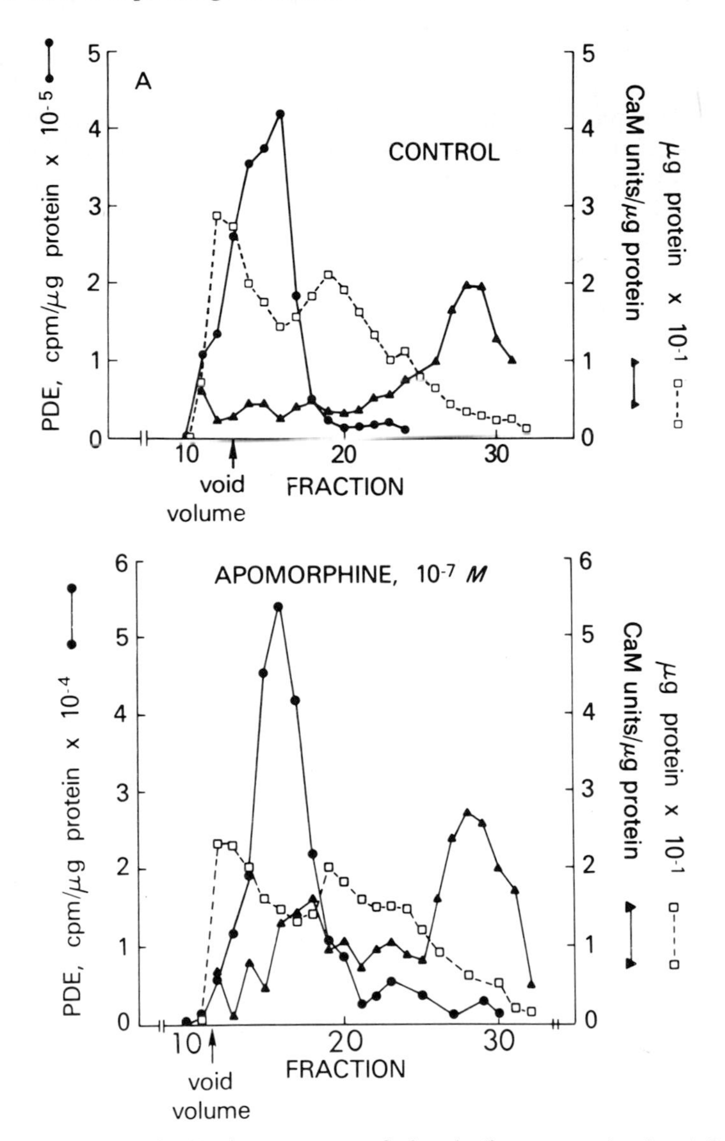

Fig. 3. Sephadex G-150 chromatogram of phosphodiesterase and calmodulin in the supernatant fraction prepared from homogenates of striatal slices. (Top) Striatal slices were incubated for 90 min in Krebs–Ringer bicarbonate buffer, pH 7.4. (Bottom) Striatal slices were preincubated for 60 min as described above followed by an incubation for 30 min in the presence of 10^{-7} *M* apomorphine. The slices were homogenized in 0.32 *M* sucrose and centrifuged at 10^5 *g* for 60 min. (Modified from Hanbauer *et al.* 1979b.)

elutes with phosphodiesterase and the total amount of soluble calmodulin increase (Fig. 3).

The development of sensitive procedures to assay calmodulin (Uzunov *et al.*, 1975; Hanbauer *et al.*, 1979a) made it possible to measure in various tissues the increase in soluble and the decrease in particulate-bound calmodulin. In brain homogenates the calmodulin content in the particulate fraction enriched with synaptic membranes is higher than in the supernatant fraction (Table IV). These findings initiated a series of studies which showed that the stimulus-coupled release of calmodulin from postsynaptic membranes into the cytosol activates phosphodiesterase (Hanbauer *et al.*, 1979a,b). Incubation of striatal slices with apomorphine (10^{-7} *M*) for at least 30 min increases the calmodulin content in the supernatant fraction (Table III) (Hanbauer *et al.*, 1979b). This increase in the content of soluble calmodulin is associated with a decrease in the apparent K_m for cAMP of the phosphodiesterase present in the same supernatant fractions. Filtration on Sephadex G-150 of the supernatant fraction of homogenates from striatal slices resolves two pools of calmodulin (Fig. 3, top). In one pool, calmodulin appears to be associated with phosphodiesterase; in the other pool, calmodulin appears to be dissociated from any large molecular weight protein. Following incubation of striatal slices with apomorphine, the amount of calmodulin associated with phosphodiesterase is increased (Fig. 3, bottom). The association of calmodulin with the enzyme is Ca^{2+}-dependent since the complex could be resolved by filtration over Sephadex G-150, which was equilibrated and eluted in the presence of 1 m*M* EGTA. These findings support the view that the activation of dopamine receptors triggers the release of membrane-bound calmodulin which complexes with phosphodiesterase and Ca^{2+}, thereby changing the catalytic properties of the phosphodiesterases. It should be noted that these studies were possible because of the introduction of an immunochemical method to measure calmodulin (Hanbauer *et al.*, 1979a). Previously, such a study was not possible because calmodulin was measured indirectly by the degree of activation of purified Ca^{2+}-dependent phosphodiesterase. This enzymatic method is relatively inaccurate because an endogenous inhibitor of calmodulin may cause artifacts thereby decreasing the range of linearity (Lucchelli *et al.*, 1978). A series of studies carried out with striatal slices showed that both calmodulin and calmodulin-activatable phosphodiesterase are located in neurons which are innervated by dopaminergic axons (Hanbauer *et al.*, 1979b). Brain hemitranssection which includes transections of the nigrostriatal fiber bundle fails to change the kinetic profile of phosphodiesterase or the content of soluble calmodulin. Moreover, incubation of slices prepared from denervated striata with apomorphine (10−7 *M*) elicits a decrease in the apparent K_m of phosphodiesterase

TABLE IV

Cellular Distribution of Calmodulin in Rat Brain[a]

Subcellular fractionation at 100,000 *g*	Calmodulin (μg/mg protein)[b]		
	Caudate nucleus	Cerebellum	Hippocampus
Supernatant	1.6 + 0.32	2.1 + 0.29	1.4 + 0.21
Pellet	5.9 + 0.38	3.8 + 0.56	3.2 + 1.13

[a] Calmodulin was determined by an enzyme-linked immunosorbent assay.
[b] $p < 0.05$.

for cAMP (Hanbauer *et al.*, 1979b). In addition, following hemitransection the amount of calmodulin stored in membranes is increased and the membrane-bound adenylate cyclase becomes supersensitive to dopamine stimulation. The increase in calmodulin content and the supersensitivity of adenylate cyclase can be blocked if, after surgery, the rats are injected twice a day with apomorphine (Luchelli *et al.*, 1978).

VI. CALMODULIN PARTICIPATION IN DOPAMINE RECEPTOR PHARMACOLOGY

Various neuronal loops participate in the modulation of striatal dopaminergic function (Hong *et al.*, 1977; Gale *et al.*, 1977, 1978). The striatal nigral pathway that feeds back on dopaminergic cell bodies located in the substantia nigra includes γ-aminobutyric acid (GABA) and substance P containing neurons. Another regulatory pathway consists of small interneurons intrinsic to the caudate nucleus which innervates the dopaminergic axons and contains enkephalin. The existence of these axo-axonic synapses between small striatal enkephalinergic interneurons and the long dopaminergic neurons have been proposed by Schwartz and co-workers (1978). They showed that the destruction of dopaminergic nerve endings decreases the number of opiate binding sites in the caudate nucleus. The firing rate of nigral dopamine neurons (Iwatsubo and Clouet, 1977) and the metabolism of striatal dopamine (Clouet and Rattner, 1970; Costa *et al.*, 1973) increase following morphine injections. It was speculated that this increase in dopamine metabolism following opiate receptor stimulation brings about an extraneuronal release of dopamine. In fact, direct behavioral evidence in support of such an activation is equivocal: morphine elicites a transient increase followed by a long-lasting decrease of motor activity associated with rigidity and catatonia which is compatible with a possible blockade of dopaminergic transmission. In order to study how

TABLE V

Changes in Apparent K_m for cAMP of Striatal Phosphodiesterase Elicited by Modulation of Opiate and Dopamine Receptors[a]

Pretreatment of rats	Addition to: Preincubation medium	Addition to: Incubation medium	Number of K_m forms	K_m apparent values of PDE (μM) ± SEM	
None	None	None	2	14 ± 1.5	100 ± 15
None	None	Morphine 5×10^{-7} M	1	31 ± 4.6	
None	Naltrexone 10^{-7} M	Morphine 5×10^{-7} M	2	11 ± 2.0	83 ± 9
None	Haloperidol 10^{-7} M	Morphine 5×10^{-7} M	2	15 ± 2.0	76 ± 10
Reserpine, 5 mg/kg ip	None	None	2	7.6 ± 3.0	55 ± 5
Reserpine, 5 mg/kg ip	None	Morphine 5×10^{-7} M	2	7.4 ± 2.5	56 ± 4
Brain hemitransection	None	None	2	14 ± 2.6	112 ± 8.0
Brain hemitransection	None	Morphine 5×10^{-7} M	2	15 ± 1.0	87 ± 8.0

[a] Rat brain was hemitransected 3 weeks and reserpine was injected 3.5 hr before the experiment. Striatal slice were incubated in Krebs–Ringer bicarbonate buffer pH 7.4 (95% O_2 + 5% CO_2). The preincubation period lasted 60 min and the incubation period lasted 30 min. The slices were homogenized in 0.32 M sucrose and centrifuged at 10^5 g_{av} for 60 min. The apparent K_m values were derived from double reciprocal plots of the initial velocity of phosphodiesterase versus cAMP concentrations of three such experiments.

dopamine receptor function is modified by persistent stimulation of opiate receptors, we investigated the changes in calmodulin content and the kinetic properties of the phosphodiesterase present in the supernatant fraction of homogenates of striatal slices incubated with morphine. The results shown in Table V indicate that persistent stimulation of the opiate receptors simultaneously increases the content of calmodulin in the supernatant fraction of homogenates of striatal slices and reduces the apparent K_m of phosphodiesterase for cAMP (Fig. 2, center) (Hanbauer *et al.*, 1979c). In addition, chromatography on Sephadex G-150 of the supernatant fraction shows that stimulation of the opiate receptors by morphine changes the profile of the calmodulin distribution in the column eluate. The amount of calmodulin that comigrates with high molecular weight proteins that include phosphodiesterase was increased following stimulation of the opiate receptors. In order to establish whether opiate and/or dopamine receptors participate in the mediation of the changes in calmodulin content and in the apparent K_m of phosphodiesterase for cAMP the effects of morphine on both parameters were studied in the presence and absence of opiate and dopamine receptor antagonists. Table V shows that pretreatment of striatal slices with naltrexone (10^{-7} M) prevents the change in the apparent K_m of phosphodiesterase for cAMP elicited by morphine. These findings indicate an interaction between enkephalinergic and dopaminergic neurons. This interaction is further supported by experiments showing that morphine fails to change the apparent K_m for cAMP of phosphodiesterase in the supernatant fraction of the homogenates prepared from striatal slices in which the dopamine axons were destroyed by transection of the nigra striatal fibers or the dopamine content was depleted by reserpine (Hanbauer *et al.*, 1979c).

Several stimulant drugs which in small doses increase motor activity and in high doses elicit sterotype behavior were tested for their effect on striatal calmodulin content. (+)-Amphetamine which was shown to elevate the turnover rate of dopamine and the cAMP content of striatum (Carenzi *et al.*, 1975) also increase the calmodulin content in the supernatant fraction of striatal homogenates (Table VI). In contrast, another central stimulant, cocaine, which similar to (+)-amphetamine enhances the dopamine turnover rate in striatum (Costa *et al.*, 1972), increases the calmodulin content in the particulate fraction prepared from striatal slices, but fails to change the calmodulin content in the supernatant fraction of striatal homogenates. A similar increase in the calmodulin content of the striatal membrane pellet was obtained during the withdrawal from a chronic treatment with haloperidol (Gnegy *et al.*, 1977b) and after denervation of the striatum by brain hemitransection (Lucchelli *et al.*, 1978). The increase in content of calmodulin coincided with supersensitivity of

TABLE VI

The Effect of Central Stimulants on the Calmodulin Content in Striatum

Drug (mg/kg, ip)	Calmodulin (% of control)[a] Soluble	Calmodulin (% of control)[a] Membrane-bound
None	100 ± 15	100 ± 19
(+)-Amphetamine (1.5)	238 + 51[b]	92 ± 9
Cocaine (30)	82 ± 14	156 ± 8[b]

[a] The calmodulin content was measured by an enzyme-linked immunosorbent assay in striatal supernatant fraction and synaptosomal pellet extract. $n = 6$.
[b] $p < 0.05$.

the dopamine receptor. These experiments show for the first time a difference in the action of the cocaine and (+)-amphetamine on membranes. These results will be completely understood when it is known how the synthesis and the storage of calmodulin in the membrane is regulated.

The translocation of calmodulin from its storage site in synaptic membranes into the cytosol can also be assessed by studying the kinetic properties of phosphodiesterase. These considerations initiated studies on the changes of the phosphodiesterase activity in striatum caused by a treatment with (+)-amphetamine, cocaine, and apomorphine. The biphasic double reciprocal plot of phosphodiesterase activity versus the cAMP concentration is shifted to a monophasic plot when the slices are prepared from striata of rats treated with apomorphine (1 mg/kg ip) or (+)-amphetamine (1.5 mg/kg ip). In contrast, the treatment of rats with cocaine (30 mg/kg ip) failed to change the biphasic double reciprocal plot of phosphodiesterase activity versus cAMP content (I. Hanbauer, unpublished results). The molecular mechanisms whereby calmodulin accumulates in striatal membranes or is preferentially released into the cytosol are not completely understood. As discussed above these mechanisms may depend on the activation of cAMP dependent or Ca^{2+}-dependent protein kinase, which phosphorylates calmodulin binding sites in the membrane. A possible mechanism which can explain this phenomenon could involve either an increase in the synthesis rate or a decrease in the release rate of this protein. Preliminary studies on the turnover rate of calmodulin in striatal membranes indicate that long-term treatment with haloperidol prolongs the half-life of calmodulin while it increases its content (I. Hanbauer, unpublished). Calculation of the turnover rate of calmodulin shows that the turnover is not changed. Since the turnover rate is not changed, it can be inferred that the long-term treatment with haloperidol could elicit a change in those factors that facilitate calmodulin binding to the membrane.

VII. CONCLUSIONS

Calmodulin appears to be involved in the regulation of dopamine receptors by acting at the adenylate cyclase and phosphodiesterase level. Activation of striatal dopamine receptors promotes translocation of calmodulin from membranes into the cytosol. This translocation can be curtailed by prior dopamine receptor blockade. Studies on drugs which activate either directly or indirectly striatal dopamine receptors introduced evidence that the release of calmodulin elicited by persistent stimulation of dopamine receptors may be rate-limiting in the "down-regulation" of cAMP. The biochemical process thereby calmodulin is released from synaptosomal membranes appears to be catalyzed by cAMP- and/or Ca^{2+}-dependent protein kinases. Increasing evidence for the regulatory role of calmodulin in dopamine receptor function implies that studies on drug actions in relation to calmodulin translocation may be a valuable biochemical characterization reflecting differences in the pharmacological profiles of drugs acting upon dopamine receptors.

REFERENCES

Blume, A. (1978). Interaction of ligands with the opiate receptors of brain membranes. Regulation of ions and nucleotides. *Proc. Natl. Acad. Sci. U.S.A.* **75,** 1713–1717.

Brostrom, C. O., Huang, Y. C., Breckenridge, B. McL., and Wolff, D. J. (1975). Identification of a calcium-binding protein as a calcium dependent regulator of brain adenylate cyclase. *Proc. Natl. Acad. Sci. U.S.A.* **72,** 64–68.

Brostrom, C. O., Brostrom, M. A., and Wolff, D. J. (1977). Calcium-dependent adenylate cyclase from rat cerebral cortex. Reversible activation by sodium fluoride. *J. Biol. Chem.* **252,** 5677–5685.

Brostrom, M. A., Brostrom, C. O., Breckenridge, B. McL., and Wolff, D. J. (1978). Calcium-dependent regulation of brain adenylate cyclase. *Adv. Cyclic Nucleotide Res.* **9,** 85–99.

Brown, J. H., and Makman, M. H. (1972). Stimulation by dopamine of adenylate cyclase in retinal homogenates and of adenosine, 3′,5′-cyclic monophosphate formation in intact retina. *Proc. Natl. Acad. Sci. U.S.A.* **69,** 539–543.

Campbell, M. T., and Oliver, I. T. (1972). 3′,5′-cyclic-nucleotide phosphodiesterase in rat tissues. *Eur. J. Biochem.* **28,** 30–37.

Carenzi, A., Cheney, D. L., Costa, E., Guidotti, A., and Racagni, G. (1975). Action of opiates, antipsychotics, amphetamines and apomorphine on dopamine receptors in rat straitum: *In vivo* changes of 3′,5′-cyclic AMP content and acetylcholine turnover rate. *Neuropharmacology* **14,** 927–940.

Cheung, W. Y. (1970). Cyclic 3',5'-nucleotide phosphodiesterase: Demonstration of an activator. *Biochem. Biophys. Res. Commun.* **38,** 533–538.

Cheung, W. Y., Bradham, L. S., Lynch, T. J., Lin, Y. M., and Tallant, E. A. (1975). Protein activator of cyclic 3',5'-nucleotide phosphodiesterase of bovine or rat brain also activates its adenylate cyclase. *Biochem. Biophys. Res. Commun.* **66,** 1055–1062.

Clement-Cormier, Y. C., Kebabian, J. W., Petzold, G. L., and Greengard, P. (1974). Dopamine-sensitive adenylate cyclase in mammalian brain: A possible role of action of antipsychotic drugs. *Proc. Natl. Acad. Sci. U.S.A.* **71,** 1113–1117.

Clouet, D., and Rattner, M. (1970). Catecholamines biosynthesis in brains of rats treated with morphine. *Science* **168,** 854–856.

Costa, E., Gropetti, A., and Naimzada, M. K. (1972). Effects of amphetamine on the turnover rate of brain catecholamines and motor activity. *Br. J. Pharmacol.* **44,** 742–751.

Costa, E., Carenzi, A., Guidotti, A., and Revuelta, A. (1973). Narcotic analgesics and the regulation of neuronal catecholamines stored. *In* "Frontiers in Catecholamine Research" (E. Usdin, and S. Snyder, eds.), pp 1003–1010. Pergamon, Oxford.

Costa, E., Gnegy, M. E., Revuelta, A., and Uzunov, P. (1977). Regulation of dopamine-dependent adenylate cyclase by a Ca^{2+} binding protein stored in synaptic membranes. *Adv. Biochem. Pharmacol.* **16,** 403–408.

Creese, I., Burt, D. R., and Snyder, S. H. (1976). Dopamine receptor binding predicts clinical and pharmacological potencies of antischizophrenic drugs. *Science* **192,** 481–483.

Creese, I., Prosser, T., and Snyder, S. H. (1978). Dopamine receptor binding: specificity, localization and regulation by ions and guanyl nucleotides. *Life Sci.* **23,** 495–500.

Gale, K., Hong, J. S., and Guidotti, A. (1977). Presence of substance P GABA in separate striatonigoal neurons. *Brain Res.* **136,** 371–375.

Gale, K., Costa, E., Toffano, G., Hong, J. S., and Guidotti, A. (1978). Evidence for the role of nigral GABA and substance P in the haloperidol-induced activation of striatal tyrosine hydroxylase. *J. Pharmacol. Exp. Ther.* **206,** 29–37.

Gnegy, M. E., Uzunov, P., and Costa, E. (1976). Regulation of dopamine stimulation of striatal adenylate cyclase by an endogenous Ca^{2+} binding protein. *Proc. Natl. Acad. Sci. U.S.A.* **73,** 3887–3890.

Gnegy, M. E., Nathanson, J. A., and Uzunov, P. (1977a). Release of the phosphodiesterase activator by cAMP dependent ATP protein phosphotransferase from subcellular fractions of rat brain. *Biochim. Biophys. Acta* **497,** 75–85.

Gnegy, M. E., Uzunov, P., and Costa, E. (1977b). Participation of an endogenous Ca^{2+} binding protein activator in the development of drug-induced supersensitivity of striatal dopamine receptors. *J. Pharmacol. Exp. Ther.* **202,** 558–564.

Hanbauer, I., Gimble, J., Yang, H. -Y. T., and Costa, E. (1979a). The role of a Ca^{2+}-dependent protein activator purified from brain in the regulation of dopamine receptors. *In* "Peripheral Dopamine Receptors" (J. L. Imbs and J. Schwartz, eds.), pp. 289–297. Pergamon, Oxford.

Hanbauer, I., Gimble, J., and Lovenberg, W. (1979b). Changes in soluble calcium-dependent regulator following activation of dopamine receptors in rat striatal slices. *Neuropharmacology* **18,** 851–857.

Hanbauer, I., Gimble, J., Sankaran, K., and Sherard, R. (1979c). Modification of striatal cyclic nucleotide phosphodiesterase by calmodulin: Regulation by opiate and dopamine receptor activation. *Neuropharmacology* **18,** 859–864.

Hong, J. S., Yang, H. -Y. T., Racagni, G., and Costa, E. (1977). Projections of substance P containing neurons from neostriatum to substantia nigra. *Brain Res.* **122,** 541–544.

Iwatsubo, K., and Clouet, D. H. (1977). Effects of morphone and haloperidol on the electrical activity of rat nigra striatal neurons. *J. Pharmacol. Exp. Ther.* **202,** 429–436.

Kebabian, J. W., and Calne, D. B. (1979). Multiple receptors for dopamine. *Nature* (*London*) **277,** 93–96.

Kebabian, J. W., Petzold, G. L., and Greengard, P. (1972). Dopamine-sensitive adenylate cyclase in caudate nucleus of rat brain and its similarity to the dopamine receptor. *Proc. Natl. Acad. Sci. U.S.A.* **69,** 2145–2149.

Klee, C. B., and Krinks, M. H. (1978). Purification of cyclic 3′,5′-cyclic nucleotide phosphodiesterase inhibitory protein by affinity chromatography on activator protein coupled to sepharose. *Biochemistry* **17,** 120–126.

Kretsinger, R. H. (1979). The informational role of calcium in the cytosol *Adv. Cyclic Nucleotide Res.* **11,** 1–26.

Kuo, J. F., and Greengard, P. (1972). An assay method for cAMP and cGMP based on their abilities to activate cAMP and cGMP-dependent protein kinase. *Adv. Cyclic Nucleotide Res.* **2,** 41–50.

Lefkowitz, R. J., Mullikin, D., and Caron, M. G. (1976). Regulation of beta-adrenergic receptors by guanyl-5-yl-imidophosphate and other purine nucleotides. *J. Biol. Chem.* **251,** 4686–4692.

Lin, Y. M., Liu, Y. P., and Cheung, W. (1975). Cyclic 3′,5′-nucleotide phosphodiesterase: Ca^{2+}-dependent formation of bovine brain enzyme-activator complex. *FEBS Lett.* **49,** 355–360.

Lucchelli, A., Guidotti, A., and Costa, E. (1978). Striatal content of Ca^{2+} dependent regulator protein and dopaminergic receptor function. *Brain Res.* **155,** 130–135.

MacLeod, R. M. (1976). Regulation of prolactin secretion. *In* "Frontiers in Neuroendocrinology" (L. Martini and W. F. Ganong, eds.), Vol. 4, pp.169–194. Raven, New York.

Miyamoto, E., Kuo, J. F., and Greengard, P. (1969). Cyclic nucleotide dependent protein kinases. III. Purification and properties of adenosine 3′,5′-monophosphate-dependent protein kinase from bovine brain. *J. Biol. Chem.* **244,** 6395–6402.

Neer, E. J. (1978). Multiple forms of adenylate cyclase. *Adv. Cyclic Nucleotide Res.* **9,** 69–83.

Randrup, A., and Munkvad, I. (1970). Biochemical, anatomical and psychological investigations of sterotyped behavior induced by amphetamines. *In* "Amphetamines and Related Compounds" (E. Costa and S. Garattini, eds.), pp. 695–713. Raven, New York.

Revuelta, A., Uzunov, P., and Costa, E. (1976). Release of phosphodiesterase activator from particulate fractions of cerebellum and striatum by putative neurotransmitters. *Neurochem. Res.* **1,** 217–228.

Schwarcz, R., Creese, I., Coyle, J. T., and Snyder, S. H. (1978). Dopamine receptors localized on cerebral cortical afferents to rat corpus striatum. *Nature* (*London*) **271,** 766–768.

Schwartz, J. C., Pollard, H., Llorens, C., Malfray, B., Gros, C., Pradelles, P., and Dray, F. (1978). Endorphin and endorphin receptors in striatum: Relationships with dopaminergic neurons. *Adv. Biochem. Psychopharmacol.* **18,** 245–263.

Singer, S. L., and Nicholson, G. L. (1972). The fluid mosaic model of the structure of cell membranes. *Science* **175,** 720–731.

Sutherland, E. W., and Rall, T. W. J. (1957). The properties of an adenine ribonucleotide produced with cellular particles, ATP Mg^{++} and epinephrine or glucagon. *J. Am. Chem. Soc.* **79,** 3608–3613.

Teo, T. S., and Wang, J. H. (1973). Mechanism of action of a cyclic 3′,5′-monophosphate phosphodiesterase from bovine heart by calcium ions: Identification of the protein activator as a Ca^{2+} binding protein. *J. Biol. Chem.* **248,** 5950–5955.

Thompson, W. J., and Appleman, M. M. (1971). Characterization of cyclic nucleotide phosphodiesterase of rat tissues. *J. Biol. Chem.* **246,** 3145–3150.

Uzunov, P., and Weiss, B. (1972). Separation of multiple molecular forms of cyclic nucleotide phosphodiesterase in rat cerebellum by polyacrylamide gel electrophoresis. *Biochim. Biophys. Acta* **284,** 220–226.

Uzunov, P., Revuelta, A., and Costa, E. (1975). A role for the endogenous activator of 3′,5′-nucleotide phosphodiesterase in rat adrenal medulla. *Mol. Pharmacol.* **11,** 506–510.

Uzunov, P., Gnegy, M. E., Lehne, R., Revuelta, A., and Costa, E. (1976). A neurobiological role for a protein activator of cyclic nucleotide phosphodiesterase. *Adv. Biochem. Psychopharmacol.* **15,** 283–301.

Waisman, D. M., Singh, T. J., and Wang, J. H. (1978). The modulator-dependent protein kinase. *J. Biol. Chem.* **253,** 3387–3390.

Wang, J. H., and Desai, R. (1976). A brain protein and its effect on the Ca^{2+} and protein modulator activated cyclic nucleotide phosphodiesterase. *Biochem. Biophys. Res. Commun.* **72,** 926–932.

Weiss, B., and Costa, E. (1967). Adenylcyclase activity in the rat pineal gland: Effects of norepinephrine and chronic denervation. *Science* **156,** 1750–1752.

Weiss, B., and Costa, E. (1968). Selective stimulation of adenylate cyclase of rat pineal gland by pharmacologically active catecholamines. *J. Pharmacol. Exp. Ther.* **161,** 310–319.

Wolff, D. J., and Brostrom, C. O. (1979). Properties and functions of the calcium dependent regulator protein. *Adv. Cyclic Nucleotide Res.* **11,** 27–88.

Chapter 13

Immunocytochemical Localization of Calmodulin in Rat Tissues

JEFFREY F. HARPER
WAI YIU CHEUNG
ROBERT W. WALLACE
STEVEN N. LEVINE
ALTON L. STEINER

I. INTRODUCTION

Calmodulin is an important regulatory protein which mediates many of the intracellular effects of ionic calcium. Various properties of the protein

CALCIUM AND CELL FUNCTION, VOL. I

ISBN0-12-171401-2

are well described in other chapters of this volume. Biochemical studies have provided much information about the nature of the calcium–calmodulin interaction and calmodulin's ability to regulate many enzymatic activities. Relatively little is known, however, about the cellular location of this important molecule; even less is understood about the factors which regulate its cellular localization. Such regulation could provide a point of control of calmodulin-mediated cellular processes.

Our approach to the understanding of calmodulin's cellular regulation utilizes immunocytochemistry, a technique which allows microscopic detection of tissue antigens. In theory (usually true in practice) any tissue-bound antigen to which one possesses an antiserum can be localized in slices cut from frozen tissue. The presence of a particular antigen can be established for any tissue, for any cell type within a tissue, and even for subcellular structures. Information about subcellular "pools" can thus be obtained. The technique is particularly useful for the demonstration of alterations in cellular antigen distribution which occur upon perturbation of the system. Thus, the possibility of hormone regulation of the antigen's location within a cell can be explored. There are limitations to the technique. The lack of localization of an antigen does not necessarily indicate a lack of antigen and could be due to lack of antigen recognition. This would happen if antigen determinant sites are "hidden" by the tissue. Even with antigen recognition, immunofluorescence studies yield data which are qualitative rather than quantitative. Nevertheless, clues obtained with immunocytochemical techniques can be used as a guide for future biochemical investigations. The use of both immunocytochemical and biochemical techniques should provide more meaningful information than can either approach alone.

Several studies designed to localize intracellular calmodulin have been performed; results of some are described in this volume. We will primarily present data obtained in our laboratories in this chapter. Our data show that calmodulin is localized in cellular compartments such as nuclei and glycogen particles. Endocrine regulation of calmodulin localization is also demonstrated. The possible biochemical significance of these findings is discussed.

II. METHODS

A. Production of Antisera to Calmodulin

Immunofluorescence investigations require one to possess an antiserum specific for the antigen to be localized. Native calmodulin appears to be a weak antigen. Calmodulin derivatized with an average of three dinitrophenyl residues is, however, antigenic in rabbits (Wallace and Cheung, 1979). Briefly, calmodulin was purified to homogeniety from bovine brain prior to dinitrophenylation. Derivatization was accomplished by incubating calmodulin (0.18 m*M*) with 1-fluoro-2, 4-dinitrobenzene (1.2 m*M*) in 50 m*M* $NaHCO_3$ for 15 min at room temperature. Dinitrophenyl–calmodulin was separated from unreacted dinitrobenzene by Sephadex G-25 column chromatography. The number of dinitrophenyl groups attached per calmodulin was monitored at 365 nm. Antiserum to calmodulin was raised in two rabbits by injection of 0.84 mg dinitrophenyl–calmodulin in Freund's complete adjuvant on days 1, 17, 33, and 64. The immunoglobulin fraction obtained from serum collected on day 72 was used for these immunocytochemical studies.

B. Immunocytochemistry Techniques

Tissue was usually obtained from male rats (150–300 g, Charles River CD®) which had been killed by cervical dislocation. Adrenal tissue was obtained from rats that had received dexamethason (0.5 mg ip) 21, 16, and 11 hr and, where indicated, ACTH repository gel (20 U sc) 16, 11, and 1 hr prior to sacrifice. Skeletal muscle was removed from barbiturate-anesthetized animals.

Tissue slices 4 μm thick were cut from blocks of unfixed tissue frozen in Optimal Cutting Temperature (OCT) compound (Miles Laboratories) as previously described (Spruill and Steiner, 1979). Immunofluorescent staining was performed using an indirect "sandwich" technique. Briefly, tissue slices were overlaid with anti-calmodulin serum (30 min room temperature), rinsed (three times for 3 min) in phosphate-buffered saline (PBS), then incubated with fluorescein-conjugated goat anti-rabbit IgG (30 min). Tissue was mounted with 50% glycerol in PBS following a final wash. In some experiments, tissue slices were treated with *Bacillis subtilis* α-amylase (Sigma) for 30 min prior to staining, a procedure designed to hydrolyze glycogen exposed within the slices.

Photographs of all immunofluorescence studies were taken to provide a semiquantitative record of staining. Each series was exposed and printed under identical conditions. Differences of contrast in the photographs are, therefore, indicative of a difference in fluorescent light intensity observed under the microscope.

The unlabeled antibody–peroxidase technique was used to localize skeletal muscle calmodulin in some experiments. Gastrocnemius muscle clamped at its resting length was fixed in 2% Paraformaldehyde. The tissue blocks were equilibrated in sucrose and finally 10% glycerol, frozen in

OCT, and sectioned at 6 μm thickness in a cryostat. The tissue sections were dried onto glass slides, then incubated at room temperature with the following series: normal goat serum (1%, v/v, 30 min), rabbit anti-calmodulin immunoglobulin (60 min), goat anti-rabbit IgG (60 min), peroxidase–rabbit anti-peroxidase complex (60 min), diaminobenzidine (30 min), and finally 2% OsO_4 (20 min). After each treatment the slices were rinsed in PBS, and after the final treatment were dehydrated in steps to 100% xylene, then mounted in Permount.

C. Assessment of Antiserum Specificity

A potential problem with all immunofluorescence studies is the possibility that nonspecific interactions be interpreted as specific indications of the presence of an antigen. Many controls need to be performed, and the desired result is that only minimal background fluorescence should be detected. The controls performed with each tissue include substitution of the specific anti-calmodulin serum with (1) PBS in the absence of serum, (2) serum from an unimmunized rabbit, (3) antiserum previously absorbed with purified calmodulin at the proper antigen–antibody ratio (empirically derived), and (4) serum depleted of anti-calmodulin by passage over a calmodulin-Sepharose affinity column. Each control was performed with the tissues reported herein; greatly reduced staining was observed in each case. Representative photographs of the controls are included in each figure, although obviously not all of the dark photographs are reproduced. We have found that generally little background staining is due to the anti-calmodulin serum. Some of the faint background fluorescence can be accounted for by the nonspecific staining of the fluorescien-conjugated goat anti-rabbit IgG (Miles Laboratories) regardless of the presence of first antiserum. We did not purify the commercially prepared fluorescein-conjugated antiserum.

The anti-calmodulin serum was tested for specificity as an additional control. Troponin C, a calcium-binding protein with approximately 70% homology to calmodulin (Watterson *et al.*, 1976), showed no significant interaction with the anti-calmodulin serum assessed either by radioimmunoassay (Wallace and Cheung, 1979) or upon attempted antibody absorption with troponin C (not shown). Immunofluorescence for tissue calmodulin is thus not interfered with by the presence of troponin C.

III. LOCALIZATION OF CALMODULIN

A. Immunofluorescence Techniques

1. General Considerations

The antiserum used in these studies has been well characterized by radioimmunoassay (Wallace and Cheung, 1979, and unpublished experiments). The antiserum, although raised against dinitrophenyl-calmodulin, recognizes native calmodulin with strong avidity. The antibody is also highly selective, exhibiting little if any cross-reactivity with troponin C. Sensitivity and selectivity are necessary but not sufficient criteria for specificity of immunocytochemical staining. Control experiments performed as described in Section II, C demonstrate that our antiserum indeed localizes tissue calmodulin specifically.

The results obtained with immunocytochemistry depend upon tissue preparation as much as on antiserum specificity. Most experiments were performed using unfixed frozen sections of rat tissue which were thawed and dried onto glass slides. This method works well with antigens bound to cellular structures, but soluble antigens are probably not retained. They are removed during the necessary washing steps of the immunofluorescent staining procedure (Sternberger, 1979). We did not estimate the percentage of calmodulin that was retained within the tissue. Presumably this percentage varies in the different subcellular compartments. For instance, glycogen particles contain calmodulin which can be released by α-amylase but which normally survives immunofluorescent staining (Sections III, A, 3 and 4). Little calmodulin seems to be removed from that compartment. However, retention of calmodulin in the nucleus and cytoplasm may be dependent upon specific binding of calmodulin to receptor proteins in some cellular structure. If this is true, the calmodulin localized in those sites would consist of a specific subclass of total cellular calmodulin, e.g., that associated with receptor proteins which are not removed upon washing. Nearly all known interactions between calmodulin and its receptor proteins are dependent upon the presence of Ca^{2+}. It is possible that immunofluorescence localizes only Ca^{2+}-replete calmodulin, which may be the fraction responsible for modulation of enzyme activities. The localization of calmodulin by immunofluorescence may go beyond the mere cataloging of intracellular location of this important regulatory protein; it may indicate its cellular sites of action.

2. Localization of Calmodulin in Intestine

a. Mucosa. Duodenal tissue cut in cross section exhibits intense calmodulin staining in several areas. A portion of one villus is shown in each

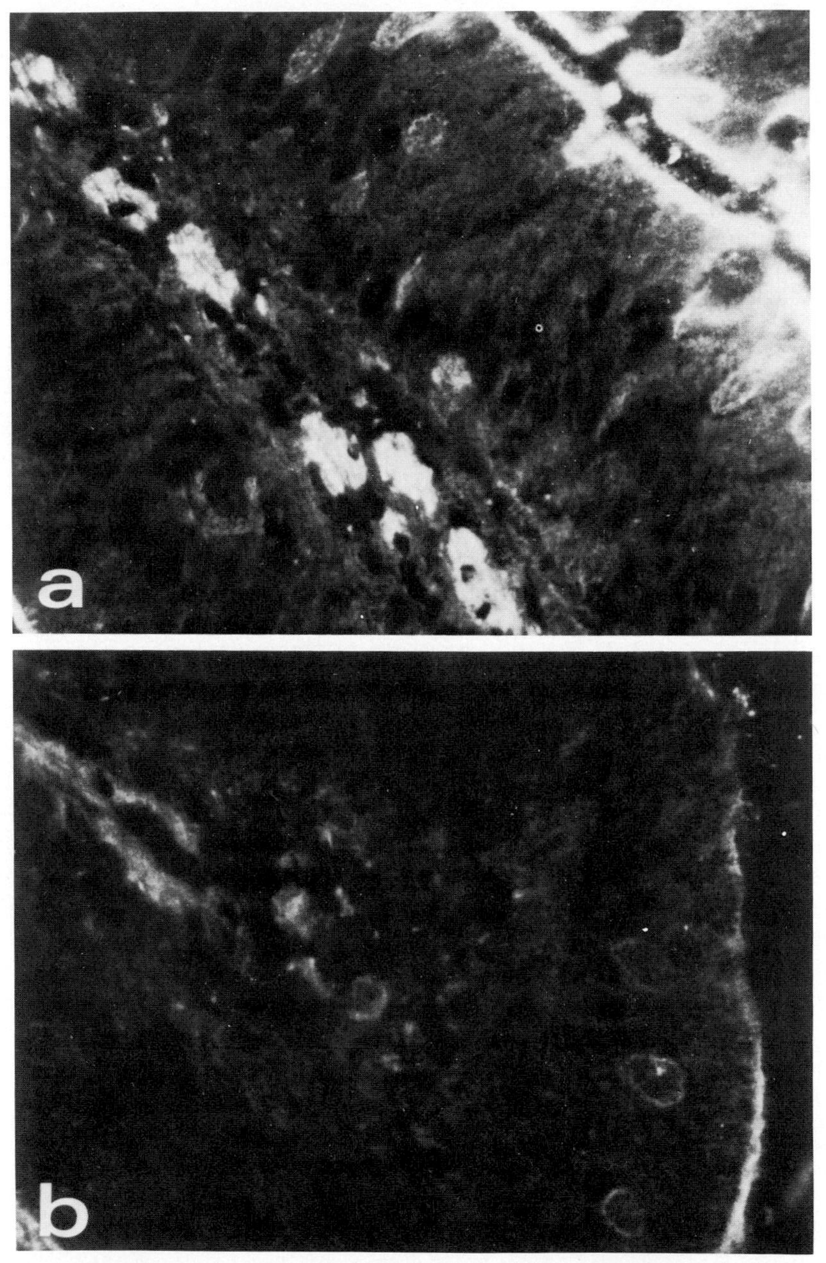

Fig. 1. Localization of calmodulin in intestine. Slices of rat duodenum were cut in cross section and stained for calmodulin using indirect immunofluorescence. (a) Tissue stained with anti-calmodulin serum; (b) tissue stained with anticalmodulin previously absorbed with calmodulin. $\times 900$.

panel of Fig. 1. The brush border [seen in the upper right corner of panel (a) and along the right side of the control preparation in panel (b)] exhibits intense specific staining for calmodulin. Some discrete areas of the lamina propria, found in the center of each photograph, are also brightly stained. The lamina propria is a structure in the core of each villus which contains a number of cell types; the stained cells have not been identified. Some immunofluorescence is found in the cytoplasm of the columnar absorbing cells, which span the distance from the lamina propria to the edge of the villus. Good specificity of staining at all sites is observed (Fig. 1b).

b. Smooth Muscle. Calmodulin's presence in smooth muscle is indicated with rather intense immunofluorescent staining (Fig. 2). The entire cytoplasm of the smooth muscle cell exhibits as much specific immunofluorescence for calmodulin as is found in the brush border of the intestinal mucosa. Staining specificity for calmodulin in smooth muscle is demonstrated in Fig. 2b using absorbed antiserum.

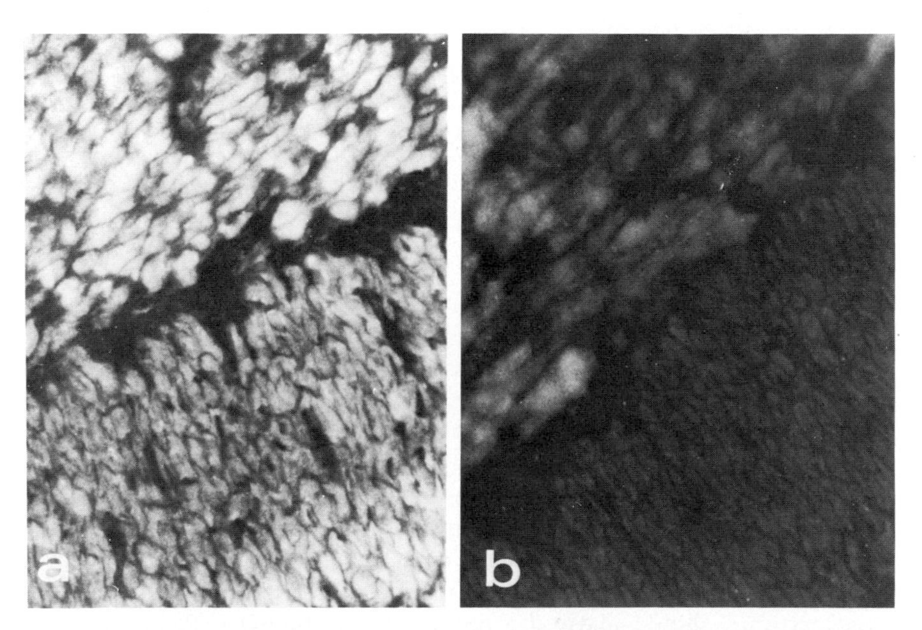

Fig. 2. Calmodulin in smooth muscle. Slices of rat duodenum were cut in cross section and stained for calmodulin; longitudinal and circular smooth muscle fields were chosen for photography. (a) Tissue stained with anti-calmodulin serum; (b) anti-calmodulin serum absorbed with calmodulin. ×600.

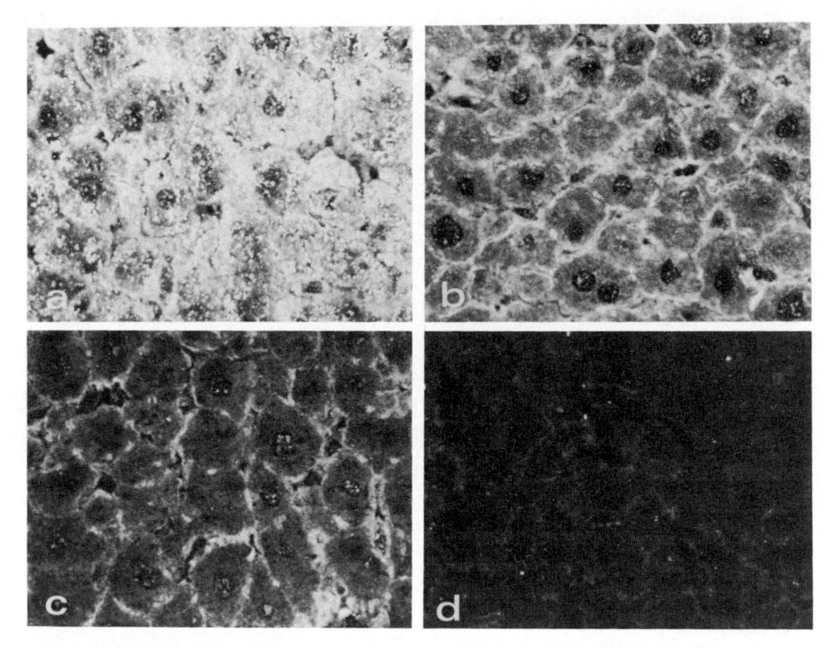

Fig. 3. Localization of calmodulin in rat liver. Rat liver slices (4 μm thick) were stained for calmodulin. (a) Liver taken from an untreated, fed animal; (b) liver from an animal fasted overnight but given water *ad libitum;* (c) liver taken from a fed rat; prior to staining the slice was treated for 30 min with α-amylase (5 units in 50 μl); (d) liver slice stained with antiserum absorbed with calmodulin. ×415. (Reprinted from Harper *et al.* 1980, by permission.)

3. Liver

Calmodulin is localized in several hepatocyte compartments (Fig. 3). Nuclear elements are stained in a reticular pattern of small irregularly shaped areas. The cell membrane also exhibits specific calmodulin staining. The most noticable staining however, is on "dots" within the cytoplasm. The granular cytoplasmic pattern is probably associated with glycogen particles; this conclusion is based upon two types of experiments (Fig. 3b and c). Glycogen content of livers from fasted rats is quite low, and, as shown in Fig. 3b, no granular cytoplasmic staining is found in livers from these animals. Glycogen can also be removed *in vitro* by incubation of tissue slices with α-amylase (Lillie, 1947). Disruption of glycogen particles should allow the previously contained calmodulin to be washed from the tissue slice during staining procedures, or become diffused throughout the cytoplasm. Figure 3 shows that the removal of glycogen by these techniques specifically removes glycogen particle calmodulin and does not disturb immunofluorescence in cell membranes or nuclei. Some nongranule cytoplasmic staining is also apparent. These findings indicate that only one of several calmodulin pools is associated with glycogen granules. Figure 3d demonstrates that the immunofluorescent staining in liver slices is blocked using antiserum to which calmodulin had been absorbed.

4. *Skeletal Muscle*

The rat gastrocnemius muscle contains calmodulin localized in a highly ordered fashion (Fig. 4). Intense staining of one striation band is apparent, as is a faint longitudinal localization. The longitudinal staining apparently demonstrates intermyofibrillar calmodulin possibly associated with the sarcoplasmic reticulum.

The location of calmodulin associated with the band structure was investigated using both immunofluorescence and unlabeled antibody techniques. Staining with fluorescent label can be used to demonstrate band location. The tissue is photographed with both emitted fluorescent light and polarized light; matching the band patterns of corresponding halves of the same field in each photograph suggest that calmodulin is localized predominantly at the I band, since the bright immunofluorescent staining matches the dark (isotropic) bands in the polarization micrograph (Fig. 5). The localization of calmodulin at the I band is corroborated by the immunoperoxidase technique, which allows band location without resorting to the photographic matching necessary with immunofluorescence. The peroxidase label on the antiserum is localized by oxidation of diaminoben-

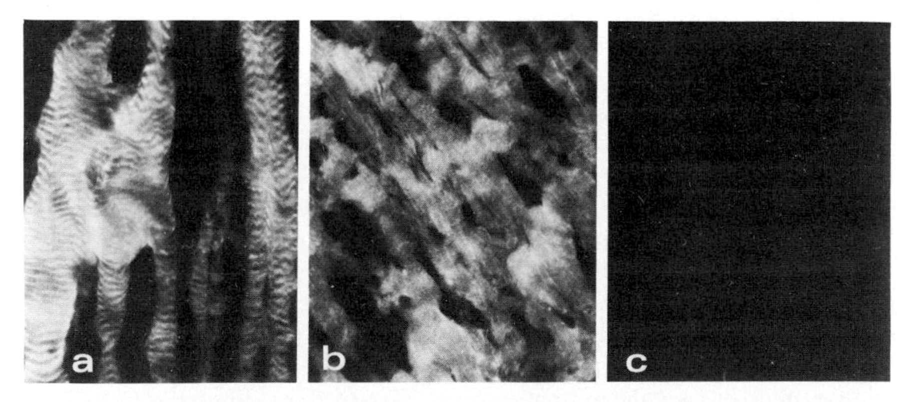

Fig. 4. Localization of calmodulin in rat skeletal muscle. Slices taken from the rat gastrocnemious muscle were stained for calmodulin. (a) Untreated slice; (b) slice treated for 30 min with α-amylase (500 units in 50 μl); (c) untreated tissue slice stained with antiserum absorbed with calmodulin prior to use. $\times$ 395. (Reprinted from Harper *et al.*, 1980, by permission.)

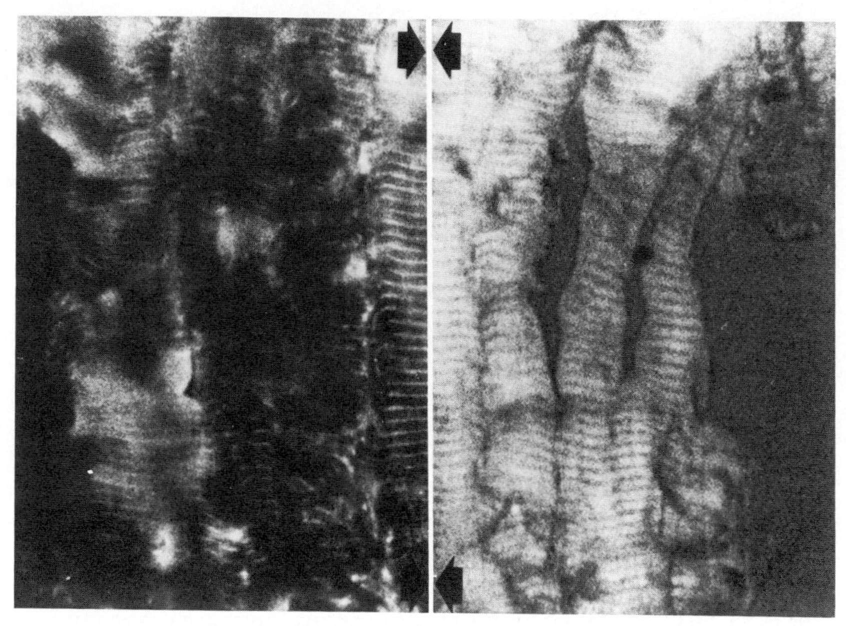

Fig. 5. Demonstration of calmodulin at I bands in skeletal muscle. A slice of gastrocnemius muscle was stained for calmodulin by indirect immunofluorescence. One microsocpic field was photographed with emitted fluorescent light and again with polarized light. The figure shows a pastiche made from the two photographs. The left side was exposed with fluorescent light, and the right side with a polarized source. The individual photographs were cut along one fiber and the band patterns matched (arrows). ×865.

zidene, the product of which is a dark precipitate. Polarized light microscopy can then be performed to reveal that the dark precipitate is aligned with the dark polarizable I bands (not shown).

The relative distribution of calmodulin on the striations can be estimated with a densitometric tracing of a photograph of the immunoperoxidase staining pattern (Fig. 6). The area of the photograph scanned (shown bound by arrows) is aligned with the densitometer trace. A 15 μm grid was used for the baseline; the specific trace shows spikes corresponding with the arrow points. Calmodulin, localized on I bands, exists with a periodicity of 2.4 μm (resting length). The entire I band, including the Z line, stains for calmodulin. Some calmodulin is localized on A bands, but the relative peak sizes indicate that at least 75% of the calmodulin localized on skeletal muscle is found associated with I bands.

The terminal cisternae of the sarcoplasmic reticulum and glycogen are both found along I bands (Porter and Palade, 1957; Mancini, 1948). Since calmodulin was found to be associated with glycogen in liver, the possibility that calmodulin is associated with glycogen in muscle was tested using the α-amylase preincubation technique. This treatment greatly reduced staining on the I band (Fig. 4b). The α-amylase treatment appears specific for I band and presumably glycogen-associated calmodulin; exposure to α-amylase did not alter intermyofibrillar staining. Cyclic GMP, localized by immunofluorescence to A bands (Ong and Steiner, 1977), was similarly unaffected by preincubation of tissue with α-amylase (not shown).

5. Adrenal Cortex

Immunofluorescence reveals the presence of calmodulin in adrenal cortex cells (Fig. 7). Figure 7a shows the pattern seen in adrenals taken from

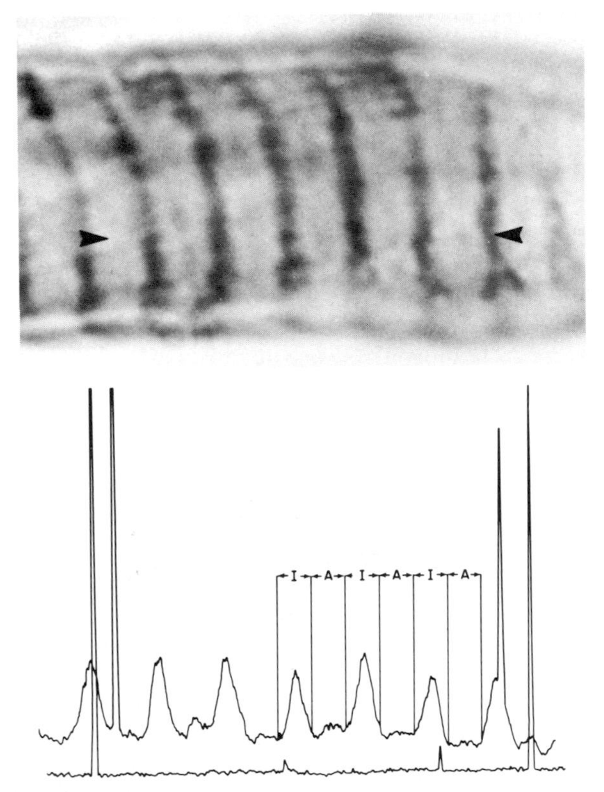

Fig. 6. Quantitation of staining on skeletal muscle. Skeletal muscle fixed at its resting length was stained for calmodulin by the immunoperoxidase technique. A densitometer trace of the muscle band pattern is shown below the corresponding photograph of the area analyzed. The baseline spikes are calibrated at 15 μm, while the spikes in the trace correspond to the points marked by the arrows in the photograph. The areas of I and A bands are shown. Band areas were defined by polarization microscopy. ×3600. (Reprinted from Harper *et al.*, 1980, by permission.)

rats which had been treated with dexamethasone to inhibit ACTH secretion. Some immunofluorescence is found in nuclei and the cytoplasm; no specific structural pattern is apparent in the cytoplasm. Dark vacuoles, which are probably cholesterol droplets, are seen in the cytoplasm. The administration of ACTH to rats previously treated with dexamethasone produces profound changes in the pattern of immunofluorescence. Figure 7b shows the effects of ACTH gel administered during the final 16 hr of treatment. Intense immunofluorescence is found in nuclei, appearing as bright discrete areas each containing subnuclear (perhaps nucleolar) patches. Cytoplasmic calmodulin staining appears to be relatively unchanged by the hormone treatment. The hormone also reduces the size of the vacular droplets, probably as a result of increased steroid hormone production from the stored cholesterol. Control incubations show that the level of nonspecific staining in adrenal tissue is negligible (Fig. 7c).

B. Biochemical Localization

Calmodulin was originally discovered as an activator of cyclic nucleotide phosphodiesterase (Cheung, 1967, 1970). It has since been found to regulate a variety of soluble and membrane-bound enzymes, and indeed has been found in both compartments (Smoake *et al.*, 1974). Gnegy *et al.* (1977) determined that calmodulin was present in both microsomes and cytosol of rat brain homogenized in sucrose; calmodulin detected in nuclear and mitochondrial fractions could be accounted for by contamination from the other fractions. Beale *et al.* (1977) homogenized Sertoli cell-

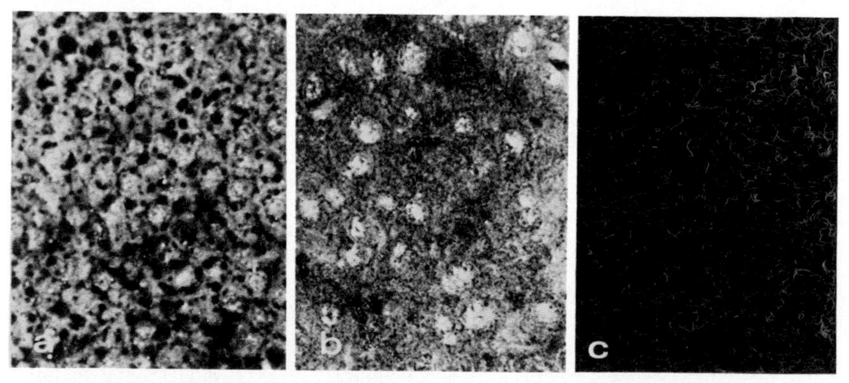

Fig. 7. Effect of ACTH on the distribution of calmodulin in adrenal cortex. Rat adrenals were removed after drug injections defined in text (Section II). (a) Adrenal tissue removed from a dexamethasone-treated rat; (b) tissue removed from a rat which had been treated with both dexamethasone and ACTH; (c) tissue stained with serum from an unimmunized rabbit. ×380. (Reprinted from Harper *et al.*, 1980, by permission.)

enriched testis in saline containing 1 m*M* EDTA. They noted that calmodulin was localized largely in the same manner as triose phosphate isomerase, a soluble enzyme, with about 9% of the cellular calmodulin associated with the nucleus.

A major difference between these two sets of localization data lies in the absence of microsomal calmodulin observed by Beale *et al.* (1977). Sertoli cells may, in fact, have no microsomal calmodulin. The respective homogenization media used provide an alternative explanation for the lack of microsomal calmodulin. Beale *et al.* (1977) homogenized testicular tissue in a buffer containing EDTA, which has since been shown to affect calmodulin distribution in other tissues. Teshima and Kakiuchi (1978) and Vandermeers *et al.* (1978) both found that membranes obtained with a homogenization medium containing Ca^{2+} had nearly twice as much calmodulin as membranes homogenized in the presence of EGTA, with reciprocal changes in supernatant calmodulin. Vandermeers *et al.* (1978) showed that binding of [^{125}I]calmodulin to guinea pig brain membranes occurred in the presence of Ca^{2+}, but was rapidly and completely reversible upon chelation of Ca^{2+} by EGTA. Both groups found that calmodulin bound to a small number of saturable receptor sites on the membranes.

The Ca^{2+}-dependent binding of calmodulin to membranes may be a mechanism which achieves regulation of calmodulin action. Alterations of the intracellular Ca^{2+} concentration could regulate the exchange of membrane-associated calmodulin into the cytoplasm, affecting the calmodulin concentration in both compartments. This view is supported by several findings. Vandermeers *et al.* (1978) observed the same Ca^{2+} dependence for binding of [^{125}I]calmodulin to brain membranes and stimulation of adenylate cyclase by native calmodulin.

Farrance and Vincenzi (1977) showed that erythrocyte membranes prepared in the presence of EGTA contained less Ca^{2+}/Mg^{2+}-ATPase activity than did membranes prepared in the presence of Ca^{2+}. Reduced Ca^{2+}/Mg^{2+}-ATPase activity was presumably due to removal of membrane calmodulin, since readdition of Ca^{2+} and calmodulin to the EGTA-treated membranes restored ATPase activity. Another potential mechanism for regulation of calmodulin's intracellular location has been reported by Gnegy *et al.* (1976). They showed that calmodulin could be released from rat striatum membranes by the specific actions of cyclic AMP-dependent protein kinase. Neither calmodulin nor adenylate cyclase was phosphorylated during this process, but one specific fraction of striated membrane protein did accept phosphate from [γ-^{32}P]ATP. The released calmodulin came predominately from the microsomal subfractions containing synaptic membranes and vesicles (Gnegy *et al.*, 1977). Phosphorylation-mediated calmodulin release from these membranes appears to correlate well with the reduction of dopamine- and fluoride-stimulated adenylate cyclase activities (Gnegy *et al.*, 1976). Basal adenylate cyclase activity was not affected by calmodulin release. Recent evidence shows that the calmodulin released by cyclic AMP-dependent protein kinase is derived mostly from the Ca^{2+}-dependent pool. Some calmodulin not released by EGTA can still be released by protein kinase, however (Lau and Gnegy, 1979). This body of work is reviewed by Hanbauer and Costa (this volume, Chapter 12).

C. Correlation of Immunocytochemical and Biochemical Localization Studies

Assignment of biochemical functions to calmodulin localized by immunofluorescence is not easy. Preincubation of tissue slices with α-amylase affords us a specific method of assignment for the calmodulin pool within glycogen granules. Exposure of tissue slices to α-amylase does not remove all cytoplasmic staining, nor does it alter staining in liver membranes and nuclei. The cytoplasmic calmodulin remaining after α-amylase treatment could be responsible for regulation of any of the known enzymes modulated by this protein. Some calmodulin localized by immunofluorescence may be involved in as yet unrecognized functions. The role of calmodulin in cell nuclei, especially that in the adrenal cortex under control of ACTH, cannot at present be explained biochemically.

IV. QUESTIONS RAISED BY IMMUNOCYTOCHEMICAL LOCALIZATION

A. Hormonal Regulation of Calmodulin Distribution

Our experiments show that hormones regulate the amount of localized nuclear calmodulin in the adrenal cortex. Experiments in uterus suggest that hormones also affect the distribution of calmodulin in those cells (Harper *et al.*, in press). Our experiments do not indicate the percentage of total cellular calmodulin detected by immunofluorescence. Soluble calmodulin may be removed from the tissue slice during the various rinse procedures, so that only a fraction of the cellular calmodulin is visualized. The localized calmodulin may represent that fraction bound to cellular structures, which implies that Ca^{2+}-replete calmodulin might be preferentially localized (Section III, A, 1). Immunocytochemical localization of calmodulin may thus prove to be a sensitive probe of biologically active calcium ion as well as the cellular site of calmodulin action.

The finding that ACTH alters nuclear calmodulin localization raises several questions. Is the change due to increased "active" calmodulin in nuclei, with or without increased total nuclear or cellular calmodulin? Is the increased pool of active calmodulin in turn due to increased nuclear Ca^{2+}, release of calmodulin from nuclear membrane into nucleoplasm, or even translocation into nucleoplasm from cytoplasm? At a molecular level, what is the function of nuclear calmodulin? Previous studies raise the possibility that it may participate in the regulation of RNA synthesis. It could act through a novel mechanism, or through activation of enzymes such as a protein kinase or cyclic nucleotide phosphodiesterase which would in turn produce the final effects. Fuhrman and Gill (1974) have found that ACTH produces incresed nucleolar RNA polymerase I activity 16 hr after injection into guinea pigs; we show increased calmodulin immunofluorescence in what appears to be nucleoli at that time. Johnson and Hadden (1975) noted that lymphocyte RNA polymerase I is fully activated *in vitro* upon addition of Ca^{2+} and cyclic GMP; their findings have not yet been confirmed in other tissues. The answers to the above questions, raised because of observations made with immunocytochemistry, require biochemical investigations. Nevertheless, our experiments indicate the potential advances that can be accomplished through the combined studies of biochemistry and immunocytochemistry.

B. Calmodulin and Cyclic GMP

Several points of interaction are known to exist between calmodulin and cyclic GMP, at least *in vitro*. The degradation of cyclic GMP is catalyzed by a calmodulin-regulated low K_m cyclic GMP phosphodiesterase (Kakiuchi *et al.*, 1973). Soluble guanylate cyclase, an enzyme which catalyzes cyclic GMP synthesis, is activated by high concentrations of calcium (Hardman *et al.*, 1971) and may be calmodulin independent: Particulate guanylate cyclase from skeletal muscle is inhibited by calcium at concentrations low enough to suggest interaction with calmodulin (Levine *et al.*, 1979). Intracellular Ca^{2+} concentration may be regulated by calmodulin-activated calcium transport (Hinds *et al.*, 1978; Katz and Remtulla, 1978). Other interaction schemes are also possible. Glycogen synthase, for instance, is known to be phosphorylated by two kinases, one regulated by calmodulin (phosphorylase kinase) and another by cyclic AMP (cyclic AMP-dependent protein kinase) (Rylatt *et al.*, 1979; Srivastava *et al.*, 1979; Huijing and Larner, 1966). Similar dual regulation of protein phosphorylation by calmodulin-dependent and cyclic GMP-dependent kinases may likewise exist.

Presumptive evidence for interactions between cyclic GMP and calmodulin is provided by our immunofluorescence studies. A number of tissues have nearly identical staining patterns whether stained with antisera specific for calmodulin or cyclic GMP. Both antisera are found in adrenal cortex cell cytoplasm (diffuse) and nuclei (nucleolar-like) (Whitley *et al.*, 1975; Fig. 7). Significantly, ACTH stimulates a large increase in nuclear staining for both antigens (Spruill *et al.*, unpublished results n.d.; Fig. 7b). Both antisera stain hepatocyte plasma membranes and nuclei (reticular pattern) (Ong *et al.*, 1975; Fig. 1), but only the anti-calmodulin serum stains glycogen granules. Both antisera decorate skeletal muscle bands; again, only calmodulin is associated with glycogen. Thus, different bands are stained preferentially by the antisera (Koide *et al.*, n.d.; Fig. 5). Both antisera stain the intestinal brush border, lamina propria, and smooth muscle cytoplasm (Ong *et al.*, 1975, and Figs. 1 and 2), although anti-calmodulin does not stain intestinal cell nuclei. The striking colocalization of calmodulin and cyclic GMP in nearly every subcellular site of each tissue studied is suggestive of interrelated functions. This speculation, of course, has not been proved. The questions raised by this observation are perhaps answerable with biochemical techniques.

V. CONCLUDING REMARKS

Our findings obtained with indirect immunofluorescence microscopy demonstrate that calmodulin is associated with glycogen particles and membranes; this was predictable from previous biochemical studies, though calmodulin regulation of liver glycogen metabolism had not been reported. Relatively large amounts of calmodulin staining in nuclei is demonstrable, and in adrenal cortex the localization of this nuclear calmodulin is dependent upon the hormonal activity of the animal. This is a novel observation, unanticipated from previous work with undividing cells. These results suggest that hormones may act not only through the regulation of intracellular calcium ions but also through regulation of the intracellular location of calmodulin.

ACKNOWLEDGMENTS

We thank Shu-Hui Ong and Hui-Lan Huang for their excellent technical assistance, Dr. T. C. Vanaman and G. Jamieson for assistance with affinity purification of calmodulin, and Celeste Layton for her aid in the preparation of this manuscript. These studies, which were benefited by helpful discussions with Dr. O. M. Rosen during the initial stages, were supported by U.S. Public Health Service Awards AM05992 (JFH), AM07129 (SNL), AM17439 (ALS), AM05789 (RWW) and NS08059 (WYC), and by ALSAC (WYC).

REFERENCES

Beale, E. G., Dedman, J. R., and Means, A. R. (1977). Isolation and regulation of the protein kinase inhibitor and the calcium-dependent cyclic nucleotide phosphodiesterase regulator in the Sertoli cell-enriched testis. *Endocrinology* **101,** 1621–1634.

Cheung, W. Y. (1967). Cyclic 3′,5′-nucleotide phosphodiesterase: Pronounced stimulation by snake venom. *Biochem. Biophys. Res. Commun.* **29,** 478–482.

Cheung, W. Y. (1970). Cyclic 3′,5′-nucleotide phosphodiesterase. Demonstration of an activator. *Biochem. Biophys. Res. Commun.* **38,** 533–538.

Farrance, M. L., and Vincenzi, F. F. (1977). Enhancement of (Ca^{2+} + Mg^{2+})-ATPase activity of human erythrocyte membranes by hemolysis in isosmotic imidazole buffer II. Dependence on calcium and a cytoplasmic activator. *Biochim. Biophys. Acta* **471,** 59–66.

Fuhrman, S. A., and Gill, G. N. (1974). Hormonal control of adrenal RNA polymerase activities. *Endocrinology* **94,** 691–700.

Gnegy, M. E., Uzunov, P., and Costa, E. (1976). Regulation of dopamine stimulation of striatal adenylate cyclase by an endogenous Ca^{++}-binding protein. *Proc. Natl. Acad. Sci. U.S.A.* **73,** 3887–3890.

Gnegy, M. E., Nathanson, J. A., and Uzunov, P. (1977). Release of the phosphodiesterase activator by cyclic AMP-dependent ATP: Protein phosphotransferase from subcellular fractions of rat brain. *Biochim. Biophys. Acta* **497,** 75–85.

Hardman, J. G., Beavo, J. A., Gray, J. P., Chrisman, T. D., Patterson, W. D., and Sutherland, E. W. (1971). The formation and metabolism of cyclic GMP. *Ann. N.Y. Acad. Sci.* **185,** 27–35.

Harper, J. F., Cheung, W. Y., Wallace, R. W., Huang, H.-L., Levine, S. N., and Steiner, A. L. (1980). Localization of calmodulin in rat tissues. *Proc. Natl. Acad. Sci. U.S.A.* **77,** 366–370.

Harper, J. F., Flandroy, L., Wallace, R. W., Cheung, W. Y., and Steiner, A. L. (in press). ACTH- and estradiol-stimulated changes in the immunocytochemical localization of cyclic nucleotides, protein kinases, and calmodulin. *Adv. Cyclic Nucleotide Res.* **13,** in press.

Hinds, T. R., Larsen, F. L., and Vincenzi, F. F. (1978). Plasma membrane Ca^{2+} transport: Stimulation by soluble proteins. *Biochem. Biophys. Res. Commun.* **81,** 455–461.

Huijing, F., and Larner, J. (1966). On the effect of adenosine 3′,5′-cyclophosphate on the kinase of UDPG: α-1,4-glycan α-4-glycosyl transferase. *Biochem. Biophys. Res. Commun.* **23,** 259–263.

Johnson, L. D., and Hadden, J. W. (1975). Cyclic GMP and lymphocyte proliferation: Effects on DNA-dpendent RNA polymerase I and II activities. *Biochem. Biophys. Res. Commun.* **66,** 1498–1505.

Kakiuchi, S., Yamazaki, R., Teshima, Y., and Uenishi, K. (1973). Regulation of nucleoside cyclic 3′:5′-monophosphate phosphodiesterase activity form rat brain by a modulator and Ca^{2+}. *Proc. Natl. Acad. Sci. U.S.A.* **70,** 3526–3530.

Katz, S., and Remtulla, M. A. (1978). Phosphodiesterase protein activator stimulates calcium transport in cardiac microsomal preparations enriched in sarcoplasmic reticulum. *Biochem. Biophys. Res. Commun.* **83,** 1373–1379.

Koide, Y., Spruill, W. A., Kapoor, C. L., Huang, H., Levine, S. N.. Ong, S. H., Beavo, J. A., and Steiner, A. L. (n.d.). Immunocytochemical localization of components of cAMP-dependent protein kinase in rat skeletal muscle, liver and adrenal. (Unpublished results.)

Lau, Y. S., and Gnegy, M. E. (1979). Calmodulin release from membrane preparations of rat striatum. *Pharmacologist* **21,** 253 (abstr.).

Levine, S. N., Steiner, A. L., Earp, H. S., and Meissner, G. (1979). Particulate guanylate cyclase of skeletal muscle. Effects of Ca^{2+} and other divalent cations on enzyme activity. *Biochim. Biophys. Acta* **566,** 171–182.

Lillie, R. D. (1947). Malt diastase and ptyalin in place of saliva in the identification of glycogen. *Stain Technol.* **22,** 67–70.

Mancini, R. E. (1948). Histochemical study of glycogen in tissues. *Anat. Rec.* **101,** 148–156.

Ong, S. H., and Steiner, A. L. (1977). Localization of cyclic GMP and cyclic AMP in cardiac and skeletal muscle: Immunocytochemical demonstration. *Science* **195,** 183–185.

Ong, S. H., Whitley, T. H., Stowe, N. W., and Steiner, A. L. (1975). Immunohistochemical localization of 3′:5′-cyclic AMP and 3′:5′-cyclic GMP in rat liver, intestine and testis. *Proc. Natl. Acad. Sci. U.S.A.* **72,** 2022–2026.

Porter, K. R., and Palade, G. E. (1957). Studies on the endoplasmic reticulum. III. Its form and distribution in striated muscle cells. *J. Biophys. Biochem. Cytol.* **3,** 269–300.

Rylatt, D. B., Embi, N., and Cohen, P. (1979). Glycogen synthase kinase-2 from rabbit skeletal muscle is activated by the calcium-dependent regulator protein. *FEBS Lett.* **98,** 76–80.

Smoake, J. A., Song, S.-Y., and Cheung, W. Y. (1974). Cyclic 3′,5′-nucleotide phosphodiesterase. Distribution of the enzyme and its protein activator in mammalian tissues and cells. *Biochim. Biophys. Acta* **341,** 402–411.

Spruill, W. A., and Steiner, A. L. (1979). Cyclic nucleotide and protein kinase immunocytochemistry. *Adv. Cyclic Nucleotide Res.* **10,** 169–186.

Spruill, W. A., Koide, Y., Huang, H., Levine, S. N., Ong, S., Steiner, A. L., and Beavo, J. A. (n.d.). Cyclic GMP dependent protein kinases: Characterization of antisera and immunocytochemical localization in several rat tissues. (Unpublished results).

Srivastava, A. K., Waisman, D. M., Brostrom, C. O., and Soderling, T. R. (1979). Stimulation of glycogen synthase phosphorylation by calcium-dependent regulator protein. *J. Biol. Chem.* **254,** 583–586.

Sternberger, L. A. (1979): "Immunocytochemistry," 2nd ed. Wiley, New York.

Teshima, Y., and Kakiuchi, S. (1978). Membrane-bound forms of Ca^{2+}-dependent protein modulator: Ca^{2+}-dependent and independent binding of modulator protein to the particulate fraction from brain. *J. Cyclic Nucleotide Res.* **4,** 219–231.

Vandermeers, A., Robberecht, P., Vandermeers-Piret, M.-C., Pathe, J., and Christophe, J. (1978). Specific binding of the calcium-dependent regulator protein to brain membranes from the guinea pig. *Biochem. Biophys. Res. Commun.* **84,** 1076–1081.

Wallace, R. W., and Cheung, W. Y. (1979). Calmodulin: Production of an antibody in rabbit and development of a radioimmunoassay. *J. Biol. Chem.* **254,** 6564–6571.

Watterson, D. M., Harrelson, W. G., Keller, P. M., Sharief, F., and Vanaman, T. C. (1976). Structural similarities between the Ca^{2+}-dependent regulatory proteins of 3′,5′-cyclic nucleotide phosphodiesterase and actomyosin ATPase. *J. Biol. Chem.* **251,** 4501–4513.

Whitley, T. H., Stowe, N. W., Ong, S. H., Ney, R. L., and Steiner, A. L. (1975). Control and localization of rat adrenal cyclic guanosine 3′,5′-monophosphate. Comparison with adrenal cyclic adenosine 3′,5′-monophosphate. *J. Clin. Invest.* **56,** 146–154.

Chapter 14

Immunocytochemical Studies of the Localization of Calmodulin and CaM-BP_{80} in Brain

JOHN G. WOOD
ROBERT W. WALLACE
WAI YIU CHEUNG

I. INTRODUCTION

Calmodulin is a calcium-binding protein with very broad tissue and species distribution. The protein regulates a variety of brain enzymes in a calcium-dependent manner, including cyclic nucleotide phosphodiesterase (Cheung, 1967, 1970; Kakiuchi and Yamazaki, 1970), adenylate cyclase (Brostrom *et al.*, 1976; Lynch *et al.*, 1976, 1977), synaptosomal mem-

CALCIUM AND CELL FUNCTION, VOL. I

ISBN 0-12-171401-2

brane Ca^{2+}-ATPase (Soube *et al.*, 1979), myosin light chain kinase (Dabrowska and Hartshorne, 1978), and enzyme systems which may be responsible for phosphorylating synaptic membranes (Schulman and Greengard, 1978a,b). Further, the brain contains several calmodulin-binding proteins whose functions have not been identified; they may be additional calmodulin-regulated enzymes. One of these proteins has been extensively characterized. It has a molecular weight of 80,000 and has been referred to as modulator-binding protein (Wang and Desai, 1977; Klee and Krinks, 1978), inhibitor protein, or CaM-BP_{80} (Wallace *et al.*, 1978, 1979).

Since calmodulin appears to influence several enzymes associated with neurotransmission, it is important to establish its localization in the brain. One approach is to use immunocytochemical methodology. In this communication we describe the distribution of calmodulin and one of its binding proteins (CaM-BP_{80}) in various regions of rat brain.

II. IMMUNOCYTOCHEMICAL METHODS

A. Antisera

The required first step in immunocytochemistry is the preparation of monospecific antisera to highly purified antigens. Antisera to calmodulin and CaM-BP_{80} have been prepared in rabbits and their specificities established (Wallace and Cheung, 1979; Wallace *et al.*, 1979, 1980).

B. Fixation

To localize proteins by immunocytochemical methods it is necessary to define conditions of fixation for the tissue to achieve adequate morphological preservation with a minimal loss of antigenic activity of the tissue protein. We find that fixatives containing very low levels of glutaraldehyde (0.1–0.3%) generally provide adequate morphology with very little loss of antigenicity (Wood *et al.*, 1976). A low level of glutaraldehyde appears sufficient to cross-link tissue proteins (Wood, 1973) without forming large heteropolymers leading to altered conformational and antigenic properties. In our studies, the rat is perfused through the heart for 15 min with a fixative containing 4.0% paraformaldehyde and 0.1% glutaraldehyde in 0.12 *M* Millonig's phosphate buffer, pH 7.2 (Millonig, 1961). After perfusion, the brain is carefully removed and stored overnight at 4°C in phosphate buffered 4.0% paraformaldehyde without glutaraldehyde. This procedure improves the morphological quality of the tissue without causing significant loss of antigenic reactivity of calmodulin or CaM-BP_{80}.

C. Preparation of Tissue

Brain tissue which has been stored overnight in paraformaldehyde is rinsed briefly in phosphate-buffered saline (PBS), pH 7.2. Appropriate regions are dissected into 1–3 mm cubes with a razor blade using a dissecting microscope for orientation. The cubes of tissue are sectioned with a Sorvall Tissue Chopper, an Oxford Vibratome, or an International Cryostat. The best combination of thinness and flatness of slice are obtained with the Cryostat, but the freezing and thawing steps necessary with this procedure produce undesirable artifacts at the electron microscopic level. The Vibratome is preferred over the Tissue Chopper because the slices obtained are flatter and hence easier to section for electron microsopy. The cubes of tissue are mounted on sections of balsa wood (1–1.5 cm cubed) using 5.0% agar to secure the tissue to the wood and to provide a matrix through which the knife passes before reaching the tissue. The slices obtained are 40–50 μm thick, and only the best slices are chosen for immunocytochemistry.

D. Incubation Procedures

Any adhering agar is teased away from the slices with a broken wood applicator stick. The slices are incubated for 30 min at room temperature with constant gentle agitation in 1:200 (anti-calmodulin) or 1:400 (anti-CaM-BP_{80}) dilutions of sera. Controls are either slices incubated with the same dilutions of nonimmune sera or slices incubated with immune sera which have been absorbed with an excess of the antigen. The slices are washed for at least 3 hr with five to seven changes of PBS and then incubated for 30 min with a 1:600 or 1:1000 dilution of peroxidase-labeled Fab fragments of purified goat anti-rabbit IgG (Avrameas and Ternynck, 1971; Mendell *et al.*, 1973; Mendell and Whitaker, 1978). The slices are washed again for at least 3 hr and then gently agitated for 12 min at 0°C in a solution containing 30 mg 3,3′-diaminobenzidine-4 HCl (DAB) and 10 μl of 30% hydrogen peroxide in 50 ml of PBS. The slices are washed for 3 min in PBS, and those to be used for electron microscopy are dissected into pieces about 1-mm square, and postfixed for 30 min in 2% osmium tetroxide in 0.12 *M* Millonig's buffer prior to processing for embedding in Epon-Araldite. Superficial ultrathin sections are obtained *en face* to the original cut face, exposed to antibody reagents, and examined in the electron microscope with or without prior staining in uranyl acetate and lead citrate.

The slices to be used for light microscopy are washed overnight in PBS and treated with 0.001% osmium tetroxide in PBS for 10 sec prior to dehydration and mounting on slides in a drop of Permount. The slides mounted for light microscopy are studied carefully to identify areas of interest to be examined at the electron microscopic level.

III. TISSUE LOCALIZATION

A. Light Microscopic Level

Calmodulin and CaM-BP_{80} localization have been studied in mouse and rat basal ganglia, cerebellum, and cerebral cortex. In the basal ganglia both antisera are bound within the soma but not the nucleus of large cells and within processes emanating from these cells (Wood *et al.*, 1980). Labeled punctate profiles which appear to be cell processes cut in transverse section are distributed throughout the neuropil. The islands of white matter which are characteristic of the basal ganglia do not contain detectable reaction product except for labeled punctate profiles at the edge of the white matter islands.

The most thorough investigation of calmodulin and CaM-BP_{80} localization in cerebral cortex has been performed on coronal sections taken approximately midway through the level of the basal ganglia. In this region the localization pattern with both antisera appears to be identical. The label is located within the soma and processes of large and small cells (Figs. 1 and 2). Many of these cells, such as the large pyramidal cells can be identified as neurons at the light microscopic level (Figs. 1 and 2). On the basis of size and morphological characteristics, most of the processes can be identified as dendrites traversing in a vertical plane relative to the pial surface (Figs. 1 and 2). Throughout the neuropil labeled punctate profiles are observed which may be dendritic processes cut in transverse sections. These punctate profiles are much less numerous than in coronal sections of basal ganglia. The staining of cell somata in the cerebral cortex (as well as the basal ganglia) is not uniform in a given small region. Some somata stain more intensely than others, suggesting differences in antigen concentration or accessibility of antigen sites to the antibodies.

All of the identifiable neurons in the cerebellum, including Purkinje cells, basket and stellate cells, granule and Golgi II cells, and neurons of the deep cerebellar nuclei, contain reaction product, indicating the presence of calmodulin and CaM-BP_{80} within the cell soma. In cases where dendrites can be clearly identified, such as Purkinje cell and deep cerebellar neuron dendrites, reaction product is evident (Figs. 3 and 4). In general, the staining in cerebellar neurons appears less intense than that in

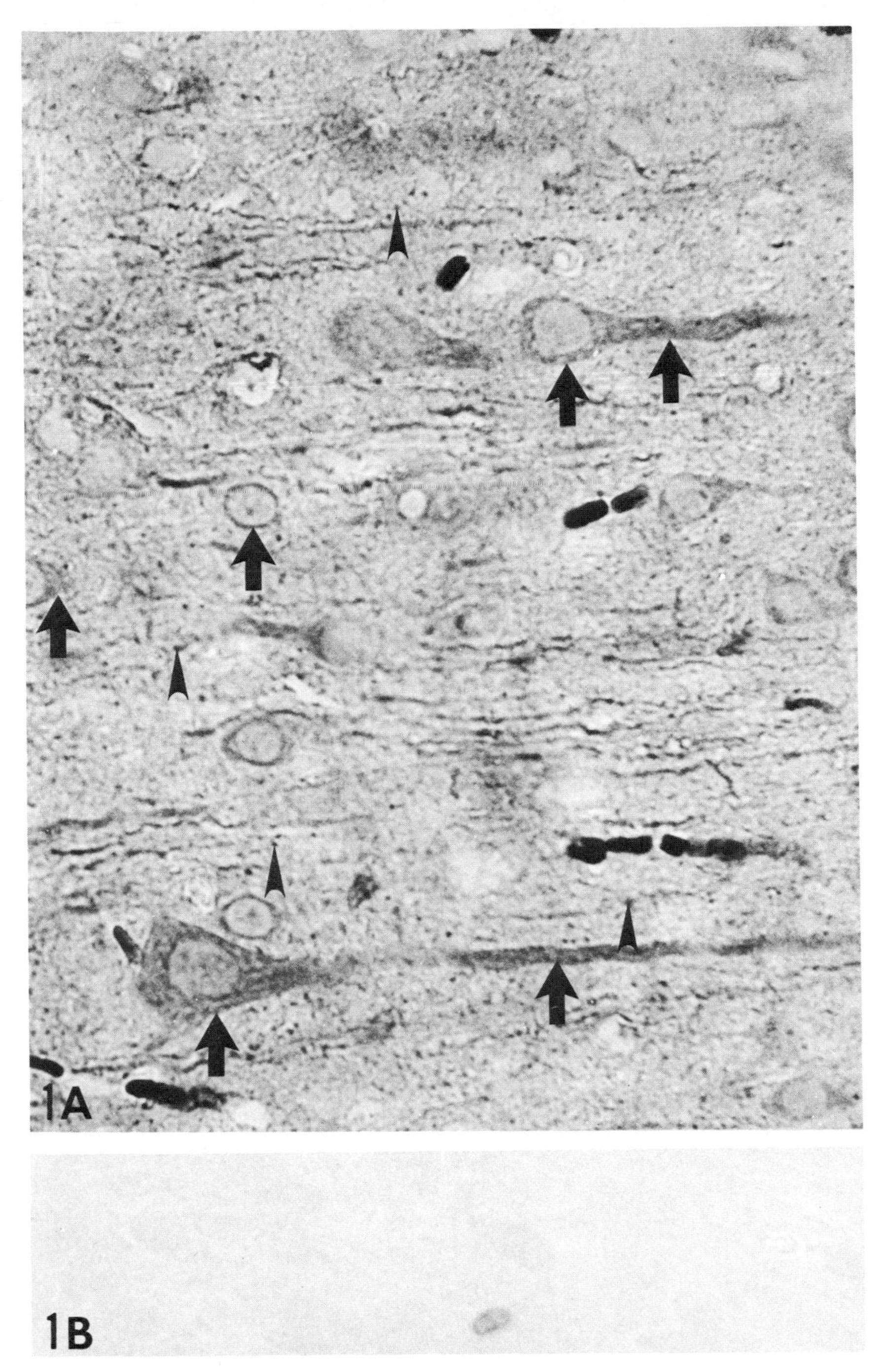

Fig. 1. Light micrographs of rat cerebral cortex slice treated with anti-calmodulin (A) or absorbed anti-calmodulin (B) sera. The soma and dendrites of large cells (arrows) are stained as well as a number of punctate profiles (arrowheads) in the neuropil. No specific staining is seen in control tissue using absorbed antiserum (see text) (B). ×650.

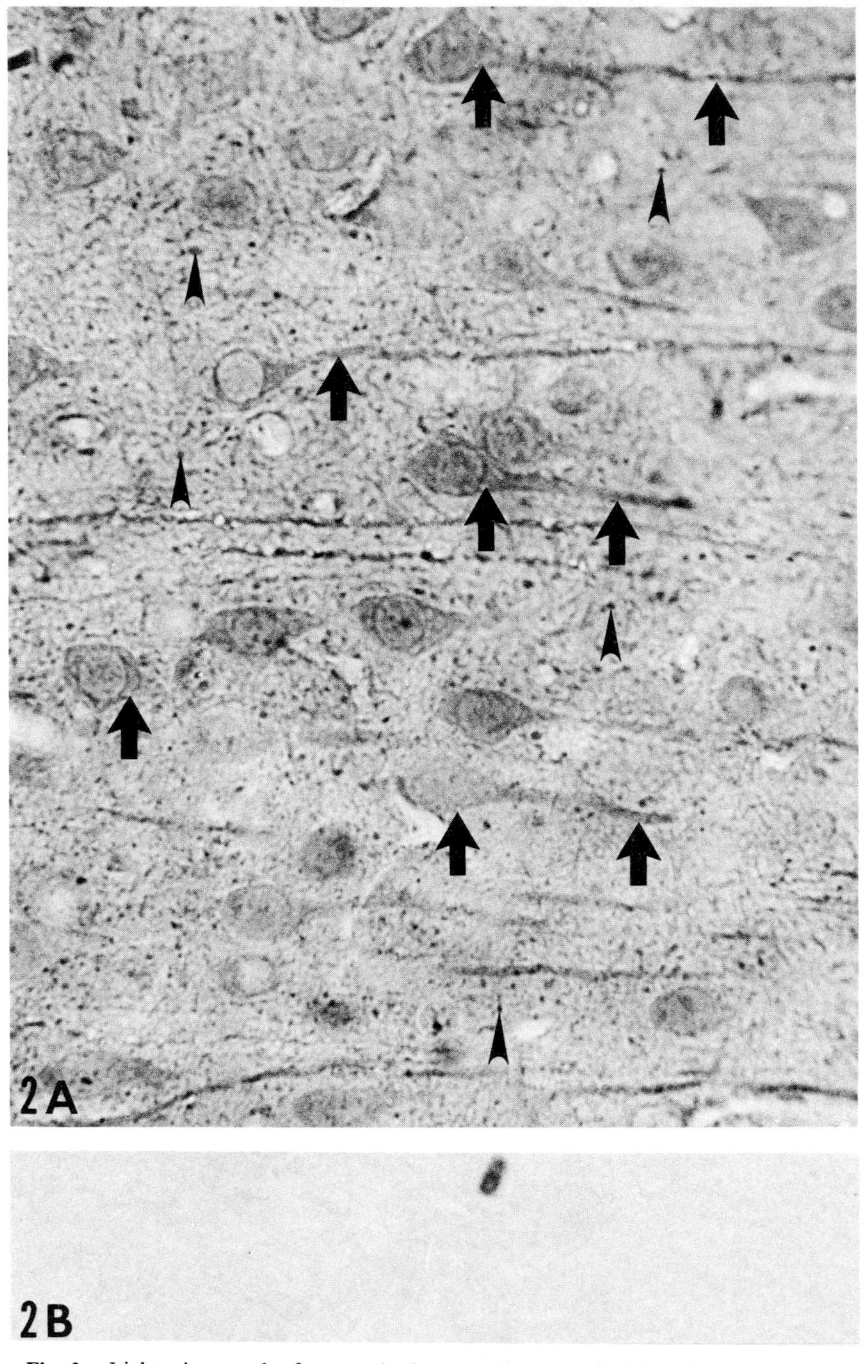

Fig. 2. Light micrograph of rat cerebral cortex slice treated with anti-CaM-BP_{80} (A) or absorbed anti-CaM-BP_{80} sera. Description same as Fig. 1. ×650.

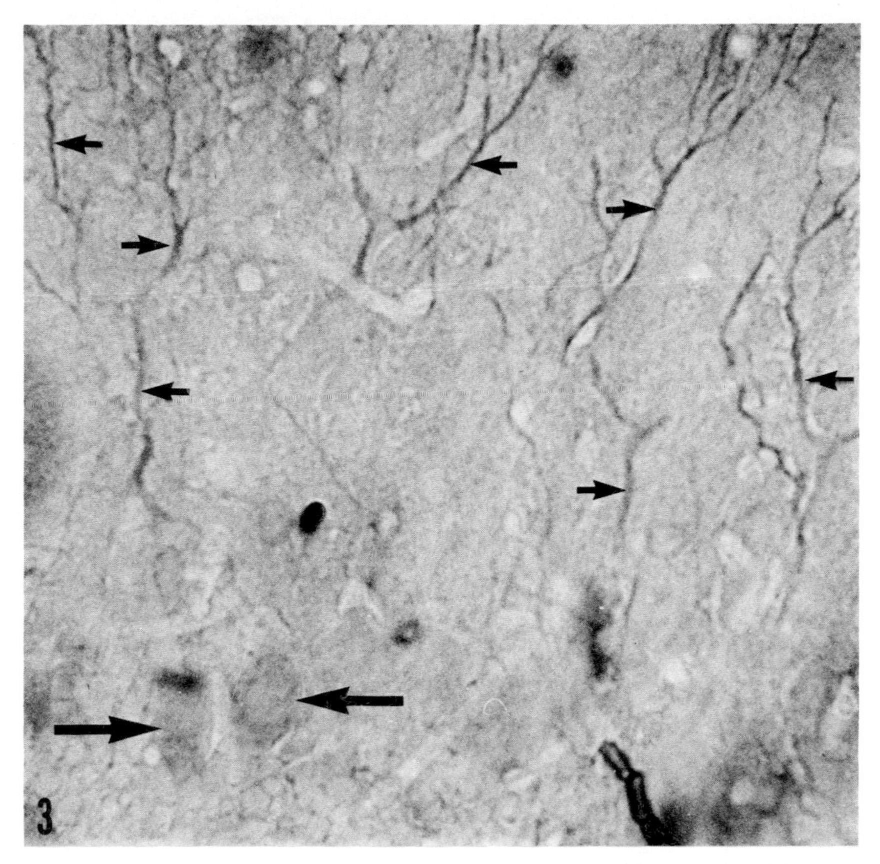

Fig. 3. Light micrograph of rat cerebellum slice treated with anti-calmodulin serum. Purkinje cell dendrites (small arrows) are more intensely stained than Purkinje cell soma (large arrows). ×800.

cortical or basal ganglia neurons. The staining of Purkinje cells in the soma is less than that in the dendrite, and the smaller branches of these dendrites frequently stain most intensely (Figs. 3 and 4). Purkinje cell somata receive a large number of inhibitory synapses from basket cells whereas small Purkinje cell dendritic branches receive a large number of excitatory synapses from granule cells.

B. Electron Microscopic Level

Detailed ultrastructural analysis of the immunocytochemical localization of calmodulin and CaM-BP_{80} have been performed in the mouse cau-

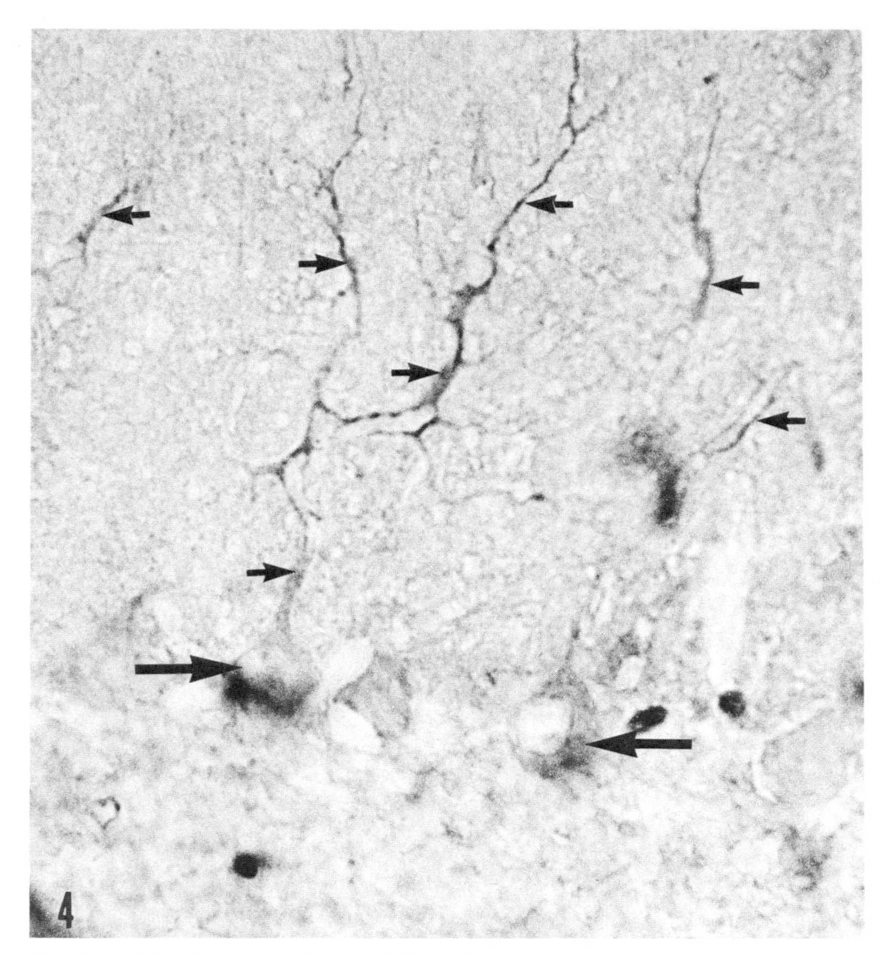

Fig. 4. Light micrograph of rat cerebellum slice treated with anti-CaM-BP_{80} serum. Description same as Fig. 3. ×750.

date-putamen only. We have not yet been able to ascertian whether these results may be generalized to include other brain regions. In the caudate-putamen the label is found in the cell somata and dendrites of neurons (Wood *et al.*, 1980). No label is observed in astrocytes or oligodendroglia. Although it is difficult to rule out the possibility that some of the neuronal processes are unmyelinated axons, most of the labeled profiles can be identified as dendrites on the basis of fine structural features and the presence of presynaptic terminals contacting them. Identified axons and presynaptic terminals do not contain reaction product. Within the neuronal

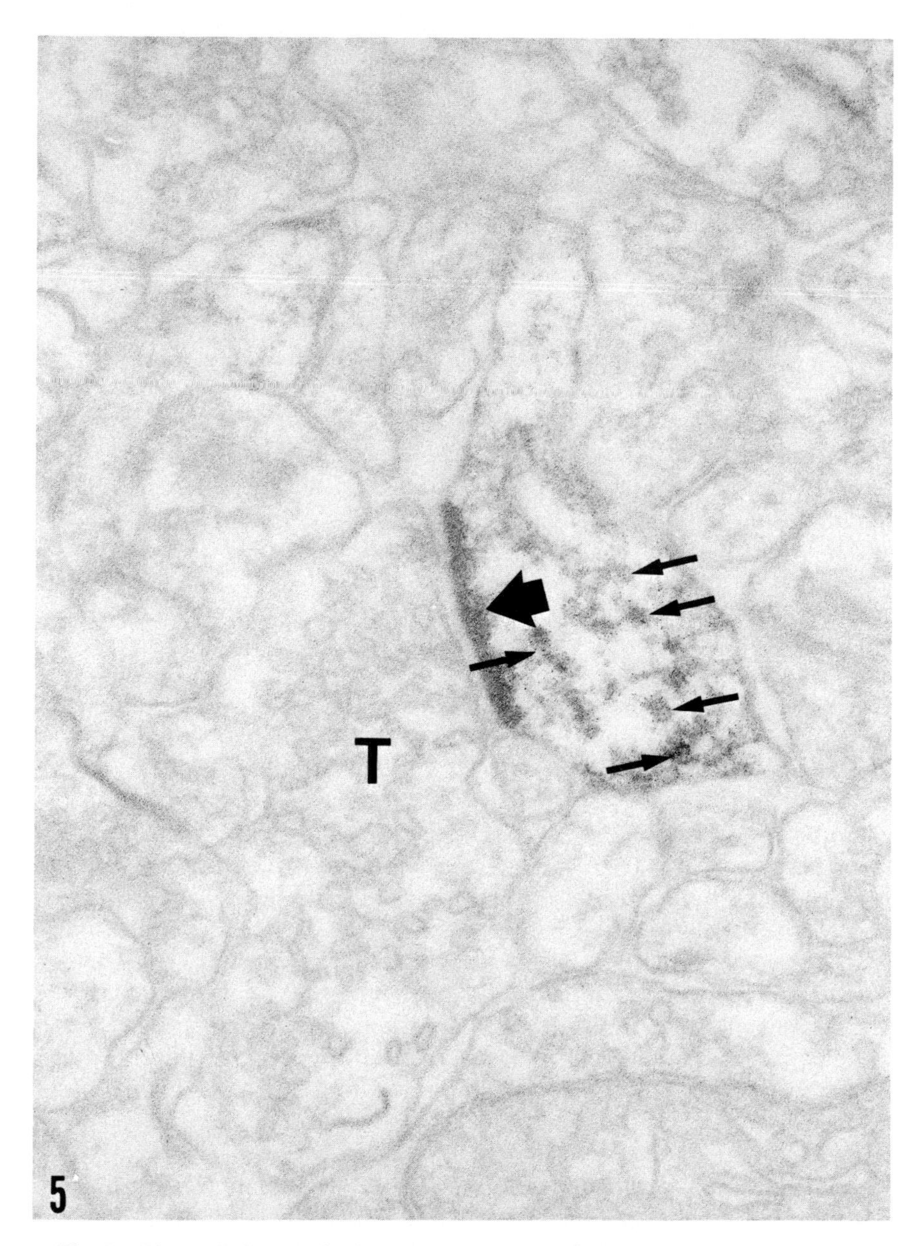

Fig. 5. Electron micrograph of rat caudate putamen treated with anti-calmodulin serum showing staining of the post synaptic density (large arrow) and dendritic microtubules (small arrows). T, presynaptic terminal. ×95,000.

somata, the reaction product is present around certain organelles and in the cytoplasm (Wood *et al.*, 1980). The somal labeling is concentrated on microtubules and ribosomes. The immunocytochemical labeling within dendrites is associated with the postsynaptic density (PSD) and microtubules; a trace of label is sometimes present around the mitochondria (Fig. 5). In lightly labeled profiles the reaction product associated with the PSD and individual microtubules is particularly striking (Fig. 5) (Wood *et al.*, 1980).

IV. FUNCTIONAL CONSIDERATIONS

Our results indicate that calmodulin and CaM-BP_{80} are localized within neurons in the mouse basal ganglia, hippocampus, cerebellum, and cerebral cortex. In all areas examined, the proteins are concentrated in neuronal somata and processes, many of which can be identified as dendrites. The localization pattern of the two proteins appears to be identical in each brain area and they may share a similar role in neurotransmission. Our results do not rule out other sites of localization of the two proteins; they may be present in presynaptic or glial structures at levels below detection by our method, or they may have masked antigenic sites.

At the electron microscopic level in the mouse caudate-putamen, calmodulin and CaM-BP_{80} are found on all cellular organelles within the soma, and primarily on the PSD and microtubules of dendrites. The somal labeling is most intense on free ribosomes and polysomes sites which would be consistant with a role for these cell structures in the biosynthesis of both proteins. The label around somal microtubules is expected since these structures are so prominently stained in the dendrite (see below). The label deposited on mitochondrial membranes, the outer nuclear envelope membrane and the cytoplasmic face of the rough and smooth endoplasmic reticulum as well as the cytoplasm itself may represent additional sites of localization of the proteins. Alternatively, these proteins may be precipitated onto these structures during the process of fixation or handling of the tissue for immunocytochemistry. It is not possible at this time to identify the somal sites at which calmodulin and CaM-BP_{80} are located *in vivo*.

The localization of calmodulin and CaM-BP_{80} within dendrites appears much more discrete than that in the soma. The reaction product is confined to the PSD, microtubules and, with considerable variability, the outer mitochondrial membrane. We have previously considered that the mitochondrial label may be artifactual (Wood *et al.*, 1980).

The localization of calmodulin and CaM-BP_{80} at the PSD and dendritic microtubules implies a functional role for the proteins at these sites. Calmodulin was originally discovered as an activator of cyclic nucleotide phosphodiesterase (cf. Cheung, 1967, 1970). Although cytochemical studies indicate that phosphodiesterase is associated with the PSD (Florendo *et al.*, 1971), its activity does not appear to be regulated by calmodulin (Grab *et al.*, 1979). It is conceivable that tissue protease may have acted on phosphodieterase during the preparation of PSD, rendering phosphodiesterase insensitive to calmodulin (Cheung, 1970). The synapse does contain a calmodulin-regulated protein kinase which is associated, at least in part, with the PSD (Grab *et al.*, 1979; Schulman and Greengard, 1978a,b).

The localization of calmodulin and CaM-BP_{80} along dendritic microtubules may be compared to the work of Welsh *et al.* (1978); they demonstrated by immunofluorescence that calmodulin is associated with microtubules of the mitotic spindle of dividing cells. The same group also show that calmodulin interferes with the assembly of microtubules *in vitro* (Marcum *et al.*, 1978). Collectively, these results support a role for calmodulin in microtubular function; the exact nature of this role will require extensive further experimentation.

ACKNOWLEDGMENTS

Supported in part by USPHS Grants NS 12590 (JGW), NS-08059 (WYC), AM 05689 (RWW), CA 21765 (WYC), ALSAC (WYC), and the Sloan Foundation (JGW).

REFERENCES

Avrameas, S., and Ternynck, T. (1971). Peroxidase labeled antibody and Fab conjugates with enhanced intracellular penetration. *Immunochemistry* **8,** 1175–1179.

Brostrom, M. A., Brostrom, C. O., Breckenridge, B. McL., and Wolff, D. J. (1976). Regulation of adenylate cyclase from glial tumor cells by calcium and a calcium-binding protein. *J. Biol. Chem.* **251,,** 4744–4750.

Cheung, W. Y. (1967). Cyclic 3′,5′-nucleotide phosphodiesterase: Pronounced stimulation by snake venom. *Biochem. Biophys. Res. Commun.* **29,** 478–482.

Cheung, W. Y. (1970). Cyclic 3²,5′-nucleotide phosphodiesterase: Demonstration of an activator. *Biochem. Biophys. Res. Commun.* **38,** 533–538.

Dabrowska, R., and Hartshorne, D. J. (1978). A Ca^{2+}- and modulator-dependent myosin light chain kinase from non-muscle cells. *Biochem. Biophys. Res. Commun.* **85,** 1352–1359.

Florendo, N. T., Barnett, R. J., and Greengard, P. (1971). Cyclic 3′,5′-nucleotide phosphodiesterase: Cytochemical localization in cerebral cortex. *Science* **173,** 745–748.

Grab, D. J., Berzins, K., Cohen, R. S. and Siekevitz, P. (1979). Presence of Calmodulin is postsynaptic densities isolated from canine cerebral cortex. *J. Biol. Chem.* **254,** 8690–8696.

Kakiuchi, S., and Yamazaki, R. (1970). Calcium dependent phosphodiesterase activity and its activating factor (PAF) from brain. *Biochem. Biophys. Res. Commun.* **41,** 1104–1110.

Klee, C. B., and Krinks, M. H. (1978). Purification of cyclic 3',5'-nucleotide phosphodiesterase inhibitory protein by affinity chromatography on activator protein coupled to sepharose. *Biochemistry* **17,** 120–126.

Lin, Y. M., Liu, Y. P., and Cheung, W. Y. (1974). Cyclic 3':5'-nucleotide phosphodiesterase: Purification, characterization, and active form of an activator from bovine brain. *J. Biol. Chem.* **249,** 4943–4954.

Lynch, T. J., Tallant, E. A., and Cheung, W. Y. (1976). Ca^{++}-dependent formation of brain adenylate cyclase-protein activator complex. *Biochem. Biophys. Res. Commun.* **68,** 616–625.

Lynch, T. J., Tallant, E. A., and Cheung, W. Y. (1977). Rat brain adenylate cyclase: Further studies on its stimulation by a Ca^{2+}-binding protein. *Arch. Biochem. Biophys.* **182,** 124–133.

Marcum, J. M., Dedman, J. R., Brinkley, B. R., and Means, A. R. (1978). Control of microtubule assembly-disassembly by calcium-dependent regulator protein. *Proc. Natl. Acad. Sci. U.S.A.* **75,** 3771–3775.

Mendell, J. R., and Whitaker, J. N. (1978). Immunocytochemical localization studies of myelin basic protein. *J. Cell Biol.* **76,** 502–511.

Mendell, J. R., Whitaker, J. N., and Engel, W. K. (1973). The skeletal muscle binding site of antistriated muscle antibody in myasthemia gravis: An electron microscopic immunohistochemical study using peroxidase conjugated antibody fragments. *J. Immunol.* **111,** 847–856.

Millonig, G. (1961). Advantages of a phosphate buffer for OsO_4 solutions in fixation. *J. Appl. Phys.* **32,** 1637 (abstr.).

Schulman, H., and Greengard, P. (1978a). Stimulation of brain protein phosphorylation by calcium and an endogenous heat-stable protein. *Nature* (*London*) **271,** 478–479.

Schulman, H., and Greengard, P. (1978b). Ca^{++} dependent protein phosphorylation system in membranes from various tissues, and its activation by "calcium dependent regulation." *Proc. Natl. Acad. Sci. U.S.A.* **75,** 5432–5436.

Soube, K., Ichida, S., Yoshida, H., Yamazaki, R., and Kakiuchi, S. (1979). Occurrence of a Ca^{2+}- and modulator protein-activatable ATPase in the synaptic plasma membranes of brain. *Biochem. Biophys. Res. Commun.* **99,** 199–202.

Wallace, R. W., and Cheung, W. Y. (1979). Calmodulin: Production of an antibody in rabbit and development of a radioimmunoassay. *J. Biol. Chem.* **254,** 6564–6571.

Wallace, R. W., Lynch, T. J., Tallant, E. A., and Cheung, W. Y. (1978). An endogenous inhibitor protein of brain adenylate cyclase and cyclic nucleotide phosphodiesterase. *Arch. Biochem. Biophys.* **187,** 328–334.

Wallace, R. W., Lynch T. J., Tallant, E. A., and Cheung, W. Y. (1979). Purification and characterization of an inhibitor protein of brain adenylate cyclase and cyclic nucleotide phosphodiesterase. *J. Biol. Chem.* **254,** 377–382.

Wallace, R. W., Tallant, E. A., and Cheung, W. Y. (1980). High levels of a heat-labile calmodulin-binding protein ($CaM\text{-}BP_{80}$) in bovine neostriatum. (Submitted for publication.)

Wang, J. H., and Desai, R. (1977). Modulator binding protein: Bovine brain protein exhibiting the Ca^{2+}-dependent association with the protein modulator of cyclic nucleotide phosphodiesterase. *J. Biol. Chem.* **252,** 4175–4184.

Welsh, M. J., Dedman, J. R., Brinkley, B. R., and Means, A. R. (1978). Calcium-dependent regulator protein: Localization in mitotic apparatus of eukaryotic cells. *Proc. Natl. Acad. Sci. U.S.A.* **75,** 1867–1871.

Wood, J. G. (1973). The effects of glutaraldehyde and osmium on the proteins and lipids of myelin and mitochondria. *Biochim. Biophys. Acta* **329,** 118–127.

Wood, J. G., McLaughlin, B. J., and Vaughn, J. E. (1976). Immunocytochemical localization of GAD in electron microscopic preparations of rodent CNS. *In* "GABA in Nervous System Function" (Roberts, Chase, and Tower, eds.).

Wood, J. G., Wallace, R. W., Whitaker, J. N., and Cheung, W. Y. (1980). Immunocytochemical localization of calmodulin and a heat-labile calmodulin binding protein in basal ganglia of mouse brain. *J. Cell Biol.* **84,** 66–76.

Chapter 15

Calmodulin-Binding Proteins

JERRY H. WANG
RAJENDRA K. SHARMA
STANLEY W. TAM

I. INTRODUCTION

Wang and Desai (1976) originally reported the existence in bovine brain of a heat-labile inhibitor protein of cyclic nucleotide phosphodiesterase. In a subsequent study on the mechanism of the enzyme inhibition, the inhibitor protein was demonstrated to undergo Ca^{2+}-dependent association with calmodulin (Wang and Desai, 1977). Klee and Krinks (1978) discovered and purified a similar phosphodiesterase inhibitor protein which displays specific association with calmodulin, and Wallace *et al.* (1978) de-

CALCIUM AND CELL FUNCTION, VOL. I

ISBN 0-12-171401-2

scribed a smiliar brain protein which inhibited both adenylate cyclase and phosphodiesterase. Since calmodulin regulates many of the calmodulin-dependent enzymes by a general mechanism involving Ca^{2+}-dependent and reversible association with the enzymes, the possibility that the bovine brain inhibitor protein of phosphodiesterase represents another calmodulin-regulated protein rather than a physiological inhibitor has been raised (Wang and Desai, 1977). However, in the absence of a positive identification of its putative physiological function, the protein has been designated as modulator (calmodulin)-binding protein (Wang and Desai, 1977).

During the last 2 years, it has become clear that many other proteins, in addition to known calmodulin-dependent enzymes and the calmodulin-binding protein, are capable of specific association with calmodulin. Sharma *et al.* (1978a,b) have discovered and purified a heat-stable phosphodiesterase inhibitor which also undergoes Ca^{2+}-dependent association with calmodulin. Using affinity chromatography on a calmodulin–Sepharose 4B column, Klee and Krinks (1978) have isolated a fraction of bovine brain proteins which are capable of Ca^{2+}-dependent association with the affinity column. This fraction contains several proteins besides the known calmodulin-dependent enzymes and the calmodulin-binding protein. In addition, a large amount of calmodulin binding activity has been detected in particulate fraction of rat brains (Kakiuchi *et al.*, 1978). Thus, it seems more appropriate to use the term "calmodulin-binding proteins" for a group of proteins. As a working definition, calmodulin-binding proteins are proteins which display specific association with calmodulin, possess inhibitory activity against calmodulin-dependent enzymes, and have no other known function. For the identification of individual calmodulin-binding proteins which have been sufficiently characterized in terms of molecular properties the abbreviated value of their molecular weight may be used as a systemic designation e.g., $CaM\text{-}BP_{80}$, $CaM\text{-}BP_{70}$, etc.*

Among calmodulin-binding proteins, two have been purified to homogeneous state from bovine brain. One is a heat-labile protein of molecular weight about 85,000 (Klee and Krinks, 1978; Wallace *et al.*, 1979; Sharma *et al.*, 1979). This protein may be designated $CaM\text{-}BP_{80}$ (Wang and Waisman, 1979). The other protein, a heat-stable protein of about 70,000 daltons, originally called heat-stable inhibitor protein of phosphodiesterase (Sharma *et al.*, 1978,a,b) may be designated $CaM\text{-}BP_{70}$. As additional calmodulin-binding proteins are purified and sufficiently characterized, they may be designated accordingly. On the other hand, when physiological functions for the individual calmodulin-binding proteins are discovered, the proteins will be renamed according to their functions.

* $CaM\text{-}BP_{80}$ has also been referred to as the inhibitor protein (Wang and Desai, 1976; Klee and Krinks, 1978), modulator binding protein (Wang and Desai, 1977), calmodulin binding protein I (Wang and Waisman, 1979) and calcineurin (Klee, C. B., Crouch, T. H., and Krinks, M. H., 1979, *Proc. Natl. Acad. Sci., U.S.A.* **76,** 6270–6273).

II. HEAT-LABILE CALMODULIN-BINDING PROTEIN ($CaM-BP_{80}$)

A. Discovery

Partially purified bovine brain calmodulin-dependent cyclic nucleotide phosphodiesterase has consistently been found to require higher concentrations of calmodulin for activation than similar preparations of the enzyme from bovine heart (Fig. 1). This observation has led to the discovery in the brain enzyme preparation of a phosphodiesterase inhibitor protein which can be separated from the enzyme by gel filtration on Sephadex G-200 column (Wang and Desai, 1976). The presence of this inhibitor protein in the brain enzyme preparation accounts for the increased requirement of calmodulin for the enzyme activation.

Independently, Klee and Krinks (1978) and Wallace *et al.* (1978) have demonstrated the existence of an inhibitor protein in phosphodiesterase preparations. From the thermal stability and the elution volume from gel filtration columns, the inhibitor proteins of cyclic nucleotide phosphodiesterase described by the various investigators appear to be the same protein—$CaM-BP_{80}$

B. Characteristics of the Inhibition

As is illustrated in Fig. 2, calmodulin-dependent phosphodiesterase activity is continuously inhibited with increasing concentrations of $CaM-BP_{80}$. The maximally inhibited enzyme shows only the basal activity, i.e., the enzyme activity obtained in the absence of calmodulin or Ca^{2+}. In contrast to the calmodulin-dependent phosphodiesterase, the calmodulin-independent enzyme and the trypsin-desensitized phosphodiesterase are not inhbited by $CaM-BP_{80}$ (Fig. 2). Thus, this calmodulin-binding protein specifically inhibits the calmodulin activation of phosphodiesterase. The inhibition of cyclic nucleotide phosphodiesterase by the calmodulin-binding protein is not reversed by increasing concentrations of Ca^{2+}, indicating that the inhibitor does not exert its effect on the enzyme by chelating Ca^{2+} (Wang and Desai, 1976; Wallace *et al.*, 1978).

The calmodulin-binding protein interacts antagonistically with calmodulin in the cyclic nucleotide phosphodiesterase reaction. Inhibition of the enzyme by the binding protein is pronounced at low concentrations of calmodulin, but can be completely abolished at high concentrations of calmodulin. As it is shown in Fig. 3A, when a higher concentration of the binding protein is present in the enzyme reaction, higher concentrations of calmodulin would be required to activate the enzyme. There appears to be a linear relationship between the concentration of the binding protein

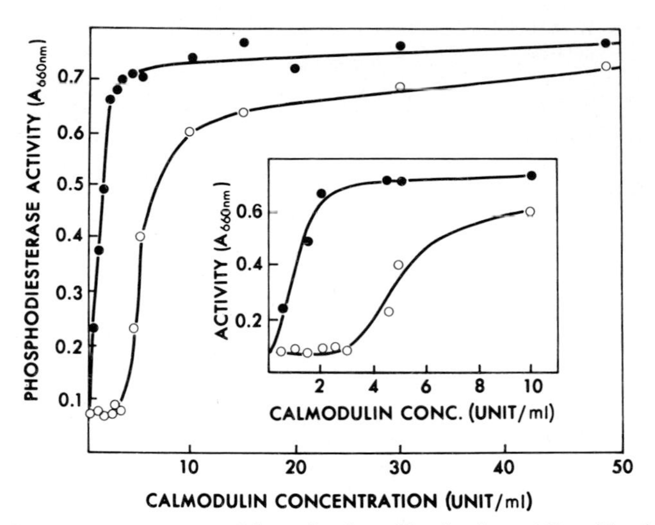

Fig. 1. Dose-response curves of the activation of bovine heart (●) and bovine brain cyclic nucleotide phosphodiesterase (○) by calmodulin. Data at low concentrations of calmodulin drawn at an expanded scale. (From Wang and Desai, 1976.)

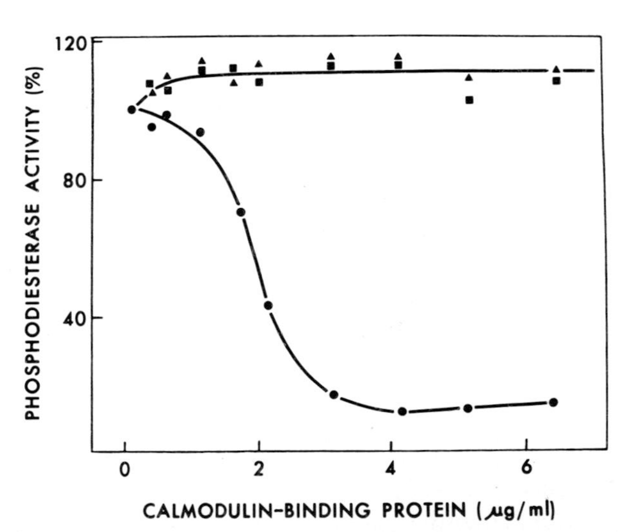

Fig. 2. Effect of CaM-BP_{80} on the calmodulin-independent phosphodiesterase (▲), trypsin-desensitized phosphodiesterase (■), or calmodulin-dependent phosphodiesterase (●). (From Wang and Desai, 1977.)

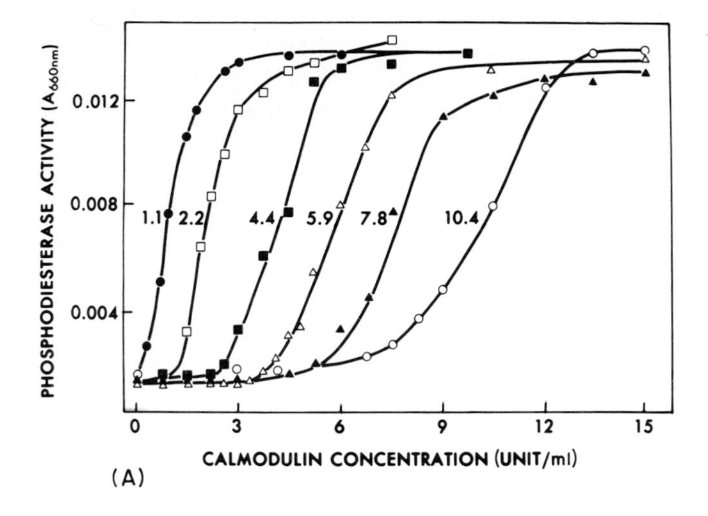

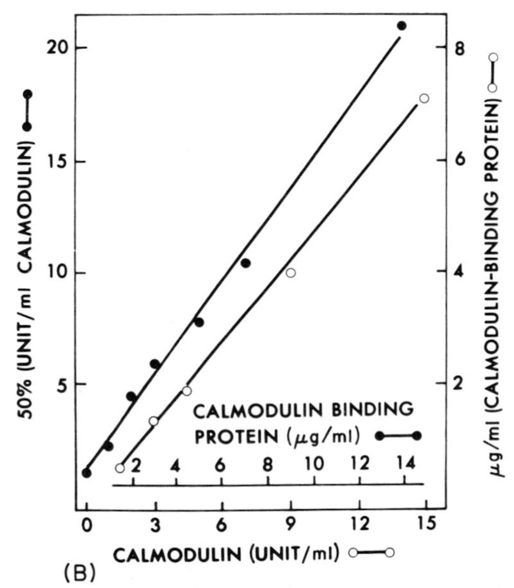

Fig. 3. (A) Dose-response curves of the activation of phosphodiesterase by calmodulin in the absence of (●) or presence of 1.0 (□), 2.0 (■), 3.0 (△), 5.0 (▲), and 7.0 μg/ml of a partially purified CaM-BP_{80}(○). Numbers in the figure referring to the concentration of calmodulin in unit/ml at 50% activation of the enzyme. (B) The dependence of the calmodulin concentration at 50% of phosphodiesterase activation upon the concentration of CaM-BP_{80} in the assay. (From Wang and Desai, 1977.)

in the reactions and the amount of calmodulin required for 50% maximal activation (Fig. 3B). Similarly, when higher concentrations of calmodulin are used for the activation of cyclic nucleotide phosphodiesterase, higher concentrations of CaM-BP_{80} would be needed for the enzyme inhibition (Wang and Desai, 1977). These results suggest that calmodulin and CaM-BP_{80} interact stoichiometrically in the phosphodiesterase reaction.

In addition to cyclic nucleotide phosphodiesterase, other calmodulin-regulated enzymes including brain adenylate cyclase (Wallace *et al.*, 1978; Wescott *et al.*, 1979), myosin light chain kinase (Waisman *et al.*, 1978a), and erythrocyte Ca^{2+}-Mg^{2+} ATPase (Larsen *et al.*, 1978), as well as the calmodulin-stimulated Ca^{2+} transport of erythrocyte membrane (Larsen *et al.*, 1978) are inhibited by CaM-BP_{80}. The inhibition of these reactions is similar to that of phosphodiesterase in that it is specific against the calmodulin-stimulated activities with no effect on the basal activities. The inhibition by CaM-BP_{80} in the reactions is completely abolished at high concentrations of calmodulin. As an example, the inhibition of brain adenylate cyclase by CaM-BP_{80} and its reversal by calmodulin are depicted in Fig. 4.

The antagonistic interactions between CaM-BP_{80} and calmodulin in the various enzyme reactions has been demonstrated using calmodulins from various sources which include bovine brain, bovine heart (Wang and

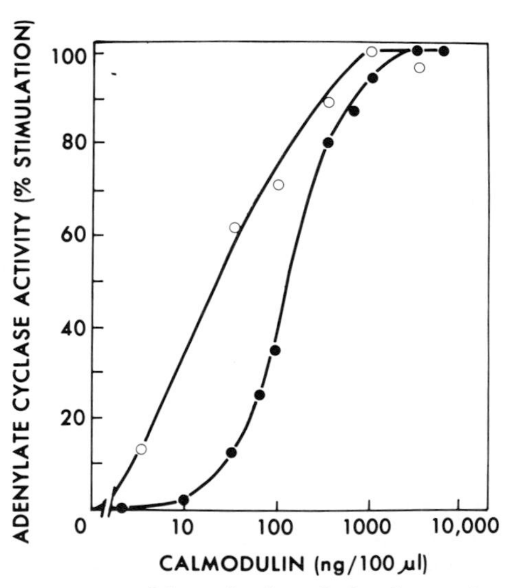

Fig. 4. Dose-response curves of the activation of adenylate cyclase by calmodulin in the presence (●) or absence of CaM-BP_{80} (○). (From Wallace *et al.*, 1978.)

Desai, 1976), human erythrocyte (Larsen *et al.*, 1978), earthworm (Waisman *et al.*, 1978b), and higher plants (Taylor *et al.*, 1980). The result suggests that calmodulin lacks tissue and species specificity in its interactions with CaM-BP_{80}.

C. Mechanism of Action

The specific and antagonistic interaction between calmodulin and CaM-BP_{80} is consistent with the notion that the two proteins associate to neutralize each other's effects in the various reactions. However, the possibility that both proteins interact directly with the enzymes in a mutually exclusive manner cannot be excluded from the kinetic characterizations. These alternative mechanisms have been distinguished by studies on the direct interaction between the enzyme and CaM-BP_{80} and between calmodulin and CaM-BP_{80}.

The interaction between calmodulin and CaM-BP_{80} has been examined by many investigators using various techniques. These include gel filtration (Wang and Desai, 1977), affinity chromatography on calmodulin–Sepharose 4B conjugate (Klee and Krinks, 1978; Wallace *et al.*, 1978), anion exchange chromatography, disc gel electrophoresis (Sharma *et al.*, 1979), density gradient centrifugation, and chemical cross-linking (Richman and Klee, 1978). In all these studies, results are in agreement with the notion that CaM-BP_{80} and calmodulin undergo Ca^{2+}-dependent and reversible association with each other.

The demonstration of Ca^{2+}-dependent association between calmodulin and CaM-BP_{80} using the technique of gel filtration is reproduced in Fig. 5 (Wang and Desai, 1977). When a mixture of calmodulin and CaM-BP_{80} was chromatographed on a Sephadex G-200 column using a buffer containing EGTA, the two proteins were eluted from the column well separated at elution volumes corresponding to those of the free proteins (Fig. 5A). On the other hand, when the protein mixture was chromatographed with a buffer containing Ca^{2+}, a protein fraction containing both proteins was eluted from the column. This fraction had an elution volume smaller than those of the free proteins (Fig. 5B).

In contrast to calmodulin, calmodulin-dependent cyclic nucleotide phosphodiesterase does not appear to associate with CaM-BP_{80}, either in Ca^{2+}- or EGTA-containing buffers. When mixtures of the two proteins were chromatographed on a Sephadex G-200 column, elution profiles of the two proteins were similar to those of the free proteins (Wang and Desai, 1977). The gel filtration technique has also been used to examine the possible formation of a ternary protein complex of calmodulin, calmodulin-dependent phosphodiesterase, and CaM-BP_{80} in the presence of

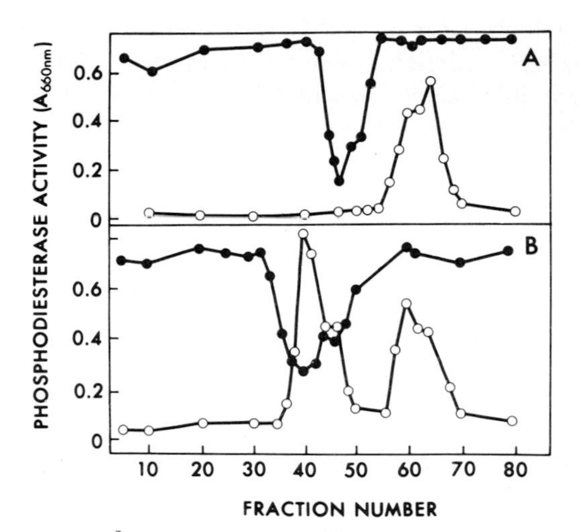

Fig. 5. The demonstration of Ca^{2+}-dependent association of calmodulin and CaM-BP_{80} by gel filtration. The mixture of calmodulin and CaM-BP_{80} was chromatographed on G-200 Sephadex column in buffers containing either EGTA (A) or Ca^{2+} (B). Fractions were analyzed for calmodulin (○) and calmodulin-binding protein activities (●). (From Wang and Desai, 1977.)

Ca^{2+}. No evidence for the formation of the ternary complex was obtained (Wang and Desai, 1977).

A mechanism of inhibition of cyclic nucleotide phosphodiesterase by CaM-BP_{80} has been proposed and may be represented by the following reaction scheme (Scheme I).

$$Ca^{2+} + CaM \rightleftharpoons Ca^{2+}\text{—}CaM \rightleftharpoons Ca^{2}\text{—}CaM^{*} \quad (1)$$

$$Ca^{2}\text{—}CaM^{*} + PDE \rightleftharpoons Ca^{2+}\text{—}\overset{\overset{PDE}{|}}{CaM^{*}} \rightleftharpoons Ca^{2+}\text{—}\overset{\overset{PDE^{*}}{|}}{CaM^{*}} \quad (2)$$

$$Ca^{2+}\text{—}CaM^{*} + CaM\text{-}BP_{80} \rightleftharpoons Ca^{2+}\text{—}CaM^{*}\text{—}CaM\text{-}BP_{80} \quad (3)$$

Scheme I

where PDE, CaM, and CaM-BP_{80} stand for phosphodiesterase, calmodulin and calmodulin-binding protein, respectively, and PDE* and CaM* denote active forms of the respective proteins. The proposed mechanism is consistent with the generally accepted mechanism of calmodulin regulation of phosphodiesterase (Wang *et al.*, 1975) and the kinetic characterization of the enzyme inhibitions by CaM-BP_{80} (Section II,B). In addition, the proposal takes into consideration the observations that calmodulin and CaM-BP_{80} undergo Ca^{2+}-dependent association and that there are no direct interactions between CaM-BP_{80} and phosphodiesterase and no ternary protein complex of phosphodiesterase, calmodulin, and CaM-BP_{80}. Although the inhibition of other calmodulin-dependent enzymes by CaM-BP_{80} has not been investigated as extensively as that of phosphodiesterase, the similarity in the kinetic characteristics of the enzyme inhibitions suggest that Scheme I may also apply to the other systems.

D. Purification and Molecular Properties

CaM-BP_{80} was purified from bovine brain first by Klee and Krinks (1978) and later by Wallace *et al.* (1979) and Sharma *et al.* (1979). All the purification procedures take advantage of the Ca^{2+}-dependent, reversible association of calmodulin with CaM-BP_{80}. Table I summarizes results of a typical protein preparation by the purification procedure currently used in the authors' laboratory. The procedure provides about 300-fold purification of the protein from crude brain extracts with an overall yield about 41%. The final protein preparation appears homogeneous upon disc gel electrophoresis and analytical ultracentrifugal analysis. The fact that homogeneous protein is obtained with a 300-fold purification suggests that the protein constitutes 0.3% protein of the total brain extract. This value is tentative since the assay of CaM-BP_{80} in crude tissue extracts is semiquantitative at best. However, it has been clearly shown that CaM-BP_{80} is the most abundant among bovine brain proteins capable of Ca^{2+}-dependent association with calmodulin (Klee and Krinks, 1978; Sharma *et al.*, 1979).

CaM-BP_{80} is a globular protein with a molecular weight of about 85,000. Some of the physical parameters of the protein are summarized in Table

TABLE I

Purification of CaM-BP_{80} from Bovine Brain[a]

Step	Total units	Specific activity (unit/mg)	Recovery (%)	Purification (fold)
Homogenate	1,280,000	33	100	1
DEAE-cellulose	1,008,900	200	79	6
Affi-Gel Blue	846,000	350	66	11
Calmodulin-Sepharose-4B gel	666,660	8,960	52	271
Sephadex G-200	530,850	10,850	41	329

[a] Three kilograms of bovine brain used for purification.

TABLE II

Physical Properties of CaM-BP_{80}

Molecular weight	
From sedimentation and gel filtration	84,000
From sedimentation equilibrium	86,000
Sedimentation constant, $s_{20,w}$	4.96
Stokes radius (Å)	40.5
Absorbance ($A_{278\,nm}$, 1%)	9.8

II. Although small discrepancies in some of the reported values are seen, there is general agreement in these values to indicate unequivocally that the same protein was being investigated in the different laboratories.

Gel electrophoresis of CaM-BP_{80} in the presence of SDS shows the existence of two distinct polypeptides with mobilities corresponding to molecular weights of about 60,000 and 15,000 (Fig. 6) (Klee and Krinks, 1978; Wallace *et al.*, 1979; Sharma *et al.*, 1979). That these polypeptides represent subunits of CaM-BP_{80} rather than cross-contamination of unre-

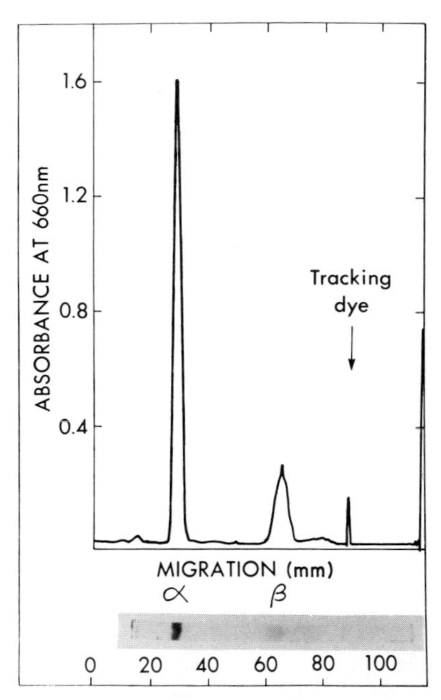

Fig. 6. Analysis of purified CaM-BP_{80} by sodium dodecyl sulfate-polyacrylamide gel electrophoresis. (From Sharma *et al.*, 1979.)

lated proteins is supported by several lines of evidence. First, disc gel electrophoresis and ultracentrifugal analysis of the protein under nondenaturating conditions indicated that the protein was homogeneous. Second, relative amounts of the two polypeptides appeared constant throughout the CaM-BP_{80} peak eluted from the Sephadex G-200 column, the final step of the protein purification (Sharma *et al.*, 1979). Third, CaM-BP_{80} prepared in different laboratories using different purification procedures all contained the two polypeptides. Fourth, although the large polypeptide was purified alone and shown to possess phosphodiesterase inhibition activity (Sharma *et al.*, 1979), this polypeptide was very unstable and could be stabilized by the small polypeptide (R. K. Sharma, R. Desai, and J. H. Wang, unpublished observation). The two subunits of CaM-BP_{80} have been designated α (60,000 daltons) and β (14,500 dalton) units (Sharma *et al.*, 1979).

The mass ratio of α unit to β unit in CaM-BP_{80} has been determined by densitometric tracing of the SDS-electrophoretic gel and by a determination of the protein recoveries after the separation of the two peptides on Sephadex G-100 column in the presence of 0.2% SDS or 6 *M* urea. Average values obtained for the mass ratio is 2.3. Using this value and molecular weights determined for the native protein and the separated subunits α and β, Sharma *et al.* (1979) have proposed a subunit structure of $\alpha\beta_2$ for CaM-BP_{80}. The data used for the determination of the subunit structure in this study are summarized in Table III. Other studies using mainly the values of molecular weight of the protein and its subunits have proposed a subunit structure of $\alpha\beta$ for CaM-BP_{80} (Klee and Krinks, 1978; Wallace *et al.*, 1979). Due to uncertainties in the determination of molecular weight and protein mass, further studies are needed to establish unambiguously the subunit structure of the protein. In addition, the possibility that two types of CaM-BP_{80}, $\alpha\beta$ and $\alpha\beta_2$ types may be purified from bovine brain should also be considered.

TABLE III

Subunit Structure of CaM-BP_{80}

Subunit	Molecular weight	Mass ratio (% of CaM-BP_{80})[a]	Molar ratio (mole/mole CaM-BP_{80})[b]
α	60,000	69	0.98
β	14,500	31	1.82

[a] Calculated on the basis of a mass ratio of subunits α/β of 2.22.

[b] Molecular weight of calmodulin binding protein of 85,000 was used for the calculation.

E. Interaction with Calmodulin

It has already been indicated previously (Section II,C), the Ca^{2+}-dependent and reversible interaction between calmodulin and CaM-BP_{80} has been demonstrated by using various techniques. However, the stoichiometry and the mechanism of the interaction are not clearly established.

Richman and Klee (1978) have studied the interaction using the glycerol density centrifugation. The interaction was found to occur only if the buffer contained Ca^{2+}. When the complex of calmodulin and CaM-BP_{80} isolated by the centrifugation was analyzed by SDS-gel electrophoresis, the stoichiometry of calmodulin and CaM-BP_{80} has been reported to be 0.7 to 1. Huang *et al.* (1978) have studied the interaction of calmodulin and CaM-BP_{80} by fluorescence spectroscopy. While CaM-BP_{80} is a tryptophan-containing protein (Klee and Krinks, 1978; Huang *et al.*, 1978), calmodulin does not contain tryptophan (Stevens *et al.*, 1976; Watterson *et al.*, 1976). The fluorescence due to tryptophan residues of CaM-BP_{80} is quenched upon mixing of the two proteins in medium containing Ca^{2+}. The stoichiometry of the interaction between calmodulin and CaM-BP_{80} as monitored by the fluorescence quenching is reported to be 1 to 2 (Huang *et al.*, 1978). The discrepancy in the reported stoichiometry of interaction of the two proteins indicates the need for further study on this reaction.

The observation that upon binding of calmodulin, the fluorescence of CaM-BP_{80} is quenched suggests that calmodulin induces a change in conformation of CaM-BP_{80} (Huang *et al.*, 1978). Thus, it seems that an additional step: a change in CaM-BP_{80} conformation after its association with calmodulin should be proposed in Scheme I (Section II,C). Wallace *et al.* (1978) have observed a marked stabilization effect of calmodulin on CaM-BP_{80} against thermal inactivation. This stabilization effect is observed only in medium containing Ca^{2+}. The result also suggests that the specific and Ca^{2+}-dependent association of calmodulin and CaM-BP_{80} is accompanied with a change in conformation of CaM-BP_{80}.

The existence of two distinct subunits of CaM-BP_{80} has prompted studies on the identification of the calmodulin-binding subunit of this protein. Richman and Klee (1978) have approached this problem by using dimethyl suberimidate to cross-link calmodulin and CaM-BP_{80}. Upon cross-linking of an active ^{125}I-monoiodinated calmodulin to CaM-BP_{80} followed by SDS-gel electrophoresis of the protein sample, two radioactive protein bands with mobilities corresponding to proteins of molecular weights about 80,000 and 93,000 can be detected on the gel. The result suggests that the large subunit α binds calmodulin. The lack of radioactive band with mobility corresponding to proteins of molecular weight about

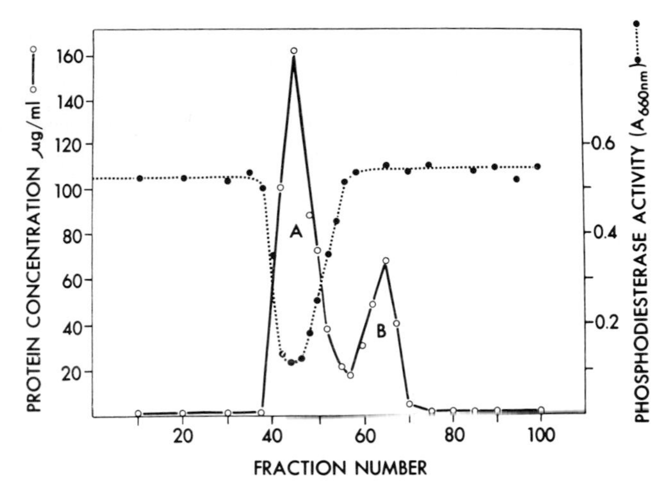

Fig. 7. Separation of subunits of CaM-BP_{80} on Sephadex G-100 in the presence of 6 *M* urea. (From Sharma *et al.*, 1979.)

32,000 is indicative of the lack of direct association of β subunit and calmodulin (Richman and Klee, 1978). Sharma *et al.* (1979) have also shown that α subunit is the calmodulin-binding subunit of CaM-BP_{80} using isolated α subunits. In the presence of 6 *M* urea in the buffer, α and β subunits of CaM-BP_{80} can be separated from each other by gel filtration on a Sephadex G-100 column. The cyclic nucleotide phosphodiesterase inhibitory activity is found to be exclusively associated with the α fraction (Fig. 7). Fractions containing β subunit neither inhibits nor affects the inhibitory activity of the subunits (Sharma *et al.*, 1979).

F. Physiological Function

During the brief period since its discovery, CaM-BP_{80} has been considered as being a physiological inhibitor protein of calmodulin, the common subunit of calmodulin regulated enzymes, or a calmodulin-regulated protein having its own biological activity. It seems most likely that this protein represents a calmodulin-regulated protein, however, its putative biological activity has not yet been identified.

CaM-BP_{80} was originally discovered as an inhibitor of cyclic nucleotide phosphodiesterase and adenylate cyclase. However, the ability to inhibit calmodulin action is not unique to this protein. Many other proteins capable of association with calmodulin have also been shown to inhibit calmodulin-activated phosphodiesterase, among them are a few other calmodulin-dependent enzymes. For example, chicken gizzard myosin light chain kinase has been shown to inhibit cyclic nucleotide phosphodiesterase in a manner similar to CaM-BP_{80} (Dabrowska *et al.*, 1978). Tam *et al.* (1980) have purified a few inhibitor proteins of cyclic nucleotide phosphodiesterase from rabbit skeletal muscle and shown that they are protein kinases: myosin light chain kinases and phosphorylase kinase. Thus, the observation that CaM-BP_{80} inhibits calmodulin dependent enzymes does not necessarily mean that this protein is a physiological inhibitor of calmodulin.

Calmodulin has been shown to exist in great excess over calmodulin-regulated enzymes in many animal tissues (Cheung, 1971; Smoake *et al.*, 1974; Waisman *et al.*, 1975). Although CaM-BP_{80} is the most abundant bovine brain protein among those which are capable of Ca^{2+}-dependent association with calmodulin, its amount in bovine brain extract is still far less than that of calmodulin. Sharma *et al.* (1979) have estimated the molar ratio of CaM-BP_{80} to calmodulin in bovine brain to be about 1 to 10. Thus, CaM-BP_{80} is not capable of inhibiting other calmodulin regulated enzymes in crude bovine brain extracts.

The postulate that CaM-BP_{80} represents a common subunit of calmodulin-regulated protein is a very attractive hypothesis, but is not supported by recent experimental results. Several other proteins which show Ca^{2+}-dependent reversible association with calmodulin have been purified to homogeneity and characterized. These proteins, which include CaM-BP_{70} (Sharma *et al.*, 1978b); cyclic nucleotide phosphodiesterase from bovine brain (Morrill *et al.*, 1979) or bovine heart (La Porte *et al.*, 1979); myosin light chain kinase from chicken gizzard (Dabrowska *et al.*, 1978), turkey gizzard (Adelsteine *et al.*, 1978), and rabbit skeletal muscle (Yagi *et al.*, 1978; Tam *et al.*, 1980), do not appear to contain subunits of molecular weights about 60,000 and 15,000. In addition, purified rabbit skeletal muscle phosphorylase kinase which contains tightly bound calmodulin does not contain subunits of 60,000 and 15,000 daltons (Cohen *et al.*, 1978).

Although it seems most probable that CaM-BP_{80} represents another calmodulin-regulated protein, its putative biological activity has not yet been established. Several enzymatic activities have been tested using pure samples of the protein, and none of the tests gave positive results. These include ATPase, GTPase, cyclic AMP and cyclic GMP phosphodiesterase, 5′-nucleotidase, adenylate cyclase, guanylate cyclase, and protein kinase using histone, casein, and phosphorylase as substrates (Wang and Desai, 1977; Klee and Krinks, 1978). However, there are many other enzymatic activities which should be tested. In addition, the protein may very well have a function which is noncatalytic. The physiological function of CaM-BP_{80} clearly requires further studies.

III. HEAT-STABLE CALMODULIN-BINDING PROTEIN ($CaM\text{-}BP_{70}$)

A. Identification and Separation from $CaM\text{-}BP_{80}$

During the purification of $CaM\text{-}BP_{80}$, Sharma *et al.* (1978a) have discovered another inhibitor protein of calmodulin-dependent cyclic nucleotid phosphodiesterase. The new inhibitor protein can be separated from $CaM\text{-}BP_{80}$ by DEAE-cellulose column chromatography which is used as an early step in the purification of $CaM\text{-}BP_{80}$. As is illustrated in Fig. 8, two fractions of inhibitor activities of cyclic nucleotide phosphodiesterase are obtained upon chromatography of a crude preparation of $CaM\text{-}BP_{80}$. $CaM\text{-}BP_{80}$ is eluted from the DEAE-cellulose column at the NaCl concentrations ranging from about 0.1 to 0.25 *M*. This fraction also contains most of the calmodulin-dependent phosphodiesterase. The additional inhibitory activity which usually represents about 10 to 20% of the total activity is eluted at much lower NaCl concentrations: 0 to 0.05 *M*.

Like $CaM\text{-}BP_{80}$, the new inhibitor is specific toward calmodulin-dependent phosphodiesterase. It shows no effect on calmodulin-independent enzyme nor on the basal activity of the calmodulin-dependent phosphodiesterase (Sharma *et al.*, 1978a). The most striking difference between the two inhibitor fractions from DEAE-cellulose column is the difference

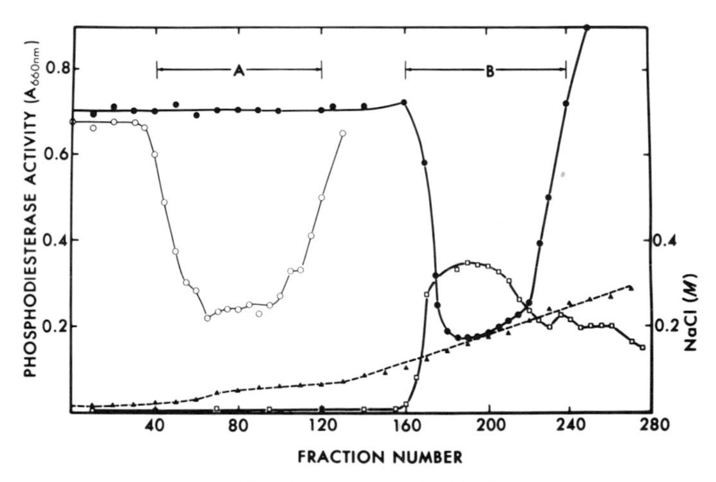

Fig. 8. Separation of two bovine brain calmodulin-binding proteins by DEAE-cellulose chromatography. Column eluates were analyzed for phosphodiesterase activity (□), and calmodulin-binding protein activity using 20 μl (●) or 100 μl (○) aliquots, NaCl concentration (▲). (From Sharma *et al.*, 1978a.)

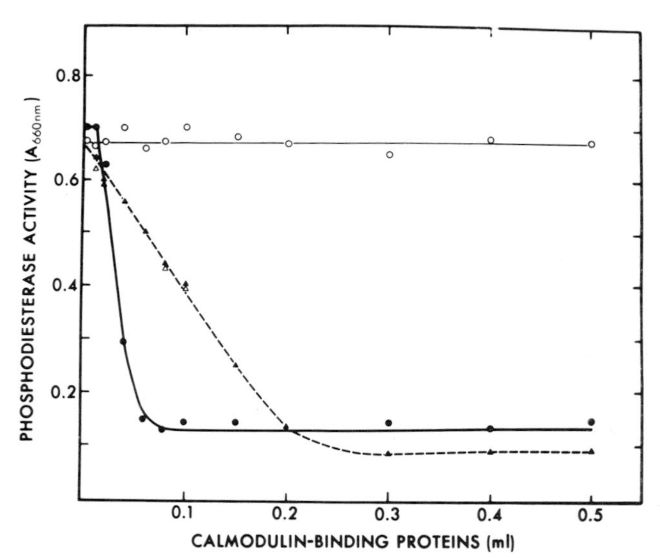

Fig. 9. Inhibition of calmodulin-dependent cyclic nucleotide phosphodiesterase by increasing amounts of CaM-BP_{80} (●), heat-treated CaM-BP_{80} (○), CaM-BP_{70} (△), or heat-treated CaM-BP_{70} (▲). (From Sharma *et al.*, 1978.)

in thermal stability. CaM-BP_{80} is a heat-labile protein; it loses its phosphodiesterase inhibitory activity after 1 min incubation in a boiling water bath. In contrast, fractions containing the new inhibitor protein can be incubated in a boiling water bath for over 5 mins without any decrease in its inhibitory activity toward phosphodiesterase (Fig. 9). Due to its remarkable heat stability, the protein was designated originally the heat-stable inhibitor protein of cyclic nucleotide phosphodiesterase (Sharma *et al.*, 1978a).

Prior to the discovery of CaM-BP_{70} in bovine brain, Dumler and Etingof (1976) reported the existence in bovine retina a heat-stable inhibitor protein of cyclic AMP phosphodiesterase. However, it is not clear whether the retina inhibitor protein is specific toward calmodulin-activated phosphodiesterase. The molecular weight of the inhibitor protein from retina determined by gel filtration method is 40,000 (Dumler and Etingof, 1976). This value is significantly lower than the molecular weight of CaM-BP_{70} determined using the same method: 70,000 (Sharma *et al.*, 1978a,b) (also see Section III,C).

B. Mechanism of Action

The kinetic characterization of the inhibition of cyclic nucleotide phosphodiesterase by CaM-BP_{70} is similar to that of the enzyme inhibition by

$CaM-BP_{80}$. The enzyme inhibition is pronounced at low concentrations of calmodulin and completely reversed by high concentrations of calmodulin. When higher concentrations of $CaM-BP_{70}$ are used in the reaction, higher amounts of calmodulin are required for the reversal of enzyme inhibition (Sharma *et al.*, 1978a). Thus, the mechanism of inhibition of cyclic nucleotide phosphodiesterase by $CaM-BP_{70}$ is similar to that by $CaM-BP_{80}$. The reaction may be described by Scheme I (Section II,C) with the substitution of $CaM-BP_{70}$ for $CaM-BP_{80}$.

The association of $CaM-BP_{70}$ with calmodulin can be clearly demonstrated in buffers containing Ca^{2+} by either gel filtration on a G-75 Sephadex column or affinity chromatography on calmodulin-Sepharose 4B conjugate (Sharma *et al.*, 1978a,b). In the presence of EGTA while $CaM-BP_{80}$ displays no interaction with calmodulin (Wang and Desai, 1977), gel filtration analysis suggests that $CaM-BP_{70}$ still interacts, albeit very weakly, with calmodulin (Sharma *et al.*, 1978a).

C. Purification and General Properties

$CaM-BP_{70}$ has been purified to apparent homogeneity as judged by SDS-gel elecrophoresis. The purification procedure depends mainly on a calmodulin-Sepharose 4B affinity chromatography step (Sharma *et al.*, 1978b). Data for a typical preparation are summarized in Table IV. The protein is seen to be purified by over 200,000-fold from the crude brain extract. The fact that $CaM-BP_{70}$ requires such a high degree of enrichment to attain homogeneity indicates that this protein exists in very minute amounts in bovine brain.

Although the amount of $CaM-BP_{70}$ in bovine brain is very low, it is very potent in its inhibitory activity toward calmodulin-activated phosphodiesterase. Figure 10 shows that the protein is more than 20 times more potent than $CaM-BP_{80}$ in the inhibition of phosphodiesterase. The ability and potency of the protein in the inhibition of other calmodulin-dependent enzymes have not been studied.

There are several heat-stable inhibitor proteins which show specific inhibitory activities toward certain regulatory enzymes. For example, heat-stable inhibitor proteins for cyclic AMP-dependent protein kinases (Walsh *et al.*, 1971) and for phosphoprotein phosphatases (Brandt *et al.*, 1975; Khandelwal and Zinman, 1978) have been described and purified. Beale *et al.* (1977) have discovered a heat-stable inhibitor which acts on both cyclic AMP-dependent protein kinases and cyclic AMP phosphodiesterase. Calmodulin-binding protein, however, appears to be specific for calmodulin-dependent phosphodiesterase; it displays no inhibitory activity toward cyclic AMP-dependent protein kinases on a purified phosphoprotein phosphatase (Sharma *et al.*, 1978b).

TABLE IV

Purification of CaM-BP_{70} from Bovine Brain[a]

Step	Total units	Specific activity (U/mg)	Recovery (%)	Purification (fold)
Homogenate	40,810	0.35	100	1
$(NH_4)_2SO_4$, 0–55%	41,000	2.0	100	6
DEAE-cellulose	40,950	10.0	100	29
Heat treatment	34,618	12.0	84	34
CM-cellulose	30,000	91.0	73	260
Calmodulin-Sepharose 4B gel	13,800	19,714.0	34	56,325
Calmodulin-Sepharose 4B gel	13,340	83,375.0	33	238,214

[a] For purification, 2.5 kg of bovine brain used.

Molecular weight of CaM-BP_{70} has been determined by both gel filtration and SDS-gel electrophoresis techniques. The values obtained by two methods are similar: about 70,000 (Sharma *et al.*, 1978a,b). Thus, CaM-BP_{70} appears to be a monomeric protein. Due to the very low amount of pure protein available, molecular properties of this protein have not been studied in any detail.

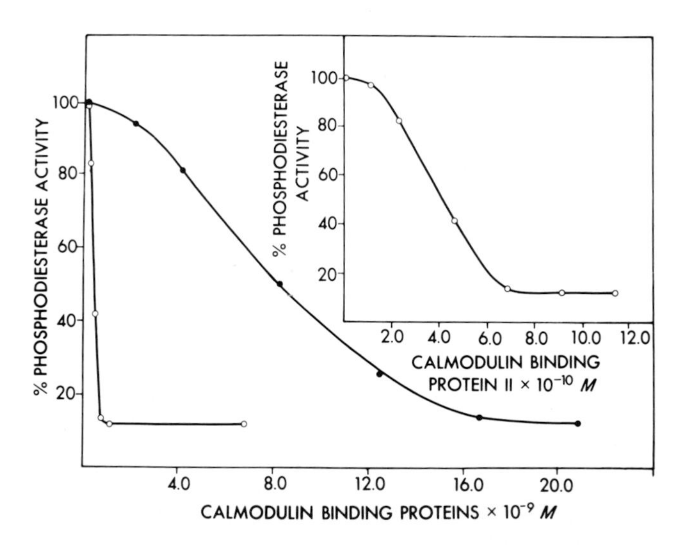

Fig. 10. Dose-response curves of the inhibition of phosphodiesterase by CaM-BP_{80} (●) and CaM-BP_{70} (○). (From Sharma *et al.*, 1979.)

The possibility that CaM-BP_{70}, like CaM-BP_{80}, represents a calmodulin-regulated proteins with its own biological activity has been considered. Several enzyme activities, including ATPase, GTPase, cyclic AMP phosphodiesterase, 5′-nucleotidase, and protein kinase (using histones as the substrate), have been tested using a pure sample of the protein. The protein does not seem to possess any of these activities (Sharma *et al.*, 1978b).

IV. OTHER CALMODULIN-BINDING PROTEINS

A. Other Soluble Brain Calmodulin-Binding Proteins

Klee and Krinks (1978) have described, in addition to calmodulin-dependent cyclic nucleotide phosphodiesterase and CaM-BP_{80}, the existence in bovine brain extract of other proteins which are specifically absorbed onto calmodulin-Sepharose 4B column in the presence of Ca^{2+} and dissociated from the column in buffers containing EGTA. The calmodulin affinity column fraction of bovine brain has been shown to contain multiple polypeptides (Klee and Krinks, 1978; Watterson and Vanaman, 1976). In addition to those representing the phosphodiesterase and CaM-BP_{80}, there are polypeptides with mobilities corresponding to molecular weights of 225,000, 150,000, 80,000, 41,000, and 38,000. The purification and characterization of any of those proteins have not been described. Recently, we have obtained a protein preparation from bovine brain which consists of predominantly the high molecular weight polypeptides (MW 200,000) (R. Desai, R. K. Sharma, and J. H. Wang, unpublished). The fraction does not exhibit any of the various enzyme activities tested, including those of ATPase, GTPase, cyclic AMP or cyclic GMP phosphodiesterase, hisone kinase, phosphorylase kinase, and myosin light chain (using skeletal muscle light chains) kinase. The protein possesses inhibitory activity toward calmodulin activated phosphodiesterase.

B. Particulate Calmodulin-Binding Proteins

Several of the calmodulin regulated enzymes are found predominantly in the particulate fraction of the cell fractionation, such as adenylate cyclase, Ca^{2+}-Mg^{2+}-ATPase (Gopinath and Vincenzi, 1977; Jarrett and Penniston, 1977; Sobue *et al.*, 1979) and the membrane calmodulin-dependent protein kinase (Schulman and Greengard, 1978a,b). In addition, the calmodulin-dependent cyclic nucleotide phosphodiesterase (Kakiuchi *et al.*, 1978; Teshima and Kakiuchi, 1978) and glycogen phosphorylase kinase

(Hammermeister *et al.*, 1965) appear to have a bimodal distribution: partly particulate and partly soluble. Thus, it is not surprising that calmodulin may be partly soluble and partly membrane bound in the cells.

Cheung (1971) originally showed that calmodulin existed both in the soluble and particulate fractions of bovine brain. The study of intracellular distribution of calmodulin in mammalian brains has been extended by Kakiuchi and co-workers (Teshima and Kakiuchi, 1978; Kakiuchi *et al.*, 1978). Using hypotonic solutions to disrupt synaptosomes, these investigators have obtained membrane preparations from rat brain which is largely free of trapped cytosol fractions. The amount of calmodulin found in these membrane preparations depends on the medium used for their preparations. The preparations contain about 20% of the total calmodulin activity when prepared with a medium containing EGTA. With a medium containing Ca^{2+}, the membrane preparation may bind about 40% of the total calmodulin activity (Kakiuchi *et al.*, 1978; Teshima and Kakiuchi, 1978). The results seem to indicate that there are two types of membrane bindings of calmodulin, a Ca^{2+}-dependent and a Ca^{2+}-independent calmodulin binding to the membranes.

The Ca^{2+}-dependent association of calmodulin with the rat brain microsomal fraction prepared with an EGTA-containing buffer has been directly demonstrated and characterized (Teshima and Kakiuchi, 1978). The binding of calmodulin to the membrane is saturable, reversible by EGTA and temperature dependent. As is illustrated in Fig. 11, at saturating amount of calmodulin, 4.5 μg of calmodulin may bind to a microsomal fraction containing 1 mg microsomal proteins. That microsomal proteins, rather than other components, are responsible for the specific calmodulin

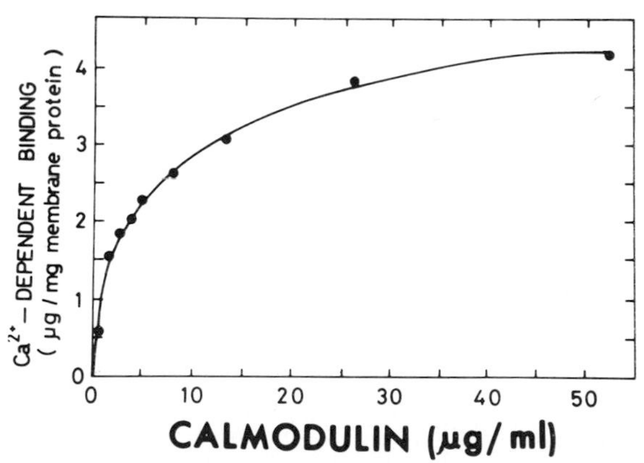

Fig. 11. Saturation curve of the Ca^{2+}-dependent association of calmodulin with microsomal fraction. (From Teshima and Kakiuchi, 1978.)

binding is suggested from the observation that trypsin treatment of the microsome fraction completely abolished the Ca^{2+}-dependent association of calmodulin.

The observed Ca^{2+}-dependent association of calmodulin to the membrane fraction is at least partly due to the Ca^{2+}-dependent association of the protein to the membrane-bound enzymes (Lynch *et al.*, 1976; Wallace *et al.*, 1979). The suggestion is supported by the observation that the binding of calmodulin to synaptic membrane displays the same Ca^{2+} and Mg^{2+} dependence as that of adenylate cyclase activation by the protein (Vandermeers *et al.*, 1978). However, there appear to be in addition to the known calmodulin-dependent enzymes, other membrane proteins which are capable of binding to calmodulin in the presence of Ca^{2+}. For example, Teshima and Kakiuchi (1978) have shown that heat treatment of the membrane for 5 min at 70°C results in 40% decrease in the Ca^{2+}-dependent binding of calmodulin but a complete loss in cyclic nucleotide phosphodiesterase and adenylate cyclase activities.

The Ca^{2+}-dependent calmodulin binding is not restricted to the membrane preparations from brain tissue. Sobue *et al.* (1979) have examined a dozen rat tissues and found that all the membrane preparations are capable of Ca^{2+}-dependent binding of calmodulin. Among these preparations, those of cerebrum, adrenal, and cerebella have the highest binding activities. The calmodulin-binding activity is found to be especially rich with the synaptic membrane fraction of cerebrum (Sobue *et al.*, 1979).

The existence of particulate calmodulin-binding proteins have been inferred from the binding activity of the membrane preparation. No calmodulin-binding protein has been purified free of known calmodulin-regulated enzymes. Thus, much more work is needed to establish the nature of these membrane-bound calmodulin-binding sites.

REFERENCES

Adelstein, R. S., Conti, M. A., Hathaway, D. R., and Klee, C. B. (1978). Phosphorylation of smooth muscle myosin light chain kinase by the catalytic subunit of adenosine 3²:5² monophosphate-dependent protein kinase. *J. Biol. Chem.* **253,** 8347–8350.

Beale, E. G., Dedman, J. R., and Means, A. R. (1977). Isolation and characterization of a protein from rat testies which inhibits cyclic AMP-dependent protein kinase and phosphodiesterase. *J. Biol. Chem.* **252,** 6322–6327.

Brandt, H., Lee, E. Y. C., and Killilea, S. D. (1975). A protein inhibition of rabbit liver phosphorylase phosphatase. *Biochem. Biophys. Res. Commun.* **63,** 950–956.

Cheung, W. Y. (1971). Cyclic 3²,5²-nucleotide phosphodiesterase: Evidence for and properties of protein activator. *J. Biol. Chem.* **246,** 2859–2869.

Cohen, P., Burchell, A., Foulkes, J. G., Cohen, P. T. W., Vanaman, T. C., and Nairn, A. A.

(1978). Identification of the Ca^{2+}-dependent modulator protein as the fourth subunit of rabbit skeletal muscle phosphorylase kinase. *FEBS Lett.* **92,** 287–293.
Dabrowska, R., Sherry, J. M. F., Aromatorio, D. K., and Hartshorne, D. J. (1978). Modulator protein as a component of the myosin light chain kinase from chicken gizzard. *Biochemistry* **17,** 253–258.
Dumler, I. L., and Etingof, F. N. (1976). Protein inhibitor of cyclic adenosine 3²:5²-monophosphate phosphodiesterase in retina. *Biochim. Biophys. Acta* **429,** 474–484.
Gopinath, R. M., and Vincenzi, F. F. (1977). Phosphodiesterase protein activator mimics and red blood cell cytoplasmic activator of (Ca^{2+}-Mg^{2+}) ATPase. *Biochem. Biophys. Res. Commun.* **77,** 1203–1209.
Hammermeister, K. E., Yunis, A. A., and Krebs, E. G. (1965). Studies on phosphorylase activation in the heart. *J. Biol. Chem.* **240,** 986–991.
Huang, C. Y., Chock, P. B., Sharma, R. K., and Wang, J. H. (1978). Interaction of bovine brain Ca^{2+}-regulation cAMP phosphodiesterase modulator and its binding protein. *Fed. Proc., Fed. Am. Soc. Exp. Biol.* **37,** 1427.
Jarret, H. W., and Penniston, J. T. (1977). Partial purification of the Ca^{2+}-Mg^{2+} ATPase activator from human erythrocytes: Its similarity to the activator of 3²:5²-cyclic nucleotide phosphodiesterase. *Biochem. Biophys. Res. Commun.* **77,** 1210–1216.
Kakiuchi, S., Yamazaki, R., Teshima, Y., Uenishi, K., Yasuda, S., Kashiba, A., Sobue, K., Ohshima, M., and Nakajima, T. (1978). Membrane-bound protein modulator and phosphodiesterase. *Adv. Cyclic Nucleotide Res.* **9,** 253–264.
Khandelwal, R., and Zinman, S. M. (1978). Purification and properties of a heat-stable protein inhibitor phosphoprotein phosphatase from rabbit liver. *J. Biol. Chem.* **253,** 560–565.
Klee, C. B., and Krinks, M. H. (1978). Purification of cyclic 3²,5²-nucleotide phosphodiesterase inhibitor protein by affinity chromatography on activator protein coupled Sepharose. *Biochemistry* **17,** 120–126.
LaPorte, D. C., Toscano, W. A., and Storm, D. R. (1979). Cross-linking of iodine-125-labeled, calcium-dependent regulatory protein to the Ca^{2+}-sensitive phosphodiesterase purified from bovine heart. *Biochemistry* **18,** 2820–2825.
Larsen, F. L., Raess, B. U., Hinds, T. R., and Vincenzi, F. F. (1978). Modulator binding protein antoganizes activation of (Ca^{2+} + Mg^{2+})-ATPase and Ca^{2+} transport of red blood cell membranes. *J. Supramol. Struct.* **9,** 269–274.
Lynch, T. J., and Cheung, W. Y. (1979). Human erythrocyte Ca^{2+}-Mg^{2+}-ATPase: Mechanism of stimulation by Ca^{2+}. *Arch. Biochem. Biophys.* **194,** 165–170.
Lynch, T. J., Tallant, E. A., and Cheung, W. Y. (1976). Ca^{2+}-dependent formation of brain adenylate cyclase-protein activator complex. *Biochem. Biophys. Res. Commun.* **68,** 616–625.
Morrill, M. E., Thompson, S. T., and Stellwagen, E. (1979). Purification of a cyclic nucleotide phosphodiesterase from bovine brain using blue dextran-Sepharose chromatography. *J. Biol. Chem.* **254,** 4371–4374.
Richman, P. G., and Klee, C. B. (1978). Interaction of ^{125}I-labeled Ca^{2+}-dependent regulation protein with cyclic nucleotide phosphodiesterase and its inhibitory protein. *J. Biol. Chem.* **253,** 6323–6326.
Schulman, H., and Greengard, P. (1978a). Stimulation of brain membrane protein phosphorylation by calcium and an endogenous heat-stable protein. *Nature (London)* **271,** 478–479.
Schulman, H., and Greengard, P. (1978b). Ca^{2+}-dependent protein phosphorylation system in membranes from various tissues, and its activation by "calcium-dependent regulator." *Proc. Natl. Acad. Sci. U.S.A..* **75,** 5432–5436.
Sharma, R. K., Wirch, E., and Wang, J. H. (1978a). Inhibition of Ca^{2+}-activated cyclic nu-

cleotide phosphodiesterase reaction by a heat-stable inhibitor protein from bovine brain. *J. Biol. Chem.* **253,** 3575–3580.

Sharma, R. K., Desai, R., Thompson, T. R., and Wang, J. H. (1978b). Purification of the heat-stable inhibitor protein of the Ca^{2+}-activated cyclic nucleotide phosphodiesterase by affinity chromatography. *Can. J. Biochem.* **56,** 598–604.

Sharma, R. K., Desai, R., Waisman, D. M., and Wang, J. H. (1979). Purification and subunit structure of bovine brain modulator binding protein. *J. Biol. Chem.* **254,** 4276–4282.

Smoake, J. A., Song, S. Y., and Cheung, W. Y. (1974). Cyclic 3²,5²-nucleotide phosphodiesterase. Distribution and developmental changes of the enzyme and its protein activator in mammalian tissues and cells. *Biochim. Biophys. Acta* **341,** 402–411.

Sobue, K., Muramoto, Y., Yamazaki, R., and Kakiuchi, S. (1979). Distribution in rat tissues of modulator binding protein of particulate nature. *FEBS Lett.* **105,** 105–109.

Stevens, F. C., Walsh, M., Ho, H. C., Teo, T. S., and Wang, J. H. (1976). Comparison of calcium-binding proteins: Bovine heart and brain protein activators of cyclic nucleotide phosphodiesterase and rabbit skeletal muscle troponin C. *J. Biol. Chem.* **251,** 4495–4500.

Tam, S. W., Waisman, D. M., and Wang, J. H. (1980). In preparation.

Taylor, W. A., Waisman, D. M., Sharma, R. K., and Wang, J. H. (1980). Distribution of calmodulin in plants. *Can. J. Biochem.* (submitted for publication).

Teshima, Y., and Kakiuchi, S. (1978). membrane-bound forms of Ca^{2+}-dependent protein modulator: Ca^{2+}-dependent and independent binding of modulator protein to the particulate fraction from brain. *J. Cyclic Nucleotide Res.* **4,** 219–231.

Vandermeers, A., Robberecht, P., Vandermeers-Piret, M. C., Pathé, J., and Christophe, J. (1978). Specific binding of the calcium-dependent regulator protein to brain membranes from the guinea pig. *Biochem. Biophys. Res. Commun.* **84,** 1076–1081.

Waisman, D. M., Stevens, F. C., and Wang, J. H. (1975). The distribution of the Ca^{2+}-dependent protein activator of cyclic nucleotide phosphodiesterase in invertebrates. *Biochem. Biophys. Res. Commun.* **65,** 975–982.

Waisman, D. M., Singh, T. J., and Wang, J. H. (1978a). A modulator-dependent protein kinase from rabbit skeletal muscle. *Fed. Proc.*, **37,** 1303.

Waisman, D. M., Stevens, F. C., and Wang, J. H. (1978b). Purification and characterization of a Ca^{2+}-binding protein in Lumbricus terresties. *J. Biol. Chem.* **253,** 1106–1113.

Wallace, R. W., Lynch, T. J., Tallant, E. A., and Cheung, W. Y. (1978). An endogenous inhibitor protein of brain adenylate cyclase and cyclic nucleotide phosphodiesterase. *Arch. Biochem. Biophys.* **187,** 328–334.

Wallace, R. W., Lynch, T. J., Tallant, E. A., and Cheung, W. Y. (1979). Purification and characterization of an inhibitor protein of brain adenylate cyclase and cyclic nucleotide phosphodiesterase. *J. Biol. Chem.* **254,** 377–382.

Walsh, D. A., Ashby, C. D., Gonzalez, C., Calleins, D., Fischer, E. H., and Krebs, E. G. (1971). Purification and characterization of a protein inhibitor of adenosine 3²,5²-monophosphate-dependent protein kinase. *J. Biol. Chem.* **246,** 1977–1985.

Wang, J. H., and Desai, R. (1976). A brain protein and its effect on the Ca^{2+}- and protein modulator-activated cyclic nucleotide phosphodiesterase. *Biochem. Biophys. Res. Commun.* **72,** 926–932.

Wang, J. H., and Desai, R. (1977). Modulator binding protein: Bovine brain protein exhibiting the Ca^{2+}-dependent association with the protein modulator of cyclic nucleotide phsophodiesterase. *J. Biol. Chem.* **252,** 4175–4184.

Wang, J. H., and Waisman, D. M. (1979). Calmodulin and its role in the second messenger system. *Curr. Top. Cell. Regul.* **15,** 47–107.

Wang, J. H., Teo, T. S., Ho, H. C., and Stevens, F. C. (1975). Bovine heart protein activator of cyclic nucleotide phosphodiesterase. *Adv. Cyclic Nucleotide Res.* **5,** 179–194.

Watterson, D. M., and Vanaman, T. C. (1976). Affinity chromatography purification of a cyclic nucleotide phosphodiesterase using immobilized modulator protein, a troponin C-like protein from brain. *Biochem. Biophys. Res. Commun.* **73,** 40–46.

Watterson, D. M., Harrelson, W. G., Keller, P. M., Sharief, F., and Vanaman, T. C. (1976). Structural similarities between the Ca^{2+}-dependent regulatory proteins of 3^2:5^2-cyclic nucleotide phosphodiesterase and actomyosin ATPase. *J. Biol. Chem.* **251,** 4501–4513.

Westcott, K. R., LaPorte, D. C., and Storm, D. R. (1979). Resolution of adenylate cyclase sensitive and insensitive to Ca^{2+} and calcium-dependent regulatory protein (CDR) by CDR-sepharose affinity chromatography. *Proc. Natl. Acad. Sci. U.S.A.* **76,** 204–208.

Yagi, K., Yazawa, M., Kakiuchi, S., Oshima, M., and Uenishi, K. (1978). Identification of an activator protein for myosin light chain kinase as the Ca^{2+}-dependent modulator protein. *J. Biol. Chem.* **253,** 1338–1340.

Chapter 16

Mechanisms and Pharmacological Implications of Altering Calmodulin Activity

BENJAMIN WEISS
THOMAS L. WALLACE

CALCIUM AND CELL FUNCTION, VOL. I

ISBN 0-12-171401-2

I. INTRODUCTION

A continuing goal of pharmacologists is to develop chemical agents that will selectively restore abnoramlly functioning biological processes to normal. One sequence that can be followed to accomplish this goal is to discover how these processes are regulated and then to search for chemicals that may mimic or interfere with these processes. The discovery of calmodulin, an endogenous calcium-binding protein, that can regulate the activity of a number of enzyme systems provides a new, and as yet relatively unexplored, field for uncovering novel and potentially more effective pharmacological agents.

The history of this area of research follows a not uncommon progression of events in that the first hint that there may be an endogenous regulator of phosphodiesterase activity was found serendipitously (see Cheung, this volume, Chapter 1) in studies of regulatory properties of phosphodiesterase (Cheung, 1967). Shortly thereafter, it was reported that this regulator was a heat-stable protein and that the ability of this protein to activate phosphodiesterase was dependent on the presence of calcium (Cheung, 1970; Kakiuchi and Yamazaki, 1970). The demonstration that phosphodiesterase was not a simple molecular entity but rather existed in multiple molecular forms (Thompson and Appleman, 1971; Uzunov and Weiss, 1972a) was followed by the finding that one of these forms was calcium dependent (Uzunov and Weiss, 1972a; Kakiuchi *et al.*, 1972, 1973). This led to the observation that calmodulin did not activate all the phosphodiesterases indiscriminately, but selectively activated one of these forms of the enzyme, the one that depended upon calcium for its activity (Uzunov and Weiss, 1972a).

The discovery that drugs could selectively interfere with the activity of calmodulin was also made serendipitously while studying the influence of antipsychotic agents on the metabolism of cyclic nucleotides. Early studies showed that antipsychotic agents inhibited the activity of adenylate cyclase and that this inhibition was related to the clinical antipsychotic effectiveness of these drugs, thus supporting the hypothesis that antipsychotic activity may be associated with an inhibition of the formation of cyclic AMP (Uzunov and Weiss, 1971). Further studies, however, showed that antipsychotic agents not only blocked adenylate cyclase activity but also inhibited the activity of phosphodiesterase (Uzunov and Weiss, 1971). The reason why this group of drugs inhibited both of these enzyme systems was an intriguing question, one which could not lay unexplored. The answer came from the following sequence of experiments. Studies of the actions of antipsychotics on each of the multiple forms of phosphodiesterase in brain showed that the drugs preferentially inhibited the same form of phosphodiesterase as that which was activated by calmodulin (Weiss *et al.,* 1974). Further, these drugs inhibited the calmodulin-induced activation of phosphodiesterase in concentrations that did not alter the basal phosphodiesterase activity, suggesting that the antipsychotics were acting on calmodulin rather than on phosphodiesterase (Weiss *et al.,* 1974; Levin and Weiss, 1976; Wolff and Brostrom, 1976; Filburn *et al.,* 1979). Subsequent studies in which the binding of antipsychotics to calmodulin and to phosphodiesterase was measured directly showed that the specific mechanism by which antipsychotics inhibited the action of calmodulin on phosphodiesterase was by binding to calmodulin (Levin and Weiss, 1977, 1979; Weiss and Levin, 1978).

It is now known that calmodulin not only activates specific forms of phosphodiesterase but specific forms of adenylate cyclase (Brostrom *et al.,* 1975; Cheung *et al.,* 1975a) and a number of other enzymes as well (see below). Since antipsychotic drugs can bind to and, thereby, inhibit calmodulin, one may predict that they would inhibit the calmodulin-induced activation of each of these enzymes. In all cases studied this has now shown to be true. For example, antipsychotic drugs have been reported to specifically inhibit the calmodulin-induced activation of adenylate cyclase (Brostrom *et al.,* 1977; Brostrom *et al.,* 1978a), ATPase (Kobayashi *et al.,* 1979; Levin and Weiss, 1980) and other calmodulin-dependent enzymes (see below). These results suggest that the inhibitory effect of antipsychotic drugs on the calmodulin-induced stimulation of adenylate cyclase and phosphodiesterase may be explained by a common mechanism. In fact, several of the biochemical actions of antipsychotic drugs may be explained by this same mechanism, namely, by their direct binding to and inhibition of calmodulin.

Although numerous questions remain, including whether any of the known biochemical effects of antipsychotics can explain their pharmacological and clinical actions, these studies do suggest that certain of their effects may be caused by modifying the actions of calmodulin. They suggest further that new classes of drugs may be developed based on their selective inhibition or activation of calmodulin. What follows is a brief review of these studies, a discussion of the pharmacological implications of inhibiting the activity of calmodulin, and some directions for future research.

II. DISTRIBUTION OF CALMODULIN

A. Species Distribution

Calmodulin is an ubiquitous protein having been found in every species of animal so far examined. Originally detected in mammals (Cheung, 1970, 1971; Kakiuchi and Yamazaki, 1970), it has now been found in other classes of Chordata, including Agnatha (Childers and Siegel, 1975) and Aves (Head *et al.*, 1977; Dabrowska *et al.*, 1977; Hidaka *et al.*, 1979b), and in many other phyla, including Protozoa (Suzuki *et al.*, 1979), Echinodermata, Mollusca, Nemathelminthes, and Porifera (Waisman *et al.*, 1975). It has also been found in higher plants (Anderson and Cormier, 1978; Charbonneau and Cormier, 1979).

Although most studies show that the calmodulin isolated from different tissues and species is nearly identical (see Wolff and Brostrom, 1979), some chemical differences have been reported. For example, the calmodulin isolated from rat testis (Dedman *et al.*, 1978), bovine brain (Vanaman *et al.*, 1977; Watterson *et al.*, 1980), human erythrocytes (Jarrett and Penniston, 1978), and canine postsynaptic densities (Grab *et al.*, 1979) was reported to have varying amounts of at least six different amino acids. Whether these represent differences between species or tissues is still unclear. These observations do, however, suggest that calmodulin may exist in several molecular forms.

It also is not yet completely clear if the calmodulin that activates phosphodiesterase, for example, is identical in every respect to that which activates adenylate cyclase or ATPase. This is an important issue for pharmacologists because if there are different forms of calmodulin and if these forms are tissue selective or have different functions within the cell, it may be possible to develop relatively selective pharmacological agents that alter the activity of specific calmodulins in different tissues.

B. Tissue Distribution

Not only is calmodulin found in every species examined but in all tissues as well (Smoake *et al.*, 1974; Kakiuchi *et al.*, 1975a,b; Egrie *et al.*, 1977; Drabikowski *et al.*, 1978; Grand *et al.*, 1979). The concentration of calmodulin, however, varies markedly among tissues. For example, the concentration of calmodulin in the testes and brain is about 10 times greater than that found in other tissues (Smoake *et al.*, 1974). The concentration of calmodulin also varies between different parts of an organ. In the brain, high levels of calmodulin are found in the corpus striatum and frontal cerebral cortex; intermediate levels in the cerebellum, corpus callosum, hippocampus, hypothalamus, and thalamus; and low levels in the medulla and pons (Egrie *et al.*, 1977). Calmodulin is also present in certain specialized tissues and cells; it is found in superior cervical ganglia (Boudreau and Drummond, 1975), retina (Liu and Schwartz, 1978; Liu *et al.*, 1979), the Sertoli cells of the testes (Beale *et al.*, 1977), the islets of Langerhans of the pancreas (Sugden *et al.*, 1979; Valverde *et al.*, 1979), and the proximal tubules of the kidney (Filburn and Sacktor, 1976), and it has been demonstrated in isolated lymphocytes (Hait and Weiss, 1977; Burgess *et al.*, 1978), ova, spermatozoa (Jones *et al.*, 1978), and fibroblasts (Van Eldik and Watterson, 1979; LaPorte *et al.*, 1979).

C. Subcellular Distribution

Calmodulin has been found in most subcellular fractions of tissues. Generally the cytoplasmic fraction contains high amounts of calmodulin and the nuclear, mitochondrial, and microsomal fractions contain lower amounts (Cheung *et al.*, 1975b; Beale *et al.*, 1977; Egrie *et al.*, 1977; Gnegy *et al.*, 1977c). The mitochondrial fraction of liver, however, is an exception in that it has been reported to contain no calmodulin (Smoake *et al.*, 1974).

Calmodulin has been reported to be associated with specific cellular components of brain, including synaptic (Sobue *et al.*, 1979) and vesicular membranes (DeLorenzo *et al.*, 1979), postsynaptic densities (Grab *et al.*, 1979), and microtubules (Runge *et al.*, 1979). There is also a report that calmodulin is associated with the mitotic apparatus of eukaryotic cells, suggesting a role in chromosome movement (Welsh *et al.*, 1978). The activator protein has also been reported to be associated with glycogen particles in the cytoplasm of liver and with the I band of skeletal muscle (Harper *et al.*, 1980).

One must be careful in interpreting these studies since various conditions can alter the distribution of calmodulin between membranes and cytoplasm. The addition of calcium to the medium has been shown to increase the amount of calmodulin associated with membranes, and, conversely, removal of calcium has been shown to increase its concentration in the cytoplasm (Vanaman *et al.*, 1976; Watterson *et al.*, 1976; Teshima and Kakiuchi, 1978). Moreover, chemical agents, such as biogenic amines and reserpine, have been reported to increase the concentration of calmodulin in cytoplasm by releasing it from membranes (Revuelta *et al.*, 1976; Hanbauer *et al.*, 1979a,b), and ACTH has been shown to increase the amount of the activator protein in nuclei of adrenal cortical cells (Harper *et al.*, 1980).

III. BIOCHEMICAL ACTIONS OF CALMODULIN

A. Activation of Phosphodiesterase

Cyclic nucleotide phosphodiesterase exists in multiple molecular forms. These forms have different molecular weights, electrophoretic mobilities, kinetic properties, substrate affinities, ionic requirements, and heat stabilities (for reviews of these properties, see Weiss, 1975; Weiss and Fertel, 1977; Wells and Hardman, 1977; Weiss and Greenberg, 1978; Strada and Thompson, 1978). These forms also have different susceptibilities to drugs, a property of interest to pharmacologists since it suggests the means of selectively altering the intracellular concentration of cyclic nucleotides in discrete tissues (Weiss and Hait, 1977).

Each tissue, and perhaps each type of cell, appears to have its own unique pattern and ratio of the different molecular forms of phosphodiesterase. This has been observed in numerous tissues, including cerebrum (Uzunov *et al.*, 1974) cerebellum (Uzunov and Weiss, 1972a), caudate nucleus (Fertel and Weiss, 1974), lung (Fertel and Weiss, 1976), pancreas (Weiss and Fertel, 1977), liver, kidney, heart, skeletal muscle, and adipose tissue (Thompson and Appleman, 1971) and in cloned astrocytoma and neuroblastoma cells (Uzunov *et al.*, 1974).

At least one of these forms of phosphodiesterase is activated by calmodulin. Figure 1, for example, shows the effect of varying concentrations of calmodulin on the activation of the different molecular forms of phosphodiesterase isolated from bovine brain by polyacrylamide gel electrophoresis. As can be seen, calmodulin increased phosphodiesterase activity of peak II by about 15-fold, but had no effect on peaks I or IV. The relatively small degree of activation of peak III has probably due to a contamination with peak II since calmodulin did not activate peak III when

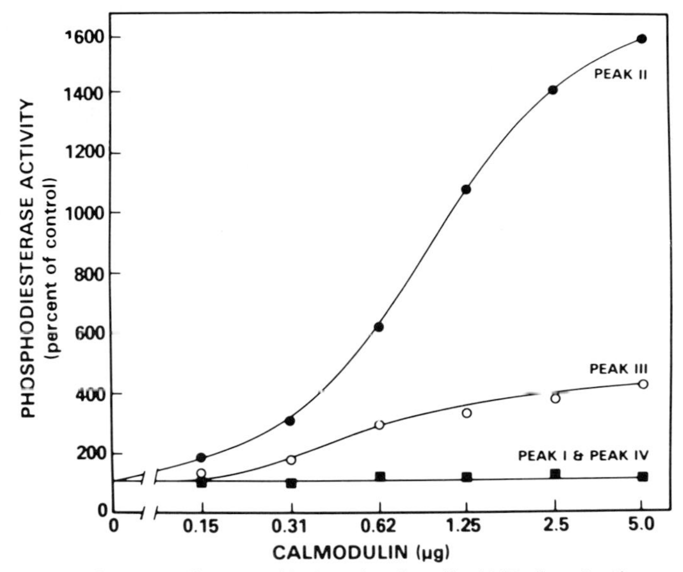

Fig. 1. Effect of calmodulin on purified peaks of cyclic AMP phosphodiesterase of beef cerebrum. The soluble supernatant fraction of beef cerebral homogenates was subjected to polyacrylamide gel electrophoresis, and the cyclic AMP phosphodiesterase activity was determined. The phosphodiesterase activity of peaks I, II, III, and IV was determined in the presence of varying quantities of calmodulin. (Taken from Weiss, 1975, by permission.)

prepared from tissues in which peak III was present but which contained no peak II.

The properties of the calmodulin-dependent phosphodiesterase have been studied intensively by several investigators. It hydrolyzes both cyclic AMP and cyclic GMP (Brostrom and Wolff, 1976), having a higher Michaelis constant (K_m) for cyclic AMP than for cyclic GMP (see Wolff and Brostrom, 1979). The addition of calmodulin has been reported to decrease the K_m value of the enzyme (Uzunov *et al.,* 1976; Brostrom and Wolff, 1976). Activation of the calmodulin-sensitive form of phosphodiesterase is dependent on the presence of calcium and magnesium, although certain other divalent cations can at least partially substitute for these ions (Kakiuchi *et al.,* 1972; Cheung *et al.,* 1975b). Calcium, which binds to calmodulin in a ratio of up to 4 calcium ions to 1 molecule of calmodulin (Watterson *et al.,* 1976), converts calmodulin to a more helical structure (Liu and Cheung, 1976). Apparently, this conversion is required for the binding to and activation of the calmodulin-sensitive enzymes (Wang *et al.,* 1975; Liu and Cheung, 1976; Kuo and Coffee, 1976; Klee, 1977; Walsh and Stevens, 1977; Wolff *et al.,* 1977; Drabikowski *et al.,* 1977;

Seamon, 1979; Walsh *et al.*, 1979). Magnesium serves as a cofactor for the enzyme (Sutherland and Rall, 1958) and perhaps also for calmodulin. Magnesium markedly increases the affinity of phosphodiesterase for calmodulin (Brostrom and Wolff, 1976), possibly by directly binding to divalent cation sites on calmodulin (Wolff *et al.*, 1977). As a result of their studies Wolff *et al.*, (1977) proposed that in its physiological state calmodulin may have 3 calcium and 1 magnesium ion bound to the cation sites.

The distribution of the calmodulin-sensitive phosphodiesterase has also been extensively studied. Often, the relative activity of this form of the enzyme parallels the concentration of calmodulin in that tissue. Smoake *et al.* (1974) and Kakiuchi *et al.* (1975a,b) reported such a correlation in the adrenals, whole brain, kidneys, and liver, and Egrie *et al.* (1977) and Greenberg *et al.* (1978) showed that regions of the brain having high levels of calmodulin also had high activities of the calmodulin-dependent phosphodiesterase. However, not all tissues show a correlation between calmodulin and the calmodulin-sensitive phosphodiesterase. For example, testes, which have the highest concentration of calmodulin, have one of the lowest levels of phosphodiesterase activity (Smoake *et al.*, 1974). This is in agreement with the studies of Beale *et al.* (1977) who reported that the amount of calmodulin present in testes is in excess of that needed for maximal stimulation of phosphodiesterase activity. Further, although calmodulin is present in lungs, there is little or no detectable calmodulin-dependent phosphodiesterase in this tissue (Fertel and Weiss, 1976). Little or no calmodulin-sensitive phosphodiesterase has also been observed in other tissues that contain calmodulin, including adrenal medulla (Egrie and Siegel, 1975; Levin and Weiss, 1978b) and lymphocytes (Hait and Weiss, 1977). These findings suggested that calmodulin has other functions besides stimulating phosphodiesterase. Indeed, it is now known that calmodulin serves as a regulatory protein for many calcium-dependent enzymes. It may be mentioned, however, that in brain, a correlation does exist between the activities of calmodulin and phosphodiesterase, suggesting that at least in this tissue one of the major roles of calmodulin may be to regulate phosphodiesterase activity.

In measuring the relative abundance of the calmodulin-sensitive phosphodiesterase in tissues, certain precautions must be taken to avoid erroneous conclusions. In many cases calmodulin remains associated with the enzyme even after chromatographic procedures. If this occurs the enzyme may already be maximally activated so no further activation is apparent. This problem can be overcome by homogenizing the tissue with the calcium chelator, ethyleneglycol-bis(β-aminoethyl ether)-*N*-*N*,*N*'*N*'-tetraacetic acid (EGTA). By chelating calcium, calmodulin dissociates from the calmodulin-dependent phosphodiesterase (Teo *et al.*, 1973; Teshima and Kakiuchi, 1974; Pledger *et al.*, 1975) and may then be separated from the enzyme by chromatographic or electrophoretic techniques. A similar problem was encountered in attempts to demonstrate a calmodulin-dependent activation of adenylate cyclase, and a similar procedure was proposed for its solution (Brostrom *et al.*, 1975).

Another technique one can use to demonstrate a calmodulin-sensitive phosphodiesterase is to take advantage of the different heat stabilities of the phosphodiesterases. Heating a soluble supernatant fraction of rat cerebral homogenates at 50°C for 30 min results in an increased proportion of activatable phosphodiesterase (Fig. 2). These findings can be explained by earlier studies showing that the activatable form of phosphodiesterase (peak II) was more stable than the nonactivatable form of phosphodiesterase (peak III) (Uzunov *et al.*, 1974).

Numerous questions in this area still remain. For example, since it is difficult to dissociate calmodulin completely from phosphodiesterase, it is still uncertain whether or not tissues contain a calcium-dependent, calmodulin-independent form of phosphodiesterase. The demonstration that calcium activates phosphodiesterase may be explained either by a direct activation of the enzyme by calcium or by the presence of a tightly coupled calmodulin–phosphodiesterase complex that is activated further by calcium.

Another unanswered question is the number of forms of phosphodiesterase that actually exist in tissue. It has been shown that the forms of phosphodiesterase identified in tissue are dependent to a great extent on the

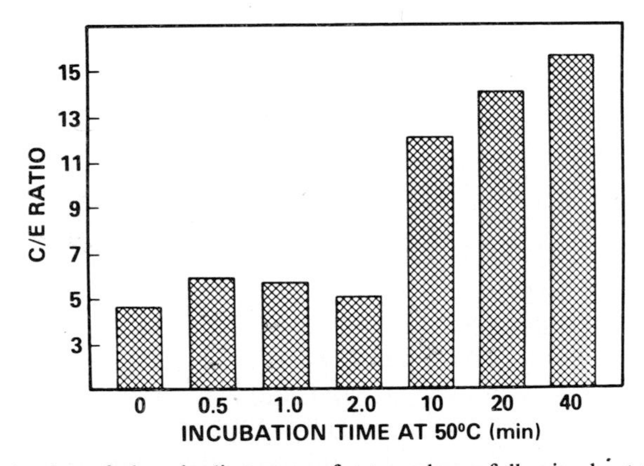

Fig. 2. Activation of phosphodiesterase of rat cerebrum following heat treatment. The 100,000 *g* soluble fraction of rat cerebral homogenates was incubated at 50°C for varying periods of time. The samples were then assayed at 37°C for phosphodiesterase activity in the presence of 10 units of calmodulin or 400 μM EGTA. The ratio of the phosphodiesterase activity in the presence of calmodulin to that in the presence of EGTA(C/E ratio) is a measure of the degree of activation of the enzyme.

techniques used to isolate them and that these forms of the enzyme may even be interconvertible (Chassy, 1972; Strada and Thompson, 1978). Nevertheless, the majority of the evidence clearly supports the view that multiple forms of phosphodiesterase do exist, that there are different relative activities of these forms in different tissues, and that one form of the enzyme can be selectively activated by calmodulin.

B. Activation of Adenylate Cyclase and Guanylate Cyclase

A few years after the demonstration that calmodulin activates phosphodiesterase, this calcium-binding protein was reported to activate adenylate cyclase. Like phosphodiesterase, adenylate cyclase also exists in at least two forms, only one of which is dependent on calmodulin for activity (C. O. Brostrom *et al.*, 1975; Cheung *et al.*, 1975a, 1978; Lynch *et al.*, 1976; M. A. Brostrom *et al.*, 1978a). Thus far, the calmodulin-dependent form of adenylate cyclase has been reported in brain (Brostrom *et al.*, 1978a; Cheung *et al.*, 1978), glioma (Brostrom *et al.*, 1976) and neuroblastoma (Cheung *et al.*, 1978). Other nonneural tissues, including heart, kidney and erythrocytes, do not appear to have this form of the enzyme (Cheung *et al.*, 1978).

The distribution of calmodulin-dependent adenylate cyclase varies in the different regions of the brain: medulla, hippocampus, thalamus, and hypothalamus have high activities; cerebrum, cerebellum, olfactory bulb, and caudate nucleus intermediate activities; and pons a low activity of calmodulin-sensitive adenylate cyclase (Cheung *et al.*, 1978).

In addition to activating adenylate cyclase, calmodulin has recently been reported to activate guanylate cyclase from *Tetrahymena* (Nagao *et al.*, 1979). Whether a calmodulin-sensitive guanylate cyclase is in other tissues is still unclear. These findings suggest that, at least in some tissues, calmodulin can modify the biosynthesis and biodegradation of both cyclic AMP and cyclic GMP.

An important issue that has yet to be resolved is whether calmodulin is involved in the stimulation of adenylate cyclase induced by the catecholamines. In favor of this possibility is the evidence that the catecholamine-stimulated formation of cyclic AMP is dependent on calcium (von Hungen and Roberts, 1973; Schwabe and Daly, 1977; Schwabe *et al.*, 1978; Brostrom *et al.*, 1979); in brain the accumulation of cyclic AMP induced by-α-adrenergic agonists apparently is totally dependent on the presence of calcium, while that elicited by β-adrenergic agonists is partially dependent on calcium (Schwabe and Daly, 1977). Further, norepinephrine-stimulated cyclic AMP accumulation in glial tumor cells is reduced by depleting intracellular calcium, and restored by readding calcium (Brostrom *et al.*, 1979). Using a different approach, it was reported (Gnegy *et al.*, 1976; Costa *et al.*, 1977) that the release of calmodulin from membranes decreases the responsiveness of adenylate cyclase to dopamine in the corpus striatum. Finally, phenothiazine antipsychotics block the activation of adenylate cyclase induced by norepinephrine (Weiss and Kidman, 1969; Weiss, 1970; Uzunov and Weiss, 1971, 1972b; Sulser and Robinson, 1978) and dopamine (Clement-Cormier *et al.*, 1974; Sulser and Robinson, 1978) as well as that induced by calmodulin (C. O. Brostrom *et al.*, 1977; M. A. Brostrom *et al.*, 1978a). Evidence against calmodulin having a role in dopamine-stimulated adenylate cyclase is provided by experiments showing that dopamine increases adenylate cyclase activity even in the presence of EGTA (Clement-Cormier *et al.*, 1975). Therefore, the answer to this important question of whether calmodulin is involved in the catecholamine-induced stimulation of adenylate cyclase must await further experiments.

Investigations into the possible influence of calmodulin on the stimulation of adenylate cyclase induced by guanine nucleotides also yielded inconclusive results. Some studies suggested that calmodulin is required for the activation of adenylate cyclase by guanine nucleotides (Brostrom *et al.*, 1978b; Moss and Vaughn, 1977; Toscano *et al.*, 1979) while others suggested that it is not (Cheung *et al.*, 1978). A resolution of this question may also shed light on the role of calmodulin in the catecholamine-stimulated adenylate cyclase since guanine nucleotides apparently are necessary for the activation of adenylate cyclase by catecholamines (see Levitzki and Helmreich, 1979).

C. Activation of ATPase

ATPase also exists in multiple molecular forms, these forms having different substrate affinities (Robinson, 1976) and ionic requirements (Robinson, 1976; Scharff, 1976; Rosenblatt *et al.*, 1976; Gubitz *et al.*, 1977). The classification of the ATPases is based on their sensitivity to cations; these include ($Na^+ + K^+$)-, ($Ca^{2+} + Mg^{2+}$)-, and (Mg^{2+})-dependent forms. Of these, calmodulin has been shown to specifically activate the ($Ca^{2+} + Mg^{2+}$)-ATPase. This was demonstrated first in erythrocyte membranes (Bond and Clough, 1973; Jarrett and Penniston, 1977; Gopinath and Vincenzi, 1977; Vincenzi and Farrance, 1977; Lynch and Cheung, 1979; Niggli *et al.*, 1979; Jarrett and Kyte, 1979; Levin and Weiss, 1980) and subsequently in synaptic plasma membranes (Sobue *et al.*, 1979; Kuo *et al.*, 1979). As an example, Fig. 3 shows the effects of calmodulin on the ($Ca^{2+} + Mg^{2+}$)-ATPase from erythrocyte membranes. Calmodulin elic-

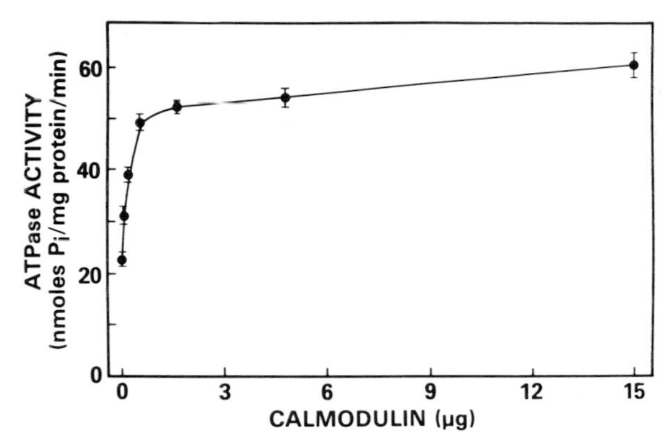

Fig. 3. Effect of calmodulin on calcium-magnesium ATPase activity of rat erythrocytes. The effect of calmodulin on ATPase activity of rat erythrocyte membranes was determined in the presence of 4 m*M* Mg^{2+} and 0.05 m*M* Ca^{2+}. Each point represents the mean of four determinations. Vertical brackets represent the standard error. (Taken from Levin and Weiss, 1980, by permission.)

ited about a three fold activation of this form of ATPase but had no effect on the (Mg^{2+})-dependent form of the enzyme (Levin and Weiss, 1980).

The mechanism by which calmodulin activates (Ca^{2+} + Mg^{2+})-ATPase appears to be similar to that by which it activates adenylate cyclase and phosphodiesterase. That is, calmodulin reacts with calcium, then the calmodulin–calcium complex interacts with the enzyme causing its activation (Lynch and Cheung, 1979).

Since (Ca^{2+} + Mg^{2+})-ATPase appears to play an important part in regulating the intracellular concentrations of calcium (Schatzman and Vincenzi, 1969; Lee and Shin, 1969; Olson and Cazort, 1969), it was not surprising to find that calmodulin produced changes in calcium transport. Calmodulin-stimulated calcium uptake into inside-out vesicles of erythrocyte membranes (Larsen and Vincenzi, 1979), stimulated calcium transport in microsomal preparations enriched in cardiac sarcoplasmic reticulum (Katz and Remtulla, 1978), and altered the ATP-dependent calcium uptake into synaptic plasma membranes (Kuo *et al.*, 1979). Obviously pharmacological blockade of calmodulin's action on (Ca^{2+} + Mg^{2+})-ATPase would have profound biological consequences.

D. Activation of Kinases

Calmodulin has been reported to activate a variety of protein kinases, including myosin light chain kinase (Dabrowska *et al.*, 1977, 1978; Yagi *et al.*, 1978; Yazawa *et al.*, 1978; Yazawa and Yagi, 1978; Walsh *et al.*,

1980), phosphorylase kinase (Cohen *et al.*, 1978; Waisman *et al.*, 1978; Depaoli-Roach *et al.*, 1979; Shenolikar *et al.*, 1979), and glycogen synthase kinase (Rylatt *et al.*, 1979; Srivastava *et al.*, 1979). Thus it appears that calmodulin activates several kinases that are involved in stimulating glycogenolysis in skeletal muscle.

Besides activating the kinases in muscle, calmodulin has also been shown to activate NAD kinase in plants (Anderson and Cormier, 1978). Calmodulin also appears to be required for the calcium-dependent phosphorylation of synaptosomal membranes (Schulman and Greengard, 1978a; Greengard, 1978; Sieghart *et al.*, 1980) and brain cytosol (Yamauchi and Fujisawa, 1979). Calcium-dependent phosphorylation of membrane proteins from lung, spleen, skeletal muscle, vas deferens, heart, and adrenal has also been shown to be calmodulin-dependent (Schulman and Greengard, 1978b).

The calmodulin-dependent protein kinase system of synaptic vesicles may be involved in the Ca^{2+}-stimulated release of neurotransmitters (DeLorenzo *et al.*, 1979). Depolarization of synaptosomes causes the release of norepinephrine and the phosphorylation of synaptosomal membranes (DeLorenzo and Freedman, 1978; DeLorenzo *et al.*, 1979). An endogenous factor present in synaptosomes was shown to be responsible for these effects and was identified as calmodulin. The addition of calmodulin resulted in the release of norepinephrine from synaptic vesicles and caused the phosphorylation of proteins in these vesicles (DeLorenzo *et al.*, 1979). The report that the release of acetylcholine and dopamine from vesicles is also dependent on calmodulin (DeLorenzo *et al.*, 1979) suggests that this calcium-binding protein may have a general role in releasing neurotransmitters from nerve endings.

E. Other Actions of Calmodulin

In addition to the actions already mentioned, calmodulin has been reported to stimulate the activity of phospholipase A2 in platelets (Wong and Cheung, 1979). It also appears to play a role in intestinal secretion (Ilundain and Naftalin, 1979), insulin secretion (Sugden *et al.*, 1979), disassembly of microtubules (Marcum *et al.*, 1978; Nishida *et al.*, 1979; Kumagi and Nishida, 1979), and smooth muscle contraction (Hidaka *et al.*, 1979a).

Clearly, the list of actions of calmodulin is not yet complete. The challenge to the pharmacologist is to determine how calmodulin's activity can be modified and, in particular, to determine if any of the diverse actions of calmodulin can be altered selectively. This question will be addressed in Section IV.

IV. ALTERATION OF CALMODULIN ACTIVITY

Since calmodulin has such an ubiquitous role in biology, it is important to learn how its activity is regulated. So far, several materials have been shown to influence calmodulin activity. These include divalent cations, endogenous proteins and psychotropic drugs.

A. Inhibition of Calmodulin Activity by Calcium Chelators

Calcium chelators such as EGTA inhibit the formation of the Ca^{2+}–calmodulin complex that is essential for calmodulin's activity (Teo and Wang, 1973; Wolff and Brostrom, 1974; Lin *et al.*, 1975). The consequence of this is that calmodulin can no longer activate calmodulin-sensitive enzymes. For example, Fig. 4 shows that calmodulin increased the activity of a calmodulin-sensitive form of phosphodiesterase (peak II) by about eightfold; this activation was completely inhibited by EGTA. By contrast, EGTA had little or no effect on peak II phosphodiesterase when

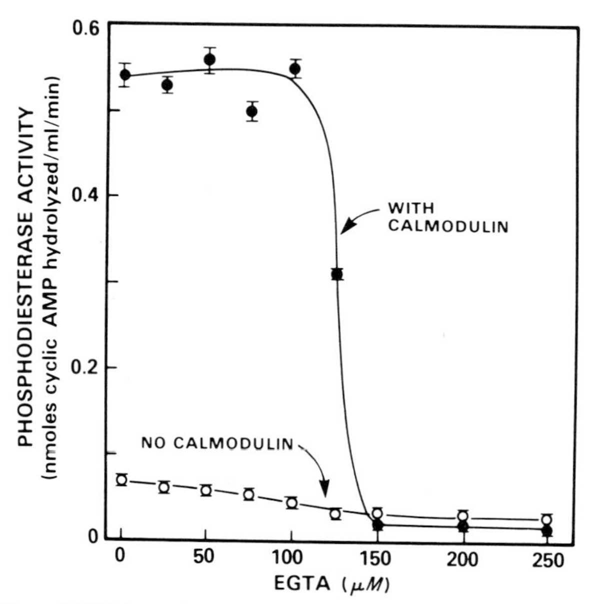

Fig. 4. Effect of EGTA on the calmodulin-induced activation of cyclic AMP phosphodiesterase of rat cerebrum (peak II). Peak II phosphodiesterase was isolated by polyacrylamide gel electrophoresis. Enzyme activity was determined using 400 μM cyclic AMP as substrate in the presence and absence of 10 units of calmodulin and varying concentrations of EGTA. Each point represents the mean value of four determinations. Vertical brackets represent the standard error.

measured in the absence of calmodulin, or on any of the calmodulin-insensitive phosphodiesterases.

B. Inhibition of Calmodulin-Sensitive Enzymes by Drugs

1. Inhibition of Phosphodiesterase Activity

Early studies on the influence of psychotropic drugs on the metabolism of cyclic nucleotides in brain showed that antipsychotics inhibited phosphodiesterase from cerebrum to a far greater extent than they inhibited the enzyme from cerebellum (Uzunov and Weiss, 1971). The reason for this differential inhibition of phosphodiesterase became clear from the following observations. Brain contains multiple forms of phosphodiesterase, and the proportion of one of these forms (designated peak II) is much greater in cerebrum (Uzunov *et al.*, 1974) than in cerebellum (Uzunov and Weiss, 1972a). This form of phosphodiesterase (peak II) is selectively activated by calmodulin (Uzunov and Weiss, 1972a) and selectively inhibited by phenothiazine antipsychotics (Weiss *et al.*, 1974). The explanation for the selectivity by which these drugs inhibited peak II phosphodiesterase came from experiments showing that the phenothiazines blocked the calmodulin-induced activation of phosphodiesterase in concentrations that failed to inhibit a nonactivated peak II phosphodiesterase or other

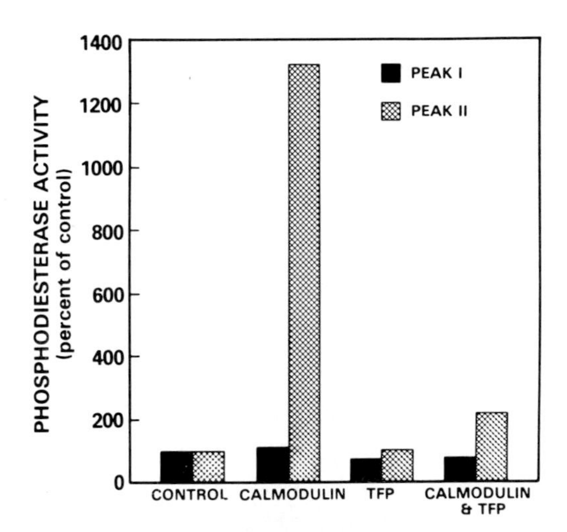

Fig. 5. Effects of trifluoperazine and calmodulin on peak I and peak II phosphodiesterase activity of rat cerebrum. Each purified peak of phosphodiesterase was assayed for cyclic AMP phosphodiesterase activity in the absence and presence of calmodulin and trifluoperazine (TFP) (25 μM). The calmodulin was added in amounts that produced a maximum increase in enzyme activity. (Taken from Weiss and Greenberg, 1975, by permission.)

forms of the enzyme (Weiss *et al.*, 1974; Levin and Weiss, 1976). An example of this effect is shown in Fig. 5 which illustrates the interaction of calmodulin and trifluoperazine on isolated forms of phosphodiesterase from rat cerebrum (peaks I and II). Calmodulin had no effect on peak I but produced more than a tenfold increase in the activity of peak II. Trifluoperazine blocked the calmodulin-induced activation of peak II phosphodiesterase but had no effect on its basal activity or on peak I phosphodiesterase measured in the presence or absence of calmodulin.

The questions that now presented themselves were what other drugs blocked the calmodulin-induced activation of phosphodiesterase and whether these drugs fit into a particular pharmacological class. The results showed that of a variety of pharmacological agents examined, the most potent compounds were clinically effective antipsychotic drugs (Levin and Weiss, 1976). Figure 6, for example, shows that whereas trifluoperazine inhibited the calmodulin-sensitive form of phosphodiesterase (peak II), other phenothiazines having weak antipsychotic activity had little inhibitory effects on this enzyme. These latter compounds included trifluoperazine sulfoxide, a metabolite of trifluoperazine, and promethazine, a phenothiazine that has antihistaminic effects but little antipsychotic actions.

Interestingly, the calmodulin-sensitive phosphodiesterase was inhibited, by several different chemical classes of antipsychotics; besides phenothiazines, the butyrophenones, thioxanthenes, and diphenylbutylpiperidines also inhibited calmodulin activity. By contrast, antidepressant drugs such as amitriptyline and desipramine, and anxiolytics, such as medazepam and chlordiazepoxide, were much less potent in blocking calmodulin. Other phosphodiesterase inhibitors, such as theophylline and papaverine and several other centrally acting drugs, such as amphetamine, (+)-lysergic acid diethylamide [(+)-LSD], pentobarbital, and morphine, showed little specific inhibition of calmodulin-sensitive phosphodiesterase (Levin and Weiss, 1976).

Further confirmation that antipsychotics specifically blocked the calmodulin-sensitive phosphodiesterase came from experiments in which we measured the effects of these drugs on phosphodiesterase from different brain regions. The results showed that phosphodiesterase activity of those tissues having the highest percentage of calmodulin-sensitive phosphodiesterase was inhibited to the greatest extent by phenothiazine antipsychotics (Greenberg and Weiss, 1976).

The finding that drugs can inhibit phosphodiesterase activity by blocking the action of calmodulin suggests a novel mechanism for inhibiting this enzyme. It also suggests the possibility of developing a specific inhibitor of one of the multiple forms of phosphodiesterase; the one that can be activated by calmodulin. Since this form of phosphodiesterase is unequally

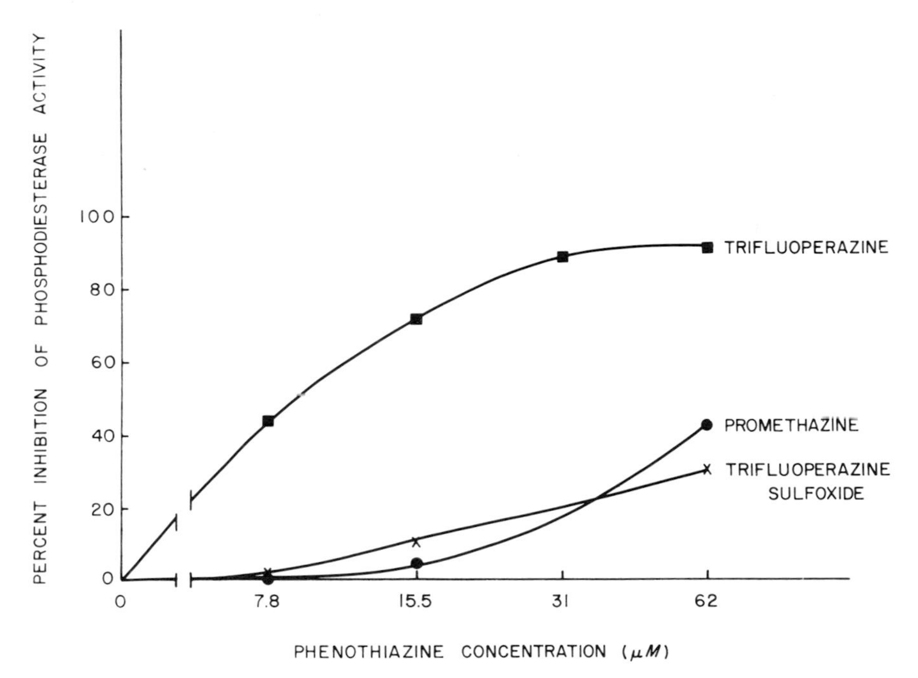

Fig. 6. Inhibition of phosphodiesterase activity of rat cerebrum (peak II) by phenothiazine derivatives. Peak II phosphodiesterase was isolated from rat cerebrum by polyacrylamide gel electrophoresis. Cyclic AMP phosphodiesterase activity was analyzed in the presence of calmodulin (5 μg of protein) and varying concentrations of trifluoperazine, trifluoperazine sulfoxide, or promethazine. Each point represents the mean value of four determinations. (Taken from Weiss *et al.*, 1974, by permission.)

distributed in the body, being concentrated in specific regions of the brain, one may be able to alter phosphodiesterase activity, and therefore the metabolism of cyclic nucleotides, in selected brain areas (see Weiss, 1975; Weiss and Greenberg, 1975, 1978; Weiss and Fertel, 1977).

Although the studies cited above suggest that the inhibition of the calmodulin-sensitive phosphodiesterase is relatively selective for antipsychotic drugs, it should be emphasized that the types of drugs studied thus far are limited. In fact, a recent report shows that vinblastine can also inhibit the activation of cyclic AMP phosphodiesterase induced by a calcium-dependent protein activator (Watanabe *et al.*, 1980). In addition, several vascular relaxing agents have been shown to inhibit other calmodulin-activated enzymes, including (Ca^{2+} + Mg^{2+})-ATPase (Kobayashi *et al.*, 1979; Hidaka *et al.*, 1980) and myosin light chain kinase (Hidaka *et al.*, 1979b). These agents include prenylamine, N^2-dansyl-L-arginine-4-*t*-butylpiperidine amide, and *N*-(6-aminohexyl)-5-chloro-1-naphthalenesulfonamide. Whether these compounds possess antipsychotic activity is not known.

2. *Inhibition of Adenylate Cyclase Activity*

The activity of the catecholamine-sensitive adenylate cyclase may be altered by agents acting at several different sites on the enzyme–receptor complex. These sites include the catalytic moiety on adenylate cyclase, the adrenergic receptors linked to the catalytic moiety, and various regulatory proteins associated with the complex. Purine nucleotides apparently act at the catalytic site by competing with substrate (see Weiss and Fertel, 1977), and adrenergic antagonists act at the receptors by preventing the binding of catecholamines to the receptor (Weiss and Costa, 1968). The actions of antipsychotic drugs appears to be more complex; they block both the catecholamine-induced activation of adenylate cyclase (Uzunov and Weiss, 1971; Miller *et al.*, 1974; Clement-Cormier *et al.*, 1974) and the calmodulin-induced activation of adenylate cyclase (C. O. Brostrom *et al.*, 1977; M. A. Brostrom *et al.*, 1978a).

Since the catalytic subunit of adenylate cyclase appears to be similar in all tissues, blocking this site inhibits enzymatic activity nonspecifically. Blockade of hormonal receptors, however, selectively alters adenylate cyclase activity since receptors are hormone and tissue specific. Blockade of calmodulin's action may also be a selective way of inhibiting adenylate cyclase since relatively few tissues appear to have a calmodulin-dependent form of this enzyme (Cheung *et al.*, 1978).

The mechanism by which antipsychotic drugs inhibit calmodulin-activated adenylate cyclase appears to be similar to that by which they block calmodulin-activated phosphodiesterase; like in the phosphodiesterase system these drugs inhibit the calmodulin-activated form of adenylate cyclase in concentrations that have no effect on basal enzyme activity (C. O. Brostrom *et al.*, 1977; M. A. Brostrom *et al.*, 1978a).

The observations that phenothiazines inhibit adenylate cyclase stimulated by both catecholamines and calmodulin raise the possibility that calmodulin may be involved in the catecholamine-induced activation of adenylate cyclase and that antipsychotic agents may inhibit the catecholamine- and calmodulin-induced activation of adenylate cyclase by a common mechanism. However, no studies are yet available that demonstrate directly that blockade of calmodulin by antipsychotic drugs results in inhibition of catecholamine-stimulated adenylate cyclase.

3. *Inhibition of ATPase Activity*

In what now may be considered a predictable sequence of studies, the demonstration that calmodulin could activate a specific (Ca^{2+} + Mg^{2+})-

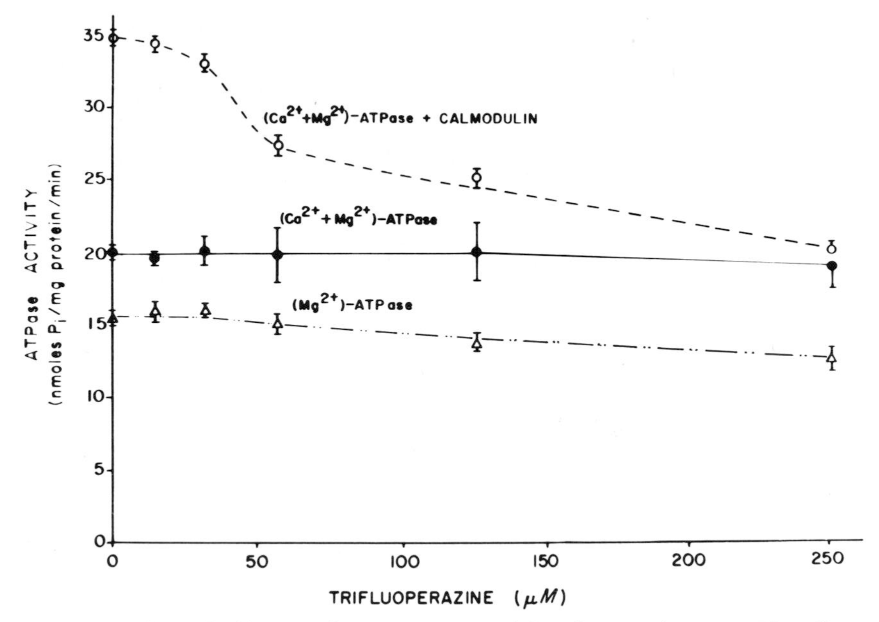

Fig. 7. Effect of trifluoperazine on ATPase activity of rat erythrocytes. The effect of varying concentrations of trifluoperazine on the ATPases of rat erythrocyte membranes was determined. $(Ca^{2+} + Mg^{2+})$-ATPase activity was measured using 0.05 m*M* Ca^{2+} and 4 m*M* Mg^{2+}. (Mg^{2+})-ATPase was measured using 4 m*M* Mg^{2+}. Calmodulin, when present, was at a concentration of 3 μg/ml. Each point represents the mean value of four determinations. Vertical brackets represent the standard error. (Taken from Levin and Weiss, 1980, by permission.)

ATPase was followed by the observation that phenothiazine antipsychotic drugs could specifically inhibit the calmodulin-induced activation of the enzyme in concentrations that failed to block other forms of ATPase (Kobayashi *et al.*, 1979, Levin and Weiss, 1980). For example, experiments in which ATPase of rat erythrocytes was studied revealed that trifluoperazine inhibited the calmodulin-induced activation of $(Ca^{2+} + Mg^{2+})$-ATPase at concentrations that produced no significant inhibition of nonactivated $(Ca^{2+} + Mg^{2+})$-ATPase or of (Mg^{2+})-ATPase (Fig. 7). The concentration necessary for 50% inhibition (IC_{50}) of $(Ca^{2+} + Mg^{2+})$-ATPase by trifluoperazine was 50 μM. Concentrations of drugs up to 200 μM failed to significantly inhibit ATPase activity when measured in the presence of either Mg^{2+} alone or in the presence of Mg^{2+} plus Ca^{2+} in the absence of calmodulin. Increasing the concentrations of calmodulin antagonized the trifluoperazine-induced inhibition of ATPase activity (Levin and Weiss, 1980). Thus, these studies are similar to the previous ones on phosphodiesterase and adenylate cyclase and suggest that trifluoperazine, and possibly other antipsychotics, inhibit calmodulin-stimulated ATPase by directly blocking calmodulin's action.

The recent finding of a calmodulin-dependent ATPase in synaptic membranes (Sobue *et al.*, 1979a) suggests further that some of the actions of antipsychotics on nerve terminals may be explained by a blockade of this enzyme. Antipsychotics are known to prevent the release of catecholamines from presynaptic terminals (Seeman and Lee, 1975, 1976). This release, in turn, may be dependent on the calmodulin-sensitive (Ca^{2+} + Mg^{2+})-ATPase (Kuo *et al.*, 1979).

4. Inhibition of Other Enzymes and Processes

In addition to inhibiting calmodulin-dependent phosphodiesterase, adenylate cyclase and (Ca^{2+} + Mg^{2+})-ATPase, antipsychotic drugs have now been shown to inhibit a number of other enzymes that are activated by calmodulin. These include calmoduln-activated phospholipase A2 in platelets (Wong and Cheung, 1979), calmodulin-activated myosin light chain kinase in gizzards (Hidaka *et al.*, 1979b), and calmodulin-activated phosphorylase kinase (Shenolikar *et al.*, 1979) and glycogen synthase kinase (Srivastava *et al.*, 1979) in skeletal muscle. In all these studies the antipsychotics were shown to inhibit the calmodulin-induced activation of these enzymes in concentrations that failed to inhibit the basal enzyme activity.

Antipsychotic drugs also have been reported to inhibit certain physiological effects that involve calmodulin, including chloride secretion in the gut (Ilundain and Naftalin, 1979), glucose-stimulated insulin release in the pancreas (Sugden *et al.*, 1979), and contraction of aortic smooth muscle (Hidaka *et al.*, 1979a).

C. Inhibition of Calmodulin Activity by Endogenous Substances

A number of proteins that have the ability to inhibit the calmodulin-induced activation of enzymes have recently been isolated from brain. At least two of these proteins are soluble and one is particulate. A major distinguishing feature of the soluble proteins is their relative stabilities to heat. The heat-labile protein has been shown to block the stimulation of calcium transport induced by calmodulin (Larsen *et al.*, 1978) and to inhibit the calmodulin-induced stimulation of phosphodiesterase (Wang and Desai, 1977; Klee and Krinks, 1978; Wallace *et al.*, 1978), adenylate cyclase (Wallace *et al.*, 1978), and ATPase (Larsen *et al.*, 1978).

It is still not certain whether all the heat-labile proteins isolated by the different investigators are identical. Some of the discrepancies in the reported molecular weights may be due to differences in isolation techniques and to the presence of subunits of this protein. For example, one heat-labile protein has been shown to consist of at least two subunits (Wallace *et al.*, 1978; Sharma *et al.*, 1979), only one of which has inhibitory properties (Sharma *et al.*, 1979).

A heat-stable protein has also been isolated from brain and has been shown to inhibit the activation of phosphodiesterase induced by calmodulin (Sharma *et al.*, 1978a,b).

Finally, a third protein, which also may function as an endogenous inhibitor of calmodulin, has recently been described (Sobue *et al.*, 1979b). Unlike those previously mentioned, this one is particulate. High concentrations of this protein are found in neural tissues such as the cerebrum and cerebellum, and low concentrations are found in nonneural tissues.

These findings of endogenous substances in brain that modify the activity of calmodulin suggests other possible sites for drug action. Since it is inherent in biology that the greater the complexity of the system, the more likely one can achieve specificity of action, the existence of these proteins may permit even greater specificity by which drugs can inhibit calmodulin's actions. Speculating further, the evidence that there are endogenous proteins that bind to and inhibit calmodulin, coupled with the evidence that clinically effective antipsychotic drugs also bind to and inhibit calmodulin, raises the possibility that endogenous peptides exist in brain that may possess antipsychotic or possibly psychotogenic activity. The discovery of agents that block or mimic these hypothesized endogenous substances may provide a novel means of treating certain forms of mental disease.

V. MECHANISM OF THE DRUG-INDUCED INHIBITION OF CALMODULIN ACTIVITY

There are at least three possible mechanisms by which one could explain the selective inhibition of the calmodulin-dependent enzymes by antipsychotic drugs. These include chelation of calcium (Rajan *et al.*, 1974), interaction of the drug with the enzyme, and interaction with calmodulin. The first possibility was eliminated by experiments showing that excess calcium could not reverse the phenothiazine-induced inhibition of phosphodiesterase activity (Levin and Weiss, 1976). The possibility that the phenothiazine antipsychotics bind directly to phosphodiesterase seemed unlikely since these drugs blocked the calmodulin-induced activation of several enzyme systems, and, therefore, one would have to hypothesize that the antipsychotics bind to several different proteins. Moreover, studies in which radiolabeled trifluoperazine was added to the purified peak II phosphodiesterase, the form of phosphodiesterase that is activated by calmodulin, failed to reveal any binding between the drug and the enzyme. However, studies in which radiolabeled trifluoperazine was added to a purified preparation of calmodulin showed that there was a high degree of selective binding between the drug and this calcium-binding protein (Levin and Weiss, 1977, 1978a, 1979). Some of the characteristics of this binding are summarized below.

A. Characteristics of the Binding of Trifluoperazine to Calmodulin

1. *Methods of Procedure*

An activatable form of phosphodiesterase (peak II) was prepared from brain by subjecting the soluble supernatant fraction of brain homogenates to polyacrylamide gel electrophoresis (Uzunov and Weiss, 1972a). Calmodulin was prepared and isolated to homogeneity according to the procedures of Teo *et al.* (1973) and Watterson *et al.* (1976). Phosphodiesterase activity was measured by the luciferin–luciferase method (Weiss *et al.*, 1972).

For the binding studies, [^{3}H]trifluoperazine and calmodulin (or other proteins) were dialyzed to equilibrium against Tris buffer, pH 7.0, containing either 0.1 m*M* calcium or 0.3 m*M* EGTA (Levin and Weiss, 1977). Displacement studies were performed utilizing varying concentrations of the radiolabeled ligand and unlabeled drug.

2. *Binding as a Function of Calmodulin Concentration*

The binding of trifluoperazine to calmodulin was linear with increasing concentrations of calmodulin when measured in the presence of calcium or EGTA (Levin and Weiss, 1977). However, the binding was six times greater in the presence of calcium than in the presence of EGTA.

3. *Saturability of Binding*

Experiments in which the binding of trifluoperazine to calmodulin was measured in the presence or absence of calcium showed that the calcium-specific binding sites were saturated at lower concentrations of trifluoperazine than the calcium-independent binding sites (see Section V,A,7).

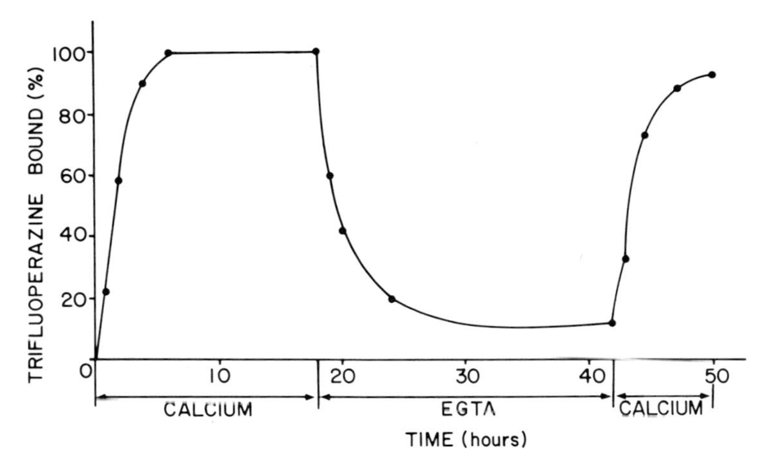

Fig. 8. Reversibility of the binding of trifluoperazine to calmodulin. At the beginning of the experiment, the dialysis bath contained 5 m*M* Tris, pH 7.0, 1 m*M* Mg^{2+}, and 0.1 m*M* Ca^{2+}. Several dialysis membranes containing either water or calmodulin (50 μg/ml) were placed in the bath. [^{3}H]Trifluoperazine was added to the bath to give a final concentration of 0.5 μM. The samples were dialyzed for 18 hr in the presence of Ca^{2+} (0.1 m*M*). EGTA (0.5 m*M*) was then added to the incubation bath. Dialysis was continued for an additional 24 hr at which time sufficient Ca^{2+} was added to achieve a concentration of free Ca^{2+} of 0.4 m*M*. The samples were then dialyzed for 8 hr more. Periodically over the course of the experiment, samples of both calmodulin and water were removed from the dialysis membranes and the radioactivity was determined. The binding of trifluoperazine to calmodulin is given as the percent of maximum binding at the times indicated. (Taken from Weiss and Levin, 1978, by permission.)

The binding of other antipsychotics to calmodulin was also saturable and calcium specific. Significant calcium-specific binding occurred at concentrations of chlorpromazine as low as 10 p*M* (Weiss and Levin, 1978).

4. Reversibility of Binding

The calcium-dependent binding of trifluoperazine to calmodulin could be reversed by the addition of EGTA and restored by the readdition of sufficient concentrations of calcium to overcome the chelating effects of EGTA (Fig. 8).

5. pH Dependence of Binding

The binding of trifluoperazine to calmodulin when measured in the presence of EGTA did not change over a pH range from 6.5 to 8.5. In the presence of calcium, the binding of trifluoperazine to calmodulin was relatively constant between pH 5 and 7.5, but was significantly reduced above pH 7.5 (Levin and Weiss, 1977).

6. *Ion Requirements for Binding*

A variety of divalent cations were examined for their ability to affect the binding of trifluoperazine to calmodulin. Of cations tested, calcium was the most effective in promoting binding of the drug to calmodulin; strontium, nickel, cobalt, zinc, and manganese were progressively less effective, and barium and magnesium were totally ineffective in promoting binding (Levin and Weiss, 1977). The monovalent cation lithium neither promoted the binding of trifluoperazine to calmodulin nor inhibited its binding when measured in the presence of calcium. None of these ions influenced the binding of trifluoperazine to bovine serum albumin.

7. *Kinetics of the Binding of Trifluoperazine to Calmodulin*

A kinetic analysis of the binding of trifluoperazine to calmodulin showed that at concentrations of trifluoperazine of 1 μM or less the bind-

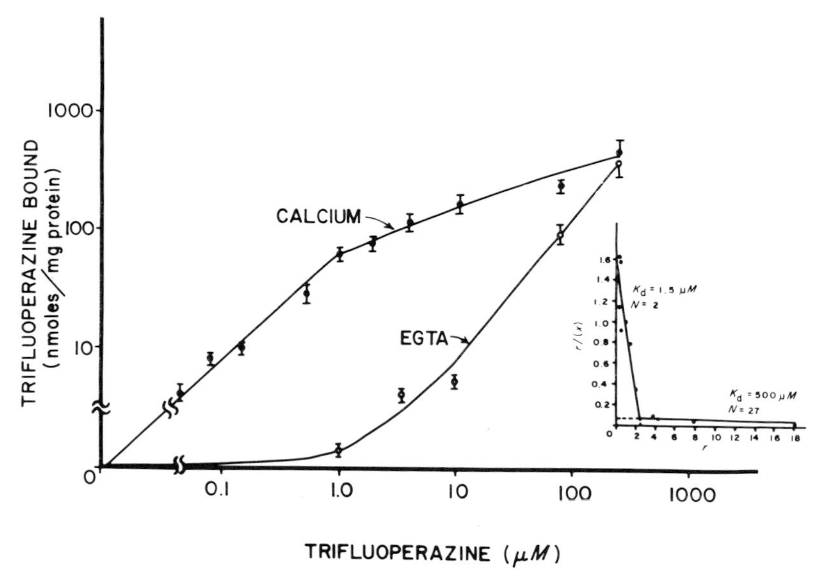

Fig. 9. Binding of trifluoperazine to human brain calmodulin as a function of the concentration of trifluoperazine. The dialysis baths contained 200 ml of 5 mM Tris, pH 7.0, 1 mM Mg^{2+}, 0.05–250 μM [^{3}H]trifluoperazine, and either 0.1 mM Ca^{2+} or 0.3 mM EGTA. The dialysis membranes contained 0.4 ml of human brain calmodulin at a concentration of 25 μg/ml. The samples were dialyzed for 18 hr at 4°C, at which time the radioactivity of the contents were determined. Each value represents the mean of four samples. Vertical bars indicate the standard error. The Scatchard plot (Scatchard, 1949) of the binding of trifluoperazine to human brain calmodulin in the presence of calcium is shown in the insert. r is the moles of trifluoperazine bound per mole of calmodulin; (x) is the concentration of free trifluoperazine at equilibrium. (Taken from Levin and Weiss, 1978a, by permission.)

ing was over 30 times greater in the presence of calcium than in the absence of calcium (Fig. 9). As the concentration of trifluoperazine increased, the ratio of calcium-dependent binding to calcium-independent binding decreased until there was no longer any difference between the binding measured in the presence or absence of calcium. Plotting the data according to the Scatchard equation (Fig. 9, insert) demonstrated two sets of binding sites for trifluoperazine: one set of high-affinity sites with an apparent K_d of 1.5 μM and two sites per molecule of calmodulin, and a second set of low-affinity sites with a K_d of 500 μM and approximately 27 sites per molecule. The binding of trifluoperazine to calmodulin in the absence of calcium displayed only the low-affinity, high-capacity binding (Levin and Weiss, 1977, 1978a).

B. Specificity of the Binding of Trifluoperazine to Calmodulin

To examine the specificity by which trifluoperazine binds to calmodulin, the binding of the phenothiazine to a variety of proteins was deter-

TABLE I

Calcium-Specific Binding of (^{3}H)Trifluoperazine to Various Proteins

Protein	Trifluoperazine bound (nmoles/mg protein)
Bovine brain calmodulin	19
Chick embryo fibroblast calmodulin	20
Human brain calmodulin	15
Rabbit brain calmodulin	12
Rat brain calmodulin	16
Aldolase	NS
Bovine serum albumin	NS
Catalase	NS
Chymotrypsinogen	NS
Cytochrome C	NS
Myosin light chain	NS
Phospholipase A	NS
Porcine brain S-100	NS
Troponin-C	1.3

Samples of various proteins (2.5 to 150 μg/ml) were dialyzed to equilibrium against (^{3}H)trifluoperazine (0.25 μM) in the presence of 0.1 mM calcium or 0.3 mM EGTA. The samples were then removed and the radioactivity determined. Values are the average of at least three determinations.

The calcium-specific binding is defined as the difference in trifluoperazine bound between that measured in the presence and absence of calcium.

NS, No significant calcium-specific binding was detectable.

Taken from Levin and Weiss, 1977, 1978a, by permission.

mined. These included calmodulin prepared from different species and tissues, several calcium-binding proteins, and a number of purified proteins of varying molecular weights (Table I). All samples of calmodulin, regardless of their source, evidenced calcium-specific binding sites for trifluoperazine. The magnitude of this binding differed somewhat among species; however, these differences may have been due to the relative purity of the samples since in each case the binding was proportional to the ability of the preparation to stimulate the activity of phosphodiesterase. Of the other calcium-binding proteins studied, only troponin C displayed calcium-specific binding of trifluoperazine. However, at low concentrations of trifluoperazine (0.2 μM) the calcium-specific binding to troponin C was less than 10% that to calmodulin (Levin and Weiss, 1978a).

Further demonstration of the specificity of antipsychotic drugs for calmodulin has come from studies showing that calmodulin was the only detectable, soluble protein in mammalian brain that binds phenothiazines (M. J. Cormier, personal communication). In fact, this binding is sufficiently specific to allow the isolation of calmodulin from tissues by phenothiazine-linked affinity chromatography (Charbonneau and Cormier, 1979; Jamieson and Vanaman, 1979).

C. Characteristics and Specificity of the Binding of Various Drugs to Calmodulin

1. Specificity of the Binding of Various Drugs to Calmodulin

The binding of a number of compounds to calmodulin was measured in the hope of determining whether this binding is associated with a particular pharmacological class of drugs (Levin and Weiss, 1979). The results showed that several antipsychotic agents belonging to different chemical classes all had a high degree of calcium-specific binding to calmodulin. These drugs, which included phenothiazines, diphenylbutylpiperidines, and butyrophenones, are all thought to produce their antipsychotic effects by blocking the actions of catecholamines postsynaptically. The tricyclic antidepressants and antianxiety agents had a low degree of calcium-specific binding, and a number of other potent centrally acting compounds that are devoid of antipsychotic activity, including (+)-amphetamine, (+)-LSD, morphine, and pentobarbital displayed no calcium-specific binding to calmodulin (Table II).

A number of neurotransmitters and agents that presumably bind to neurotransmitter receptors also showed little or no calcium-specific binding to calmodulin. These included dopamine, histamine, apomorphine, and the β-adrenergic receptor antagonist dihydroalprenolol (Levin and Weiss, 1979). The results with dopamine were particularly interesting

TABLE II

Calcium-Specific Binding of Various Drugs to Calmodulin[a]

Drug	Drug Bound (nmoles bound/mg protein)
Penfluridol	105 ± 11
Pimozide	65 ± 5
Trifluoperazine	56 ± 6
Chlorpromazine	23 ± 1
Haloperidol	13 ± 1
Amitriptyline	9.0 ± 0.7
Imipramine	6.8 ± 0.6
Chlordiazepoxide	5.5 ± 0.2
Diazepam	4.7 ± 0.1
Dihydroalprenolol	1.5 ± 0.2
(+)-Amphetamine	NS
(+)-LSD	NS
Dopamine	NS
Histamine	NS
Morphine	NS
Pentobarbital	NS

[a] Labeled drugs were added to the bath to give a final concentration of 1 μM, and the binding of these agents in the absence or presence of 0.1 mM calcium was determined by equilibrium dialysis. The calcium-specific binding was determined from the difference between the binding measured in the absence and presence of calcium. Each value is the mean of three to six experiments ± standard error. (Taken from Levin and Weiss, 1979, by permission.)

since dopamine and antipsychotics are generally thought to act on the same receptor. Our results showing that dopamine does not bind to calmodulin raises the possibility that some of the biochemical actions of antipsychotic drugs can be explained by their interaction with a non-dopamine-receptor (Weiss *et al.*, 1979).

2. *Kinetics of the Binding of Antipsychotic Drugs to Calmodulin*

All chemical classes of antipsychotic drugs studied so far exhibited a high-affinity, calcium-specific binding to calmodulin. The actual number of calcium-specific binding sites for these drugs on calmodulin appears to vary between 1 and 3 depending in part on the particular drug studied and on the batch of calmodulin used. For example, a Scatchard analysis of the binding of several antipsychotic drugs to calmodulin revealed two calcium-specific binding sites on calmodulin for trifluoperazine (K_d=1.5 μM) (Fig. 9), three for chlorpromazine (K_d=5 μM), two for haloperidol (K_d=9 μM), and one site for pimozide (K_d=0.8 μM) (Levin and Weiss, 1979).

3. Competition between Different Antipsychotic Drugs for Binding Sites on Calmodulin

The questions that were now asked were, what type of compound will displace trifluoperazine from calmodulin and do the different chemical classes of antipsychotics, which were shown previously to bind to calmodulin, bind to the same sites on calmodulin as trifluoperazine does?

To answer the first question, preparations of calmodulin were dialyzed to equilibrium against [^{3}H]trifluoperazine and various concentrations of the drug to be examined. The [^{3}H]trifluoperazine that remained bound to calmodulin was then determined. The results of these studies showed that, in general, the ability of compounds to displace trifluoperazine from calmodulin correlated with their clinical antipsychotic activity.

Table III shows that the displacement of labeled trifluoperazine from calmodulin was a characteristic common to several antipsychotic drugs of different structural classes. The antipsychotic agents, pimozide, trifluoperazine, flupenthixol, penfluridol, chlorpromazine, thiothixene, and clozapine all displaced [^{3}H]trifluoperazine from calmodulin with IC_{50} values of 20 μM or less. The phenothiazine derivatives that have little or no antipsychotic activity, such as promethazine, trifluoperazine sulfoxide, and chlorpromazine sulfoxide, were significantly less potent than the antipsychotics in displacing [^{3}H]trifluoperazine. The antipsychotic drug reserpine, which acts presynaptically by depleting catecholamines from synaptic terminals rather than by blocking the action of catecholamines at postsynaptic sites, failed to displace trifluoperazine from calmodulin. The antianxiety agent, chlordiazepoxide, and the antidepressants, amitriptyline and nortriptyline, displayed moderate potency in displacing [^{3}H]trifluoperazine. Agents that were relatively ineffective in displacing [^{3}H]trifluoperazine from calmodulin included the α-adrenergic blocking agent, phentolamine, the phosphodiesterase inhibitors, theophylline and papaverine, and the cholinomimetic agent, methacholine.

Of particular interest was the data showing that perlapine and metoclopramide did not displace [^{3}H]trifluoperazine from calmodulin. These two compounds have been reported to behave pharmacologically as though they were blocking dopamine receptors (Burki *et al.*, 1975; Peringer *et al.*, 1976) and were, therefore, thought to be antipsychotic. However, clinical trials of these compounds failed to demonstrate antipsychotic activity (Stille *et al.*, 1973; Tarsy and Baldessarini, 1974; Nakra *et al.*, 1975).

These results suggest that the ability of drugs to displace trifluoperazine from calmodulin may offer another *in vitro* test for screening potential antipsychotic agents. However, the important issue concerning the relative lack of stereospecificity with which the cis and trans isomers of flupenthixol and thiothixene displace trifluoperazine from calmodulin is yet to be re-

TABLE III

Displacement of [^{3}H]Trifluoperazine from Calmodulin by Various Compounds[a]

Compound	IC_{50} for displacement (μM)
Pimozide	1.2
Trifluoperazine	1.5
trans-Flupenthixol	2.5
Penfluridol	3.8
cis-Flupenthixol	4.0
Chlorpromazine	8.0
trans-Thiothixene	10
cis-Thiothixene	17
Clozapine	20
TFP-sulfoxide	45
Promethazine	60
(+)-Butaclamol	75
Nortriptyline	80
Chlordiazepoxide	90
Amitriptyline	100
CPZ-sulfoxide	250
Phentolamine	350
(−)-Butaclamol	350
Reserpine	>200
Perlapine	>200
Metoclopramide	>200
Molindone	>200
Papaverine	>250
Theophylline	>1000
Methacholine	>1000

[a] The displacement of [^{3}H]trifluoperazine from calmodulin by various compounds was determined by equilibrium dialysis. IC_{50} is the concentration of drug required to displace 50% of trifluoperazine from calmodulin. (Taken in part from Weiss *et al.*, 1980, by permission.)

solved. This question is discussed more fully in Section VI,A,1.

To study the question whether different chemical classes of antipsychotic agents compete for the same binding sites on calmodulin, we measured the effect of the diphenylbutylpiperidines, penfluridol and pimozide, on the calcium-specific binding of [^{3}H]trifluoperazine to calmodulin (Levin and Weiss, 1979). Penfluridol caused a concentration-dependent shift in the binding curve for [^{3}H]trifluoperazine to calmodulin. At low concentrations of trifluoperazine, penfluridol produced a marked displacement of trifluoperazine from calmodulin. However, at high concentrations of trifluoperazine, penfluridol failed to displace the phenothiazine

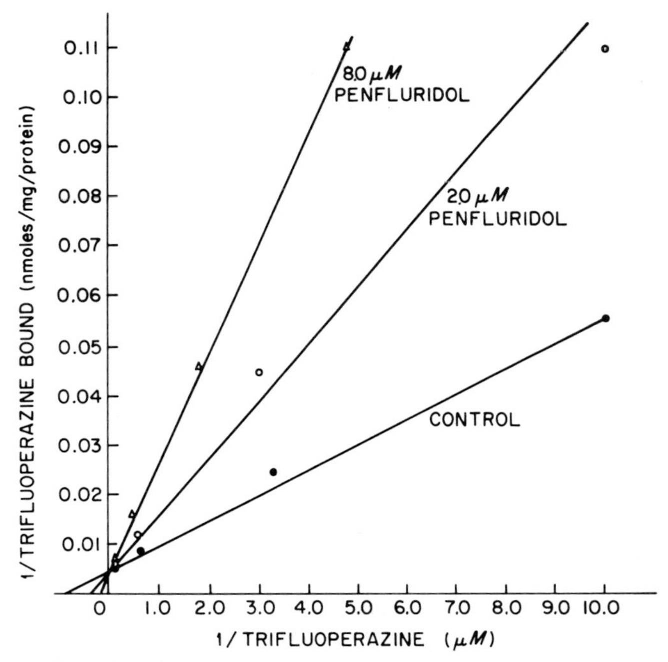

Fig. 10. Effect of penfluridol on the binding of trifluoperazine to calmodulin. [^{3}H]Trifluoperazine was added to baths containing 0, 2.0, or 8.0 μM penfluridol, and the samples were dialyzed to equilibrium. The free trifluoperazine concentrations ranged from 0.1 to 10 μM. The data are plotted as the reciprocals of the bound versus free trifluoperazine. Each point is the mean of three determinations. Lines are fitted by linear regression analysis. (Taken from Levin and Weiss, 1979, by permission.)

significantly. Kinetic analysis of the data suggests that penfluridol competitively inhibits the binding of trifluoperazine to the high-affinity binding sites on calmodulin (Fig. 10). Analysis of the influence of pimozide on the binding of [^{3}H]trifluoperazine demonstrated that pimozide, like penfluridol, also competed for the trifluoperazine binding sites on calmodulin.

4. *Irreversible Binding of Phenothiazine Antipsychotic Drugs to Calmodulin*

The initial studies on the binding of trifluoperazine to calmodulin showed that this binding was readily reversible. However, the reports that photochemical activation of halogenated phenothiazine derivatives results in an irreversible binding of these agents to certain proteins such as ATPase (Gubitz *et al.*, 1973) suggested that phenothiazines might also bind irreversibly to calmodulin.

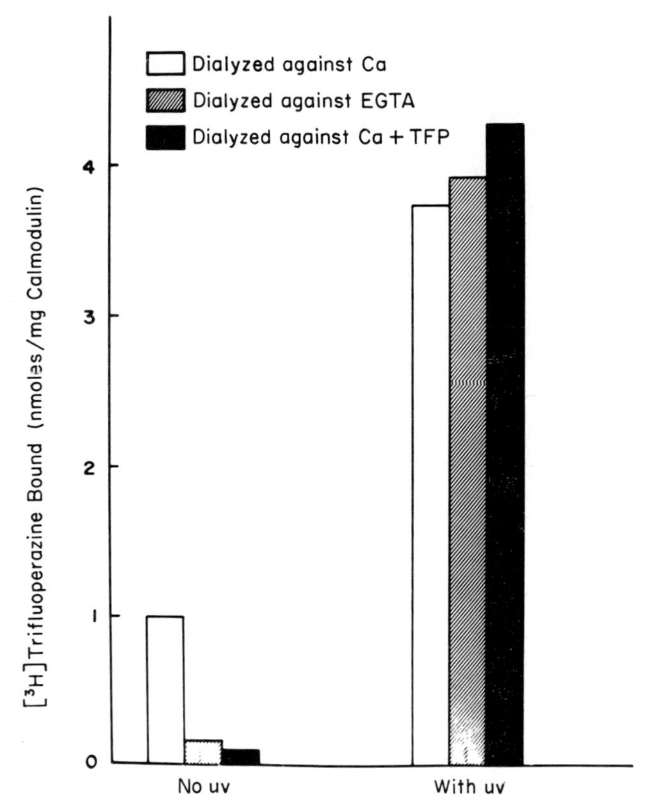

Fig. 11. Irreversible binding of trifluoperazine to calmodulin following irradiation with ultraviolet light. Samples containing 5 m*M* Tris-HCl buffer, pH 7.0, 1 m*M* Mg^{2+}, 0.1 m*M* Ca^{2+}, 40μg calmodulin/ml and 1μ*M* [^{3}H]trifluoperazine were incubated at 4°C for 30 min in the presence or absence of uv light (254 nm). Samples were then dialyzed for 16 hr against Tris buffer containing either 0.1 m*M* Ca^{2+}, 0.3 m*M* EGTA, or 0.1 m*M* Ca^{2+} and 0.1 m*M* unlabeled trifluoperazine. The amount of [^{3}H]trifluoperazine that remained bound to calmodulin was then determined. Each value is the mean of four experiments. (Taken from Weiss *et al.*, 1980, by permission.)

Figure 11 shows that ultraviolet (uv) irradiation caused an irreversible binding of trifluoperazine to calmodulin. Samples of [^{3}H]trifluoperazine and calmodulin that were not irradiated could be readily dissociated by dialyzing the samples against EGTA or by dialyzing them against high concentrations of nonradiolabeled trifluoperazine. However, if [^{3}H]trifluoperazine and calmodulin were irradiated with uv light, the binding was not reversed under the same conditions of dialysis.

The irreversible binding of trifluoperazine to calmodulin shared several of the characteristics seen in studies of the reversible binding; the irreversible binding was calcium-dependent, showed 50% maximal binding at about 5 μM and was relatively specific for calmodulin since trifluoperazine did not exhibit any Ca^{2+}-specific binding to troponin C or to bovine serum albumin. Further, other centrally active drugs, such as apomorphine and diazepam, showed little or no irreversible binding to calmodulin (Cimino *et al.*, 1979; Weiss *et al.*, 1980). However, there were important differences between the reversible and irreversible binding of drugs to calmodulin. For example, whereas all antipsychotics examined, regardless of their structure, exhibited reversible binding, not all showed irreversible binding. Nonphenothiazine antipsychotics, such as haloperidol and penfluridol, did not bind to calmodulin irreversibly. This may be due to their relative inability to be photochemically activated.

To examine the biological consequence of binding phenothiazines to calmodulin irreversibly, samples of calmodulin and trifluoperazine were irradiated and examined for their ability to activate a calmodulin-dependent phosphodiesterase. These irradiated samples failed to activate phosphodiesterase even after extensive dialysis against buffer, whereas similarly dialyzed preparations of calmodulin and trifluoperazine that had not been irradiated increased phosphodiesterase activity about fivefold (Cimino *et al.*, 1979). These studies indicate that the irreversible binding of phenothiazine antipsychotics to calmodulin causes an irreversible inhibition of calmodulin's biological activity.

Although the irreversible binding of phenothiazines to calmodulin has thus far been demonstrated only *in vitro*, it may also occur *in vivo*. It is well known that metabolites of phenothiazine antipsychotics remain in the body for extremely long periods of time following the cessation of drug treatment, indicating that these metabolites might be tightly bound to tissue constituents (Breyer and Gaertner, 1974). Perhaps the *in vivo* metabolism of phenothiazines generates reactive intermediates such as free radicals or epoxides (Forrest *et al.*, 1958) which bind irreversibly to calmodulin.

D. Correlation between Binding of Drugs to Calmodulin and Inhibition of Calmodulin Activity

Our studies indicate that there is a clear relationship between two biochemical events: the binding of drugs to calmodulin and their ability to inhibit the calmodulin-dependent activation of certain enzymes. This relationship, which has been demonstrated using a wide variety of drugs and which even extends to the stereoisomers of antipsychotics, is depicted

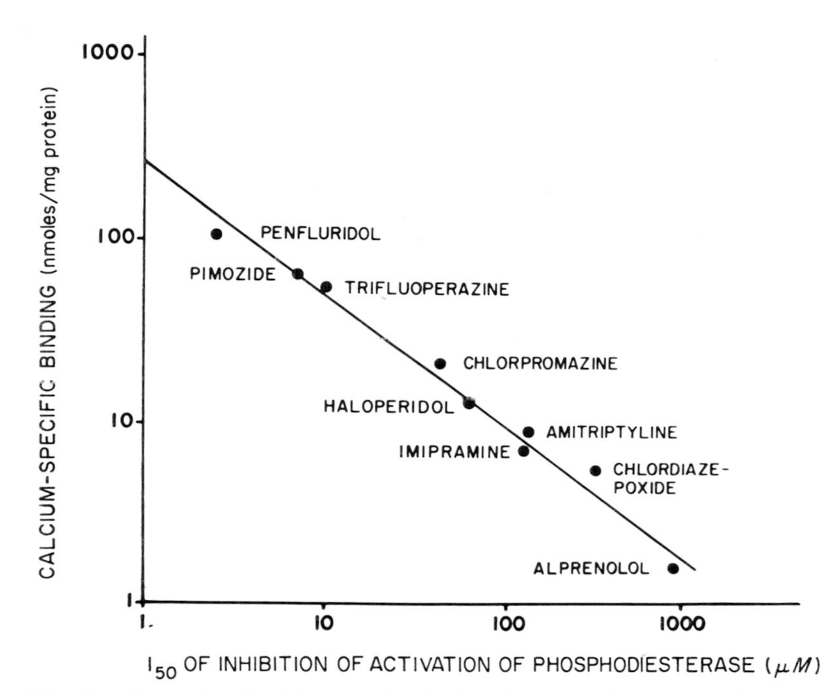

Fig. 12. Correlation of calcium-specific binding of neurotropic agents to calmodulin with their ability to inhibit the calmodulin-induced activation of phosphodiesterase. The binding of drugs to calmodulin was determined by equilibrium dialysis. The I_{50} is the concentration of drugs required to inhibit by 50% the calmodulin-induced stimulation of a calmodulin-sensitive phosphodiesterase prepared from rat brain. The line of best fit was determined by regression analysis after logarithmic transformation. The correlation coefficient was 0.99. (Taken from Levin and Weiss, 1979, by permission.)

graphically in Fig. 12. This figure shows that the degree to which drugs bind to calmodulin is directly related to their ability to inhibit the activation of phosphodiesterase. Antipsychotic agents showed the highest degree of binding to calmodulin and the greatest inhibition of calmodulin-induced activity, while antidepressants and anxiolytics demonstrated significantly less binding to calmodulin and less inhibition of phosphodiesterase activation. Agents displaying no calcium-specific binding to calmodulin showed no specific inhibition of the activation of phosphodiesterase. Thus, these studies demonstrate that the ability of agents to inhibit phosphodiesterase activation is directly correlated to their ability to bind to calmodulin. They show further that agents exhibiting the greatest potency in both parameters were clinically effective antipsychotic drugs.

VI. PHARMACOLOGICAL IMPLICATIONS OF INHIBITING THE ACTIVITY OF CALMODULIN

A. Acute Effects of Antipsychotic Drugs on Calmodulin

The studies reviewed in this chapter indicate that antipsychotic agents inhibit the activation of several calmodulin-sensitive enzyme systems by binding to and blocking the action of calmodulin. Indeed, all calmodulin-activated enzymes examined so far have been shown to be inhibited by antipsychotic drugs, suggesting that other enzymes or processes activated by calmodulin or calcium may also be inhibited by these agents. One of the important questions that remains is whether these biochemical actions of antipsychotics can explain their pharmacological and clinical effects. Although this is always an extremely difficult question to address, there is some suggestive, albeit tenuous, evidence that certain of the clinical effects of antipsychotic drugs, such as antipsychotic activity, orthostatic hypotension, and endocrinological disturbances may, in fact, be explained by their inhibition of calmodulin-activated enzymes.

1. Blockade of Calmodulin as a Basis for Explaining Antipsychotic Activity

The evidence suggesting that blockade of calmodulin may be involved in the antipsychotic actions of these drugs may be briefly summarized as follows: (a) A variety of clinically effective antipsychotic drugs of diverse chemical structure all bind to calmodulin in a high-affinity, calcium-dependent manner, whereas several other centrally active agents exhibiting no antipsychotic activity do not, and (b) calmodulin regulates the metabolism of cyclic nucleotides. These nucleotides, which mediate the actions of catecholaminergic neurotransmitters, are widely believed to play a role in certain diseases of the central nervous system, including schizophrenia (see Snyder *et al.,* 1974; Weiss and Greenberg, 1975; Nathanson, 1977; Carlsson, 1978). Inhibiting the biosynthesis of cyclic nucleotides may explain, at least in part, the pharmacological activity of antipsychotics, since many clinically effective antipsychotic agents have in common the ability to prevent the catecholamine-induced elevation of cyclic nucleotides (Uzunov and Weiss, 1972b; Miller *et al.,* 1974; Clement-Cormier *et al.,* 1974; Sulser and Robinson, 1978).

Although our studies showed that the relative ability of pharmacological agents to displace [^{3}H]trifluoperazine from calmodulin is, in general, consistent with the concept that the affinity of these agents for the activator protein may be related to their antipsychotic activity, an important question that remains to be resolved is the apparent lack of stereo-specific binding of drugs to calmodulin. Other *in vitro* actions of antipsychotics, such as their blockade of the rise of cyclic AMP induced by norepinephrine (Uzunov and Weiss, 1972b; Sulser and Robinson, 1978), their blockade of dopamine-sensitive adenylate cyclase (Miller *et al.*, 1974; Lippman *et al.*, 1975), and their displacement of the binding of dopamine (Burt *et al.*, 1976), have been shown to exhibit stereo specificity. Possible explanations for our inability to demonstrate stereospecific binding are (a) that during the purification and isolation of calmodulin, it may have undergone sufficient conformational changes to lose its stereospecificity, or (b) perhaps stereo-specific binding of antipsychotics to calmodulin is conferred by adjacent cellular constituents or surrounding cellular membranes.

Another question that should be addressed when postulating a role for calmodulin as a receptor for antipsychotic agents is the comparison between the therapeutic concentrations of antipsychotics found in tissues and the dissociation constant (K_d) between these drugs and calmodulin. It may be argued that in clinical doses these drugs are not present in therapeutic concentrations sufficient to significantly block calmodulin activity. However, these arguments are based on the concentration of antipsychotic drugs in plasma, which apparently bears little or no relationship either to the therapeutic effectiveness of these drugs (Davis *et al.*, 1978) or to the amount of drug actually present in the brain (Maickel *et al.*, 1974; Alfredsson *et al.*, 1977). For example, the concentration of chlorpromazine in brain was reported to be about 100 times that found in serum (Alfredsson *et al.*, 1977), probably reflecting the extremely high lipid–water partition coefficient of these drugs (Seeman, 1972). Further, several studies have shown that the concentration of chlorpromazine and its active metabolites in the whole rat brain are from 5 μmoles/kg (Maickel *et al.*, 1974) to 15 μmoles/kg tissue (Alfredsson *et al.*, 1977) following treatment with doses sufficient to produce pharmacological effects. These levels of chlorpromazine in brain are similar to the dissociation constant of chlorpromazine (5 μ*M*) for calmodulin (Levin and Weiss, 1979). These observations are consistant with the proposition that calmodulin may function as a pharmacologically important receptor for antipsychotic drugs.

2. *Blockade of Calmodulin as a Basis for Explaining Other Pharmacological Effects*

Orthostatic hypotension is a prominent side effect of antipsychotic therapy. Although the mechanism for this hypotensive effect is generally believed to be due to a blockade of adrenergic receptors (Byck, 1975), it may also be explained by a blockade of calmodulin. Calmodulin apparently is involved in the contraction of smooth muscle induced by various neurotransmitters, and this action is specifically blocked by antipsychotic drugs (Hidaka *et al.*, 1979a). These observations suggest that antipsychotics may cause orthostatic hypotension, in part at least, by inhibiting the calmodulin-activated contraction of vascular smooth muscle.

Another frequently observed side effect of antipsychotic therapy is an alteration of endocrine function. Antipsychotic agents increase the concentration of prolactin and decrease the concentration of growth hormone, corticotropin-releasing hormone, and neurohypophyseal hormones in plasma and reduce the concentration of gonadotropins in urine (Byck, 1975). The release of several of these hormones, as well as the release of neurotransmitters, is generally believed to occur through a common, calcium-dependent mechanism. Omission of Ca^{2+} from the extracellular fluid blocks the release of these substances; reintroduction of Ca^{2+} increases their release (see Trifaró, 1977). The evidence that calmodulin is involved in so many calcium-dependent processes suggests that it may also be involved in the calcium-dependent secretion of hormones and neurotransmitters. In support of this, it has been demonstrated that the calcium-stimulated release of norepinephrine is dependent on calmodulin (DeLorenzo *et al.*, 1979), and it is well known that antipsychotics block the release of catecholamines from nerve terminals (Seeman and Lee, 1975, 1976). Although the evidence is far from conclusive, perhaps antipsychotic agents alter the release of neurotransmitters as well as pituitary and hypothalamic hormones by blocking calmodulin-dependent enzymes in these tissues.

The smooth muscle of the intestinal wall is another site at which these drugs might act. A recent study (Ilundain and Naftalin, 1979) showed that chloride secretion in the gut was regulated by calmodulin, probably through an intermediary mechanism involving calcium transport. Trifluoperazine inhibited the secretion of chloride, raising the possibility that the antidiarrheal effects of drugs structurally related to the antipsychotics may act by blocking calmodulin.

These results taken together suggest that antipsychotic drugs exert several of their biochemical and perhaps pharmacological actions through a common mechanism, namely, by a selective binding to and inhibition of calmodulin. A scheme depicting this mechanism is shown in Fig. 13.

B. Long-Term Actions of Antipsychotic Drugs on Calmodulin

It is now clear that the short-term biochemical actions of antipsychotic drugs may not adequately explain all of their therapeutic and toxic effects; their long-term actions must also be considered. In fact, the adaptive

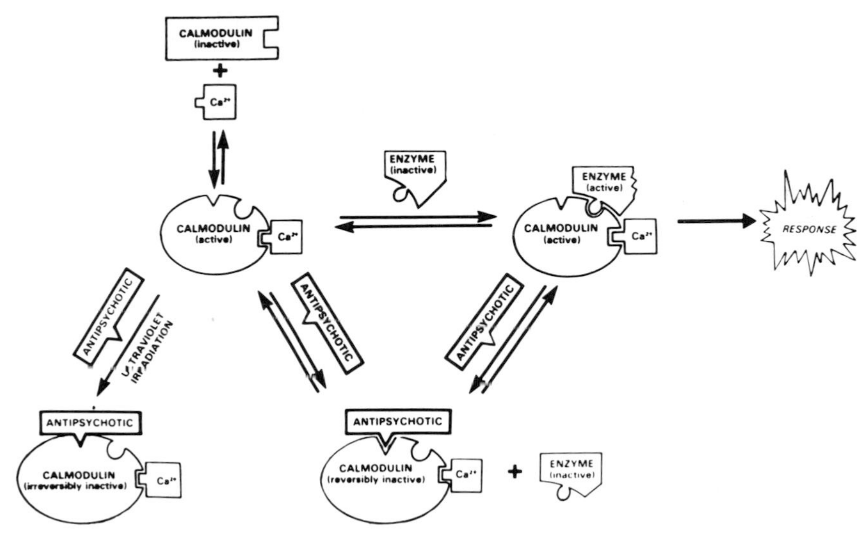

Fig. 13. Mechanism by which antipsychotic agents inhibit the action of calmodulin. The scheme describes our hypothesis for the mechanism by which antipsychotic agents block the calmodulin-induced activation of enzymes. The first step involves the interaction of calcium with calmodulin. Calcium binds to calmodulin (up to 4 calcium ions to 1 molecule of calmodulin) (Watterson *et al.*, 1976). This results in a conformational change in calmodulin (Klee, 1977; Liu and Cheung, 1976). The calcium–calmodulin complex then interacts with a specific calmodulin-sensitive form of the enzyme, causing the formation of the active calmodulin–calcium–enzyme complex (Brostrom and Wolff, 1974; Lin *et al.*, 1974; Teshima and Kakiuchi, 1974) and, subsequently, the biochemical response. The calcium–calmodulin complex may also bind to antipsychotic agents, resulting in a calmodulin–calcium–antipsychotic complex (Levin and Weiss, 1977). This binding between antipsychotic drugs and calmodulin may be reversible (Levin and Weiss, 1977) or, if irradiated with uv light, irreversible (Cimino *et al.*, 1979). In either case this complex cannot activate the calmodulin-sensitive form of enzymes (Weiss *et al.*, 1974; Levin and Weiss, 1976). In the absence of calcium, there is no interaction of calmodulin with either the enzyme or the antipsychotics (Weiss *et al.*, 1974; Levin and Weiss, 1976, 1977). The addition of antipsychotics to an enzyme that is already activated will reverse its activation. This inhibition cannot be overcome by adding more calcium or by adding more enzyme (Levin and Weiss, 1976). We suggest that a scheme similar to the one described here not only explains the mechanism by which antipsychotics inhibit the specific molecular forms of calmodulin-sensitive enzymes but may even explain how antipsychotics exert some of their diverse pharmacological actions.

changes in catecholaminergic receptors and calmodulin that take place in response to the long-term use of these drugs might provide a more accurate basis for understanding their pharmacological actions (see Weiss *et al.*, 1980; Weiss and Greenberg, 1980). Insofar as calmodulin is concerned, it has been reported (Gnegy *et al.*, 1977a,b; Lucchelli *et al.*, 1978) that the activity of calmodulin is increased following chronic treatment of rats with haloperidol. Since, as mentioned earlier, calmodulin may be involved in the response of adenylate cyclase to catecholamine stimulation, it is possible that antipsychotic drugs might be blocking the catecholamine's effects by interfering with the actions of calmodulin. To carry this hypothesis further, chronic exposure to the antipsychotics might lead to compensatory increases in the levels of calmodulin. This, in turn, may account for the tolerance that develops to some of the effects of antipsychotic therapy, and for certain long-term motor dysfunctions such as the tardive dyskinesias (Rubovitz and Klawans, 1972). Perhaps by studying the interaction of antipsychotics with calmodulin and on the catecholamine-induced alterations of cyclic nucleotides, one may obtain a better understanding of the therapeutic and toxic actions of these drugs.

VII. SUMMARY

Calmodulin is an ubiquitous, calcium-binding protein that regulates the activity of several calcium-dependent enzymes, including specific forms of phosphodiesterase, adenylate cyclase, and ATPase. Modification of the activity of calmodulin, therefore, would have profound biological consequences. A number of endogenous and exogenous substances have already been shown to alter calmodulin's activity. The exogenous agents that interfere with calmodulin's action share similar biochemical and pharmacological profiles. Biochemically, they have in common the ability to inhibit the calmodulin-induced activation of phosphodiesterase, adenylate cyclase, and ATPase, and pharmacologically, many produce effects that are characteristic of antipsychotic agents.

The specific mechanism by which antipsychotic drugs block the action of calmodulin apparently is by a direct, calcium-dependent binding to calmodulin. This proposition is supported by the demonstration that the ability of drugs to bind calmodulin is directly related to their effectiveness in inhibiting calmodulin-dependent enzymes.

The binding of drugs to calmodulin is saturable and of high affinity and is relatively specific for agents having clinical antipsychotic activity. Indeed the specificity by which antipsychotic drugs bind to calmodulin suggest that this protein may function as an endogenous receptor for antipsychotics.

In addition to exogenous compounds, a number of endogenous substances have been described which bind to and inhibit the action of calmodulin. These findings raise the possibility that there may be endogenous ligands in brain which function as endogenous psychotogens or endogenous antipsychotic materials.

The large number of actions already attributed to calmodulin will surely increase as will the types of drugs that modify its activity. The critical question to pharmacologists is whether calmodulin's activity can be altered selectively. If this can be accomplished, future research may well provide new classes of pharmacological agents that act by modifying the activity of calmodulin.

ACKNOWLEDGMENTS

We acknowledge with thanks the excellent technical assistance of Ms. Becky Simon and we thank Ms. Jose Weycis for her help with the manuscript. This work was supported by Grant #MH30096 awarded by the National Institute of Mental Health, DHEW, and by a predoctoral fellowship awarded by the Scottish Rite Schizophrenia Research Program to T. L. W.

REFERENCES

Alfredsson, G., Wiesel, F. A., and Skett, P. (1977). Levels of chlorpromazine and its active metabolites in rat brain and the relationship to central monoamine metabolism and prolactin secretion. *Psychopharmacology* **53,** 13–18.

Anderson, J. M., and Cormier, M. J. (1978), Calcium-dependent regulator of NAD kinase in higher plants. *Biochem. Biophys. Res. Commun.* **84,** 595–602.

Beale, E. G., Dedman, J. R., and Means, A. R. (1977). Isolation and regulation of the protein kinase inhibitor and the calcium-dependent cyclic nucleotide phosphodiesterase regulator in the sertoli cell-enriched testis. *Endocrinology* **101,** 1621–1634.

Bond, G. H., and Clough, D. L. (1973). A soluble protein activator of (Mg^{2+} + Ca^{2+})-dependent ATPase in human red cell membranes. *Biochim. Biophys. Acta* **323,** 592–599.

Boudreau, R. J., and Drummond, G. I. (1975). The effect of Ca^{2+} on cyclic nucleotide phosphodiesterases of superior cervical ganglion. *J. Cyclic Nucleotides Res.* **1,** 219–228.

Breyer, U., and Gaertner, H. J. (1974). Tissue accumulation of metabolites during chronic administration of piperazine-substituted phenothiazine drugs. *Adv. Biochem. Psychopharmacol.* **9,** 167–173.

Brostrom, C. O., and Wolff, D. J. (1974). Calcium-dependent cyclic nucleotide phosphodiesterase from glial tumor cells. *Arch. Biochem. Biophys.* **165,** 715–727.

Brostrom, C. O., and Wolff, D. J. (1976). Calcium-dependent cyclic nucleotide phosphodiesterase from brain: Comparison of adenosine 3′,5′-monophosphate and guanosine 3′,5′-monophosphate as substrates. *Arch. Biochem. Biophys.* **172,** 301–311.

Brostrom, C. O., Huang, Y.-C., Breckenridge, B. McL., and Wolff, D. J. (1975). Identification of a calcium-binding protein as a calcium-dependent regulator of brain adenylate cyclase. *Proc. Natl. Acad. Sci. U.S.A.* **72,** 64–68.

Brostrom, C. O., Brostrom, M. A., and Wolff, D. J. (1977). Calcium-dependent adenylate cyclase from rat cerebral cortex. *J. Biol. Chem.* **252,** 5677–5685.

Brostrom, M. A., Brostrom, C. O., Breckenridge, B. McL., and Wolff, D. J. (1976). Regulation of adenylate cyclase from glial tumor cells by calcium and a calcium-binding protein. *J. Biol. Chem.* **251,** 4744–4750.

Brostrom, M. A., Brostrom, C. O., Breckenridge, B. McL., and Wolff, D. J. (1978a). Calcium-dependent regulation of brain adenylate cyclase. *Adv. Cyclic Nucleotide Res.* **9,** 85–99.

Brostrom, M. A., Brostrom, C. O., and Wolff, D. J. (1978b). Calcium-dependent adenylate cyclase from rat cerebral cortex: Activation by guanine nucleotides. *Arch. Biochem. Biophys.* **191,** 341–350.

Brostrom, M. A., Brostrom, C. O., and Wolff, D. J. (1979). Calcium dependence of hormone-stimulated cyclic AMP accumulation in intact glial cells. *J. Biol. Chem.* **254,** 7548–7557.

Burgess, W. H., Howlett, A. C., Kretsinger, R. H., and Gilman, A. G. (1978). S49 lymphoma wild type and variant clones contain normal calcium dependent regulator *J. Cyclic Nucleotide Res.* **4,** 175–181.

Burki, H. R., Ruch, W., and Asper, H. (1975). Effects of clozapine, thioridazine, perlapine and haloperidol on the metabolism of the biogenic amines in the brain of the rat. *Psychopharmacologia* **41,** 27–33.

Burt, D. R., Creese, I., and Snyder, S. H. (1976). Properties of (^{3}H)-haloperidol and (^{3}H)-dopamine binding associated with dopamine receptors in calf brain membranes. *Mol. Pharmacol.* **12,** 800–812.

Byck, R. (1975). Drugs and the treatment of psychiatric disorders. *In* "The Pharmacological Basis of Therapeutics" (L. S. Goodman and A. Gilman, eds.), pp. 152–200. Macmillan, New York.

Carlsson, A. (1978). Antipsycotic drugs, neurotransmitters and schizophrenia. *Am. J. Psychiatry* **135,** 164–173.

Charbonneau, H., and Cormier, M. J. (1979). Purification of plant calmodulin by fluphenazine-Sepharose affinity chromatography. *Biochem. Biophys. Res. Commun.* **90,** 1039–1047.

Chassy, B. M. (1972). Cyclic nucleotide phosphodiesterase in *Dictyostelium discoideum:* Interconversion of two enzyme forms. *Science* **175,** 1016–1018.

Cheung, W. Y. (1967). Cyclic 3′,5′-nucleotide phosphodiesterase: Pronounced stimulation by snake venom. *Biochem. Biophys. Res. Commun.* **29,** 478–482.

Cheung, W. Y. (1970). Cyclic 3′,5′-nucleotide phosphodiesterase. Demonstration of an activator. *Biochem. Biophys. Res. Commun.* **38,** 533–538.

Cheung, W. Y. (1971). Cyclic 3′,5′-nucleotide phosphodiesterase. Evidence for and properties of a protein activator. *J. Biol. Chem.* **246,** 2859–2869.

Cheung, W. Y. (1980). Calmodulin plays a pivotal role in cellular regulation. *Science* **207,** 19–27.

Cheung, W. Y., Bradham, L. S., Lynch, T. J., Lin, Y. M., and Tallant, E. A. (1975a). Protein activator of cyclic 3′,5′-nucleotide phosphodiesterase of bovine or rat brain also activates its adenylate cyclase. *Biochem. Biophys. Res. Commun.* **66,** 1055–1062.

Cheung, W. Y., Lin, Y. M., Liu, Y. P., and Smoake, J. A. (1975b). Regulation of bovine brain cyclic 3′,5′-nucleotide phosphodiesterase by its protein activator. *In* "Cyclic Nucleotides in Disease" (B. Weiss, ed.), pp. 321–350. Univ. Park Press, Baltimore, Maryland.

Cheung, W. Y., Lynch, T. J., and Wallace, R. W. (1978). An endogenous Ca^{2+}-dependent activator protein of brain adenylate cyclase and cyclic nucleotide phosphodiesterase. *Adv. Cyclic Nucleotide Res.* **9,** 233–251.
Childers, S. R., and Siegel, F. L. (1975). Isolation and purification of a calcium-binding protein from electroplax of Electrophorus electricus. *Biochim. Biophys. Acta* **405,** 99–108.
Cimino, M., Prozialeck, W., and Weiss, B. (1979). Irreversible binding of trifluoperazine to calmodulin by photoaffinity labeling. *Pharmacology* **21,** 240.
Clement-Cormier, Y. C., Kebabian, J. W., Petzold, G. L., and Greengard, P. (1974). Dopamine-sensitive adenylate cyclase in mammalian brain: A possible site of action of antipsychotic drugs. *Proc. Natl. Acad. Sci. U.S.A.* **71,** 1113–1117.
Clement-Cormier, Y. C., Parrish, R. G., Petzold, G. L., Kebabian, J. W., and Greengard, P. (1975). Characterization of a dopamine-sensitive adenylate cyclase in the rat caudate nucleus. *J. Neurochem.* **25,** 143–149.
Cohen, P., Burchell, A., Foulkes, J. G., Cohen, P. T. W., Vanaman, T. C., and Nairn, A. C. (1978). Identification of the Ca^{2+}-dependent modulator protein as the fourth subunit of rabbit skeletal muscle phosphorylase kinase. *FEBS Lett.* **92,** 287–293.
Costa, E., Gnegy, M., Revuelta, A., and Uzunov, P. (1977). Regulation of dopamine-dependent adenylate cyclase by a Ca^{2+}-binding protein stored in synaptic membranes. *Adv. Biochem. Psychopharmacol.* **9,** 403–408.
Dabrowska, R., Aromatorio, D., Sherry, J. M. F., and Hartshorne, D. J. (1977). Composition of the myosin light chain kinase from chicken gizzard. *Biochem. Biophys. Res. Commun.* **78,** 1263–1272.
Dabrowska, R., Sherry, J. M. F., Aromatorio, D. K., and Hartshorne, D. J. (1978). Modulator protein as a component of the myosin light chain kinase from chicken gizzard *Biochemistry* **17,** 253–258.
Davis, J. M., Erickson, S., and Dekirmenjian, H. (1978). Plasma levels of antipsychotic drugs and clinical response. *In* "Psychopharmacology: A Generation of Progress" (M. A. Lipton, A. DiMascio, and K. F. Killam, eds.), pp. 905–915. Raven, New York.
Dedman, J. R., Jackson, R. L., Schreiber, W. E., and Means, A. R. (1978). Sequence homology of the Ca^{2+}-dependent regulator of cyclic nucleotide phosphodiesterase from rat testis with other Ca^{2+}-binding proteins. *J. Biol. Chem.* **253,** 343–346.
DeLorenzo, R. J., and Freedman, S. D. (1978). Calcium dependent neurotransmitter release and protein phosphorylation in synaptic vesicles. *Biochem. Biophys. Res. Commun.* **80,** 183–192.
DeLorenzo, R. J., Freedman, S. D., Yohe, W. B., and Maurer, S. C. (1979). Stimulation of Ca^{2+}-dependent neurotransmitter release and presynaptic nerve terminal protein phosphorylation by calmodulin and a calmodulin-like protein isolated from synaptic vesicles. *Proc. Natl. Acad. Sci. U.S.A.* **76,** 1838–1842.
Depaoli-Roach, A. A., Gibbs, J. B., and Roach, P. J. (1979). Calmodulin and calmodulin activation of muscle phosphorylase kinase. *FEBS Lett.* **105,** 321–324.
Drabikowski, W., Kuznicki, J., and Grabarek, Z. (1977). Similarity in Ca^{2+}-induced changes between troponin-C and protein activator of 3′:5′-cyclic nucleotide phosphodiesterase and their tryptic fragments. *Biochim. Biophys. Acta* **485,** 124–133.
Drabikowski, W., Kuznicki, J., and Grabarek, Z. (1978). Distribution of troponin-C and protein activator of 3′,5′-cyclic nucleotide phosphodiesterase in vertebrate tissues. *Comp. Biochem. Physiol. C* **60,** 1–6.
Egrie, J. C., and Siegel, F. L. (1975). Adrenal medullary cyclic nucleotide phosphodies-

terase: Lack of activation by the calcium-dependent activator. *Biochem. Biophys. Res. Commun.* **67,** 662–669.

Egrie, J. C., Campbell, J. A., Flangas, A. L., and Siegel, F. L. (1977). Regional, cellular and subcellular distribution of calcium-activated cyclic nucleotide phosphodiesterase and calcium-dependent regulator in porcine brain. *J. Neurochem.* **28,** 1207–1213.

Fertel, R., and Weiss, B. (1974). A microassay for guanosine 3′,5′-monophosphate phosphodiesterase activity. *Anal. Biochem.* **59,** 386–398.

Fertel, R., and Weiss, B. (1976). Properties and drug responsiveness of cyclic nucleotide phosphodiesterases of rat lung. *Mol. Pharmacol.* **12,** 678–687.

Filburn, C. R., and Sacktor, B. (1976). Cyclic nucleotide phosphodiesterases of rabbit renal cortex. Characterization of brush border membrane activities. *Arch. Biochem. Biophys.* **174,** 249–261.

Filburn, C. R., Colpo, F. T., and Sacktor, B. (1979). Mechanism of phenothiazine inhibition of Ca^{2+}-dependent guanosine 3′,5′-(cyclic)monophosphate phosphodiesterase of brain. *Mol. Pharmacol.* **15,** 257–262.

Forrest, I. S., Forrest, F. M., and Berger, M. (1958). Free radicals as metabolites of drugs derived from phenothiazine. *Biochim. Biophys. Acta* **29,** 441–442.

Gnegy, M. E., Uzunov, P., and Costa, E. (1976). Regulation of dopamine stimulation of striatal adenylate cyclase by an endogenous Ca^{+}-binding protein. *Proc. Natl. Acad. Sci. U.S.A.* **73,** 3887–3890.

Gnegy, M. E., Lucchelli, A., and Costa, E. (1977a). Correlation between drug-induced supersensitivity of dopamine dependent striatal mechanisms and increase in striatal content of Ca^{2+}-regulated protein activator of cAMP phosphodiesterase. *Naunyn-Schmiedeberg's Arch. Pharmacol.* **301,** 121–127.

Gnegy, M., Uzunov, P., and Costa, E. (1977b). Participation of an endogenous calcium binding protein activator in the development of drug-induced supersensitivity of striatal dopamine receptors. *J. Pharmacol. Exp. Ther.* **202,** 558–564.

Gnegy, M. E., Nathanson, J. A., and Uzunov, P. (1977c). Release of the phosphodiesterase activator by cyclic AMP-dependent ATP: Protein phosphotransferase from subcellular fractions of rat brain. *Biochim. Biophys. Acta* **497,** 75–85.

Gopinath, R. M., and Vincenzi, F. F. (1977). Phosphodiesterase protein activator mimics red blood cell cytoplasmic activator of $(Ca^{2+} - Mg^{2+})$ ATPase. *Biochem. Biophys. Res. Commun.* **77,** 1203–1209.

Grab, D. J., Berzins, K., Cohen, R. S., and Siekevitz, P. (1979). Presence of calmodulin in postsynaptic densities isolated from canine cerebral cortex. *J. Biol. Chem.* **254,** 8690–8696.

Grand, R. J. A., Perry, S. V., and Weeks, R. A. (1979). Troponin-C-like proteins (calmodulins) from mammalian smooth muscle and other tissues. *Biochem. J.* **177,** 521–529.

Greenberg, L. H., and Weiss, B. (1976). Activatable phosphodiesterases in various areas of rat brain. *Trans. Am. Soc. Neurochem.* **7,** 115.

Greenberg, L. H., Troyer, E., Ferrendelli, J. A., and Weiss, B. (1978). Enzymatic regulation of the concentration of cyclic GMP in mouse brain. *Neuropharmacology* **17,** 737–745.

Greengard, P. (1978). Cyclic nucleotides, phosphorylated proteins, and neuronal function. *In* "Distinguished Lecture Series of the Society of General Physiologists," Vol. 1. Raven, New York.

Gubitz, R. H., Akera, T., and Brody, T. M. (1973). Comparative effects of substituted phenothiazines and their free radicals on $(Na^{+}-K^{+})$-activated adenosine triphosphatase. *Biochem. Pharmacol.* **22,** 1229–1235.

Gubitz, R. H., Akera, T., and Brody, T. M. (1977). Control of brain slice respiration by ($Na^+ - K^+$) activated adenosine triphosphatase and the effects of enzyme inhibitors. *Biochim. Biophys. Acta* **459,** 263–277.

Hait, W. N., and Weiss, B. (1977). Characteristics of the cyclic nucleotide phosphodiesterase of normal and leukemic lymphocytes. *Biochim. Biophys. Acta* **497,** 86–100.

Hanbauer, I., Gimble, J., and Lovenberg, W. (1979a). Changes in soluble calmodulin following activation of dopamine receptors in rat striatal slices. *Neuropharmacology* **18,** 851–857.

Hanbauer, I., Gimble, J., Sankaran, K., and Sherard, R. (1979b). Modulation of striatal cyclic nucleotide phosphodiesterase by calmodulin: Regulation by opiate and dopamine receptor activation. *Neuropharmacology* **18,** 859–864.

Harper, J. F., Cheung, W. Y., Wallace, R. W., Huang, H.-L., Levine, S. N., and Steiner, A. L. (1980). Localization of calmodulin in rat tissues. *Proc. Natl. Acad. Sci. (U.S.A.)* **77,** 366–370.

Head, J. F., Mader, S., and Kaminer, B. (1977). Troponin-C-like protein from vertebrate smooth muscle. *In* "Calcium Binding Proteins and Calcium Function" (R. H. Wasserman, R. A. Corradino, E. Carafoli, R. H. Kretsinger, D. H. MacLennan, and F. L. Siegel, eds.), pp. 73–75. North-Holland Publ., New York.

Hidaka, H., Yamaki, T., Totsuka, T., and Asano, M. (1979a). Selective inhibitors of Ca^{2+}-binding modulator of phosphodiesterase produce vascular relaxation and inhibit actin-myosin interaction. *Mol. Pharmacol.* **15,** 49–59.

Hidaka, H., Naka, M., and Yamaki, T. (1979b). Effect of novel specific myosin light chain kinase inhibitors on Ca^{2+}-activated Mg^{2+}-ATPase of chicken gizzard actomyosin. *Biochem. Biophys. Res. Commun.* **90,** 694–699.

Hidaka, H., Yamaki, T., Naka, M., Tanaka, T., Hayashi, H., and Kobayashi, R. (1980). Calcium-regulated modulator protein interacting agents inhibit smooth muscle calcium-stimulated protein kinase and ATPase. *Mol. Pharmacol.* **17,** 66–72.

Ilundain, A., and Naftalin, R. J., (1979). Role of Ca^{2+}-dependent regulator protein in intestinal secretion. *Nature (London)* **279,** 446–448.

Jamieson, G. A., Jr., and Vanaman, T. C. (1979). Calcium-dependent affinity chromatography of calmodulin on an immobilized phenothiazine. *Biochem. Biophys. Res. Commun.* **90,** 1048–1056.

Jarrett, H. W., and Kyte, J. (1979). Human erythrocyte calmodulin: Further chemical characterization and the site of its interactions with the membrane. *J. Biol. Chem.* **254,** 8237–8244.

Jarrett, H. W., and Penniston, J. T. (1977). Partial purification of the $Ca^{2+} - Mg^{2+}$ ATPase activator from human erythrocytes: Its similarity to the activator of 3′:5′-nucleotide phosphodiesterase. *Biochem. Biophys. Res. Commun.* **77,** 1210–1216.

Jarrett, H. W., and Penniston, J. T. (1978). Purification of the Ca^{2+}-stimulated ATPase activator from human erythrocytes. *J. Biol. Chem.* **253,** 4676–4682.

Jones, H. P., Bradford, M. M., McRorie, R. A., and Cormier, M. J. (1978). High levels of a calcium-dependent modulator protein in spermatozoa and its similarity to brain modulator protein. *Biochem. Biophys. Res. Commun.* **82,** 1264–1272.

Kakiuchi, S., and Yamazaki, R. (1970). Calcium dependent phosphodiesterase activity and its activating factor (PAF) from brain. *Biochem. Biophys. Res. Commun.* **41,** 1104–1110.

Kakiuchi, S., Yamazaki, R., and Teshima, Y. (1972). Regulation of brain phosphodiesterase activity: Ca^{2+} plus Mg^{2+}-dependent phosphodiesterase and its activating factor from rat brain. *Adv. Cyclic Nucleotide Res.* **1,** 455–477.

Kakiuchi, S., Yamazaki, R., Teshima, Y., and Uenishi, K. (1973). Regulation of nucleoside cyclic 3′:5′-monophosphate phosphodiesterase activity from rat brain by a modulator and Ca^{2+}. *Proc. Natl. Acad. Sci. U.S.A.* **70,** 3526–3530.

Kakiuchi, S., Yamazaki, R., Teshima, Y., Uenishi, K., and Miyamoto, E. (1975a). Multiple cyclic nucleotide phosphodiesterase activities from rat tissues and occurrence of a calcium-plus-magnesium-ion-dependent phosphodiesterase and its protein activator. *Biochem. J.* **146,** 109–120.

Kakiuchi, S., Yamazaki, R., Teshima, Y., Uenishi, K., and Miyamoto, E. (1975b). Ca^{2+}/Mg^{2+}-dependent cyclic nucleotide phosphodiesterase and its activator protein. *Adv. Cyclic Nucleotide Res.* **5,** 163–178.

Katz, S., and Remtulla, M. A. (1978). Phosphodiesterase protein activator stimulates calcium transport in cardiac microsomal preparations enriched in sarcoplasmic reticulum. *Biochem. Biophys. Res. Commun.* **83,** 1373–1379.

Klee, C. B. (1977). Conformational transition accompanying the binding of Ca^{2+} to the protein activator of 3′,5′-cyclic adenosine monophosphate phosphodiesterase. *Biochemistry* **16,** 1017–1024.

Klee, C. B., and Krinks, M. H. (1978). Purification of cyclic 3′,5′-nucleotide phosphodiesterase inhibitory protein by affinity chromatography on activator protein coupled to sepharose. *Biochemistry* **17,** 120–126.

Kobayashi, R., Tawata, M., and Hidaka, H. (1979). Ca^{2+} regulated modulator protein interacting agents: Inhibition of Ca^{2+} – Mg^{2+}-ATPase of human erythrocyte ghost. *Biochem. Biophys. Res. Commun.* **88,** 1037–1045.

Kumagai, H., and Nishida, E. (1979). The interactions between calcium-dependent regulator protein of cyclic nucleotide phosphodiesterase and microtubule proteins. II. Association of calcium-dependent regulator protein with tubulin dimer *J. Biochem. (Tokyo)* **85,** 1267–1274.

Kuo, C.-H., Ichida, S., Matsuda, T., Kakiuchi, S., and Yoshida, H. (1979). Regulation of ATP-dependent Ca-uptake of synaptic plasma membranes by Ca-dependent modulator protein. *Life Sci.* **25,** 235–240.

Kuo, I. C., and Coffee, C. J. (1976). Purification and characterization of a troponin-C-like protein from bovine adrenal medulla. *J. Biol. Chem.* **251,** 1603–1609.

LaPorte, D. C., Gidwitz, S., Weber, M. J., and Storm, D. R. (1979). Relationship between changes in the calcium dependent regulatory protein and adenylate cyclase during viral transformation. *Biochem. Biophys. Res. Commun.* **86,** 1169–1177.

Larsen, F. L., and Vincenzi, F. F. (1979). Calcium transport across the plasma membranes: Stimulation by calmodulin. *Science* **204,** 306–309.

Larsen, F. L., Raess, B. V., Hinds, T. R., and Vincenzi, F. F. (1978). Modulator binding protein antagonizes activation of (Ca^{2+} + Mg^{2+})-ATPase and Ca^{2+} transport of red blood cell membranes. *J. Supramol. Struct.* **9,** 269–274.

Lee, K. S., and Shin, B. C., (1969). Studies on the active transport of calcium in human red cells. *J. Gen. Physiol.* **54,** 713–729.

Levin, R. M., and Weiss, B. (1976). Mechanism by which psychotropic drugs inhibit adenosine cyclic 3′,5′-monophosphate phosphodiesterase in brain. *Mol. Pharmacol.* **12,** 581–589.

Levin, R. M., and Weiss, B. (1977). Binding of trifluoperazine to the calcium-dependent activator of cyclic nucleotide phosphodiesterase. *Mol. Pharmacol.* **13,** 690–697.

Levin, R. M., and Weiss, B. (1978a). Specificity of the binding of trifluoperazine to the calcium-dependent activator of phosphodiesterase and to a series of other calcium-binding proteins. *Biochim. Biophys. Acta* **540,** 197–204.

Levin, R. M., and Weiss, B. (1978b). Characteristics of the cyclic nucleotide phosphodiesterases in a transplantable pheochromocytoma and adrenal medulla of the rat. *Cancer Res.* **38,** 915–920.

Levin, R. M. Weiss, B. (1979). Selective binding of antipsychotics and other psychoactive agents to the calcium-dependent activator of cyclic nucleotide phosphodiesterase. *J. Pharmacol. Exp. Ther.* **208,** 454–459.

Levin, R. M., and Weiss, B. (1980). Inhibition by trifluoperazine of calmodulin-induced activation of ATPase activity of rat erythrocyte. *Neuropharmacology* **19,** 169–174.

Levitzki, A., and Helmreich, E. J. M. (1979). Hormone-receptor-adenylate cyclase interactions. *FEBS Lett.* **101,** 213–219.

Lin, Y. M., Liu, Y. P., and Cheung, W. Y. (1974). Cyclic 3′:5′-nucleotide phosphodiesterase. Purification, characterization, and active form of the protein activator from bovine brain. *J. Biol. Chem.* **10,** 4943–4954.

Lin, Y. M., Liu, Y. P., and Cheung, W.. Y. (1975). Cyclic 3′,5′-nucleotide phosphodiesterase Ca^{2+}-dependent formation of bovine brain enzyme-activator complex. *FEBS Lett.* **49,** 356–360.

Lippman, W., Pugsley, T., and Merker, J. (1975). Effect of butaclamol and its enantiomers upon striatal homovanillic acid and adenyl cyclase of olfactory tubercle in rats. *Life Sci.* **16,** 213–224.

Liu, Y. P., and Cheung, W. Y. (1976). Cyclic 3′:5′-nucleotide phosphodiesterase. Ca^{2+} confers more helical conformation to the protein activator. *J. Biol. Chem.* **251,** 4193–4198.

Liu, Y. P., and Schwartz, H. S. (1978). Protein activator of cyclic AMP phosphodiesterase and cyclic nucleotide phosphodiesterase in bovine retina and bovine lens. Activity, subcellular distribution and kinetic parameters. *Biochim. Biophys. Acta* **526,** 186–193.

Liu, Y. P., Krishna, G., Aguirre, G., and Chader, G. J. (1979). Involvement of cyclic GMP phosphodiesterase activator in an hereditary retinal degeneration. *Nature (London)* **280,** 62–64.

Lucchelli, A., Guidotti, A., and Costa, E. (1978). Striatal content of Ca^{2+}-dependent regulator protein and dopaminergic receptor function. *Brain Res.* **155,** 130–135.

Lynch, T. J., and Cheung, W. Y. (1979). Human erythrocyte $Ca^{2+} - Mg^{2+}$-ATPase: Mechanism of Stimulation by Ca^{2+}. *Arch. Biochem. Biophys.* **194,** 165–170.

Lynch, T. J., Tallant, E. A., and Cheung, W. Y. (1976). Ca^{2+}-dependent formation of brain adenylate cyclase-protein activator complex. *Biochem. Biophys. Res. Commun.* **68,** 616–625.

Maickel, R. P., Braunstein, M. C., McGlyn, M., Snodgrass, W. R., and Webb, R. W. (1974). Behavioral, biochemical and pharmacological effects of chronic dosages of phenothiazine tranquilizers in rats. *Adv. Biochem. Psychopharmacol.* **9,** 593–602.

Marcum, J. M., Dedman, J. R., Brinkley, B. R., and Means, A. R. (1978). Control of microtubule assembly-disassembly by calcium-dependent regulator protein. *Proc. Natl. Acad. Sci. U.S.A.* **75,** 3771–3775.

Miller, R. J., Horn, A. S. and Iverson, L. L. (1974). The action of neuroleptic drugs on dopamine-stimulated adenosine cyclic 3′,5′-monophosphate production in rat neostriatum and limbic forebrain. *Mol. Pharmacol.* **10,** 759–766.

Moss, J., and Vaughn, M. (1977). Choleragen activation of solubilized adenylate cyclase: Requirement for GTP and protein activator for demonstration of enzymatic activity. *Proc. Natl. Acad. Sci. U.S.A.* **74,** 4396–4400.

Nagao, S., Suzuki, Y., Watanabe, Y., and Nozawa, Y. (1979). Activation by a calcium-binding protein of guanylate cyclase in Tetrahymena pyriformis. *Biochem. Biophys. Res. Commun.* **90,** 261–268.

Nakra, B. R. S., Bond, A. J., and Lader, M. H. (1975). Comparative psychotropic effects of metoclopramide and prochlorperazine in normal subjects. *J. Clin. Pharmacol.* **15,** 449–454.

Nathanson, J. A. (1977). Cyclic nucleotides and nervous system function. *Physiol. Rev.* **57,** 157–256.

Niggli, V., Ronner, P., Carafoli, E., and Penniston, J. T. (1979). Effects of calmodulin on the (Ca^{2+} + Mg^{2+})-ATPase partially purified from erythrocyte membranes. *Arch. Biochem. Biophys.* **198,** 124–130.

Nishida, E., Kumagai, H., Ohtsuki, I., and Sakai, H. (1979). The interactions between calcium-dependent regulator protein of cyclic nucleotide phosphodiesterase and microtubule proteins. I. Effect of calcium-dependent regulator protein on the calcium sensitivity of microbubule assembly. *J. Biochem. (Tokyo)* **85,** 1257–1266.

Olson, E. J., and Cazort, R. J. (1969). Active calcium and strontium transport in human erythrocyte ghosts. *J. Gen. Physiol.* **53,** 311–322.

Peringer, E., Jenner, P., Donaldson, I. M., and Marsden, C. D. (1976). Metoclopramide and dopamine receptor blockade. *Neuropharmacology* **15,** 463–469.

Pledger, W. J., Thompson, W. J., and Strada, S. J. (1975). Isolation of an activator of multiple forms of cyclic nucleotide phosphodiesterase of rat cerebrum by isoelectric focusing. *Biochim. Biophys. Acta* **391,** 334–340.

Rajan, K. S., Manian, A. A., Davis, J. M., and Skripkus, A. (1974). Studies on the metal chelation of chlorpromazine and its hydroxylated metabolites. *In* "The Phenothiazines and Structurally Related Drugs" (I.S. Forrest, C. J. Carr, and E. Usdin, eds.), pp. 571–591. Raven, New York.

Revuelta, A., Uzunov, P., and Costa, E. (1976). Release of phosphodiesterase activator from particulate fractions of cerebellum and striatum by putative neurotransmitters. *Neurochem. Res.* **1,** 217–227.

Robinson, J. D. (1976). (Ca-Mg)-stimulated ATPase activity of a rat brain microsomal preparation. *Arch. Biochem. Biophys.* **176,** 366–374.

Rosenblatt, D. E., Lauter, C. J., and Trams, E. G. (1976). Deficiency of a Ca^{2+}-ATPase in brains of seizure prone mice. *J. Neurochem.* **27,** 1299–1304.

Rubovitz, R., and Klawans, H. L. (1972). Implications of amphetamine-induced stereotyped behavior as a model for tardive dyskinesias. *Arch. Gen. Psychiatry* **27,** 502–507.

Runge, M. S., Hewgley, P. B., Puett, D., and Williams, R. C., Jr. (1979). Cyclic nucleotide phosphodiesterase activity in 10-nm filaments and microtubule preparations from bovine brain. *Proc. Natl. Acad. Sci. U.S.A.* **76,** 2561–2565.

Rylatt, D. B., Embi, N., and Cohen, P. (1979). Glycogen synthase kinase-2 from rabbit skeletal muscle is activated by the calcium-dependent regulator protein. *FEBS Lett.* **98,** 76–80.

Scatchard, G. (1949). The attractions of proteins for small molecules and ions. *Ann. N.Y. Acad. Sci.* **51,** 660–672.

Scharff, O. (1976). Ca^{2+} activation of membrane bound (Ca^{2+} − Mg^{2+})-dependent ATPase from human erythrocytes prepared in the presence or absence of Ca^{2+}. *Biochim. Biophys. Acta* **443,** 206–218.

Schatzman, H. J., and Vincenzi, F. F. (1969). Calcium movements across the membrane of human red cells. *J. Physiol. (London)* **201,** 369–395.

Schulman, H., and Greengard, P. (1978a). Stimulation of brain membrane protein phosphorylation by calcium and an endogenous heat-stable protein. *Nature (London)* **271,** 478–479.

Schulman, H., and Greengard, P. (1978b). Ca^{2+}-dependent protein phosphorylation system in membranes from various tissues, and its activation by calcium dependent regulator. *Proc. Natl. Acad. Sci. U.S.A.* **75,** 5432–5436.

Schwabe, U., and Daly, J. W. (1977). The role of calcium ions in accumulation of cyclic adenosine monophosphate elicited by alpha and beta adrenergic agents in rat brain slices. *J. Pharmacol. Exp. Ther* **202,** 134–143.

Schwabe, U., Ohga, Y., and Daly, J. W. (1978). The role of calcium in the regulation of cyclic nucleotide levels in brain slices of rat and guinea pig. *Naunyn- Schmiedeberg's Arch. Pharmacol.* **302,** 141–151.

Seamon, K. (1979). Cation dependent conformations of brain Ca^{2+}-dependent regulator protein detected by nuclear magnetic resonance. *Biochem. Biophys. Res. Commun.* **86,** 1256–1265.

Seeman, P. (1972). The membrane actions of anesthetics and tranquilizers. *Pharmacol. Rev.* **24,** 583–656.

Seeman, P., and Lee, T. (1975). Antipsychotic drugs: Direct correlation between clinical potency and presynaptic action on dopamine neurons. *Science* **188,** 1217–1219.

Seeman, P., and Lee, T. (1976). Neuroleptic drugs: Direct correlation between clinical potency and presynaptic action on dopamine neurons. *In* "Antipsychotic Drugs: Pharmacodynamics and Pharmacokinetics" (G. Sedvall, B. Uvnas, and Y. Zotterman, eds.), pp. 183–191. Pergamon, Oxford.

Sharma, R. K., Desai, R., Thompson, T. R., and Wang, J. H. (1978a). Purification of the heat-stable inhibitor protein of the Ca^{2+}-activated cyclic nucleotide phosphodiesterase by affinity chromatography. *Can. J. Biochem.* **56,** 598–604.

Sharma, R. K., Wirch, E., and Wang, J. H. (1978b). Inhibition of Ca^{2+}-activated cyclic nucleotide phosphodiesterase reaction by a heat-stable inhibitor protein from bovine brain. *J. Biol. Chem.* **253,** 3575–3580.

Sharma, R. K., Desai, R., Waisman, D. M., and Wang, J. H. (1979). Purification and subunit structure of bovine brain modulator binding protein. *J. Biol. Chem.* **254,** 4276–4282.

Shenolikar, S., Cohen, P. T. W., Cohen P., Nairn, A. C., and Perry, S. V. (1979). The role of calmodulin in the structure and regulation of phosphorylase kinase from rabbit skeletal muscle. *Eur. J. Biochem.* **100,** 329–337.

Sieghart, W., Schulman, H., and Greengard, P. (1980). Neuronal localization of Ca^{2+} -dependent protein phosphorylation in brain. *J. Neurochem.* **34,** 548–553.

Smoake, J. A., Song, S.-Y., and Cheung, W. Y. (1974). Cyclic 3′,5′-nucleotide phosphodiesterase distribution and developmental changes of the enzyme and its protein activator in mammalian tissues and cells. *Biochim. Biophys. Acta* **341,** 402–411.

Snyder, S. H., Banerjee, S. P., Yamamura, H. I., and Greenberg, D. (1974). Drugs, neurotransmitters and schizophrenia. *Science* **184,** 1243–1253.

Sobue, K., Ichida, S., Yoshida, H., Yamazaki, R., and Kakiuchi, S. (1979a). Occurrence of a Ca^{2+}- and modulator protein-activatable ATPase in the synaptic plasma membranes of brain. *FEBS Lett.* **99,** 199–202.

Sobue, K., Muramoto, Y., Yamazaki, R., and Kakiuchi, S. (1979b). Distribution in rat tissues of modulator-binding protein of particulate nature. *FEBS Lett.* **105,** 105–109.

Srivastava, A. K., Waisman, D. M., Brostrom, C. O., and Soderling, T. R. (1979). Stimulation of glycogen synthase phosphorylation by calcium-dependent regulator protein. *J. Biol. Chem.* **254,** 583–586.

Stille, G., Sayers, A. C., Lavener, H., and Eichenberger, E. (1973). 6-(4-methyl-l-piperazinyl)-morphanthridine (Perlapine), a new tricyclic compound with sedative and sleep-promoting properties. *Psychopharmacologia* **28,** 325–337.

Strada, S. J., and Thompson. W. J. (1978). Multiple forms of cyclic nucleotide phosphodiesterases: Anomalies or biologic regulators. *Adv. Cyclic Nucleotide Res.* **9,** 265–283.

Sugden, M. C., Christie, M. R., and Ashcroft, S. J. H. (1979). Presence and possible role of calcium-dependent regulator (calmodulin) in rat islets of Langerhans. *FEBS Lett.* **105,** 95–100.

Sulser, F., and Robinson, S. E. (1978). Clinical implications of pharmacological differences among antipsychotic drugs (with particular emphasis on biochemical central synaptic adrenergic mechanisms). *In* "Psychopharmacology: A Generation of Progress" (M. A. Lipton, A. DiMascio, and K. F. Killam, eds.), pp. 943–954. Raven Press, New York.

Sutherland, E. W., and Rall, T. W. (1958). Fractionation and characterization of a cyclic adenine ribonucleotide formed by tissue particles. *J. Biol. Chem.* **232,** 1077–1091.

Suzuki, Y., Hirabayashi, T., and Watanabe, Y. (1979). Isolation and electrophoretic properties of a calcium-binding protein from the ciliate *Tetrahymena pyriformis. Biochem. Biophys. Res. Commun.* **90,** 253–260.

Tarsy, D., and Baldessarini, R. J. (1974). Behavioral supersensitivity to apomorphine following chronic treatment with drugs which interfere with the synaptic function of catecholamines. *Neuropharmacology* **13,** 927–940.

Teo, T. S., and Wang, J. H. (1973). Mechanism of activation of a cyclic adenosine 3′:5′-monophosphate phosphodiesterase from bovine heart by calcium ions. Identification of the protein activator as a Ca^{2+} binding protein. *J. Biol. Chem.* **248,** 5950–5955.

Teo, T. S., Wang, T. H., and Wang, J. H. (1973). Purification and properties of the protein activator of bovine heart cyclic adenosine 3′,5′-monophosphate phosphodiesterase. *J. Biol. Chem.* **248,** 588–595.

Teshima, Y., and Kakiuchi, S. (1974). Mechanism of stimulation of Ca^{2+} plus Mg^{2+} dependent phosphodiesterase from rat cerebral cortex by the modulator protein and Ca^{2+}. *Biochem. Biophys. Res. Commun.* **56,** 489–495.

Teshima, Y., and Kakiuchi, S. (1978). Membrane-bound forms of Ca^{2+}-dependent protein modulator: Ca^{2+}-dependent and -independent binding of modulator protein to the particulate fraction from brain. *J. Cyclic Nucleotide Res.* **4,** 219–231.

Thompson, W. J., and Appleman, M. M. (1971). Characterization of cyclic nucleotide phosphodiesterases of rat tissues. *J. Biol. Chem.* **246,** 3145–3150.

Toscano, W. A., Jr., Westcott, K. R., LaPorte, D. C., and Storm, D. R. (1979). Evidence for a dissociable protein subunit required for calmodulin stimulation of brain adenylate cyclase. *Proc. Natl. Acad. Sci. U.S.A.* **76,** 5582–5586.

Trifaro, J. M. (1977). Common mechanisms of hormone secretion. *Annu. Rev. Pharmacol. Toxicol.* **17,** 27–47.

Uzunov, P., and Weiss, B. (1971). Effects of phenothiazine tranquilizers on the cyclic 3′,5′-adenosine monophosphate system of rat brain. *Neuropharmacology* **10,** 697–708.

Uzunov, P., and Weiss, B. (1972a). Separation of multiple molecular forms of cyclic adenosine-3′,5′-monophosphate phosphodiesterase in rat cerebellum by polyacrylamide gel electrophoresis. *Biochim. Biophys. Acta* **284,** 220–226.

Uzunov, P., and Weiss, B. (1972b). Psychopharmacological agents and the cyclic AMP system of rat brain. *Adv. Cyclic Nucleotide Res.* **1,** 435–453.

Uzunov, P., Shein, H. M., and Weiss, B. (1974). Multiple forms of cyclic 3′,5′-AMP phosphodiesterase of rat cerebrum and cloned astrocytoma and neuroblastoma cells. *Neuropharmacology* **13,** 377–391.

Uzunov, P., Lehne, R., Revuelta, A. V., Gnegy, M. E., and Costa, E. (1976). A kinetic analysis of the cyclic nucleotide phosphodiesterase regulation by the endogenous protein activator. *Biochim. Biophys. Acta* **422,** 326–334.

Valverde, I., Vandermeers, A., Anjanyulu, R., and Malaisse, W. J. (1979). Calmodulin activation of adenylate cyclase in pancreatic islets. *Science* **206,** 225–227.

Vanaman, T. C., Sharief, F., Awramik, J. L., Mendel, P. A., and Watterson, D. M. (1976). Chemical and biological properties of the ubiquitous troponin-C-like protein from non-muscle tissues, a multifunctional Ca^{2+}-dependent regulatory protein. *In*

"Contractile Systems in Non-Muscle Tissues" (S. V. Perry *et al.*, eds.), pp. 165–176. Elsevier/North-Holland Biomedical Press, Amsterdam.

Vanaman, T. C., Sharief, F., and Watterson, M. (1977). Structural homology between modulator protein and TNCs. *In* "Calcium Binding Proteins and Calcium Function" (R. H. Wasserman, R. A. Corradino, E. Carofoli, R. H. Kretssinger, D. H. MacLennan, and F. L. Siegel, eds.), pp. 107–116. North-Holland Publ., Amsterdam.

Van Eldik, L. J., and Watterson, D. M. (1979). Characterization of a calcium-modulated protein from transformed chicken fibroblasts. *J. Biol. Chem.* **254,** 10250–10255.

Vincenzi, F. F., and Farrance, M. L. (1977). Interaction between cytoplasmic (Ca^{2+} + Mg^{2+})-ATPase activator and the erythrocyte membrane. *J. Supramol. Struct.* **7,** 301–306.

von Hungen, K., and Roberts, S. (1973). Catecholamine and Ca^{2+} activation of adenylate cyclase systems in synaptosomal fractions from rat cerebral cortex. *Nature (London), New Biol.* **242,** 58–60.

Waisman, D., Stevens, F. C., and Wang, J. H. (1975). The distribution of the Ca^{2+}-dependent protein activator of cyclic nucleotide phosphodiesterase in invertebrates. *Biochem. Biophys. Res. Commun.* **65,** 975–982.

Waisman, D. M., Singh, T. J., and Wang, J. H. (1978). The modulator-dependent protein kinase. *J. Biol. Chem.* **253,** 3387–3390.

Wallace, R. W., Lynch, T. J., Tallant, E. A., and Cheung, W. Y. (1978). Purification and characterization of an inhibitor protein of brain adenylate cyclase and cyclic nucleotide phosphodiesterase. *J. Biol. Chem.* **254,** 377–382.

Walsh, M., and Stevens, F. C. (1977). Chemical modification studies on the Ca^{2+}-dependent protein modulator of cyclic nucleotide phosphodiesterase. *Biochemistry* **16,** 2742–2749.

Walsh, M., Stevens, F. C., Oikawa, K., and Kay, C. M., (1979). Circular dichroism studies of native and chemically modified Ca^{2+}-dependent protein modulator. *Can. J. Biochem.* **57,** 267–278.

Walsh, M. P., Vallet, B., Cavadore, J.-C., and Demaille, J. G. (1980). Homologous calcium-binding proteins in the activation of skeletal, cardiac and smooth muscle myosin light chain kinases. *J. Biol. Chem.* **255,** 335–337.

Wang, J. H., and Desai, R. (1977). Modulator binding protein. Bovine brain protein exhibiting the Ca^{2+}-dependent association with the protein modulator of cyclic nucleotide phosphodiesterase. *J. Biol. Chem.* **252,** 4175–4184.

Wang, J. H., Teo, T. S., Ho, H. C., and Stevens, F. C. (1975). Bovine heart protein activator of cyclic nucleotide phosphodiesterase. *Adv. Cyclic Nucleotide Res.* **5,** 179–194.

Watanabe, K., Williams, E. F., Law, J. S., and West, W. L. (1980). Specific inhibition of a calcium dependent activation of brain cyclic AMP phosphodiesterase activity by vinblastine. *Experientia* (in press).

Watterson, D. M., Harrelson, W. G., Keller, P. M., Sharief, F., and Vanaman, T. C. (1976). Structural similarities between the Ca^{2+}-dependent regulatory proteins of 3′,5′-cyclic nucleotide phosphodiesterase and actomyosin ATPase. *J. Biol. Chem.* **251,** 4501–4513.

Watterson, D. M., Sharief, F., and Vanaman, T. C. (1980). The complete amino acid sequence of the Ca^{2+}-dependent modulator protein (calmodulin) of bovine brain. *J. Biol. Chem.* **255,** 962–975.

Weiss, B. (1970). Factors affecting adenyl cyclase activity and its sensitivity to biogenic amines. *In* "Biogenic Amines as Physiological Regulators" (J. J. Blum, ed.), pp. 35–73. Prentice-Hall, Englewood Cliffs, New Jersey.

Weiss, B. (1975). Differential activation and inhibition of the multiple forms of cyclic nucleotide phosphodiesterase. *Adv. Cyclic Nucleotide Res.* **5,** 195–211.

Weiss, B., and Costa, E. (1968). Selective stimulation of adenyl cyclase of rat pineal gland by pharmacologically active catecholamines. *J. Pharmacol. Exp. Ther.* **161,** 310–319.

Weiss, B., and Fertel, R. (1977). Pharmacological control of the synthesis and metabolism of cyclic nucleotides. *Adv. Pharmacol. Toxicol.* **14,** 189–283.

Weiss, B., and Greenberg, L. H. (1975). Cyclic AMP and brain function: Effects of psychopharmacological agents on the cyclic AMP system. *In* "Cyclic Nucleotides in Disease" (B. Weiss, ed.), pp. 269–320. Univ. Park Press, Baltimore, Maryland.

Weiss, B., and Greenberg, L. H. (1978). Physiological and pharmacological significance of the multiple forms of cyclic nucleotide phosphodiesterase. *In* "Molecular Biology and Pharmacology of Cyclic Nucleotides" (G. Folco and R. Paoletti, eds.), pp. 69–84. Elsevier/North-Holland Biomedical Press, New York.

Weiss, B., and Greenberg, L. H. (1980). Modulation of beta-adrenergic receptors and calmodulin following acute and chronic treatment with neuroleptics. *Adv. Biochem. Psychopharmacol.* **24,** 139–146.

Weiss, B., and Hait, W. N. (1977). Selective cyclic nucleotide phosphodiesterase inhibitors as potential therapeutic agents. *Annu. Rev. Pharmacol. Toxicol.* **17,** 441–477.

Weiss, B., and Kidman, A. D. (1969). Neurobiological significance of cyclic 3′,5′-adenosine monophosphate. *Adv. Biochem. Psychopharmacol.* **1,** 131–164.

Weiss, B., and Levin, R. M. (1978). Mechanism for selectively inhibiting the activation of cyclic nucleotide phosphodiesterase and adenylate cyclase by antipsychotic agents. *Adv. Cyclic Nucleotide Res.* **9,** 285–303.

Weiss, B., Lehne, R., and Strada, S. J. (1972). A rapid microassay of adenosine 3′,5′-monophosphate phosphodiesterase activity. *Anal. Biochem.* **45,** 222–235.

Weiss, B., Fertel, R., Figlin, R., and Uzunov, P. (1974). Selective alteration of the activity of the multiple forms of adenosine 3′,5′-monophosphate phosphodiesterase of rat cerebrum. *Mol. Pharmacol.* **10,** 615–625.

Weiss, B., Levin, R. M., and Greenberg, L. H. (1979). Modulation of cyclic nucleotide metabolism by antipsychotics through a non-dopamine receptor. *In* "Catecholamines: Basic and Clinical Frontiers" (E. Usdin, I. J. Kopin, and J. Barchas, eds.), pp. 529–531. Pergamon, Oxford.

Weiss, B., Prozialeck, W., and Cimino, M. (1980). Acute and chronic effects of psychoactive drugs on adrenergic receptors and calmodulin. *Adv. Cyclic Nucleotide Res.* **12,** 213–225.

Wells, J. N., and Hardman, J. G. (1977). Cyclic nucleotide phosphodiesterases. *Adv. Cyclic Nucleotide Res.* **8,** 119–144.

Welsh, M. J., Dedman, J. R., Brinkley, B. R., and Means, A. R. (1978). Calcium-dependent regulator protein: Localization in mitotic apparatus of eukaryotic cells. *Proc. Natl. Acad. Sci. U.S.A.* **75,** 1867–1871.

Wolff, D. J., and Brostrom, C. O. (1974). Calcium-binding phosphoprotein from pig brain: Identification as a calcium-dependent regulator of brain cyclic nucleotide phosphodiesterase. *Arch. Biochem. Biophys.* **163,** 349–358.

Wolff, D. J., and Brostrom, C. O. (1976). Calcium-dependent cyclic nucleotide phosphodiesterase from brain: Identification of phospholipids as calcium-independent activators. *Arch. Biochem. Biophys.* **173,** 720–731.

Wolff, D. J., and Brostrom, C. O. (1979). Properties and functions of the calcium-dependent regulator protein. *Adv. Cyclic Nucleotide Res.* **11,** 27–88.

Wolff, D. J., Poirier, P. G., Brostrom, C. O., and Brostrom, M. A. (1977). Divalent cation binding properties of bovine brain Ca^{2+}-dependent regulator protein. *J. Biol. Chem.* **252,** 4108–4117.

Wong, P. Y.-K., and Cheung, W. Y. (1979). Calmodulin stimulates human platelet phospholipase A2. *Biochem. Biophys. Res. Commun.* **90,** 473–480.

Yagi, K., Yazawa, M., Kakiuchi, S., Oshima, M., and Uenishi, K. (1978). Identification of an activator protein for myosin light chain kinase as the Ca^{2+}-dependent modulator protein. *J. Biol. Chem.* **253,** 1338–1340.

Yamauchi, T., and Fujisawa, H. (1979). Most of the Ca^{2+}-dependent endogenous phosphorylation of rat brain cytosol proteins requires Ca^{2+}-dependent regulator protein. *Biochem. Biophys. Res. Commun.* **90,** 1172–1178.

Yazawa, M., and Yagi, K. (1978). Purification of modulator-deficient myosin light chain kinase by modulator protein-sepharose affinity chromatography. *J. Biochem. (Tokyo)* **84,** 1259–1265.

Yazawa, M., Kuwayama, H., and Yagi, K. (1978). Modulator protein as a Ca^{2+}-dependent activator of rabbit skeletal myosin light-chain kinase. *J. Biochem. (Tokyo)* **84,** 1253–1258.

Index

R

S

T

U

V

Molecular Biology

An International Series of Monographs and Textbooks

HAROLD A. SCHERAGA. Protein Structure. 1961

STUART A. RICE AND MITSURU NAGASAWA. Polyelectrolyte Solutions: A Theoretical Introduction, *with a contribution by Herbert Morawetz*. 1961

SIDNEY UDENFRIEND. Fluorescence Assay in Biology and Medicine. Volume I–1962. Volume II–1969

J. HERBERT TAYLOR (Editor). Molecular Genetics. Part I–1963. Part II–1967. Part III–Chromosome Structure–1979

ARTHUR VEIS. The Macromolecular Chemistry of Gelatin. 1964

M. JOLY. A Physico-chemical Approach to the Denaturation of Proteins. 1965

SYDNEY J. LEACH (Editor). Physical Principles and Techniques of Protein Chemistry. Part A–1969. Part B–1970. Part C–1973

KENDRIC C. SMITH AND PHILIP C. HANAWALT. Molecular Photobiology: Inactivation and Recovery. 1969

RONALD BENTLEY. Molecular Asymmetry in Biology. Volume I–1969. Volume II–1970

JACINTO STEINHARDT AND JACQUELINE A. REYNOLDS. Multiple Equilibria in Protein. 1969

DOUGLAS POLAND AND HAROLD A. SCHERAGA. Theory of Helix-Coil Transitions in Biopolymers. 1970

JOHN R. CANN. Interacting Macromolecules: The Theory and Practice of Their Electrophoresis, Ultracentrifugation, and Chromatography. 1970

WALTER W. WAINIO. The Mammalian Mitochondrial Respiratory Chain. 1970

LAWRENCE I. ROTHFIELD (Editor). Structure and Function of Biological Membranes. 1971

ALAN G. WALTON AND JOHN BLACKWELL. Biopolymers. 1973

WALTER LOVENBERG (Editor). Iron-Sulfur Proteins. Volume I, Biological Properties—1973. Volume II, Molecular Properties—1973. Volume III, Structure and Metabolic Mechanisms—1977

A. J. HOPFINGER. Conformational Properties of Macromolecules. 1973

R. D. B. FRASER AND T. P. MACRAE. Conformation in Fibrous Proteins. 1973

OSAMU HAYAISHI (Editor). Molecular Mechanisms of Oxygen Activation. 1974

FUMIO OOSAWA AND SHO ASAKURA. Thermodynamics of the Polymerization of Protein. 1975

LAWRENCE J. BERLINER (Editor). Spin Labeling: Theory and Applications. Volume I, 1976. Volume II, 1978

T. BLUNDELL AND L. JOHNSON. Protein Crystallography. 1976

HERBERT WEISSBACH AND SIDNEY PESTKA (Editors). Molecular Mechanisms of Protein Biosynthesis. 1977

WAI YIU CHEUNG (Editor). Calcium and Cell Function, Volume I: Calmodulin. 1980

TERRANCE LEIGHTON AND WILLIAM F. LOOMIS, JR. (Editors). The Molecular Genetics of Development: An Introduction to Recent Research on Experimental Systems. 1980

in preparation

ROBERT B. FREEDMAN AND HILARY C. HAWKINS (Editors). The Enzymology of Post-Translational Modification of Proteins, Volume 1